Environment and Agriculture: Perspectives on Sustainability

Environment and Agriculture: Perspectives on Sustainability

Edited by **Laura Vivian**

SYRAWOOD
PUBLISHING HOUSE
New York

Published by Syrawood Publishing House,
750 Third Avenue, 9th Floor,
New York, NY 10017, USA
www.syrawoodpublishinghouse.com

Environment and Agriculture: Perspectives on Sustainability
Edited by Laura Vivian

International Standard Book Number: 978-1-68286-332-9 (Hardback)

Printed in China.

Contents

Preface

This book has been an outcome of determined endeavour from a group of educationists in the field. The primary objective was to involve a broad spectrum of professionals from diverse cultural background involved in the field for developing new researches. The book not only targets students but also scholars pursuing higher research for further enhancement of the theoretical and practical applications of the subject.

There has been a growing demand for sustainable agriculture to improve environmental quality by making use of natural and organic resources. This text comprises of various topics such as use of organic manures and pesticides, land management, sustainable agricultural practices, crop yield, etc. that would provide a comprehensive insight into these fields. Researches and case studies by internationally eminent experts have been included herein. It aims to serve as a source of reference for students and professionals alike and contribute to the growth of the discipline.

It was an honour to edit such a profound book and also a challenging task to compile and examine all the relevant data for accuracy and originality. I wish to acknowledge the efforts of the contributors for submitting such brilliant and diverse chapters in the field and for endlessly working for the completion of the book. Last, but not the least; I thank my family for being a constant source of support in all my research endeavours.

Editor

1

The genotoxic effect of lead and zinc on bambara groundnut (*Vigna subterranean*)

E. O. Oladele*, P. G. C. Odeigah and I. A. Taiwo

Department of Cell Biology and Genetics, University of Lagos, Nigeria.

The effects of lead and zinc treatments on the chromosomes of bambara groundnut was investigated. The seeds of bambara groundnut were placed in Petri dishes in three replicates and allowed to germinate for five days in different concentrations: 25, 50 and 100 mg/L of both lead and zinc nitrates while the control group had distilled water. The total aberrations were examined. The mitotic index was calculated and the results were statistically evaluated by the analysis of variance at 5% significant level. The mitotic index decreased as the concentration increased ($p<0.05$). The highest mitotic index value was 3.40±0.88 for the control while the least was 1.30±0.48 for the 100 mg/L Pb treatment. The results show the most frequent chromosomal anomalies induced by these heavy metals as stickiness and bridges. Pb is much more genotoxic than Zn, as it induced more aberrations having percentage abnormalities of 92.3% while Zn had 30.0% for the highest concentration tested. Increased metal pollution can lead to some irreversible cytogenetic effects in plants and higher organisms. The study is an attempt to corroborate the toxic effect of lead and zinc on the chromosomes of plants. These results will be useful in environmental monitoring of the cytotoxicity of metals.

Key words: Heavy metal, aberrations, genotoxic, cytogenetic, mitotic index.

INTRODUCTION

Pollution by metals varies from air, water to soil and heavy metals are found to stay much longer in soil than other parts of the biosphere (Lasat, 2002). Most industrial operations release various toxic heavy metals in their effluents which eventually find their way into water sources such as lakes, rivers and streams. The alarm of heavy metal pollution started with the effects of Minamata disease caused by consumption of mercury. Liu et al. (1994) reported that high concentrations of heavy metals have been found to be chromotoxic and mutagenic to a large number of plant species. These metals including Pb characteristically inhibit root growth and cell division in plants like *Allium cepa* (Liu et al., 1994) as well as *Zea mays* L. (Jiang and Liu, 2000). Ademoroti (1996) has shown that Pb occurs more in the stem than in the root tips of *Abelmoschus esculentus* (Okro) and bitter leaf plants. Plants grown in contaminated soil have higher levels of Pb (as well as other heavy metals) than those grown in an uncontaminated soil (Bearington, 1975, Gingell et al., 1976). Zinc pollution occurs as a result of natural and anthropogenic processes and these can lead to the deposition of Zn on land. Ademoroti (1996) and Odiete (1999) noted that Zn arises as industrial effluents and such industries that produce Zn as effluents include electroplating industries, metallurgical industries, dye, paint and pigments, pharmaceuticals and mining industries. Rajesh and Rahda (2011) found that copper and zinc altered the nucleic acid contents and the nutritional status of *Vigna mungo* (L.) Hepper. Fragašovã (2001) found that lead and zinc were very toxic to *Sinapsis alba* L. seedlings. According to Mittler (2002), the toxicity of heavy metals including Pb and Zn may arise as a result of the generation of reactive oxygen species that may cause wide-ranging damage to proteins, nucleic acids, lipids and eventually apoptosis (cell death).

The bambara groundnut is important because it fixes

*Corresponding author. E-mail: esteromitiran@yahoo.com.

atmospheric nitrogen as nitrates into the soil which improves soil fertility. It is high in methionine, an essential amino acid. The beans are eaten fresh after harvest and can be dried and stored for later use and consumption (NRC, 2006). The young fresh seeds may be boiled and eaten as a snack in a manner similar to boiled peanuts and could be made into pudding locally called moi moi or okpa (bean porridge) in some parts of Nigeria. It has been reported that in Zambia, bambara groundnut is used for bread making (Brough et al., 1993) while Poulter and Caygill (2006) also reported that it could be used for milk making. The great genetic diversity potential of bambara groundnut has been described by Wazael et al. (2004). Aremu et al. (2006) carried out an assessment of heavy metals including Pb and Zn in bambara groundnut planted along some roads in Nasarawa state, Nigeria. They found the levels of these two metals to be high especially Pb at 19.2 mg/100 g which is about 10 times the recommended daily intake. W.H.O (1972), recommends daily intake to be 0.430 mg per person. If this plant was to be consumed, it would have been harmful. Studies on the treatment effects of lead and zinc on this important legume are scarce.

The objective of the study was therefore to determine the toxicity of these two metals on the chromosomes found at the root tips of the bambara groundnut. This investigation is an attempt to corroborate the toxic effect of lead and zinc on the chromosomes of plants. The results obtained may help in understanding the possible constraints in the role of Pb and Zn as pollutants in plants.

MATERIALS AND METHODS

Dry seeds of Accessions of bambara groundnut (*Vigna subterranea*): Tvsu 102 was collected from the International Institute of Tropical Agriculture (I.I.T.A.), Ibadan. Seeds were spread uniformly in Petri dishes lined with filter paper. The Petri dishes were divided into three replicates and the seeds were divided into two sets of treatment. Equal volumes of the different concentrations of lead and zinc nitrates solutions (25, 50 and 100 mg/L), respectively were administered while the control group had distilled water. The seeds were allowed to germinate within the Petri dishes and were treated with the different concentrations of each of the metals and distilled water, respectively at a temperature of 25°C for five days. Growing root tips which were brittle, translucent and gently tapering were selected from the three plants grown in the effluents of different concentration and from the control. About 2 to 3 mm terminal root tips were cut off using a sharp blade, then placed on a clean glass slide and macerated with the aid of two dissecting needles, and the remaining portion discarded. A drop of 1 N hydrochloric acid (HCL) was added to the root tip and left for 5 min; this softens the root tissue breaking up the middle lamellae. The excess acid was sucked up with a filter paper and the softened tissue is further macerated with dissecting needles so that the cells easily absorb the stain and spread adequately for microscopic observation. Then, a drop of lactic acetic orcein stain (2%) was placed on the macerated root tip and allowed to stand for 20 min for clearer viewing of the mitotic stages using the microscope (Michelle-Frainer et al., 2006). Each slide was covered with a cover slip and pressed down to allow the tissue spread out and also to allow the excess stain seep out at the edges of the cover slip. This was removed by placing the slide between the folds of the filter paper and squashing was done with the base of the dissecting set. All slides were examined under the light microscope with high power magnification (x40 objective): then the good slides were preserved by sealing the edges of the cover slip with nail varnish to prevent the stain from evaporating. The photomicrographs of good slides were then taken under the oil immersion lens (x1000 objective) using a Wild M20 microscope with MPS 55 photoautomat attachment. The mitotic index was calculated according to Balog (1982) using the formula:

$$\text{Mitotic index (M/I)} = \frac{\text{Number of cells in mitosis per field}}{\text{Total Number of cells per field}} \times \frac{100}{1}$$

The results of the mitotic index were statistically evaluated by the analysis of variance at 5% significant level using Microcal Origin 5.0 software.

RESULTS AND DISCUSSION

The results for genotoxicity assay of the treated bambara groundnuts are shown in Figure 1. 25 and 50 mg/L of lead and zinc nitrates resulted in anaphase bridges, fragmented and scattered chromosomes while the 100 mg/L concentrations mostly resulted in stickiness and vagrant chromosomes.

Table 1 shows the total number of cells analyzed, mitotic index values and the number of chromosome aberrations observed in bambara groundnut treated with different concentrations of lead and zinc nitrates. The results show that there was decrease in mitotic index with increase in treatment concentrations. The control had 3.40 while 25, 50 and 100 mg/L of Pb had mitotic index of 2.10, 1.50 and 1.30, respectively. The aberrations observed include: anaphase bridges, stickiness, vagrants, laggards and fragments. The highest treatment concentration of 100 mg/L was observed to cause mostly stickiness and vagrants especially with lead. The most frequent chromosomal aberrations were anaphase bridges, stickiness and vagrants.

A concentration-dependent increase in chromosomal abnormalities was observed for Zn and a decrease for Pb except for the highest concentrations in both cases. At 100 mg/L of Pb, the highest chromosomal aberration was observed while Zn treatment at this concentration resulted to a decrease in percentage chromosomal aberration. There was a linear correlation between the concentration of metals and the percentage of abnormalities in the treated sets especially for Pb. The total aberration (%) increased after 50 mg/L Pb exposure and a decrease was observed after 50 mg/L Zn exposure. This was probably due to increased metal concentration and subsequent detrimental effect especially for Pb. A concentration-dependent decrease in number of dividing cells was observed. This indicates an interference with cell division. There was a highly significant difference between both metal-treated set and the control set ($P < 0.05$). Scattering, bridges, stickiness

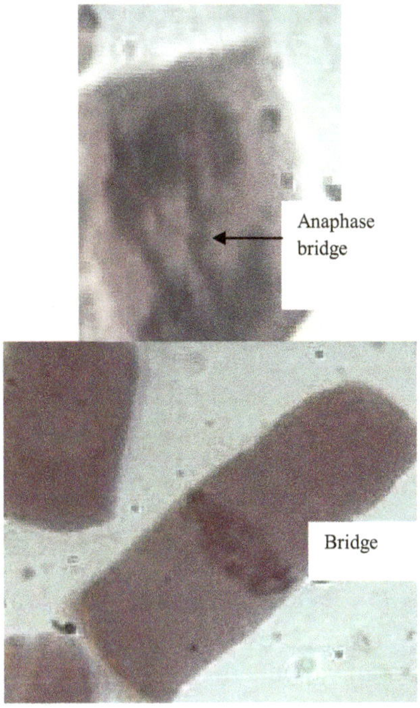

50mg/L zinc nitrate X 40

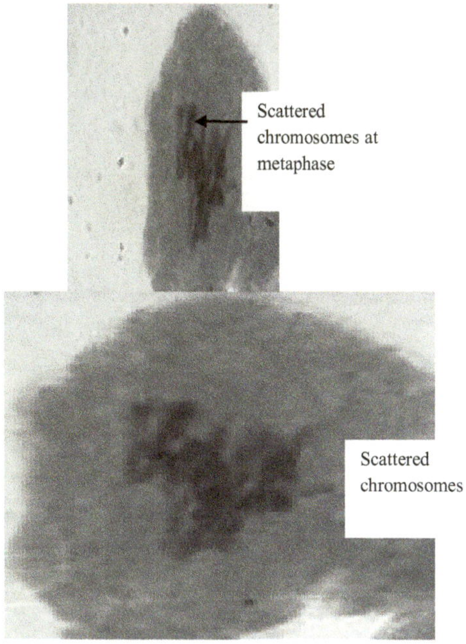

50mg/L lead nitrate X 40

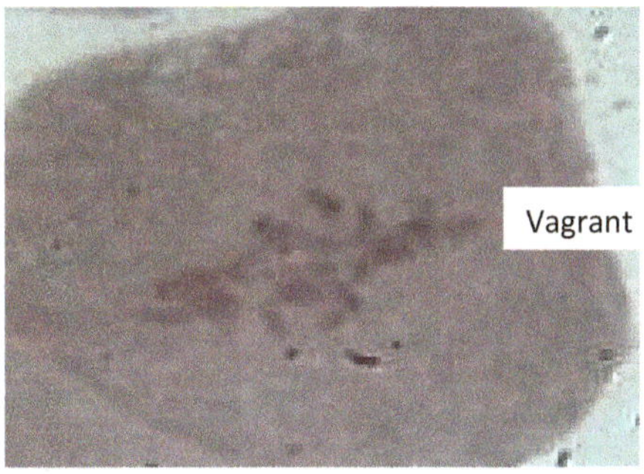

100Mg/l lead nitrate X 40

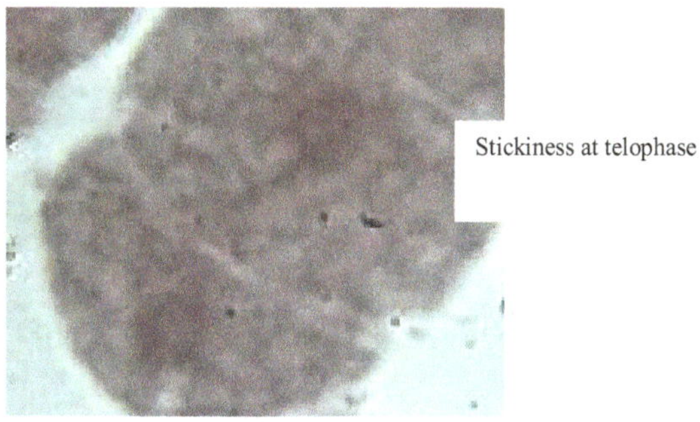

100mg/L Zinc nitrate X 40

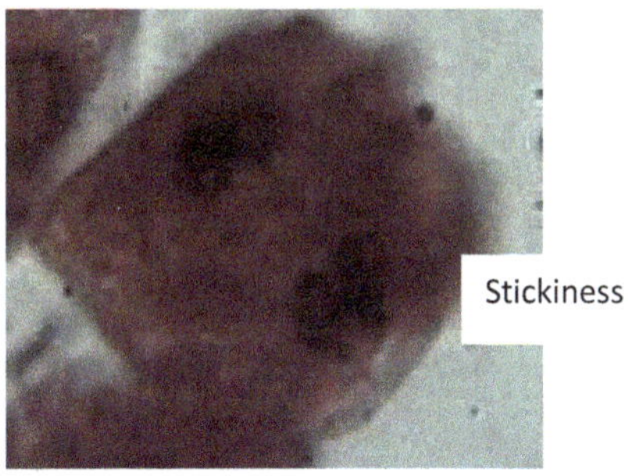

100Mg/l lead nitrate X 40

Figure 1. Chromosomal aberrations induced by Pb and Zn.

were the most frequent aberrations observed for the two heavy metals (Pb and Zn). Similar observations were shown by other studies (Fiskesjo, 1993) and (Ukaegbu and Odeigah, 2009). All the concentrations were capable of inducing different types of chromosomal abnormalities. This provides a case for comparison of the deleterious effects of these metals on the concerned plant. These aberrations were probably caused mainly by the interference of Pb and Zn with cell division at the root tips (meristems). Several researchers performed similar studies on the genotoxic effects of different chemicals on different plant materials (Ahmad and Yasmin, 1991;

Table 1. Chromosome aberrations in bambara groundnut root tips cells treated with different concentrations of lead and zinc nitrate.

Concentrations	TCN	ND	ST	CM	BF	VG	LG	MA	TA(%)	MI	MI ± SD	P-value
Control	1000	34 ($P_3M_{16}A_{12}T_4$)	0	0	0	0	0	0	-	3.40	3.40±0.88	0.7376**
25 mg/L Pb	1000	21 ($P_2M_8A_5T_6$)	0	0	3	5	0	0	38.10	2.10	2.10±0.76	0.0014*
50 mg/L Pb	1000	15 ($P_0M_5A_6T_4$)	0	0	2	0	0	0	13.3	1.50	1.50±0.63	0.0028*
100 mg/L Pb	1000	13 ($P_6M_2A_2T_3$)	3	0	4	2	3	0	92.3	1.30	1.30±0.48	<0.0001*
25 mg/L Zn	1000	26 ($P_6M_8A_6T_6$)	0	0	5	1	1	0	26.92	2.60	2.60±0.81	0.0081*
50 mg/L Zn	1000	24 ($P_0M_0A_0T_{24}$)	6	0	6	0	0	0	50.0	2.40	2.40±0.80	0.0036*
100 mg/L Zn	1000	20 ($P_1M_9A_7T_3$)	2	1	2	0	1	0	30.0	2.00	2.00±0.73	0.0010*

TCN, Total cell number; P, prophase; M, metaphase; A, anaphase; T, telophase; ND, number of dividing cells; ST, stickiness; CM, C-mitosis; BF, bridges fragment; VG, vagrant; LG, laggard; MA, multipolar anaphase; TA, total aberrations; SD, standard deviation; MI, mitotic index. Data were expressed as mean ± SD. *P<0.05, Significantly different from control; **P>0.05, not significantly different from control.

Kumar and Tripathi, 2007). Metal-induced chromosomal stickiness has been reported to interfere with cell division (Kumar and Srivastava, 2006). Kumar and Tripathi (2007) reported the interference of Pb with cell division in grass pea. Pb was also found to inhibit root growth and cell division in *Zea mays* L. (Jiang and Liu, 2000). Metal-induced chromosomal stickiness accompanied by pyknosis and chromatin degeneration has also been reported in maize by Caetano-Pereira et al. (1995), *Allium cepa* (Kumar and Tripathi, 2003) and *Helianthus annuus* (Kumar and Srivastava, 2006), these metals include Pb and Zn. Chromosome stickiness which was observed can lead to sticky metaphase and precocious separation of chromosomes. This may be due to breaking of the protein moiety of the nucleoprotein backbone by these heavy metals (Pb and Zn). Patnaik et al. (1984) had similar observations with chemicals tested on *Allium cepa*. Scattering which involves chromosomes spreading irregularly over the cell may be due to disturbance of the spindle apparatus by the metals. Anaphasic bridges formed were probably due to unequal exchange or dicentric chromosomes. Fragments observed at metaphase may be due to the failure of broken chromosomes to recombine. Fragments might be due to the stickiness of the chromosomes and the consequent failure of the arrival of chromatids at the poles. The characteristic behaviour of laggard chromosomes is that they generally lead to micronucleus formation (Kumar and Rai, 2007). Agarwal and Ansari (2001) noted that fragments caused by Pb may also be due to acentric chromosomes formed as a result of inversion. Accumulation of these abnormalities affects gamete formation and can lead to non-viable gametes, which considerably reduces plant fertility. Studies on different plant species have shown that declines in seed production are correlated with meiotic irregularities (Kumar and Rai, 2007).

From this study, it can be said that both heavy metals are capable of inducing chromosomal anomalies. Pb is much more genotoxic than Zn, as it induced more abnormalities having percentage abnormalities of 92.3% while zinc had 30.0% for the highest concentration tested. The statistical values ($P < 0.05$) obtained for Pb treated sets are much lower than those for Zn treated sets. Also, Zn treated sets had higher number of dividing cells compared to Pb treated sets. Increased soil pollution can lead to some irreversible cytogenetic effects in plants and higher organisms. Thus, mutagenic data from plant assays are very important for genetic research and for maintaining a stable ecosystem.

REFERENCES

Ademoroti CMA (1996). Environmental chemistry and toxicology. Ibadan Foludex Press Ltd., London. 215 p.

Agarwal R, Ansari MYK (2001). The effect of aniline on root tip cells of *Vicia faba* L. J. Cytol. Genet. 2(NS) 129-134.

Ahmad S, Yasmin R (1991). Effects of methyl parathion and trimiltox on the mitosis of *Allium cepa*. Cytologia 57:155-160.

Aremu MO, Olonisakin A, Ahmed SA (2006). Assessment of heavy metal content in some selected agricultural products planted along some roads in Nasarawa State, Nigeria. J. Eng. Appl. Sci. 1(3):199-204.

Balog C (1982). The mitotic index in diploid and triploid *Allium* roots. Cytologia 47:689-697.

Brough SH, Azam-Ali SN, Taylor AJ (1993). The potential of Bambara groundnut (*Vigna subterranea*) in vegetable milk production and basic protein functionality systems. Food Chem. 47:277-283.

Caetano-Pereira CM, Pagliarini MS, Brasil EM (1995). Influence of aluminium in causing chromosome stickiness in maize microsporocytes. Maydica 40:325-330.

Fargašovã A (2001). Phytotoxic effects of Cd, Zn, Pb, Cu and Fe on *Sinapsis alba* .L. seedlings and their accumulation in roots and shoots. Biologia Plantarum 44:471-473.

Fiskesjo G (1993). "The Allium test in waste water monitoring". Environ. Toxicol. Water Qual. 8:291-298.

Gingell SM, Campbell R, Martin RH (1976). The effect of Zinc, Lead and Cadmium pollution on the leaf surface microflora. J. Environ. Pollut. 11(1):25-37.

Jiang W, Liu D (2000). Effects of Pb^{2+} on root growth, cell division and nucleolus of *Zea mays* L. Bull. Envron. Contam. Toxicol. 65:786-793.

Kumar G, Rai P (2007). Genotoxic potential of mercury and cadmium in soybean. Turk. J. Biol. Sci. 31:13-15.

Kumar G, Srivastava S (2006). Mechanism of Aluminum cytotoxicity in meristematic cells of *Helianthus annus* L. J. Phytol. Res. 19:175-178.

Kumar G, Tripathi A (2003). Comparative mitotoxicity and mutagenicity of two heavy metals in *Allium cepa* L. J. Cytol. Genet. 4(NS):169-173.

Kumar G, Tripathi R (2007). Lead induced cytotoxicity and mutagenecity in Grass Pea. Turk. J. Biol. Sci. 32:73-78.

Lasat HA (2002). Phytoextraction of toxic metals: a review of biological mechanisms. J. Environ. Qual. 31(1):109–120. (PubMed)

Liu D, Jiang W, Zhao FM, Lu C (1994). Effects of lead on root growth, cell division and nucleolus of *Allium cepa*. Environ. Pollut. 86:1-4.

Michelle-Frainer K, Antonio CF, Thais S, Solange BT (2006). Effects of *Pterocaulon polystachyum* DC. (*Asteraceae)* on Onion (*Allium cepa*) Root-tip cells. Genet. Mol. Biol. 29:539-542.
Mitler R (2002).Oxidative stress, antioxidants and stress tolerance. Trends Plant Sci. 7:405-410.
National Research Council (NRC) (2006). "Bambara Bean". Lost Crops of Africa: Volume II: Vegetables. Lost Crops of Africa 2. National Academies Press. ISBN 978-0-309-10333-6.http://books.nap.edu/openbook.php?record_id=11763 and page=53. Retrieved 2008-07-15.
Odiete WO (1999). Environmental physiology of animals and pollution. Diversified Resources, Ltd., Lagos, Nigeria. 261 p.
Patnaik S, Saran BL, Patnik SM (1984). Effect of zarda (processed tobacco leaf) extract on the chromosomes of *Allium cepa*. Cytologia 49:807-814.
Rajesh D, Radha S (2011). Effect of copper and zinc on physiological and biochemical parameters in *Vigna mungo* (L.) Hepper. Int. J. Pharm. Biosci. (2)554-565.
Ukaegbu MC, Odeigah PGC (2009). Genotoxic effects of sewage effluent on *Allium cepa*. In: Report and Opinion. 1(6):36-41 (ISSN:1553-9873).
Wazael HN, Inga CB, Jorgen LC, Sven BA (2004). Analysis of genetic diversity in bambara groundnut [*Vigna subterranea* (L.) Verdc] landraces using amplified fragment length polymorphism (AFLP) markers. Afr. J. Biotechnol. 3:220-225.
World Health Organization (WHO) (1972). Evaluation of certain food additives and the contaminants: Mercury, Lead and Cadmium. WHO Tech. Rep. Ser. No. 505; FAO Nut, Meet Rep. Ser. No. 51.

2

Oral acute toxicity study of selected botanical pesticide plants used by subsistence farmers around the Lake Victoria Basin

Maud Kamatenesi-Mugisha[1]*, John Paul Buyungo[1], Patrick Engeu Ogwang[2], Alex Kasibante[2], Arop L. Deng[3], Joshua O. Ogendo[3] and Matobola J. Mihale[4]

[1]Division of Medical Ethnobotany and Product Development, Department of Biological Sciences, College of Natural Sciences, Makerere University, P. O. Box 7062, Kampala, Uganda.
[2]Natural Chemotherapeutics Research Institute, Ministry of Health, Uganda.
[3]Egerton University, P. O. Box 536, Egerton, Kenya.
[4]Open University of Tanzania, P. O. Box 31608, Dar es salaam, Tanzania.

A survey carried out around the Lake Victoria region showed evidence that people around this region use plant extracts, parts and powders to protect stored food commodities from insect pests. The widely used plants were identified and selected for biosafety assessments namely: *Ocimum gratissimum, Tithonia diversifolia*, *Eucalyptus saligna, Eucalyptus globulus* and *Cupressus lusitanica.* Wistar mice were acclimatized and divided into groups of six. Each mice group was administered with one extract at different concentrations. The extracts were administered orally and the animals were observed for 24 h. A control group was kept which received only the carrier substance orally. The LD_{50} values were determined by the use of the graphical method and regression analysis. Oral acute toxicity studies established the LD_{50} values for essential oils of *O. gratissimum*, *E. saligna* and *C. lusitanica* as 4,570, 2,290, and 3,311 mg/kg, respectively. For ethanol extracts, LD_{50} values were 12,882, 12,302, 14,996 and 11,481 mg/kg for *O. gratissimum, E. globulus*, *C. lusitanica* and *T. diversifolia*, respectively. For the aqueous extracts, the LD_{50} of *T. diversifolia* was found to be 12,302 mg/kg. For *E. globulus* and *C. lusitanica,* their aqueous LD_{50s} were beyond 15,000 mg/kg. The oral acute toxicity tests showed weak toxicities for all the plant extracts investigated in the study. The low toxicity levels exhibited by these extracts may be the reason why these plant products have been used by local communities for long without adverse effects. Chronic studies should be carried out to assess whether these extracts have serious effects on experimental animals exposed to them at small doses for a long period of time.

Key words: Oral acute toxicity, biopesticide, plant extracts, Lake Victoria Basin.

INTRODUCTION

There is a very long history of use of botanical extracts for human and veterinary medicine, as well as for the protection of field and stored crops (Berger, 1994). In the recent decades, however, due to the introduction of synthetic pesticides, the adoption of these traditional approaches of crop and post harvest protection has not been improved (Berger 1994). Today, the use of plant extracts for controlling pests has been limited to subsistence farmers, who in most cases are supported by various Non Government Organizations (NGOs) and women groups (Kamatenesi-Mugisha et al., 2008). The use of synthetic pesticides has undoubtedly increased crop production. This has been possible through reduced

*Corresponding author. E-mail: mkamatenesi@botany.mak.ac.ug.

losses caused by crop pests but many of these chemicals are hazardous to both humans and the environment.

According to Blackman and Eastop (1999), it has been estimated that hardly 0.1% of the agrochemicals used for crop protection reach the target pest leaving the remaining 99.9% to enter the environment and cause hazards to non-target organisms including humans. The situation is alarming especially in Africa and Asia where it has been reported that there is a lot of misuse of conventional pesticides due to, lack of suitable application equipment, and lack of technical training to ensure effective and safe use of pesticides (Attila, 1990).

Botanical pesticides are extracted from various plant parts (stems, seeds, roots, leaves and flower heads) of different plant species. These pesticides are known for having a broad spectrum of activity, being easy to process and use/apply, having a short residual activity and for not accumulating in the environment or in fatty tissues of warm blooded animals because they are highly biodegradable (Philip and Robert, 1998). It is, however, important to note that botanical pesticides, despite being derived from plants, may not necessarily be safe to humans and the environment. Some may be quite toxic. Toxicological studies aimed at assessing their safety should be done before they are used to avoid possible dangers to non target organisms and the environment (Belmain et al., 2001). Some plants have been scientifically tested and have been found to have good pesticidal properties. There is a concern, however, as most of the studied plants are of western origin (Jaya and Dubey, 2005). Botanical pesticides, if sufficiently exploited, can surely play a big role in reducing pollution, health risks and crop losses to pests.

Problem statement

The steadily increasing problems emanating from the use of synthetic pesticides including pest resistance, pollution of the environment and the side effects to beneficial flora and fauna demand for an urgent and intensive search for safer pesticides (Attila, 1990). On the other hand, synthetic pesticides are expensive and are in most cases not affordable by many farmers especially those with small farms around the Lake Victoria region (Mathenge, 2001). Subsistence farmers around the Lake Victoria region use botanical pesticides to protect their stored produce from storage pests; however, they have no standard dosage rates and are ignorant about the toxicity levels of these pesticides. It is important to note that, most botanical pesticides are naturally developed, some may not be safe for use by humans and may be potential pollutants of the environment. Biosafety assessments should be done before these botanicals are used (Belmain et al., 2001). Thus, this study was conducted mainly to assess the safety of plant extracts used as pesticides by the local communities around the Lake Victoria region. This study specifically determined the acute oral toxicity of the essential oils, ethanol and aqueous extracts of the selected plants using mice. In addition, the determination of the phytochemical profile of the crude ethanol and aqueous extracts of the selected plants was done.

A review of the selected plants for biosafety assessments

Ocimum gratissimum (L)

According to Katende et al. (1995), *O. gratissimum,* formerly known as *Ocimum suave* is an aromatic, perennial herb, 1 to 3 m tall. The stem is erect, round-quadrangular, much branched, glabrous or pubescent, woody at the base, often with epidermis peeling in strips. According to Hassanali et al. (1990), *O. gratissimum* is generally found growing across Africa and tropical Asia in the upland forest areas and open waste areas. It is used as a traditional medicine against stomachache, cough and influenza. The essential oil of *O. gratissimum* has antibacterial activity. Leaf extracts of *O. gratissimum* has antidiabetic properties in streptozocin-induced in diabetic rats. Other studies have shown that the oils obtained from the leaves of *O. gratissimum* contain phenols and recent studies have reported that the major components of the oil are p-cyme (59%), alpha-cyme 10%, myrcene 7% and thymol 7% (Keita et al., 2000).

Eucalyptus saligna Sm

According to Cremer (1990), *E. saligna* (Myrtaceae) is a fast growing hardwood tall tree, 30 to 55 m in height; with diameter at breast height over bark up to 2 m. Exceptional specimens may attain 60 and 2.5 m. The trunk is generally of excellent form, straight and clear of branches for half to two-thirds of tree height. Upper bark is smooth, grey-green, bluish-green or white, with a stocking of persistent, rough, brown or grey bark 1 to 4 m from ground level. The flowers are small, white in clusters of 7 to 11. Fruits are woody capsules, obconical to slightly pear-shaped, with 3 or 4 valves with thin, pointed tips which are exerted (protruding) and erect or out-curved. Seeds are brown, 1 to 2 mm long, cuboid or ovoid, with the dorsal surface pitted. It is predominantly found in moist forests on better soils of the east coast of Australia. *E. saligna* is moderately drought and frost tolerant and is widely planted in Australia and overseas (Cremer, 1990). The dry leaves of *Eucalyptus* species are frequently used in grain stores across Africa to deter feeding by *Sitophilus zeamais* and *Rhyzopertha dominica* to protect stored grains against insect attack (Hermann, 2010).

***Eucalyptus globulus* Labill.**

According to Katende et al. (1995), *E. globulus* (Myrtaceae) is usually a fast growing tall tree, grows up to 55 m high, the crown is rounded and open with straight main stems. The bark is blue-grey, smooth peeling in long strips and rough at the base. Young leaves are opposite, oval, blue-grey without stalks. The mature leaves are deep blue-green, thin and long to 30 cm, slightly curved with sharp tips, stalked and smell like camphor when crushed. The flowers have grey-green buds, wrinkled up to 2.5 cm, usually one, rarely 2 or 3 white flowers up to 4 cm across. The fruits are woody, half spheres, rough, 3 cm across with no stalks. The wood is hard, heavy and strong. The wood is usually used for poles and construction. The tree can be planted in pure stands, as an ornamental or as an avenue tree. The young leaves of this species are used to produce an essential oil used in pharmaceutical products. *E. globulus* grows naturally in the cool wet parts of S. W Australia. It prefers good-quality loams with adequate but not excessive moisture. It is suitable for areas over 2,000 m above sea level. This species does well in several uplands of Uganda. Its oil can be used for cleaning and as a natural insecticide.

***Cupressus lusitanica* (Mill)**

According to Katende et al. (1995), *C. lusitanica* (Cupressaceae) is an evergreen fast-growing tree. It has a straight trunk, generally conical but irregular in shape, branches hang down with branchlets in all directions and grow up to 35 m high. The bark is red-brown with vertical grooves which turn grey with age. The leaves are tiny, dull blue-green, in 4 ranks with spreading pointed tips. The fruit has male cones which appear like fat tips on branchlets, producing clouds of yellow pollen dust. Female cones are rounded, 1.5 cm across, brown with central pointed projections. About 75 winged seeds are released from beneath the cone scales. *C. lusitanica* grows fast on good sites and tends to be invasive at high altitudes. It is used for firewood, poles, posts, timber, ornamental purposes, shade and windbreaks and for live fencing. The leaves of this plant are used to cure some skin diseases caused by dermatophytes and have also been used to ward off insects from stored grain. In Costa Rica, a drink made by steeping a branch in alcohol is taken to alleviate coughs and cold symptoms Kuiate et al. (2006). The essential oil of *C. lusitanica* is extracted from the needles and twigs of young branches by steam distillation. This clear and fresh smelling oil is used for aromatherapy and massage to correct several body disorders. For example it restores calmness, sooths anger, improves circulation, sorts out coughs and bronchitis. It helps in varicose veins and female menstrual problem. The essential oil was screened for antimicrobial activity, and it showed antibacterial activity against *Bacillus cereus* and antifungal activity against *Aspergillus niger* (Hassanzadeh et al., 2010).

***Tithonia diversifolia* (Hemsl.)**

T. diversifolia family Asteraceae, is believed to have originated from Mexico and is commonly known as the Mexican sunflower. It is now widely spread throughout the humid and sub humid tropic regions. *T. diversifolia* is thought to have been introduced to several parts of Africa as an ornamental plant. According to Akanbi et al. (2007), it is reported to be growing in several African countries including Kenya, Nigeria, Rwanda, Zimbabwe, Cameroon, Uganda and Zambia. It is a perennial aggressive weed, a shrub, grows to about 2.5 m of height and grows well in most soils. Sun flowers are up to 10 cm across; petals are orange to yellow and 4 to 5 cm long and produce many seeds which can remain dormant for up to 4 months (Muoghalu and Chuba, 2005). It is commonly found growing in waste or abandoned lands, along roads, railway lines and on cultivated farmlands. It is considered to be an invasive plant in many parts of southern and eastern Africa, Australia and many pacific Islands. It is reported to be allelopathic where by the dense shading below the plant inhibits growth of other seedlings. Rapid vegetation reproduction and the production of many light seeds allow the plant to invade a disturbed habitat very quickly. It also competes with crops (Muoghalu and Chuba, 2005, Akanbi et al., 2007). *T. diversifolia* is known to be used for various purposes worldwide among which include the following: poultry feed, fuel, manure, fodders and land demarcation. The leaf extract is used to protect crops from termites and other insect attacks. The green biomass has been reported to be an effective nutrient source for low land rice, maize, okra, water melon and pumpkins (Akanbi et al., 2007).

MATERIALS AND METHODS

Collection of plant materials

Leaves of the selected pesticide plants (Figure 1) were collected from Mukono District, Uganda for aqueous and ethanol extraction. The essential oils were sent from the Department of Biological Sciences, Egerton University, Egerton, Kenya. Leaves were collected because the local communities use them to prepare the insecticides. The plants were identified by taxonomists in the Herbarium Department of Botany, Makerere University. This was done to ensure that the plants are rightly identified.

Preparation, extraction and concentration of plant materials

The collected plant materials (leaves) were separately exposed under shade until they became dry. Leaves of *C. lusitanica, O. gratissimum, T. diversifolia* and *E. globulus* were pounded using an

Ocimum gratissimum

Source: Floridata.com

Eucalyptus saligna

Photo taken by P.K. Bett (2011)

Eucalyptus globulus

Source: Floridata.com

Cupressus lusitanica

Source: Floridata.com

Tithonia diversifolia

Source: Floridata.com

Figure 1. Plants used in the study.

electric motor. 200 g of the pounded material were separately soaked in ethanol (95%) followed by shaking periodically for three days and then filtered. This was repeated three times to allow ethanol extract substantial quantities of the chemical constituents from the pounded plant materials. The solutions were filtered by the use of cotton wool and the filtrates were concentrated using a rotary evaporator and water bath maintained at 70°C. The concentrated ethanol extracts were put in a hot air oven maintained at 50°C and subsequently air dried. The dried extracts were kept in well labeled, dry, bijous bottles. Ethanol was used as a solvent because of its long and widespread use as a solvent for substances intended for human contact and consumption. Such substances include medicines, colorings, flavorings and scents among others.

In order to obtain the water extract, 200 g of the pounded materials were separately soaked in distilled water followed by shaking periodically for three days. The solutions were then filtered through a cheese cloth before further filtration using whatman A-1 filter paper. The filtrates were concentrated in a hot air oven maintained at 50°C for two days and were subsequently air-dried to thick residues. The obtained extracts were labeled and kept in a fridge at 4°C till when needed.

Water was used as a solvent due to its good solvent properties brought about by its polarity. Because of this, a wide range of substances are known to be soluble in water. In addition to this, the study intended to use the same or similar methods local people use to obtain or prepare botanical insecticides. A high percentage of local communities are known to use water as solvent to prepare herbal medicines and insecticides.

Extraction of the essential oils

The extraction of the essential oils of *O. gratissimum*, *C. lusitanica* and *E. saligna* was done in the Department of Biological Sciences, Egerton University, Egerton, Kenya. The essential oils were extracted by steam distillation from fresh leaves. The leaves were cut into small pieces and put into a distillation flask. Steam was allowed to pass through each batch of leaves for two hours. Essential oils were trapped in collecting tubes and put in clean, dry, well labeled bottles. The essential oils were then kept in a fridge maintained at 0°C to freeze the water which had been trapped together with the oil. The oils were decanted off since they do not freeze like water. The essential oils were kept in a fridge so as to minimize their volatile behavior which is catalyzed by relatively high temperature including room temperature.

Preparation of the test animals and determination of the acute toxicity of the plant extracts

Test animals (Wistar mice) were purchased at 4 weeks of age from the School of Veterinary Medicine, Makerere University. They were randomly group housed in stainless wire cages living enough space for clear observation of each animal. The animals were fed on conventional laboratory food with unlimited supply of drinking water (*ad libitium*). This was done for 5 days prior to dosing so as to get them acclimatized to laboratory conditions. During this period, the mice were observed to assess their health conditions based on their external appearance, nutritional conditions and general behavior.

In order to determine the preliminary acute toxicity of the different plant extracts, 4 dose levels were prepared for each plant extract. For essential oils, the following dose levels were administered: 2,500, 5,000, 7,500 and 10,000 mg/kg. Each dose level was assigned 2 test animals (mice). The animals were randomly selected and marked to permit individual identification. The animals were only given drinking water and not food for 4 h prior to dosing. This was done to avoid food-drug interactions so as to aid optimum absorption of the plant extracts from the gut. The mice were weighed and the freshly prepared test substances administered base on their weights as single doses orally (gavage method). Concentrations of the extracts were prepared in such a way that no test animal received more than 2 ml/100 g body weight for the water extracts and 1 ml/100 g body weights for the rest of the extracts. The administration of the test substances was done by using intragastric plastic tubes for the different groups of test animals. For each extract, a control group was kept and given equal amounts of the vehicle substances (distilled water for ethanol and water extracts and sunflower seed oil for the essential oils). One group was kept, it neither received plant extracts nor carrier substance throughout the experiment. The mice were observed for 24 h. The dose level that killed 50% of the members of the test group was taken as the approximate LD_{50}. The results were recorded. The approximate LD_{50} was then used to determine the correct LD_{50}.

For the correct LD_{50}, five dose levels were set within the range of the approximate LD_{50} for each extract. Different dose levels were set for the different extracts as shown in Tables 1, 2 and 3. Each group was assigned 6 members (mice). Administration of the extracts was done as explained above. The results were recorded in Tables 1, 2 and 3 for each group. Probits were determined by use of Fisher and Yates statistical tables. The LD_{50} values were determined by the use of the graphical method, plotting probits against the log dose, and thereafter running a regression analysis using a general formula:

$$y = mx + c.$$

RESULTS

The acute toxicity tests showed weak toxicities for all the plant extracts investigated in the study. However, when compared with each other, essential oils showed lower lethal doses (more toxic) followed by ethanol extracts (slightly toxic) and lastly by the aqueous extracts (relatively harmless). *E. saligna* essential oil had the lowest LD_{50} value (2,290 mg/kg) followed by *C. lustanica* (3,311 mg/kg) and then *O. gratissimum* (4,570 mg/kg). For ethanol extracts, *T. diversifolia* showed the lowest LD_{50} value (11,481 mg/kg), followed by *E. globulus* (12,302 mg/kg), *O. gratissimum* (12,882 mg/kg), and lastly *C. lusitanica* (14,996 mg/kg). For the aqueous extracts, *T. diversifolia* was found to have the lowest LD_{50} value (12,302 mg/kg). For *O. gratissimum, C. lusitanica* and *E. globulus,* their extracts did not show toxicity signs up to 15,000 mg/kg concentrations. The results are shown in the Table 4.

During the acute toxicity studies, the following general observations were made after the administration of the different essential oils to mice: Abnormal diminished body activity (hypo-activity), dizziness, rapid and deep breathing (hyperventilation), excess loss of saliva (salivation) and death of some group members. Deaths were recorded within 4 to 5 h after the administration for all essential oils. Similar observations were made for the ethanol extracts except that the dose levels at which such effects were observed were higher than those of the essential oils. It was also noted that the essential oils caused toxic signs and death within short periods as compared to the ethanol extracts. For the aqueous extracts, deaths were recorded for only *T. diversifolia*, although dizziness was observed for all extracts for sometime after which the mice recovered.

Qualitative phytochemistry

Phytochemical screening was determined asdescribed by Parekh and Chanda (2007). There were more classes of phytochemical constituents found in the aqueous extracts as compared to those found in the ethanol extracts (Table 5). Tannins, alkaloids and steroid glycosides were found to be most abundant for both the ethanol and water extracts of all the plants. Starch was not found in any of the extracts, while flavonosides were only found in the ethanol extract of *E. saligna*.

Table 1. The dose levels used in the determination of LD_{50s} of essential oils for different extracts.

Group	*Ocimum gratissimum*					*Cupressus lusitanica*					*Eucalyptus saligna*				
	1	2	3	4	5	1	2	3	4	5	1	2	3	4	5
Dose level (mg/kg)	4,500	4,800	5,000	5,500	C	2,500	3,000	4,000	4,500	C	1,500	2,000	2,500	3,000	C
Log dose	3.65	3.68	3.69	3.74		3.39	3.47	3.60	3.65		3.17	3.30	3.39	3.47	
Dead/total	1/4	3/4	3/4	4/4	0/6	2/6	3/6	2/6	5/6	0/6	0/6	1/6	2/6	6/6	0/6
Dead (%)	25	75	75	100		33	50	33	83		0	16	33	100	
Corrected (%)	25	75	75	93.75		33	50	33	83		4.16	16	33	95.8	
Probits	4.33	5.67	5.67	6.52		4.56	5.00	4.56	5.95		3.30	4.05	4.56	6.64	

C =control.

Table 2. The dose levels used in the determination of LD_{50s} of different ethanol extracts.

Group	*Eucalyptus globulus*					*Cupressus lusitanica*					*Ocimum gratissimum*					*Tithonia diversifolia*				
	1	2	3	4	5	1	2	3	4	5	1	2	3	4	5	1	2	3	4	5
Dose level (mg/kg)	C	10,0.00	12,000	14,000	15,000	C	10,000	12,000	14,000	15,000	C	10,000	12,000	14,000	15,000	C	10,000	12,000	14,000	15,000
Log dose		4	4.07	4.14	4.17		4	4.07	4.14	4.17		4	4.07	4.14	4.17		4	4.07	4.14	4.17
Dead/Total	0/6	1/6	2/6	4/6	5/6		0/6	1/6	2/6	3/6	0/6	0/6	1/6	3/6	6/6	0/6	1/6	3/6	6/6	6/6
Dead (%)		16	33	66	83		0	16	33	50		0	16	50	95.83		16	50	100	100
Corrected (%)		16	33	66	83		4.16	16	33	50		4.16	16	50	95.83		16	50	95.83	95.83
Probits		4.01	4.56	5.41	5.95		3.30	4.01	4.56	5.00		3.30	4.01	5.00	6.64		4.01	5.00	6.64	6064

C=control.

Table 3. The dose levels used in the determination of LD_{50s} of aqueous extract of *T. diversifolia*.

Group	Dose level mg/kg	Log dose	Dead/total	Dead (%)	Corrected (%)	Probits
1.	Control		0/6			
2.	12,000	4.08	3/6	50	50	5.00
3.	10,000	4.00	2/6	33	33	4.56
4.	8,000	3.90	0/6	4.16	4.16	3.30
5.	6,000	3.77	0/6	4.16	4.16	3.30

C=control.

DISCUSSION

According to the toxicity classification scale of Hodge and Sterner (1956), acute toxicity tests showed weak toxicities for all the plant extracts investigated in the study. Essential oils however, showed lower lethal doses (more toxic) followed by ethanol extracts (slightly toxic) and lastly the aqueous extracts (relatively harmless). The weak

Table 4. Showing the LD_{50s} of the different plant extracts tested in the study.

S/N	Plant name	Extract	LD_{50} Value (mg/kg)
1	*Ocimum gratissimum* (L)	Essential oil	4,570
		Ethanol	12,882
		Aqueous	>15,000
2	*Cupressus lusitanica* (Mill)	Essential oil	3,311
		Ethanol	14,996
		Aqueous	>15,000
3	*Eucalyptus saligna* (Smith)	Essential oil	2,290
4	*Eucalyptus globulus* (Labill)	Ethanol	12,302
		Aqueous	>15,000
5	*Tithonia diversifolia* (Hemsl)	Ethanol	11,481
		Aqueous	12,302

Table 5. Phytochemical analysis of plant extracts.

Parameter	*Ocimum gratissimum*		*Eucalyptus globulus*		*Tithonia diversifolia*		*Cupressus lusitanica*	
	Aqueous extract	Ethanol extract	Aqueous extract	Ethanol extract	Aqueous extract	Ethanol extract	Aqueous extract	Ethanol extract
Tannins	+ + +	-	+ + +	+ + +	+ +	-	+ + +	+ + +
Alkaloid salts	+ + +	+ +	+ +	+ + +	+ + +	+ +	+ + +	+ +
Anthracenosides	+ +	-	+ + +		-	-	+ + +	-
Coumarin derivatives	+ + +	+ +	+ + +	+ + +	-	-	+ +	+ + +
Steroid glycosides	+ + +	+ +	+ +	+ + +	+ +	+ + +	+ + +	+ + +
Flavonosides	-	-	-	+ +	-	-	-	-
Anthocyanosides	+ +	-	+ +	-	+ +	+ +	+ +	+ +
Reducing compounds	+ + +	-	+ + +	+ + +	-	-	+ + +	+ + +
Starch	-	×	-	×	-	×	-	×
Polyuronides	+ + +	×	-	×	+ + +	×	-	×
Glucides	+ + +	×	+ +	×	+ + +	×	+ + +	×
Saponins	-	×	+ + +	×	-	×	+ + +	×

The presence and qualitative abundance of phytochemical compounds IS shown: Very abundant (+++), moderate (++) , little (+), not available (-) test not done (×).

acute toxicity properties could be the reason why these plants have for long been used by local communities around the Lake Victoria region as storage pesticides.

Essential oils showed low LD_{50} values (more toxic) as compared to ethanol and aqueous extracts of the same plants. This could be attributed to their complex nature since they are normally a mixture of various compounds. Hassanzadeh et al. (2010) reported that the major components of *C. lusitanica* leaf oil are α-pinene (40 to 82%), limonene (4 to 18%), isobornyl acetate (up to 10%) and *cis*-muurola-4 (14%), 5-diene (up to 7%). For *E. saligna*, the major components are, 1, 8-cineol (eucalyptol), α-pinene (45.1%), limonene (1.5%), terpinene (8.6%), p-cymene (cymol) (22.5%), terpenene-4-ol (8.6%), terpineol (9.9%) and carvocrol (Hermann, 2010; Patricia et al., 2007). The essential oil of *O. gratissimum* has components including thymol (31.79%), euginol, myrcene (6.94%), ɣ-terpinene (12.34%), α-thujene (6.11%) and p-cymene (15.57) (Hermann, 2010; Koffi et al., 2009). Many components of essential oils are thought to be toxic when taken orally (Shaaya et al., 1991).

More phytochemical constituents were found in the aqueous extracts as compared to those found in the ethanol extracts. However, ethanol extracts showed lower LD_{50} values as compared to aqueous extracts. This could be due to the presence of toxic substances in the ethanol extracts which were not extracted by water due to their polarity characters. The results obtained from this study may be used to develop safer alternatives to fight crop pests. Based on the available literature, it is evident that plant extracts are biodegradable and thus will not cause similar environmental risks as many of the widely used synthetic agrochemicals. Biodegradability of botanicals may be an important factor which will increase demand for plant-based chemical products (Kari et al., 2011). On the other hand, it has also been found that the rapid biodegradability of these botanicals may hinder the registration process with the authorities. It is very difficult to obtain data indicating the spread of botanicals in the environment, because all the measurable components break down in soil within a few days. There is no single active ingredient or decomposition product which can be used as an indicator of the leaching risk (Kari et al., 2011). Other challenges include: the fact that most botanicals are slow in action and lack the rapid knockdown affect, brings about doubt as to whether they are as effective as the synthetic products. The issue of genetic variability of plant species in different localities is instability of the active ingredients when exposed to direct sunlight. Competition with synthetic pesticides through aggressive advertisement by commercial pesticides dealers and commercially formulated botanicals are more expensive than synthetic insecticides and are not as widely available (Yallappa et al., 2012).

Conclusions

All the extracts assessed in this study were found to have weak acute toxicity properties. The weak acute toxicity properties could be the reason why these plants have for long been used by local communities around the Lake Victoria region as storage pesticides.

RECOMMENDATIONS

The essential oils assessed in this study should be used with caution to avoid individuals from applying them for long term exposure at relatively high doses.

From this study, chronic toxicity studies should be carried out using several routes of exposure such as the oral, inhalation and skin absorption (dermal toxicity) since they are relevant occupational exposure routes.

ACKNOWLEDGEMENTS

This research was funded by Sida to Lake Victoria Research (VICRES) under Inter University Council of East Africa (IUCEA). The participating universities: Makerere University, University of Egerton and Open University of Tanzania are acknowledged for supporting the researchers and for providing the infrastructure that enabled us do the research. We thank the local communities around Lake Victoria Basin for providing their valuable indigenous knowledge and all the resource persons in the region that were valuable during the study.

REFERENCES

Akanbi WB, Adebayo FA, Togun AO, Adeyeye AJ, Olaniran OA (2007). The use of Compost Extract as Folia Spray Nutrient Source and Botanical Insecticide in *Telfairia occidentalis*. World J. Agric. Sci. 3(5):642-652.

Attila KT (1990). Study on the Production Possibilities of Botanical Pesticides in developing African countries. United Nations Industrial development Organization. 90-86421 Budapest, Hungary.

Belmain S, Neal SR, Ray GE, Golob DE (2001). Ethnobotanicals in Ghana. Insecticidal and Vertebrate toxicity associated with ethnobotanicals. J. Toxicol. 39(3):287-291.

Berger A (1994). Using natural Pesticides: Current and future perspectives. A report for the Plant Protection Improvement Programme in Botswana, Zambia and Tanzania. Swedish University of Agriculture. http://www.vaxteko.nu/html/sll/slu/rapport_vaxtsk_vetensk/RVX02/RVX02.HTM . Date accessed 3rd March 2010.

Blackman RL, Eastop VF (1999). Aphids on World's crops. John Wiley and sons. pp. 5.

Cremer KW (1990). Trees for rural Australia. CSIRO Division of Forestry and Forest Products, Inkata press, Melbourne.

Hassanali ALW, Ole–Sitayo N, Moreka L, Nokoe S, Chapya A (1990). Weevil repellent constituents of *Ocimum suave leaves* and *Eugenia caryophylla* cloves used as grain protectant in parts of East Africa. Discov. Innov. 2:91-95.

Hassanzadeh SL, Tuten JA, Vogler B, Setzer WN (2010). The chemical composition and antimicrobial activity of the leaf oil of *Cupressus lusitanica* from Monteverde, Costa Rica. Phcog Res 2010; 2:19-21 http://www.phcogres.com/text.asp?2010/2/1/19/60585

Hermann A (2010). The Chemistry and Biology of Volatiles. John Wiley & Sons Ltd. Chichester, UK.

Hodge HC, Sterner JH (1956). Combined tabulation of toxicity classes. In: Spector, W.S. Handbook of Toxicology, Philadelphia, W.B. Saunders Company.

Jaya V, Dubey NK (2005). Prospectives of botanical and microbial products as pesticides of tomorrow. Banaras Hindu University. Varanasi 221 005, India.

Kamatenesi-Mugisha M, Deng AL, Ogendo JO, Omolo EO, Mihale MJ, Otim M, Buyungo JP, Bett PK (2008). Indigenous knowledge of field insect pests and their management around Lake Victoria basin in Uganda. Afr. J. Environ. Sci. Technol. 2(8):342-348.

Kari T, Lindqvist I, Marleena H, Heikki S, Dionyssios P (2011). Use of Botanical Pesticides in Modern Plant Protection, Pesticides in the Modern World - Pesticides Use and Management, Dr. Margarita Stoytcheva (Ed.), ISBN: 978-953-307-459-7, InTech, Available from: http://www.intechopen.com/books/pesticides-in-the-modern-world-pesticides-use-and-management/use-ofbotanical-pesticides-in-modern-plant-protectionpesticides-in-modern-plant-protection

Katende AB, Ann B, Tanganass BO (1995). Useful trees and shrubs for Uganda. Caroline Agola-Nairobi.

Keita SM, Vincent C, Schmit JP, Belanger A (2000). Essential oil composition of *Ocimum basilicum*, *Ocimum gratissimum* and *Ocimum suave* in the Republic of Guinea. Flavour Fragr. J. 15:339-341.

Koffi K, Panopendou WP, Christine R, Komla S (2009). Antifungal activity of the essential oils from *Ocimum gratissimum* L. grown in Togo. J. Sci. Res. 1(1):164-171.

Kuiate JR, Bessiére JM, Vilarem G, Amvam Z (2006). Chemical composition and antidermatophytic properties of the essential oils

from leaves, flowers and fruits of *Cupressus lusitanica* Mill. from Cameroon. Flavour Fragr. J. 21:693-697.
Mathenge KL (2001). Summarized report of the international consultative and planning workshop on efficacy and side effects of the botanical pestcides/ repellents. SACDEP-KENYA.
Muoghalu JI, Chuba DK (2005). Seed Germination and Reproductive Strategies of *Tithonia diversifolia* (Hemsl.) Gray and *Tithonia rotundifolia* (P.M) Blake. Appl. Ecol. Environ. Res. 3(1):39-46. http://www.ecology.kee.hu/pdf/0301_039046.pdf . (Accessed January 2010).
Parekh J, Chanda SV (2007). *In vitro* antimicrobial activity and phytochemical analysis of some Indian medicinal plants. Turk. J. Biol. 31:53-58.
Philip GK, Robert AB (1998). Florida Pest Control Association. www.oisat.org/downloads/whatspesticide.html. (Accessed Jan. 2010)
Shaaya E, Ravid U, Paster N, Juven B, Zisman U, Pissarev V (1991). Fumigant toxicity of essential oils against four major stored-product insects. J. Chem. Ecol. 17:499-504.
Yallappa R, Nandagopal B, Thimmappa S (2012) Botanicals as Grain Protectants. Hindawi Publishing Corporation Psyche. Volume 2012, Article ID 646740, 13 pages doi:10.1155/2012/646740.

3

Effect of essential oils extracted from *Satureja calamintha*, *Mentha pulegium* and *Juniperus phoenicea* on *in vitro* methanogenesis and fermentation traits of vetch-oat hay

Arhab Rabah[1,2] , Khenaka Karima[1], Leulmi Nassima[1], Belaidi Hakim[2], Harzallah Besma[1] and Bousseboua Hacene[1]

[1]Laboratoire de Génie Microbiologique et Applications, Université Mentouri, BP 325 route de Aïn-El-Bey 25 017 Constantine. Algérie.

[2]Département des Sciences de la Nature et de la Vie, Faculté des Sciences Exactes et des Sciences de la Nature et de la Vie, Université Larbi Ben Mhidi, Oum El Bouaghi, Algérie.

The impact of inclusion of essential oils (EO) extracted from *Juniperus phoenicea, Satureja calamintha* and *Mentha pulegium* at three levels: 0, 1.66, 3.33 and 6.66 µl/ml of incubation medium, on methane production and rumen fermentation traits of vetch-oat hay was studied in *in vitro* gas production test, using 200 mg of substrate in a 60 ml graduated syringes. Gas production was recorded at 3, 6, 9, 24, 48, 72 and 96 h. Methane was measured at 3, 24 and 96 h, while pH and ammonia N were analyzed just at 24 h of incubation. At 24 h of fermentation, all doses of EO reduced significantly total gas and methane production ($P < 0.05$). At the highest concentration and after 24 h, methanogenesis was decreased by 60.98, 57.70 and 64.64% for *J. phoenicea*, *M. pulegium* and *S. calamintha*, respectively. This effect was more pronounced at the end of fermentation (96 h) where reduction in methane production was superior to 72%. However, addition of EO increased pH and decreased significantly ammonia concentration, mainly, for high EO concentrations. The results indicate that these EO had a potential to reduce methanogenesis in the rumen, but further *in vitro* and *in vivo* trials are required to search optimum dose which reduce methane production without adversely changing dietary fermentation and rumen function.

Key words: Methane, ammonia, essential oils, *Satureja calamintha (*Calament), *Mentha pulegium (fliou,* Menthe pouliot*)*, *Juniperus phoenicea (Arar,* Genévrier*)*, *in vitro* gas production test.

INTRODUCTION

Methane (CH_4) is a potent greenhouse gas and it has a global warming potential 25 times than carbon dioxide (CO_2) (Benchaar and Greathead, 2011). In fact, 33% of anthropogenic emissions of this gas are produced by ruminant livestock (Beauchemin et al., 2008). CH_4 is an end-product generating in the rumen under anaerobic conditions by methanogenic *Archaea*, using CO_2 and H_2 which are produced during the fermentation of the feed ingested by the animal. Furthermore, CH_4 represents a significant loss of dietary energy from the production system (Eckard et al., 2010); hence, reducing enteric methanogenesis may also lead to benefits for both the environment and the animal producer. Appreciable research efforts have been directed towards reducing CH_4 production by ruminants, many chemical feed additives have been tested to decrease CH_4 production in the rumen, but many of these substances appear to be toxic and some may generate chemical residues in animal derived foods (Patra and Saxena, 2010). Current research efforts are interested in using natural products, such as plant secondary metabolites such as essential oils (EO). EO are complex mixtures of lipophilic, liquid, volatile and often terpenoid compounds that are obtained by many methods of extraction (such as steam distillation, solvent extraction, supercritical CO_2 extraction) and have extensive antimicrobial activity against Gram-positive and Gram-negative bacteria (Calsamiglia et al., 2007; Benchaar and Greathead, 2011).

The objective of this study was to investigate the effectiveness of increasing doses of three EO extracted from *Satureja calamintha*, *Juniperus phoenicea* and *Mentha pulegium* in the inhibition of methanogenesis in the rumen, and their effects on *in vitro* ruminal fermentation traits of vetch-oat hay.

MATERIALS AND METHODS

Extraction procedure of essential oils

EO were extracted from aerial parts (leaves) of *S. calamintha*, *J. phoenicea* and *M. pulegium* by hydrodistillation. 100 g of each plant were mixed with 1400 ml of distilled water and boiled for 3 h. After decantation, EO were recuperated and stored in dark flasks at 4°C until tested.

In vitro fermentation

Ruminal digesta samples were obtained from three sheep (mean weight, 52 kg) and pooled together in order to achieve a homogenous inocula. Rumen digesta was collected 1 h before the morning feeding, placed in a Thermos container that was sealed immediately, and transported to the laboratory. The animals received daily 1200 g of vetch-oat hay (DM, 891.1 g/kg; ash, 50.8 g/kg DM; CP, 67.9 g/kg DM; NDF, 616.2 g/kg DM; ADF, 327.7 g/kg DM; ADL, 43.6 g/kg DM) in two equal meals at 8: 00 h and 16: 00 h. Water was available *ad libitum*. At the laboratory, the ruminal fluid was filtered through four layers of cheesecloth into pre-warmed thermos flasks.

The method of Menke et al. (1979) was used for the gas production procedure. Preparation of artificial saliva was done according to the method of Menke and Steingass (1988). The artificial saliva containing buffer solution ($NaHCO_3$, 39 g/l), macrominerals solution (Na_2HPO_4, 5.7 g; KH_2PO_4, 6.2 g and $MgSO_4.7H_2O$, 0.6 g per 1 L of distilled water), microminerals solution ($CaCl_2.2H_2O$, 13.2 g; $MnCl_2.4H_2O$, 10 g; $CoCl_2.6H_2O$, 1 g; $FeCl_3.6H_2O$, 8 g per 100 ml of distilled water) and potential redox indicator (resazurine, 0.1 g/100 ml) was prepared the day before incubation and stored at 39°C. The reducing agent ($Na_2S.9H_2O$, 0.625 g; NaOH 1 N, 4 ml; distilled water, 95 ml) was then prepared dor the incubation of the samples. Artificial saliva and ruminal fluid were mixed in a 2:1 ratio. 200 mg of vetch-oat hay (ground to pass a 1-mm screen) were weighed into 60 ml syringes and incubated at 39°C with 30 ml of inoculum. EO was added before the diluted rumen fluid at doses of 50, 100 and 200 µl/syringe which were equivalent to 1.66, 3.33 and 6.66 µl/ml.

In each trial and for each level, three syringes were used. Three syringes without EO (controls) were incubated in the same conditions at the same time.

Analytical techniques

Manual measurements of gas production were done at 3, 6, 9, 24, 48, 72 and 96 h. CH_4 in gas was determined at 3, 24 and 96 h according to the procedure described by Jouany et al. (1994) by transferring the total gas into syringes containing 4 ml of hydroxide sodium (NaOH, 10N) to absorb carbon dioxide. The gas remaining was assumed to be CH_4. After 24 h of incubation, the medium of each batch was checked for pH and analyzed for ammonia concentration as described by Chaney and Marbach (1962). 2 ml of orthophosphoric acid (50 g/l) were mixed with 10 ml of syringe contents. Samples were then centrifuged at 15,000 g for 15 min at 4°C, and the supernatant was used to determine ammonia concentration. Ammonia of blanks was subtracted from the measured $N\text{-}NH_3$ of samples to obtain the net ammonia concentration.

Calculations and statistical analysis

Cumulative gas production data were fitted to the model of Orskov and McDonald (1979) as follows: $y= a+b\ (1-e^{(-ct)})$ where a=the gas production from the immediately soluble fraction, b= the gas production from the insoluble fraction, c= the gas production rate constant for the insoluble fraction (b), t= incubation time, (a+b) = the potential extent of gas production and y= gas produced at time "t". The estimation of these parameters was made by the Neway excel software developed by Chen (1997). Data on *in vitro* gas production tests (pH, gas and methane productions, exponential parameters) were subjected to ANOVA analysis employing the SAS software (1990). The means were compared by Scott-Knott's test at the level of 5%, using the completely randomized design.

RESULTS

The effect of EO on *in vitro* gas production (ml g^{-1} DM) are presented in Table 1. EO included in the incubation medium did not affect ($P > 0.05$) gas production in the first hours of fermentation, but after 24 h, total gas production was reduced significantly ($P < 0.05$) by the addition of the three EO at all concentrations. The greatest decrease was observed with EO of *S. calamintha*. Similarly, methane production was not affected by concentration of EO during the early hours of fermentation (after 9 h of incubation, $P > 0.05$). However, a significant decrease was noted and compared with the control, at 24, 48, 72 and 96 h (Table 2). This impact was dependent upon the dose of EO administrated in syringes. The greatest decrease was observed for *S. calamintha* (65%) and the lowest for *M. pulegium* (58%). This decrease was greater than 72% for all three of the EO at the end of fermentation (96 h of incubation).

The addition of EO to the incubation medium had a significant effect ($P > 0.05$) at 24 h pH values (Table 2). Compared to the controls, addition of EO increased pH values except at the lowest concentration (1.66 µl/ml) of *J. phoenicea* and *M. pulegium*. Furthermore, inclusion of these EO in culture media reduced significantly ammonia production (Table 2) ($P < 0.05$). This effect was linearly

Table 1. Effect of increasing doses of essential oils on *in vitro* total gas production (ml/g DM).

Incubation time (h)	Control	*Juniperus phoenicea*			*Mentha pulegium*			*Satureja calamintha*			S.E.M	P
	0*	1.66*	3.33*	6.66*	1.66*	3.33*	6.66*	1.66*	3.33*	6.66*		
3	12.00	13.67	12.33	13.00	13.34	13.33	12.67	12.00	12.33	12.67	0.69	0.1531
6	13.67	15.33	15.00	14.67	15.00	14.67	13.67	13.67	15.33	14.00	0.82	0.7931
9	15.35	17.02	16.02	16.68	17.02	16.02	15.02	15.35	17.02	16.02	0.87	0.4950
24	27.35^{a}	24.68^{b}	21.01^{d}	20.01^{de}	24.68^{b}	21.01^{d}	18.68^{f}	23.35^{c}	19.35^{ef}	17.35^{g}	0.58	0.0163
48	36.02^{a}	32.35^{b}	24.68^{e}	20.68^{f}	30.35^{c}	23.02^{e}	20.35^{fg}	28.02^{d}	20.02^{fg}	18.02^{g}	1.05	0.0204
72	37.35^{a}	35.35^{b}	26.68^{d}	20.68^{f}	35.35^{b}	24.68^{e}	20.02^{f}	30.02^{c}	19.68^{f}	17.68^{g}	0.73	0.0000
96	39.35^{a}	35.35^{b}	26.68^{d}	20.68^{f}	35.35^{b}	24.68^{e}	20.02^{f}	30.02^{c}	19.68^{f}	17.68^{g}	0.78	0.0000
a**	7.39^{b}	10.65^{a}	10.84^{a}	9.86^{a}	10.85^{a}	11.38^{a}	10.41^{a}	10.12^{a}	9.74^{a}	8.95^{a}	1.98	0.0304
b**	33.42^{a}	26.58^{b}	16.31^{d}	10.87^{f}	26.66^{b}	13.43^{e}	9.74^{fg}	20.36^{c}	10.69^{f}	9.05^{fg}	1.70	0.0208
c**	0.04^{b}	0.03^{c}	0.04^{b}	0.11^{a}	0.03^{c}	0.05^{b}	0.08^{a}	0.04^{b}	0.10^{a}	0.12^{a}	0.01	0.0345

Means with different letters (a, b, c, d, e, f, and g) within the row are statistically different. S.E. M = standard error of the means. *Concentrations of essential oils (µl/ml). **Data were fitted to the equation $p = a + b(1 - e^{-ct})$ as described by Ørskov and McDonald (1979), where *p* is percentage degraded at time t, a is the zero time intercept, a + b is the potential degradability and c is the rate of degradation.

Table 2. Effect of increasing doses of essential oils on methane production (ml/g DM), pH and ammonia N concentrations.

Parameter	Incubation time (h)	Control	*Juniperus phoenicea*			*Mentha pulegium*			*Satureja calamintha*			S.E.M	P
		0*	1.66*	3.33*	6.66*	1.66*	3.33*	6.66*	1.66*	3.33*	6.66*		
	3	3.17	2.70	2.56	2.30	2.17	2.84	2.14	2.17	2.84	2.19	0.50	0.3473
CH_4 (ml/g DM)	24	7.92^{a}	3.75^{b}	3.09^{cde}	2.80^{cde}	2.42^{de}	3.35^{cd}	2.94^{cde}	2.50^{de}	2.80^{cde}	2.21^{e}	0.31	0.0015
	96	15.67^{a}	6.50^{b}	4.84^{c}	2.75^{d}	6.17^{b}	3.25^{c}	2.88^{d}	5.92^{b}	2.24^{d}	2.16^{d}	0.50	0.0114
pH	24	6.53^{c}	6.54^{c}	6.66^{b}	6.66^{b}	6.54^{c}	6.64^{b}	6.64^{b}	6.63^{b}	6.72^{a}	6.72^{a}	0.02	0.0000
NH3-N (10^2 mg/l)	24	7.00^{b}	6.94^{b}	6.86^{b}	7.65^{a}	7.68^{a}	6.64^{c}	6.78^{c}	6.93^{b}	6.45^{c}	5.44^{d}	0.02	0.0000

Means with different letters (a, b, c, d and e) within the row are statistically different. S.E.M= standard error of the means.*Concentrations of essential oils (µl/ml).

related to dose administered (r = 0.99) for *S. calamintha*. Nevertheless, a different trend was observed for *M. pulegium* and *J. phoenicea*. For the first extract, ammonia production was increased with dose of 1.66 µl/ml; however, a significant reduction was recorded with the other doses. While for *J. phoenicea*, this production was intensified in the presence of the high concentration (6.66 µl/ml) and any effect was noted for the other doses.

DISCUSSION

In this experiment, addition of EO in the culture media at all doses reduced methane production and this effect was more pronounced after 24 h of incubation; this impact was mainly attributed to their antimicrobial activity (Hart et al., 2008; Rochfort et al., 2008). In fact, these secondary compounds were constituted from a mixture of chemical molecules that exert antimicrobial activity by multiple mechanisms of action and can inhibit a broad variety of both Gram-positive and Gram-negative bacteria and other ruminal microorganisms (Calsamiglia et al., 2007). According to the literature, the prominent components of *M. pulegium* EO are menthol, menthone, piperitone, piperitenone, 1,8 cineole, α-terpineol and pulegone (Derwich et al., 2010; Mahboubi and Haghi, 2008;

Marzouk et al., 2008), the major component of *J. phoenicea* is α-Pinene and other components may coexist in low percentage such as linalool, α-phyllandrene, mycene (Bouzouita et al., 2008; Derwich et al., 2010; Ennajar et al., 2009a,b) and that of *S. calamintha* were characterized by the presence of thymol, p-cymene and γ-terpinene (Satrani et al., 2001). The impact of these EO on ruminale methanogenesis can be a result from their major compound or from the synergic action of all constituents. The same observations have been observed by several authors who studied the inhibition effect of ruminal methanogenesis by addition of individual or blends of EO. *Eucalyptus* oil inhibited CH_4 production up to 58% at 1.66 ml/L (Kumar et al., 2009) and 90.3% at 2 ml/L (Sallam et al., 2009). *M. piperita* EO at 2 ml/L reduced methanogenesis by 75.6% (Agarwal et al., 2009). A concentration of 20 mg/l of Juniper berry oil decreased CH_4 production up to 50% (Chaves et al., 2008). In other research, Macheboeuf et al. (2008) reported that thymol caused a suppression of CH_4 to the extent of 99% at 6 mM and carvacrol reduced CH_4 production by 98.4% at 5 mM.

Inclusion, EO in culture media at 3.33 and 6.66 μl/ml induced an increase in pH values compared with the control. This increase in pH can be associated with a reduction in total volatile fatty acids (VFA) concentration and/or with a high liberation of carbonate (HCO_3) ions from the buffer solution. The same results have been reported by other authors who have worked in the same conditions. For example using batch cultures, Busquet et al. (2006) noted that addition of eugenol, carvacrol, anethol, anise oil, cinnamon oil and oregano oil (at 3 mg/ml) reduced total VFA and increased pH after 24 h of fermentation. Similarly, Castillejos et al. (2006) reported that thymol, guaiacol, vanillin and eugenol at concentration of 5 mg/ml increased final pH and decreased total VFA concentration in 24 h *in vitro* batch culture incubations.

Ruminants have a low efficiency of N retention (Castillejos et al., 2005); hence, many studies have been directed to improve their ammonia N utilization. Molero et al. (2004) reported that the reduction in losses of N from the rumen can be obtained in different ways, one of them being the feeding of slowly degradable proteins and/or feed compounds that either reduce levels of rumen ammonia N and/or reduce protein degradation in the rumen. In the current study, EO had variable effects on ammonia concentrations that were partly dependent upon the EO dose. In view of the antimicrobial activity of EO, the observed decrease in ammonia N concentrations cannot be attributed to an increase in microbial growth (Macheboeuf et al., 2006). The decrease in ammonia N concentration can have many causes, such as: inhibition of proteolysis, peptidolysis or deamination. Indeed, many studies suggest that most EO reduced ammonia N by inhibition of deamination which is assured principally by ammonia hyper-producing bacteria (Castillejos et al., 2005). The decrease in ammonia concentration could also result from a negative effect on protozoa and thus a decrease in predation of bacteria. Furthermore, the increase in ammonia concentrations observed with EO of *J. phoenicea* (6.66 μl/ml) and *M. pulegium* (1.66 μl/ml) can be attributed to a selective inhibition of microorganisms that are users of ammonia N (Macheboeuf et al., 2006).

Conclusion

The three EO tested decreased CH_4 production at all doses tested and also decreased ammonia N concentrations. These secondary compounds had a potential to reduce methanogenesis in the rumen, but further *in vitro* and *in vivo* trials are required to determine the optimum dose which reduce CH_4 production without adversely affecting fermentation and rumen function. Moreover, it is important to determine their mechanisms of action.

REFERENCES

Agarwal N, Shekhar C, Kumar R, Chaudhary LC and Kamra DN (2009). Effect of peppermint (*Mentha piperita*) oil on *in vitro* methanogenesis and fermentation of feed with buffalo rumen liquor. Anim. Feed Sci. Technol. 148:321–327.

Beauchemin KA, Kreuzer M, O'Mara F and McAllister TA (2008). Nutritional management for enteric methane abatement: a review. Aust. J. Ep. Agric. 48:21–27.

Benchaar C, Chaves AV, Fraser GR, Yang Y, Beauchemin KA and McAllister TA (2007). Effects of essential oils and their components on *in vitro* rumen microbial fermentation. Can. J. Anim. Sci. 87: 413-419.

Benchaar C and Greathead H (2011). Essential oils and opportunities to mitigate enteric methane emissions from ruminants. Anim. Feed Sci. Technol. 166-167: 338-355.

Bouzouita N, Kachouri F, Ben Halima M and Chaabouni MM (2008). Composition chimique et activités antioxydante, antimicrobienne et insecticide de l'huile essentielle de *Juniperus phoenicea*. J. Soc. Chim. Tunisie. 10:119-125.

Busquet M, Calsamiglia S, Ferret A and Kamel C (2006). Plant extracts affect *in vitro* rumen microbial fermentation. J. Dairy Sci. 89:761-771.

Calsamiglia S, Busquet M, Cardozo PW, Castillejos L and Ferret A (2007). Invited Review: Essential Oils as Modifiers of Rumen Microbial Fermentation. J. Dairy Sci. 90: 2580-2595.

Castillejos L, Calsamiglia S, Ferrer A and Losa R (2005). Effects of a specific blend of essential oil compounds and the type of diet on rumen microbial fermentation and nutrient flow from a continuous culture system. Anim. Feed Sci. Technol.119: 29-41.

Castillejos L, Calsamiglia S and Ferret A (2006). Effect of essential oil active compounds on rumen microbial fermentation and nutrient flow in *in vitro* system. J. Dairy Sci. 89:2649-2658.

Castillejos L, Calsamiglia S, Ferret A and Losa R (2005). Effects of a specific blend of essential oil compounds and type of diet on rumen microbial fermentation and nutrient flow from a continuous culture system. Anim. Feed Sci. Technol. 119:2941.

Chaney AL and Marbach EP (1962). Modified reagents for determination of urea and ammonia. Clin. Chem. 8:130-132.

Chen XB (1997). Neway Exel: a utility for processing data of feed degradability and *in vitro* gas production (version 5.0). Rowett Research Institute, Aberdeen, UK.

Chaves AV, He ML, Yang WZ, Hristov AN, McAllister TA and Benchaar C (2008).Effects of essential oils on proteolytic, deaminative and methanogenic activities of mixed ruminal bacteria. Can. J. Anim. Sci., 89:97-104.

Derwich E, Benziane Z and Boukir A (2010). Chemical composition of leaf essential oil of *Juniperus phoenicea* and evaluation of its antibacterial activity. Int. J. Agric. Biol.12:199-204.

Derwich E, Benziane Z and Boukir A (2010). GC/MS Analysis and Antibacterial Activity of the Essential Oil of *Mentha Pulegium* Grown in Morocco. Res. J. Agr. Biol. Sci. 6(3):191-198.

Eckard RJ, Grainger C and De Klein CAM (2010). Options for the abatement of methane and nitrous oxide from ruminant production: a review. Livest. Sci.130: 47-56.

Ennajar M, Bouajila J, Lebrihi A, Mathieu F, Abderraba M, Raies A and Romdhane M (2009b). Chemical composition and antimicrobial and antioxidant activities of essential oils and various extracts of *Juniperus phoenicea* L. (Cupressacees). J. Food. Sci. 74 (7):71-364.

Ennajar M, Bouajila J, Lebrihi A, Mathieu F, Arlette S, Manef A, Raies A and Romdhane M (2009a). The influence of organ, season and drying method on chemical composition and antioxidant and antimicrobial activities of *Juniperus phoenicea* L. essential oils. J. Sci. Food. Agric. 90:462-470.

Hart KJ, Yáñez-Ruiz DR, Duval SM, McEwan NR and Newbold CJ (2008). Plant extracts to manipulate rumen fermentation. Anim.Feed Sci.Technol. 147:8-35.

Kumar R, Kamra DN, Agrawal N and Chaudhary LC (2009). Effect of eucalyptus (*Eucalyptus globulus*) oil on *in vitro* methanogenesis and fermentation of feed with buffalo rumen liquor. Anim. Nutr. Feed Technol. 9: 237-243.

Macheboeuf D, Morgavi DP, Papon Y, Mousset JL and Arturo-Schaan M (2008). Dose-response effects of essential oils on *in vitro* fermentation activity of the rumen microbial population. Anim. Feed Sci. Technol. 145:335-350.

Macheboeuf D, Papon Y, Artuso-Schaan M, Mousset JL and Cherel R (2006). Utilisation d'extraits végétaux (huiles essentielles et extrait de polyphénols) pour diminuer la dégradation ruminale des protéines-Étude *in vitro*. Renc. Rech. Ruminants. 13:69-72.

Mahboubi M and Haghi G (2008). Antimicrobial activity and chemical composition of *Mentha pulegium* L. essential oil. J. Ethnopharmacol. 119:325-327.

Marzouk B, Ben Hadj Fredj M, Chraief I, Mastouri M, Boukef K and Marzouk Z (2008). Chemical composition and antimicrobial activity of essential oils from Tunisian *Mentha pulegium* L. J. Food. Agr. Environ. 6:78-82.

Menke KH and Steingass H (1988). Estimation of the energetic feed value obtained by chemical analysis and *in vitro* gas production using rumen fluid. Anim. Res. Dev. 28:7-55.

Molero R, Ibars M, Calsmiglia S, Ferret A and Losa R (2004). Effects of a specific blend of essential oil compounds on dry matter and crude protein degradability in heifers fed diets with different forage to concentrate ratios. Anim. Feed Sci. Technol. 114:91-104.

Orskov ER and McDonal I (1979). The estimation of protein degradability in the rumen from incubation measurements weighted according to rate of passage. J. Agric. Sci. 92: 499-503.

Patra AK and Saxena J (2010).A new perspective on the use of plant secondary metabolites to inhibit methanogenesis in the rumen. Phytochem., 71:1198-1222.

Rochfort S, Parker AJ and Dunshea FR (2008). Plant bioactives for ruminant health and productivity. Phytochem., 69:299-322.

Sallam SMA, Bueno ICS, Brigide P, Godoy PB, Vitti DMSS and Abdalla AL (2009). Efficacy of eucalyptus oil on *in vitro* rumen fermentation and methane production. Options Meditérranéennes, 85: 267-272.

Satrani B, Abdellah F, Fechtal M, Talbi M, Blaghen M and Chaouch A (2001). Composition chimique et activité antimicrobienne des huiles essentielles de *Satureja calamintha* et *Satureja alpine* du Maroc. Ann. Fals. Exp. Chim., 94 (956):241-250.

Effects of poor sanitation on public health: Case of Yopougon town (Abidjan)

Yapo Toussaint Wolfgang[1,2]*, Mambo Veronique[1], Yapo Ossey Bernard[1], Seka M'bassiguié Arsène[1] and Houenou Pascal Valentin[1]

[1]Laboratory of Environmental Sciences, Research Group of Water Chemistry, University of Abobo-Adjamé (UAA), 02 BP 801 Abidjan 02, Côte d'Ivoire.
[2]National Institute of Public Hygiene (INHP), BPV 14 Abidjan, Côte d'Ivoire.

Effects of poor environmental hygiene on population health were studied in the precarious living quarters of Doukoure, the village of Niangon-adjame and the viable living quarters of millionaire's in Yopougon town. This study included sanitation, water sources, as well as, population health. It appears that excreta management is from self-purification works (WC with septic tank) (60%), collective purification work (gutter storm water, sewer) (26%), anarchic connections (30%) and defecation in nature relatively less important (2%) for all living quarters visited. Furthermore, almost 90% of the populations have access to the drinking water, which is rather used. In Doukouré's living quarters and Niangon-Adjamé's village, public tap water (7%) and traditional wells (3%) are used for domestic needs. Data collected in 15 health centers in the town on 125 446 cases of morbidity, revealed that malaria is much contracted by populations with 58% cases against 14% for diarrhea and 28% for acute respiratory infections. Under-five children are much affected with 24% of the patients among which 43% were cases of the diarrheic diseases. Also, malaria and respiratory infections are more important during rainy seasons, in contrast to diarrheal diseases caused by poor hygiene of people.

Key words: Public health, malaria, diarrhea, respiratory infections, sanitation, risk.

INTRODUCTION

Most large African cities are overcrowded, due to urban attraction. Little controlled urban growth leads to poor management of solid and liquid wastes produced by cities. This leads to many problems of sanitation. The sanitation in these cities is generally dominated by self purification works. They often repress wastewater that trickles down in living quarter streets emitting strong foul odors (Strauss et al., 2000). Moreover, water intended for consumption are very often contaminated because of drinking water connection with the sewage (Schwartzbrod, 1992). This fact can cause high health risks and environmental pathologies arising from it are very important (OMS, 1992). They are mostly of diarrheal diseases (Carr, 2001). In addition, poor drainage of wastewater and stormwater leads to the proliferation of ponds that are habitats of mosquito vectors of various diseases including malaria (OMS, 1985). Africa south of the Sahara is the most concerned in the sense that the majority (90%) of malaria deaths occur there. In this part of the continent, it is now recognized that malaria is both a disease caused by poverty and a cause of poverty (OMS / RBM, 2001). Côte d'Ivoire is a country where malaria is endemic (WHO, 1985, 2000). According to the report on the progress of the National Program against Malaria (PNLP) of Côte d'Ivoire in 2003, malaria is transmitted throughout the year with an increase during the rainy season (Mémain, 2003). In this country, there is resistance of *Plasmodium falciparum* to chloroquine (Djaman et al., 2001).

*Corresponding author. E-mail: toussaintwolfgang@yahoo.fr.

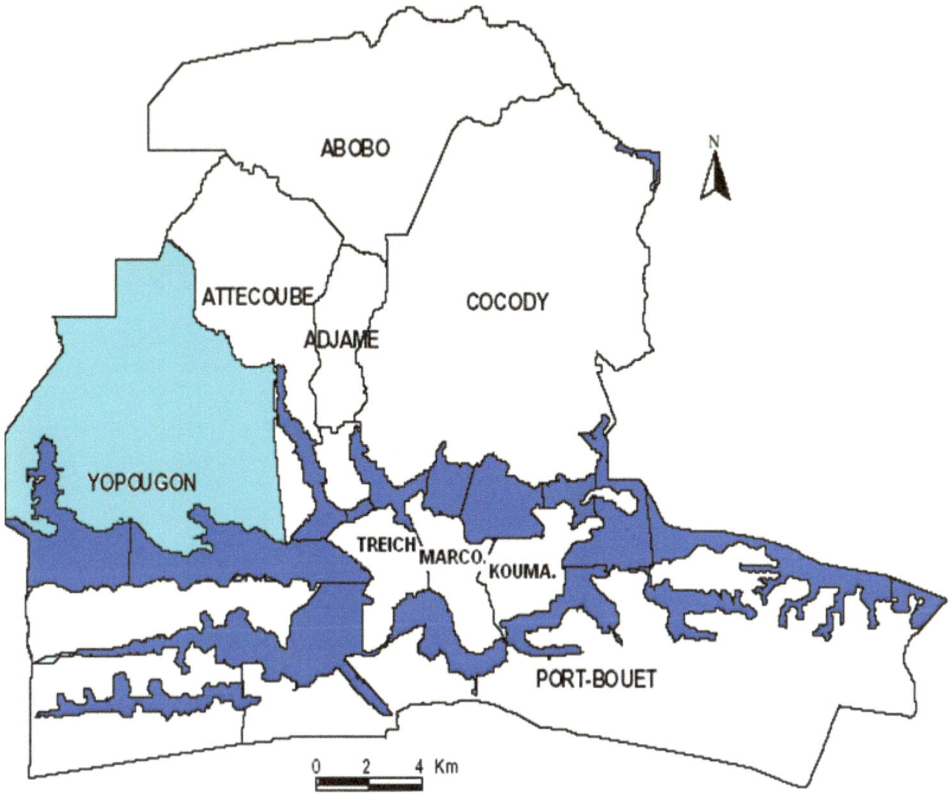

Figure 1. Map of Yopougon town.

These various pathologies are the result of many interactive phenomena between health and environmental hygiene highlight many health risks associated with poor hygiene (Hibberd et al, 1998) including vector-borne diseases and infections (Forget and Lebel, 2002).

In fact, poor quality waters, poor environmental sanitation (excreta management, drainage) and poor sanitation contribute much to deteriorating population health (Ezzati et al., 2005). Abidjan is not outside for these different problem types. Economic capital of Côte d'Ivoire, it is now a metropolis of nearly 3.6 million inhabitants. With an average annual growth rate of 3.2% between 2000 and 2005, Abidjan's population alone represents 43.8% of the urban population of Côte d'Ivoire and 19.7% of Ivorian population (United Nations, 2006). This strong growth of Abidjan's population has resulted in the production of solid and liquid waste, household as industrial. In the western part of Abidjan city, is located Yopougon town. In this town, urban wastewaters (domestic, storm water, industrial) are discharged into the lagoon through sewers and channel existing storm water drainage without prior treatment.

The objective of this work is thus, to establish a snapshot of Yopougon town in sanitation and mode of supply in drinking water in association with population health. Diseases investigated are including malaria, diarrheal diseases and acute respiratory infections.

MATERIALS AND METHODS

Study site

The study was performed in Yopougon town (Figure 1). It has an area of approximately 85 km^2. In 1998, its population was estimated at 708 155 inhabitants (National Institute of Statistics, 1998). Based on an average annual growth rate of Abidjan's population by 3.7% between 1995 and 2000 and 3.2% between 2000 and 2005 (United Nations, 2006), the population today is around 900 000 inhabitants. The town is bordered on the south by the Ebrié lagoon.

Data collection

The study took place in precarious living quarters, a village lagoon and modern living quarters of Yopougon town. Selection of living quarters was done as part of an interdisciplinary approach to understanding global health problems in the town in relation to environment, sociology and health engineering. The investigation was performed from March to October 2008 using an investigation form. It was conducted with a team of investigators who run through the various living quarters of the town. Households responding to the questionnaires were the different samples. The number of households visited in different parts of Yopougon town are divided as follows: 30 in precarious living quarters (Doukouré) (30%), 30 in the village (Niangon Adjamé) (30%) and 40 in viable area (millionaire's living quarters) (40%). The questionnaires focused on sources of drinking water, the method of managing solid and liquid waste and population health. Moreover, sanitary data were collected in fifteen health centers. The information researched concerned the recurrent diseases including malaria, diarrheal diseases and acute respiratory infections. The data were collected by

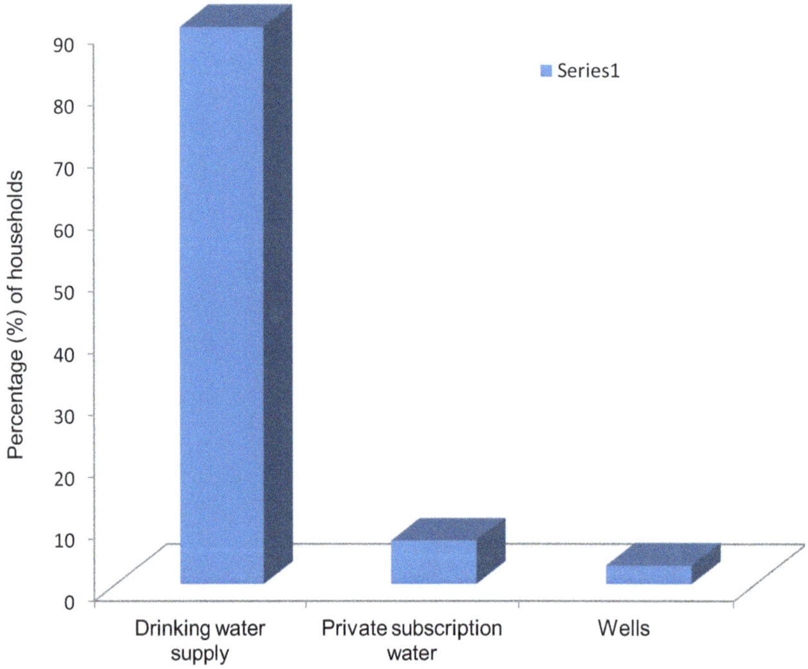

Figure 2. Sources of drinking water supply.

the information system of management, which is the monthly statement of all the activities mad in health centers. Concerning the division of the population in age, the classic range (0-4, 5-14, 15-29 and 30 years and more) used in medicine was selected.

RESULTS

With an average of 10 persons per household, we have a size of 1 000 people. Data collected in health facilities, were from 15 health centers spread over the town with a size of 124 446 people with diseases such as malaria, diarrheal diseases and acute respiratory infections.

Sources of population water supply

Drinking water from Water Supply Company in Côte d'Ivoire (SODECI), wells and public tap of SODECI were identified. It appears from our study that about 90% of investigated households used drinking water of SODECI. Contrariwise, water from public taps resold by individuals in Doukouré's living quarters (7%), as well as, wells of Niangon Adjamé'village (3%) were used to lesser degrees (Figure 2); but it happened that water quality of SODECI was harmed by connection of wastewater with drinking water system.

Sanitation

In different living quarters of the town, human excretas were evacuated either in self purification works, or in collective purification works, or in nature (bush, edges of the lagoon).

It appears from our investigations that about 60% of households used self purification works. Defecation in nature was used too in the town (2%). Anarchic connections to the sewer system were observed especially in precarious living quarters (12%). Collective purification works systems were used in approximately 26% of cases (Figure 3). However, misuse of collective purification works (anarchic connections to the sewer system, dumping of domestic waste into the drains rainwater) lead to the fact that wastewaters were hardly drained. This causes the stagnation of sewage, proliferation of larval gites and strong emissions of foul odors.

Prevalence rate of some diseases and health center attendance

The prevalence rate of malaria, diarrhea and acute respiratory infections of 124 446 people received into consultation in 15 health centers in the town indicates that among all these diseases, malaria diseases were much contracted by people with 58% of cases. After malaria, we have acute respiratory infections (ARI) with 28% of cases. Diarrheal diseases were the least contracted for entire population with 14% of cases. Moreover, we have noted that under-five children often contracted malaria with 69% of cases. Unlike the general population, diarrheal diseases were much contracted by under-five with 26% of cases. Acute respiratory infections

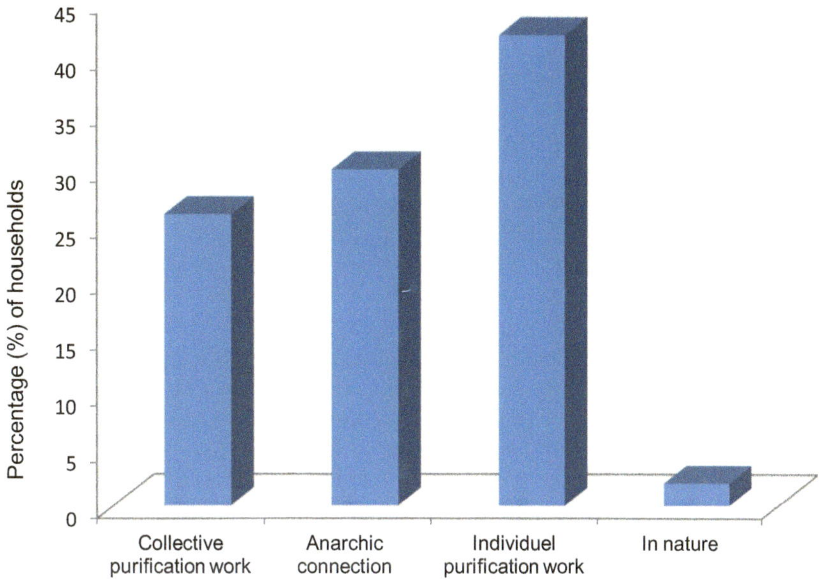

Figure 3. Management mode of wastewater.

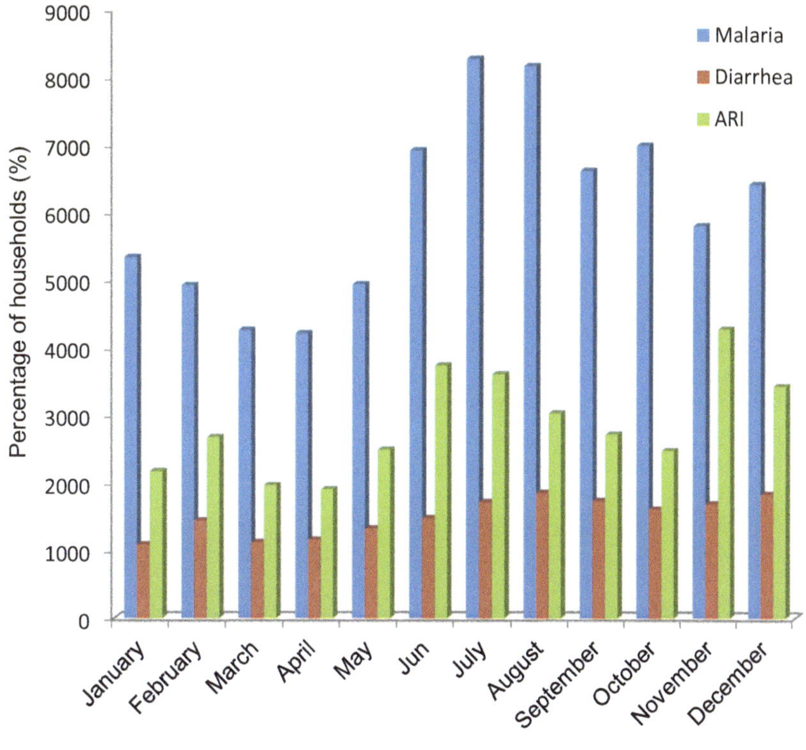

Figure 4. Evolution of pathologies in Yopougon town during the year 2006.

(ARI) come in last place with 5% of cases. Figure 4 showed the evolution of malaria, ARI and diarrheal diseases within the year. Malaria diseases were more raised during June, whereas, the ARI were much more concentrated in the period of February, July and November, respectively corresponding to dry and rainy season unlike diarrheal diseases, which don't change too much during the year. First-line treatment, most households have opted for some treatments without consultation (self-medication) (68%) to the detriment of health facilities where the attendance rate is low (32%).

DISCUSSION

This work uses an interdisciplinary approach in studying poor sanitation and health risks. It reveals that the accessibility of population with drinking water (that provided by SODECI) is very important in Abidjan city. According to the studies of Coulibaly et al. (2004), 99% of households of the municipality of Port - Bouet were connected to the networks of drinking water. This rate approaches that of our study in Yopougon town which is 90% self purification sanitation system. Excreta management is mainly (60%) using in Yopougon. This type of excreta management is found in the study of Coulibaly et al. (2004) in Port-Bouet town. Also, we should note that our work is consistent with the work of Kuitcha et al. (2008), which revealed that the self purification works are much used by Cameroonians. The lack of sanitation contributes much to the deteriorating health of populations (Dongo et al., 2008). Diarrheal diseases are caused by several parasites including *Salmonella* (most frequently occurred) and are located in unhealthy environments, lack of adequate modern sewerage system, or associated with poor environmental hygiene (Farthing, 2000). This situation is much more noticeable in most cities in developing countries where poor environmental and especially section of under five children are most affected by diarrheal diseases (Coulibaly et al., 2004). Our investigations lead to similar observations with a child morbidity by 26% compared to diarrheal diseases.

Moreover, the prevalence rate of malaria registered in health centers is more than twice that found by Dagnan et al. (2002), whose work, extended to the whole Yopougon give a rate of 23.3%. This rate is significantly higher than that found by Coulibaly et al. (46%) in the town of Port-Bouet. This could be explained by the stagnation of sewage into living quarters because of the self purification works due to the strong demographics of the city of Abidjan linked to the politico-military situation of 2002. Ultimately, the results obtained in relation to malaria are in line with rates in Côte d'Ivoire and in countries with endemic malaria transmission (OMS, 2001).

Concerning acute respiratory infections, our study shows that peaks are observed in the month of February, July and November. The peak in February coincides with the dry season while the months of July and November correspond to rainy seasons (large and small rainy season). Guedonon (1987), in Benin found that the IRA also experiences an upsurge during the hot and humid seasons. In terms of our work, the results show that the IRA of under-five children is around 5%. This rate is distinctly below 88% found in the work of Bakonde et al. (1998) in Togo. These results shows that the policies of health authorities in connection with the reduction of ARI through the expanded program on vaccination of under-five are very effective. Indeed, according to WHO recommendations (1990), vaccination against pertussis and measles in particular is needed to reduce the prevalence of ARI.

In terms of therapeutic routes, our results show that over 60% of Yopougon's population practice self-medication. The studies of Bossart (2002) in Benin show the dominating role of the self-medication as therapeutic route. In fact, some medicinal plants used by the populations in the treatment of the malaria (*Cassia occidentalis*) are scientifically proved. Also, it is necessary to note that Guava is sometimes used to treat gastrointestinal pain and diarrhea. Clinical studies show the power of action of some substances extracted from the guava in the treatment of diarrheas (Lozoya et. al., 2002). Besides, a study on 200 Indian mothers, whose children suffered from respiratory infections showed that 25% of the mothers used some honey to treat the cough of their children while 27% used of the ginger (Tona et. al., 2004). But for Kouadio et al (2006), a treatment without prior consultation may be more expensive than treatment after diagnosis of a modern specialist. Thus, ignorance of households on the choice of a suitable therapeutic route appears to be an economic risk factor for household towards the pathologies.

Therefore, in this context of poverty, the development of a good policy of sanitation with the implication of all the factors (state and municipal authorities, communities) is necessary to improve people lives. The results of this study are therefore very interesting database for future actions in the town. They will plan and validate measures to remedy sanitation deficiencies and improve population health.

Conclusion

Finally, sanitation situation in Yopougon town is very worrying. In different living quarters of the town, human excreta are disposed of either in sanitation facilities, or in collective purification works, or in nature. However, misuse of these facilities causes the stagnation of sewage in the streets of neighborhoods, with escape of strong foul odors and proliferation of larval gites. These are all problems that threaten the health and well-being.

In terms of drinking water, the populations surveyed use a lot of water of SODECI for domestic needs. Unlike the wells, they are used to lesser degrees. Regarding the health status of populations, recurrent illness in Yopougon includes malaria, diarrheal diseases and acute respiratory infections. The rates for malaria and respiratory infections are most important during the rainy season, in contrast to diarrheal diseases that are caused by poor hygiene and risk behaviors of populations.

Also it should be noted that under-five are affected by these pathologies especially malaria, which is much contracted by this section of the population. Sensitization on good hygiene practices and good policy of municipal authorities for sanitation is essential to improve people lives. The results of this study will be particularly useful to health authorities and municipalities. The town can be used as tools to raise public awareness on good hygiene practices and sanitation.

ACKNOWLEDGEMENTS

We thank officers of the National Institute of Public Health, Health District of Yopougon and students of the Master of Science and Technology of Water of the 2007-2008 session of the University of Abobo-Adjamé and those KIBIO Group for their important contribution to the success of this research.

REFERENCES

Bakonde B, Tatagan K, Kessie K, Kafechina ABL, Assimadi K, Paupe J, Scheinmann P (1998). Epidémiologie Hospitalière des Infections Respiratoires Aigues (Ira) Basses Chez Le Nourrisson et L'enfant Togolais. Médecine d'Afrique Noire 45(7).

Bossart R (2002). Social relations and the management of illness in Abidjan (Côte d'Ivoire). Urban Health Dev. Bull. 5:13-18.

Carr R (2001). Excreta-related infections and the role of sanitation in the control of transmission. Water Quality : Guidelines, Standards and Health.WHO. Ed. Frewtrell L. and Bartram J., London, UK. ISBN: 1 90022228, pp. 89-113.

Coulibaly L, Diomandé D, Coulibaly A, Gourène G (2004). Utilisation des ressources en eaux, assainissement et risques sanitaires dans les quartiers précaires de la commune de Port-Bouët (Abidjan; Côte d'Ivoire). Vertigo 3(5).

Dagnan NS, Kone B, Tiembre I, Ekra KD, Benié VJ, Ndoutabe M, Tagliante-Saracino J (2002). Etude de la prévalence du paludisme dans la commue de Yopougon Abidjan- Côte d'Ivoire. Médecine d'Afrique Noire 49(11):507-510.

Djaman JA, Yapi A, Dje MK, Diarra JN, Guede-Guina F (2001). Sensibilité *in vitro* à la chloroquine de *Plasmodium falciparum* à Abidjan. Médicine d'Afrique noire. 48:371-374.

Dongo K, Kouamé KF, Koné B (2008). Analyse de la situation de l'environnement sanitaire des quartiers défavorisés dans le tissu urbain de Yopougon a Abidjan, Côte d'Ivoire. Vertigo 3(8).

Ezzati M, Utzinger J, Cairncross S, Cohen AJ, Singer BH (2005). Environmental risks in the developing world: exposure indicators for evaluating interventions, programmes, and policies. J. Epidemiol. Community Health 59:15-22.

Farthing MJG (2000). Diarrhoea: a significant worldwide problem. Int. J. Antimicrob. Agents. 14(1):65-69.

Forget G, Lebel J (2002). An ecosystem approach to human health. Int. J. Occup. Environ. Health 7:3-36.

Guedonon A (1987). Contribution à l'étude des staphylococcies pleuropulmonaires au CHU de Cotonou. Thèse Méd. Cotonou, no. 13, pp. 111.

Hibberd P (1998). Risk factors for childhood diarrheal disease in communities involved in the CIMEP Bénin Project. Methodology et and results of the baseline survey. EHP (Environmental Health Project). USAID, p. 33.

Institut National de la Statistique (INS) (1998). Synthèse des résultats définitifs du Recensement Général de la Population et de l'Habitat (RGPH-98), p. 32.

Kouadio AS, Cissé G, Obrist B, Wyss K, Zinsstag J, Yao YJ, Tanner M (2006). Fardeau économique du paludisme sur les ménages démunis des quartiers défavorises d'abidjan, Côte d'ivoire-Vertigo 3(3).

Kuitcha D, Kamgang KBV, Sigha NL, Lienou G, Ekodeck GE (2008). L'approvisionnement en eau, assainissement et risques pour la santé à Yaoundé, Cameroun- Journal africain de Sciences et Technologies Environnementales 2(11).

Lozoya X, Reyes-Morales H, Chávez-Soto MA, Martínez-García MC, Soto-González Y, Doubova SV (2002). Intestinal anti-spasmodic effect of a phytodrug of *Psidium guajava* folia in the treatment of acute diarrheic disease. J. Ethnopharmacol. 83:19-24.

Memain D (2003). Lutte contre le paludisme en Côte d'Ivoire. Programme National de Lutte contre le paludisme, Rapport annuel sur l'état d'avancement, p. 65.

Organisation Mondiale de la Santé (OMS) (1985). Manuel de l'aménagement de l'environnement en vue de sa démoustication., Organisation Mondiale de la Santé, Génève.

Organisation Mondiale de la Santé (OMS) (1992). Evacuation des eaux de surface dans les communautés à faibles revenus. Organisation Mondiale de la Santé, Génève.

Organisation Mondiale de la Santé (OMS) (2001). Roll Back Malaria (RBM) –Recule du paludisme. Aspects économiques du paludisme, Genève.

Programme Nationale de Lutte contre le Paludisme (PNLP) (2003). Rapports annuels d'activités 2001, 2002, 2003. Document technique, p. 27.

Schwartzbrod L (1992). Virologie des eaux : Aspects épidémiologiques. Microbiologie-Aliment-Nutrition 10:213-222.

Strauss M, Heinss U, Montangero A (2000). On-site sanitation : When the pits are full. Planning for resource protection in fecal sludge management. Proceedings, Int. Conference, Bad Elster 20-24 Nov. 1998. Schriftenreihe des vereins für wasser-, boden- und Lufthygiene, 105 :Water, sanitation and Health-Resolving conflict between drinking-Water demands and pressures from society'wastes (I. Chorus, U. Ringelband, G. Schlag and O. Schmoll, eds.). IWA Publishing House and WHO Water Series. ISBN No. 3-932816-34-X.

Tona L, Cimanga RK, Mesia K, Musuamba CT, Bruyne TD, Apers S, Hernans N, Van-Miert S, Pieters L, Totte J, Vlietinck AJ (2004). In vitro antiplasmodial activity of extracts and fractions from seven medicinal plants used in the Democratic Republic of Congo. J. Ethnopharmacol. 93:27-32.

United Nations (2006). World Urbanization Prospects : The 2005 Revision. United Nations, New York. pp. 210.

5

The logistic model-generated optimal stocking levels, optimal species mix and carrying capacity of wild herbivores in southern rangelands game ranches in Kenya

P. I. D Kinyua, J. T. Njoka and Chris O. Magero

Department of Land Resource Management and Agricultural Technology, University of Nairobi, P.O. Box 29053 - 00625 Nairobi, Kenya.

This paper derives logistic models for Grant's gazelle (*Gazella granti*), Thompson's gazelle (*Gazella Thompsonii*) and plains Zebra (*Equus burchellii*), in a game ranching system. The study was conducted on an 8100 hectares savannah ecosystem in Machakos district, with an average annual rainfall of 514±159.3 mm. Modelled as discrete-time logistic equations with fixed carrying capacities, it captures the wildlife herbivore population dynamics. Time series data, covering a period of 16 years, is used to generate the fixed carrying capacities and the interaction parameters endogenously. The estimation of the logistic models involves estimation of econometric models for each herbivore species, followed by the recovery of the carrying capacities, mathematically. The model-generated carrying capacities are 144.30 AU, 75.53 AU and 54.65 AU; the equivalent number of animals is 986, 929 and 144; for Grant's gazelle, Thompson's gazelle and Zebra, respectively. The derived "maximum sustainable yield" (MSY) stocking levels is 44 AU, 24 AU and 29 AU; the equivalent animal number is 297, 393 and 76; for Grant's gazelle, Thompson's gazelle and Zebra, respectively. Similarly, the MSY off-take level is 10.42 AU, 8.39 AU and 6.37 AU for these species respectively. These results show that the discrete-time logistic models are applicable to sustainable management of game ranching enterprise.

Key words: Carrying capacity, animal units, game ranching, species mix, logistic models, MSY.

INTRODUCTION

Proper game management requires good grazing management strategies that provide effective means of manipulating the grazing game animals, so as to achieve the highest level of animal production sustainably (Pratt and Gwynne, 1977). The key issues in grazing management are the determination of optimal stocking level, optimal species mix and carrying capacity; the latter is the upper bound stocking level. The current method of estimating these parameters involves trial and error approximations, based on the availability of forage on the one hand, and the animal forage requirements on the other (Pratt and Gwynne, 1977; Stoddart et al., 1975). This method is clearly subjective. An alternative method of determining herbivore optimal stocking levels, optimal species mix and carrying capacities is based on systems analysis procedures (Njoka and Kinyua, 2006; Kinyua, 1998).

This method involves deriving a set of logistic equations for each species in a given ranch (Equations 1) (Njoka and Kinyua, 2001), coupled with recovery of carrying capacities from these equations mathematically. This approach is completely objective, and it is the underlying the objective of this study.

Under this formulation, both carrying capacity and exchange ratios are endogenously determined (Kinyua and Njoka, 2001), making it possible to empirically estimate the population growth models for Grant's gazelle, Thompson's gazelle and Zebra.

$$H_{it+1} - H_{it} = \Delta H_{it} = \beta_i H_{it}\left(1 - \frac{(H_{it} + \sum_{j=1}^{n-1}\lambda_{ij}H_{jt})}{K_i}\right) - Y_{it} \quad (1)$$

for $i = 1,\ldots,n-1$ $j = 1,\ldots,2$ and $i \neq j$.

Here H_{it+1}, measured in animal units, is the standing herbivore population at period t+1 where time step is measured in periods of six months. Y_{it+1} is a control variable, representing the off-take level of Herbivore species *i* at period *t+1* and is measured in animal units; H_{it} represents the standing biomass of game herbivore species *i* at period *t* and is measured in animal units; and K_i, also measured in animal units, represents the carrying capacity for game herbivore species *i*. The parameter β_i represents the exponential growth rate for the game herbivore species *i*. The parameter σ_{ij} represents the interaction effect of game herbivore species *j* on species i, and this interaction parameter is negative, zero or positive depending on whether the interactive relationship is complementary, supplementary or competitive, respectively. This parameter also gives an estimate of exchange ratios or grazing pressure equiva-lence among species, that is, it plays the role of converting H_{jt} animal units into the equivalent H_{it} animal units. For this particular study, "n" is equal to three animal species.

Conceptual view

The formulation of wild herbivores' discrete-time logistic equations with fixed carrying capacity is captured in Equations1 which are difference equations (Starfield and Bleloch, 1986; Anderson, 1991; Caughley and Sinclair, 1994; Njoka and Kinyua, 2006). Conceptually, the term K_i represents the maximum number of animal units of game herbivore species *i* that can be supported through period *t*. There are three cases as shown in (Equations 2, 3 and 4).

Case 1 (stocking level below carrying capacity):

$$K_i \succ (H_{it} + \sum_{j}^{n-1}\sigma_{ij}H_{jt}), \quad (2)$$

for $i = 1,\ldots,n$, $j = 1,\ldots,n-1$ and $i \neq j$.

This implies that the animal unit months (AUMs) of available forage exceeds the AUMs of forage demand by the standing population of game herbivores species *i*. The standing game herbivore population increases since there is enough forage to meet their maintenance requirements and leave a balance for growth and reproduction.

Case 2 (stocking level equal to carrying capacity):

$$K_i = (H_{it} + \sum_{j}^{n-1}\sigma_{ij}H_{jt}), \quad (3)$$

for $i = 1,\ldots,n-1$, $j = 1,\ldots,n-1$ and $i \neq j$.

This implies that the AUMs of available forage exactly matches the AUMs of forage demand by the standing population of game animal species *i*. The standing herbivore population barely meets its maintenance requirements, leaving no balance for growth and reproduction. As a result, the population change is zero, implying that H_{it} is at the maximum stocking level.

Case 3 (stocking level above carrying capacity):

$$K_i \prec (H_{it} + \sum_{j}^{n-1}\sigma_{ij}H_{jt}), \quad (4)$$

for $i = 1,\ldots,n$, $j = 1,\ldots,n-1$ and $i \neq j$.

In this case the forage demand by wild herbivore species *i* is greater than the AUMs of available forage, or the standing herbivore population is greater than the carrying capacity. The excess number of animals will die out, resulting in the decline of the standing population of herbivore species *i*.

The discreet-time game herbivore logistic models, (Equations 1) is a quadratic equations without an intercept (Beattie and Taylor, 1985) and are derived from the discrete time compounding equations (Equations 5a).

$$H_{it} = H_{i0}(1+r_i)^t - Y_{it}, \quad (5a)$$

for $i = 1,\ldots,n$.

Where, H_{it} represents the population of herbivore species *i* at period *t*; H_{i0} represents the initial population (at $t = 0$) of herbivore species *i*; r_i represents the rate of herbivore population growth per period:, Y_{it} represents the off-take level for herbivore species *i* at period t; t represents discrete time-steps of periods of six months. The term $H_{i0}(1+r_i)^t$ is the discrete-time compounding formula, capturing the population growth of herbivore species *i*. When $t = 1$, Equations (5a) become Equations (5b).

$$H_{i1} = H_{i0}(1+r_i)^1 - Y_{i1}, \quad (5b)$$

for i = 1,…, n.
When $t=2$, Equations (5b) become Equations (5c).

$$H_{i2} = H_{i0}(1+r_i)(1+r_i) - Y_{i2} = H_{i1}(1+r) - Y_{i2} \quad (5c)$$

for i = 1,…,n.

The general form of Equations (5c) is shown in Equations (5d); these equations are independent of time

$$H_{it+1} = H_{it}(1+r_i) - Y_{it+1}, \quad (5d)$$
for i = 1,…,n.

From an ecological point of view, the rate of herbivore population growth per period (r_{i}), is hypothesized to be determined by intra- and inter-species competition thus it is more realistically modelled as shown in Equations (5e).

$$r_i = \{\beta_i - (\delta_i H_{it} + \textstyle\sum_j^{n-1} \lambda_{ij} H_{jt})\}.$$
for i = 1,…,n, j = 1,…,n-1 and i ≠ j. (5e)

Here parameter β_i represents the fixed rate of population change for species *i* while parameter δ_i represents intra species competition. Parameter λ_{ij} represents inter species interaction between species *i* and species *j*; this parameter is negative, zero or positive depending on whether the interactive relationship is competitive, supplementary or complementary, respectively Substituting (r_i) in Equations (5d) with (r_i) in Equations (5e) yields Equation (5f)

$$H_{it+1} = H_{it}\{1 + [\beta_i - (\delta_i H_{it} + \sum_{j=1}^{n-1} \lambda_{ij} H_{jt})]\} - Y_{it+1}$$
for i = 1,…,n, j = 1,…,n-1 and i ≠ j. (5f)

Multiplying out Equations (5f) and rearranging terms yields Equations (6a), and like Equations (1), they are different equations. Shifting Y_{it+1} from the right to the left hand side of Equations (6a) yields Equations (6b). Equations (6b) are the econometric (statistical) models, and although they are nonlinear in variables, they are linear in parameters.

$$H_{it+1} - H_{it} = \Delta H_{it} = \beta_i H_{it} - \delta_i H^2{}_{it} + \sum_{j=1}^{n-1} \lambda_{ij} H_{it} H_{jt} - Y_{it+1}, \quad (6a)$$
for i = 1,…,n, j = 1,…,n-1 and *i ≠ j.*

$$H_{it+1} - H_{it} + Y_{it+1} = \Delta H_{it} + Y_{it+1} = \beta_i H_{it} - \delta_i H^2{}_{it} + \sum_{j=1}^{n-1} \lambda_{ij} H_{it} H_{jt}, \quad (6b)$$

for i = 1,…,n,
j = 1,…,n-1 and
i ≠ j.

MATERIALS AND METHODS

Study area

The study area is located on the David Hopcraft Ranch, 40km south-east of Nairobi, on the Athi-Kapiti Plains along the Nairobi-Mombasa road in Machakos District, Eastern Province of the Republic of Kenya. The ranch area is 8100 ha and its altitude is 1600 m to 1700 m above sea level. The ranch vegetation is typical of eco-climatic zone 4 (Pratt and Gwynne, 1977). The ranched vegetation comprises of *Themeda triandra* grassland, a mixture of *Balanites glabra-Themeda triandra* tree grassland, *Acacia drepanolobium* and *Balanites glabra* bush grassland, and *Acacia seyal* and *Acacia xanthophloea* riverine woodland. Its average annual rainfall is 514±159.3 mm. The median rainfall for the study site for the same period was 599.4 mm between 2001 and 1978 with an annual range from 177.4 to 924.5 mm.

Methods

This study focused on analysis of sustainable management and sustainable conservation of Grant's gazelle, Thompson's gazelle and Zebra, coupled with game cropping. The Zebra is a large non-ruminant grazer (267 - 318 kg) capable of consuming bulk forage especially grasses. The Thompson's gazelle (20 to 25 kg) and is a selective feeder consuming highly nutritious grasses and while the Grant's gazelle (32 to 42 Kg) is slightly stockier than the Grant's gazelle, and is a mixed feeder utilizing both high quality grass and browse (Hofmann and Stewart, 1972). The feeding habits of the Grant's gazelle would therefore overlap with that of the Thompson's gazelle and the Zebra depending on forage availability and level of interspecies and intrapecies competition. Grazing requirement is also based on the metabolic weight of the animal which is calculated from the Animal Unit Coefficient (AUC) Table 1.

Through on-the site visits to the David Hopcraft ranch, during 1995 and 1996, time series data covering the period 1981 through 1996 on Grant's gazelle Thompson's gazelle and Zebra populations and off-take levels were collected from ranch records. Animal numbers and off-take levels were then converted to animal units (AUs), effectively converting discrete data to continuous data, where one animal unit (AU) is equivalent to the metabolic body weight of 454 kg. Based on this data, the econometric population growth models, Equations (6b), were estimated for each of the three herbivore species, namely, Grant's gazelle, Thompson's gazelle and Zebra, using SPSS software (Kleinbaum and Kupper, 1978). This procedure was used to estimate livestock growth models (Njoka and Kinyua, 2006); these models can also be estimated using cross-sectional data.

After the estimation of the econometric models (Equation 6b), the logistic models (Equations 1) is recovered (Figure 1) from the former as follows:

The derived logistic models, that is,

$$\Delta H_{it} = \beta_i H_{it}\left(1 - \frac{H_{it} + \sum_{j=1}^{2} \sigma_{ij} H_{jt}}{K_i}\right) - Y_{it+1}$$

for i = 1,…,3, j = 1,…,2 and i □ j; portray three scenarios as follows:

(a) When $Y_{it+1} \succ \Delta H_{it}$,
for I = 1,…3

Table 1. Approximate exchange ratios for animals based on metabolic body weight.

Species	Approximate weight (Kg)	Animal unit coefficient
Grant's gazelle	35	0.1463
Thompson's gazelle	23	0.0813
Zebra	272	0.38

Source: (Kinyua, 1998)

$$\Delta H_{it} = \beta_i H_{it}(1 - \frac{H_{it} + \sum_{j=1}^{2} \sigma_{ij} H_{jt}}{K_i}) - Y_{it+1}$$

the game harvests are higher than the populations growth per period; as a result, the populations decline.

(b) When $Y_{it+1} = \Delta H_{it}$,

for i = 1,...,3, game harvests are equal to the t growth of the pulations per period; as a result, the populations are constant or at a steady state.

This also means that

$$\Delta H_{it} = 0$$

and that,

$$Y_{it+1} = \beta_i H_{it}(1 - \frac{H_{it} + \sum_{j=1}^{2} \sigma_{ij} H_{jt}}{K_i})$$

for i=1,...,3;

These equations are quadratic production functions without intercept. Multiplying out these equations yield their equivalent counterpart, also production functions without intercept (equations 6c), as follows:

$$Y_{it+1} = \beta_i H_{it} - \delta_i H^2{}_{it} + \sum_{j=1}^{n-1} \lambda_{ij} H_{it} H_{jt}$$

for i = 1,...,3, j = 1,...,2 and i □ *j.*

These equations are then solved for the first order conditions which, in turn, are solved simultaneously because the animal species share resources. The solutions are the "maximum sustained yield" stocking levels; by substituting the values of "maximum sustained yield" stocking levels into equations 6c, the outcomes are "maximum sustained yield" off-take levels.

(c) When

$$Y_{it+1} \prec \Delta H_{it},$$

for i = 1,..., 3, the game harvests are less than the growth of the populations per period; as a result, the populations increase.

RESULTS AND DISCUSSION

The estimation results of Grant's gazelle, Thompson's gazelle and Zebra econometric models (equations 6b) are presented in Table 2. The F-test for the econometric models for the Grant's gazelle, Thompson's gazelle and Zebra were significant ($P < 0.05$). This shows that the econometric model is suitable for capturing the population change for all three species. The exponential growth rate parameters for the three models were positive, as expected, and significant ($P < 0.05$) for all species. The intra-species competition parameters were negative, as expected, and significant ($P < 0.05$) for the three models.

The interaction parameter of Grant's gazelle and Thompson's gazelle were negative ($P<0.05$), suggesting that Thompson's gazelle is competitive with Grant's gazelle. In retrospect, this negative interaction of Thompson's gazelle with Grant's gazelle could emanate from overlap of their feeding habits. Grant's gazelle is a mixed feeder, eating both grass and browse, while Thompson's gazelle is a grazer. Both species prefer open grassland plains. By virtue of Grant's gazelle being a grazer, its feeding habit overlaps with that of Thompson's gazelle and consequently they tend to compete for the available resources.

The interactive parameter of Grant's gazelle with Zebra is negative ($P<0.1$), suggesting that the two species are competitive on the grounds that the former is a mixed feeder while the latter is a grazer. The R^2 for the model is 0.413. Thompson's gazelle's interaction parameter with Zebra is negative ($P<0.05$) depicting a competitive relationship which may explain why Zebra competes with Thompson's gazelle. Grant's gazelle interaction parameter is positive suggesting complementary relationship. However the parameter was not significant ($P>0.1$) The R^2 for Thompson's gazelle model is 0.45 which is good.

For the Zebra model, the Thompson's gazelle's interaction parameter is negative and significant ($P<0.05$), suggesting competitive relationships between the two species. Both Zebra and Thompson's gazelle are grazers preferring open grasslands plains. By virtue of their dietary requirement, Thompson's gazelle likely competes with Zebra for available feed resources. In contrast

Table 2. Parameter estimates for the econometric models. Equation (6b) (n = 3)[a]

Species	Gr	Th	Zb
Gr	0.722 (3.842)**	0.001 (0.001)	0.007 (1.193)
Gr^2	-0.005 (-2.121)**		
	-0.006 (-2.356)**	1.027 (4.335)**	-0.017 (-4.249)**
Th^2		-0.014 (-4.031)**	
Gi^2			
Zb	-0.005 (-1.851)	-0.014 (-3.836)**	1.258 (5.235)**
Zb^2			-0.023 (-4.577)**
R^2	0.413	0.45	0.560
F	(4.755)**	(5.522)**	(8.579)**

Gr=Grants gazelle; Th=Thompson's gazelle; Zb=Zebra; [a] Results are for equations identified in top row of the table, with parameters indicated in the left-hand column; F and t-statistics are in parenthesis: ** indicates significant at the 0.05 level and * at the 0.1 level.

Table 3. Parameter values for the logistic population models, Equation (1), after derivation from the econometric models in.

Species	Gr	Th	Zb
Gr	0.722	-0.074	0.006
Th	1.198	1.027	-0.328
Zb	1.0	1.0	1.258
K[b]	144.3 (986)	75.529 (929)	54.645 (144)

Gr, Grant's gazelle; Th, Thompson's gazelle and Zb, Zebra [a]Results are for equations identified in top row of table, with parameters indicated in the left-hand column; t-statistics are in parenthesis: ** indicates significant at the 0.05 level and * at the 0.1 level.
[b]Carrying capacity in AUs; figures in bracket represent carrying capacities in animal numbers.

the intra-species parameter between Zebra and grant's gazelle is slightly but not significantly positive (P>0.1), suggesting complementary relationship. This may be explained by the likelihood that Grant's gazelle does not compete for the grass favoured by Zebra.

The other components of the results are the recovery of the logistic models to estimate the species carrying capacity, the derivation of the "maximum sustained yield" stocking level and the "maximum sustained yield" off-take level.

The derivation of the logistic models, as discussed earlier, is shown in Table 3 for Grant's gazelle, Thompson's gazelle and Zebra. The carrying capacities in animal units (AUs) are 144.3, 75.529 and 54.645 for Grant's gazelle, Thompson's gazelle and Zebra, respectively. This converts to 986 Grant's gazelle, 929 Thompson's gazelle and 144 Zebra. The Carrying Capacity is the upper boundary stocking rate for the ranch.

The "maximum sustainable yield" (MSY) stocking levels are 43.504AUs, 23.840AUs and 28.785AUs, or 279, 293 and 76 animals for Grant's gazelle, Thompson's gazelle and Zebra, respectively. The "maximum sustained yield off-take" levels are 10.42AUs, 8.39AUs and 6.37AUs or 71, 103 and 17 animals for these species, respectively.

Conclusion

The key issues in the sustainable management of wild and domestic herbivores are the optimal stocking levels, the carrying capacities, the "maximum sustained yield off-take" levels and the optimal species mix. This systems analysis approach, based on the logistic model, has been applied here to solve the range manager dilemma. This is a novel ground breaking approach, which provides an alternative to showing how the key parameters can be determined in a ranching enterprise. The results obtained in this study confirmed that it is possible to derive the "fixed carrying capacity" logistic model for three wildlife species. These models have in turn been used to estimate the carrying capacity for each of the species and provide solutions for the MSY. Observations at the David Hopcraft ranch showed stocking level numbers of between 138 and 552 Grant's gazelle, 155 and 971 Thompson's gazelle and 122 and 154 Zebra over the period covered by the study, while the carrying capacity models predicted 986, 929 and 144 animal numbers respectively. In comparison the Grant's gazelle remained consistently understocked over the years; the Thompson's gazelle on average was understocked while the Zebra fluctuated within the range predicted by the model. In particular, this procedure has direct application in determining the number of animals to be harvested in a wildlife cropping policy and in wildlife culling as a management tool. It is also usable in determining the animal numbers that should be translocated for management purposes. Finally, this procedure can be used to analyse the dynamics of wild herbivores, including issues like the population doubling time.

REFERENCES

Anderson SH (1991). Managing our Wildlife Resources. Englewood, New Jersey: Prentice Hall Inc.

Beattie BR, Taylor CR (1985). The Economics of Production. New York: John Wiley & Sons.

Caughley GJ, Sinclair AR (1994). Wildlife Ecology and Management. Boston: Blackwell Scientific Publications.

Hofmann RR, Stewart DR (1972). Grazer and Browser: A Classification Based on Stomach Structure and Feeding Habit of East African Ruminant Mammalia.

Kinyua PI (1998). Game ranching in Machakos District Kenya: An applicaton of mathematical programming to the study of wildlife policy. University of British Columbia.

Kinyua PI, Njoka JT (2001). Animal Exchange Ratios: An Alternative Point of View. Afr. J. Ecol. 59 - 64.

Kleinbaum DG, Kupper L (1978). Applied Regression Analysis and Other Multivariable Methods. Belmont: Thomson Learning Academic Resource Center.

Njoka JT, Kinyua PI (2006). The logistic model - generated carrying capacities, maximum sustained off-take and optimal stocking rates for Kenya's commercial ranches. 23(2).

Pratt DJ, Gwynne MD (1977). Rangeland Management and Ecology in East Africa. Hutington, New York: Robert E. Krieger Publishing Company.

Starfield AM, Bleloch AL (1986). Building Models for Conservation and Wildlife Management. (New York and London): Macmillan.

Stoddart AL, Smith AD, Box TW (1975). Range Management. New York: McGraw - Hill, Inc.

6

Evaluation of copper and lead immobilization in contaminated soil by single, sequential and kinetic leaching tests

R. A. Wuana[1], S. G. Yiase[2], P. D. Iorungwa[1] and M. S. Iorungwa[1]

[1]Department of Chemistry, University of Agriculture, Makurdi, 970001, Benue State, Nigeria.
[2]Department of Chemistry, Benue State University, Makurdi, 970001, Benue State, Nigeria.

The effectiveness of natural clay, calcium phosphate, poultry manure and rice husks as cheap and ecologically non-invasive amendments for immobilizing Cu and Pb in contaminated soil was assessed. A moderately contaminated soil was sampled from a cultivated field in the vicinity of an active waste dump, characterized and amended with the immobilizing agents at 5 to 20% w/w. Single (calcium nitrate), sequential (optimized European Communities Bureau of Reference, BCR method) and kinetic (ethylenediaminetetraacetic acid, EDTA) leaching tests were performed before and after soil amendments. Single extractions showed that metal immobilization efficiencies, *E* were generally high ($20 \leq E \leq 81\%$), increased as the agent dose and varied in the order: phosphate > natural clay > poultry manure > rice husks. BCR extractions showed that the agents attenuated Cu and Pb mobilities and induced changes in their fractionation patterns as most of the hitherto mobile fractions became associated with the less bioavailable forms. Two metal forms: readily labile and less labile, with apparent leaching rates k_1 (10^{-2} min^{-1}) and k_2 (10^{-3} min^{-1}), respectively were identified by fitting EDTA leaching data into a two-first-order model. The immobilizing agents suppressed both k_1 and k_2 for Cu and only k_2 for Pb. These agents may reduce risks resulting from Cu and Pb contamination in soil since metal mobility is related to its bioavailability.

Key words: Bioavailability, chemical extraction, contaminated soil, immobilizing agent.

INTRODUCTION

The disturbance and acceleration of nature's slowly occurring biogeochemical cycle of metals by anthropogenic activities have made most soils of rural, urban, agricultural and industrial areas to be contaminated by one or more of the toxic heavy metals (D'amore et al., 2005; Rajaganapathy et al., 2011; Corrochano et al., 2011). The term heavy metal has been widely adopted for a group of metals and metalloids and their compounds probably due to their toxicity in biological systems as well as contamination and persistence in environmental media (Bhat and Khan, 2011). Heavy metal contamination of soil may pose risks and hazards to humans and the ecosystem through: direct ingestion or contact with contaminated soil, the food chain (soil-plant-human or soil-plant-animal-human), drinking of contaminated ground water, reduction in food quality (safety and marketability) via phytotoxicity, reduction in land usability for agricultural production causing food insecurity, land tenure problems and economic losses (McLaughlin et al., 2000a, b; Suruchi and Khanna, 2011;

Matini et al., 2011). Contamination of soil with heavy metals such as copper and lead particularly appears to be virtually permanent. Copper, for instance, is an essential heavy metal, but in high doses can cause anaemia, liver and kidney damage, stomach and intestinal irritation, while lead is a non-essential heavy metal that affects the functioning of the blood, liver, kidney and brains of human beings (Irmak, 2010; Sarma, 2011). Copper normally occurs in soil from horticultural preparations, copper pipes, irrigation water as well as from additives designed to control algal growth, while lead is a component of most industrial/domestic paints and leaded petrol in addition to untreated sewage from residential areas (Yap et al., 2008; Irmak, 2010). Increasing public concern about environmental and soil contamination has led to the promulgation of national and international regulations to reduce pollution and to remediate polluted sites. Consequently, over the last few decades a great deal of research has been carried out in order to develop strategies for the management of agricultural ecosystems (Chukwuka and Omotayo, 2008), as well as remediation of former industrial sites which have been exposed to diffuse pollution by toxic heavy metals (Finžgar et al., 2006; Belviso et al., 2010). Among the remediation technologies available for contaminated sites, *in situ* (in place) immobilization methods are relatively more advantageous since they are generally less expensive and disruptive to the natural landscape, hydrology and ecosystems than conventional excavation, treatment and disposal methods.

In situ immobilization refers to the reduction of metal bioavailability and leachability in contaminated soils using environment-friendly amendments. Since metal solubility is related to its mobility and bioavailability, metal immobilization may reduce environmental risk. Thus, the success of a strategy for in situ immobilization of metals can be evaluated by the ability to reduce contaminant solubility/extractability, which must also have a long-term effect (Alvarenga et al., 2008). Heavy metal-contaminated soils have classically been ameliorated using additives such as lime, phosphate, red mud, bone meal and organic matter and their effectiveness for field applications have already been demonstrated (Awodun, 2007; Fawzy, 2008; Gatimu et al., 2010a, b; Santos et al., 2010; Florido et al., 2011; Ogbo and Okhuoya, 2011). Most research efforts aimed at assessing the effectiveness of amendments for *in situ* immobilization have either employed one or combination of two single, sequential and kinetic extractions (Cao et al., 2003; Fawzy, 2008; Santos et al., 2010). In this study, single, sequential and kinetic extractions have been conjointly employed to assess the ability of some cheap, non-toxic agents- natural clay, phosphate salt, poultry manure and rice husks at immobilizing copper and lead in a contaminated soil. The study was carried out on a composite soil, sampled from a cultivated field in the vicinity of a waste dump at the Benue Industrial Layout, Makurdi, a rapidly growing city in the Lower Benue River Basin, a major agricultural zone in North-Central Nigeria from January to May, 2011.

MATERIALS AND METHODS

Materials

Multipurpose flask shaker (Model TT 12F, Techmel and Techmel, Texas, US), digitally operated high speed centrifuge (Model TGL-16G, Shanghai, China), pH meter (Fisher Hydrus 300 model), normal laboratory glassware (borosilicate), polyethylene vessels, electronic weighing balance (Gallenkamp 80) were used. A flame atomic absorption spectrophotometer (Buck Scientific Model 200A, Norwalk, Connecticut, US) was used for metal assay. Natural clay, phosphate salt [$Ca_3(PO_4)_2$], poultry manure and rice husks were used as immobilizing agents.

Sample collection, pre-treatment, characterization and amendment

Five top (0 to 20 cm) soil samples were collected at a cultivated field around an active waste dumpsite at the Benue Industrial Layout, Makurdi (7.44°N, 8.33°E), North-Central Nigeria. The soil samples were collected into polythene bags and taken to the laboratory, air-dried, ground and sieved to give <2 mm particle size and composited as the parent soil. Standard operating procedures were used to test soil properties: pH (1:25 soil/water ratio) (Maxted et al., 2007), bulk density (CDPRHAP, 1998), particle size distribution by the hydrometer method (Bouyoucos, 1962), organic matter by the Walkley-Black rapid acid dichromate oxidation (Schumacher, 2002), cation exchange capacity by the summation of Na, K, Ca and Mg (Jackson, 1962; Black, 1965), soil total metals by aqua regia digestion (ISO, 1995). Similarly, the immobilizing agents: natural clay, phosphate salt, poultry manure and rice husks were characterized, as the case may be, in terms of some of the afore-mentioned attributes and evaluated for their effectiveness at immobilizing Cu and Pb in contaminated soil. Different portions of a fixed mass of soil (1 kg) were separately put in polyethylene vessels and treated with 0, 52.6, 111.2 and 250 g (0, 5, 10 and 20% w/w) of the agent. Each treatment was performed in triplicate and incubated for 1 month at ambient temperature.

Single batch extractions

Portions of the unamended and amended soils were used. The single extractant was expected to leach the plant available metals. This scheme adopted the procedure of Mench et al. (1994). That is, 10 g of soil (<2 mm) was shaken with 0.1 M $Ca(NO_3)_2$ (50 mL) for 2 h at 20°C. Extracts were filtered through ashless filter paper. The 0.1 M $Ca(NO_3)_2$ extracts were acidified with 14 M HNO_3 (1.0 mL) to prevent metal adsorption.

BCR sequential extractions

The geochemical forms of Cu and Pb viz: extractable, M_1; oxide-bound, M_2; organic matter-bound, M_3 and residual, M_4 fractions in both the pre-amended and post-amended soil samples were assessed by the sequential extraction scheme proposed by the European Communities Bureau (BCR method) described by Ure et al. (1993).

Table 1. Physicochemical properties of the parent soil and immobilizing agents used for the study.

Property	Parent soil	Natural clay	Phosphate salt	Poultry manure	Rice husks
pH	5.7±1.3	6.0±0.7	8.1±0.1	6.9±0.0	6.5±0.5
Sand	34.8±2.5	34.8±2.5	-	-	-
Silt	22.0±2.2	10.0±2.0	-	-	-
Clay	31.0±3.1	55.2±4.6	-	-	-
OM (%)	5.8±0.3	0.9±0.1	-	45.3±1.3	-
CEC (meq/100 g)	10.6±2.2	13.0±0.5	-	7.9±1.0	-
Total Cu (mg/kg)	173.0±7.5	0.5±0.1	-	100.6±3.0	5.5±0.0
Total Pb (mg/kg)	133.0±6.7	0.9±0.0	-	1.5±0.0	1.5±0.1

Mean ± SD, n = 3.

Kinetic extractions

One gram portions of both the unamended soil and its sub-samples treated with the immobilizing agents (at 20% w/w) were separately transferred to 100 mL beakers provided. Aliquots of 50 mL of the extracting solution (0.05 M ethylenediaminetetraacetic acid, EDTA) were added to each of the beakers at a solid/solution ratio of 1:50 and the mixtures was stirred manually, homogenized with a vortex mixer, and then kept on a multipurpose flask shaker. At the elapse of selected time intervals: 10, 20, 30 min and 1, 2, 6, 10, 12, 16 h; the slurries were filtered through an ash less filter paper, centrifuged and kept for Cu and Pb assay.

RESULTS AND DISCUSSION

Physicochemical characteristics of parent soil and immobilizing agents

Some physicochemical properties of the parent soil and immobilizing agents are reported in Table 1. Textural analysis characterized the parent soil as sandy clay loam. The other properties were: pH (5.7), organic matter (5.8%), cation exchange capacity (10.6 meq/100 g). Soil total Cu and Pb were 171.0 and 133.0 mg kg^{-1}, respectively indicating moderate contamination with Cu and Pb. Mean values of the physicochemical attributes of natural clay were: pH (6.0), sand (34.8%), silt (10.0%), clay (55.2%), organic matter (0.9%), cation exchange capacity (13.0 meq/100 g). The only attribute measured for phosphate salt was pH value of 8.1. Poultry manure had the following attributes: pH (6.9), organic matter (45.3%), cation exchange capacity (7.9 meq/100 g) and total Cu and Pb, 100.6 and 1.5 mg/kg, respectively. The markedly high concentration of Cu in poultry manure may be attributed to its addition as a growth promoter to feed preparations (Wuana and Okieimen, 2011) and so during soil amendment, the manure needs to be scrupulously added at such doses that would not trigger short-term leaching pulses of the heavy metals in the soil. For rice husks, the parameters measured were: pH (6.5), total Cu and Pb (5.5 and 1.5 mg/kg, respectively). Rice husk is believed to be essentially composed of lots of silica, the organic compound constituent being mainly cellulose, hemicellulose and lignin (Okieimen et al., 2007). The efficiency of an agent to immobilize heavy metals in soil will be greatly influenced by the pH, organic matter content and cation exchange capacity.

Assessment of metal immobilization efficiency

The effectiveness of natural clay, calcium phosphate, poultry manure and rice husks as low-cost, non-toxic and ecologically non-invasive amendments for immobilizing Cu and Pb in contaminated soil was assessed by performing chemical leaching tests viz: single $Ca(NO_3)_2$ extractions to determine plant available metals, BCR sequential extractions and EDTA kinetic extractions before and after soil amendments.

Single calcium nitrate extractions

The results of $Ca(NO_3)_2$ extractions carried out on the unamended and post-amended soils are presented in Table 2. The immobilization efficiency, *E (%)* was calculated using Equation 1.

$$E\,(\%) = \left(\frac{M_o - M_e}{M_o} \right) \text{x}\,100 \qquad (1)$$

Where, M_e = equilibrium extractable amount of metal in the amended soil (mg/kg); M_o = initial extractable amount of metal in the unamended soil (mg/kg). The results are illustrated in Figure 1.

Prior to amendment, pseudototal Cu and Pb concentrations in the parent soil were 173.0 and 133.0 mg/kg. With increased dosage of the immobilizing agents, however, these concentrations decreased generally after one month of equilibration. For natural clay, the range of immobilization efficiencies, E for Cu and Pb were $45.8 \le E \le 75.4\%$ and $38.9 \le E \le 77.5\%$, respectively.

Table 2. Calcium nitrate-extracted Cu and Pb concentrations (mg/kg) in soil before and after amendment with various doses of immobilizing agents.

Agent	Dose (% *w/w*)	Cu	Pb
Parent soil	0	135.7±9.5	90.4±7.5
	5	73.5±7.9	55.2±9.8
Natural clay	10	61.3±6.6	40.4±9.0
	20	33.4±6.0	20.3±3.6
	5	60±5.2	40.5±8.6
Phosphate salt	10	50.9±6.3	28.2±8.0
	20	25.4±7.5	18.4±5.0
	5	90.7±6.1	60.2±7.0
Poultry manure	10	80.3±5.7	50.4±5.4
	20	40.4±4.0	25.3±2.4
	5	102.3±7.0	72.1±5.3
Rice husks	10	90.2±7.3	51.4±5.0
	20	50.0±6.0	37.6±4.5

Mean ± s.d, n = 3.

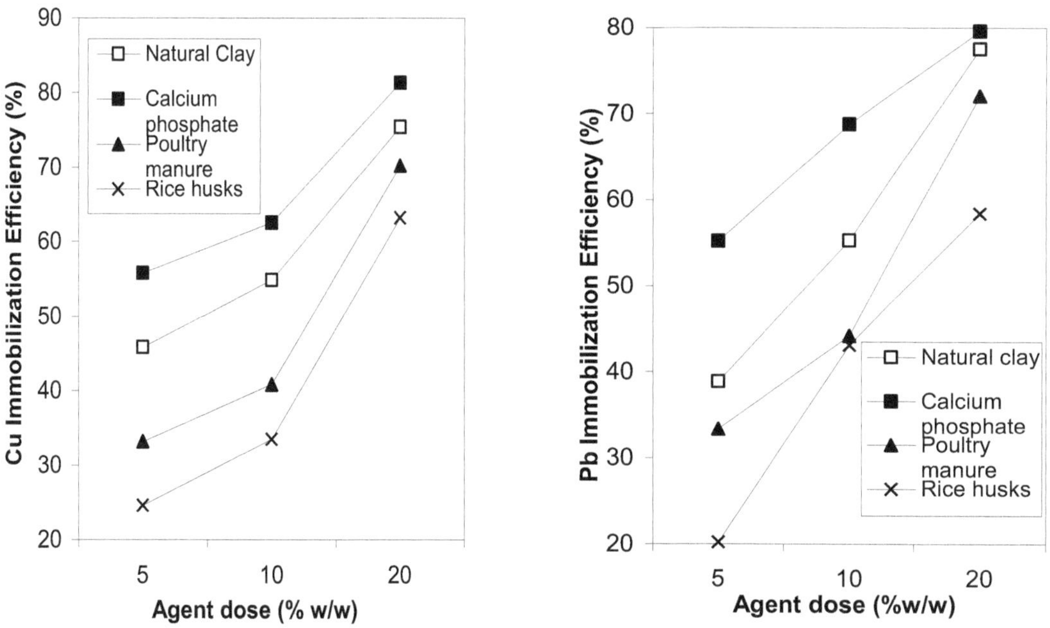

Figure 1. Efficiency of Cu and Pb immobilization in soil amended with different agents.

Natural clay consists mainly of clay mineral which is made up of silica (tetrahedral) layer alternating with one alumina (octahedral) layer held rigidly with hydrogen bonding. Clay may be generally described as 40% aluminum oxide, 46% silica oxide and 14% water. Soil amendment with clay will be more profitable at doses that bring the percentage clay to about 20% (Nara, 2010). Metal immobilization by this agent may occur through adsorption on the silica and alumina surfaces and/or through ion exchange.

Calcium phosphate showed immobilization efficiencies in the range of $55.8 \leq E \leq 81.3\%$ and $55.2 \leq E \leq 79.6\%$ for Cu and Pb, respectively. The mechanism of metal immobilization by calcium phosphate may be through adsorption onto the negative functional group on the phosphate ions (Cui et al., 2010).

The ranges of E values recorded for poultry manure were $33.2 \leq E \leq 70.2\%$ for and $33.4 \leq E \leq 72.1\%$ for Pb. The addition of poultry manure raises both the soil organic matter content and pH, increases the soil metal binding ability, thereby reducing their mobility and leachability.

Rice husks showed E values in the range of $24.6 \leq E \leq 63.2\%$ and $20.2 \leq E \leq 58.4\%$ for Cu and Pb, respectively. Metal immobilization by rice husks may also be through adsorption as it consists of silica oxide and adsorbing organic compounds such as cellulose and hemicellulose.

Overall, E values ranged between $20.2 \leq E \leq 81.3\%$ and varied in the sequence: calcium phosphate > natural clay > poultry manure > rice husks for both Cu and Pb. The E values represent good efficiencies of the agents that can be recommended in field studies for Cu and Pb immobilization in contaminated soil.

Assessment of changes in Cu and Pb fractionation in soil

The extent of Cu and Pb incorporation in the amended soil was assessed by changes in their geochemical forms relative to the unamended soil as determined by the BCR sequential extraction procedure. Metal concentrations (mg/kg) in individual soil fractions extracted by the sequential procedure are presented in Table 3. The metal concentrations were converted to extraction yields (%) by using Equation 2 and the results are illustrated in Figure 2.

$$\text{Extraction Yield (\%)} = \frac{M_i}{\sum_{i=1}^{4} M_i} \text{x}\,100 \qquad (2)$$

Where, $\sum_{i=1}^{4} M_i = M_1 + M_2 + M_3 + M_4$; and M_1, M_2, M_3 and M_4 are the extractable, oxide-bound, organic-bound and residual metal pools (mg/kg), respectively. Since the first metal fraction is the most mobile, extraction yield in this fraction has been defined as the mobility factor, M_f.

BCR sequential extractions showed that the concentrations, hence extraction yields of Cu and Pb in the various soil fractions varied considerably with the nature and level of soil amendment (Figure 2a and b).

Metal concentrations extracted by the first step of the sequential extraction procedure were somewhat comparable to the plant available fractions from calcium nitrate single extractions. Metal mobilities, M_f were reduced following amendment with the immobilizing agents. For instance, in the parent soil, M_f for Cu and Pb were 76.4 and 68.5%, respectively. Natural clay amendment at 5, 10 and 20% w/w lowered the M_f for Cu to 41.4, 34.9 and 18.2%, respectively; while corresponding values for Pb were 40.3, 30.5 and 15.9%. Calcium phosphate treatments at 5, 10, and 20% w/w gave M_f for Cu as 35.4, 30.5 and 15.6%, respectively; while corresponding values for Pb were 31.4, 20.0 and 15.0%. The M_f for Cu in soil amended with poultry manure at 5, 10 and 20% w/w were 53.5, 48.8 and 25.7%, respectively; while corresponding values for Pb were 50.1, 39.7 and 23.1%. Treatment of soil at 5, 10 and 20% w/w with rice husks reduced Cu mobility to 60.8, 55.6 and 33.9%, respectively; while corresponding values in the case of Pb were 57.2, 41.6 and 31.8%. Depending on the M_f values, the ability of the treatments to immobilize the metals varied in the sequence: calcium phosphate > natural clay > poultry manure > rice husks. Additionally, M_f values for Cu were higher than those for Pb indicating the generally low mobility of Pb in natural soils (Pueyo et al., 2008). Reduced M_f recorded for both Cu and Pb following these treatments interprets the ability of the agents to attenuate metal mobility in soil via stabilization.

Metal extraction yields in the oxide-bound fraction of the amended soil were higher relative to the unamended soil, implying that some of the hitherto mobile metals were converted to the less bioavailable Fe/Mn-oxide form. This fraction contained as much as 11.8 - 18.3%, 29.5 to 36.3%, 12.3 to 19.5% and 12.1 to 22.5% of Cu in soils treated with natural clay, calcium phosphate, poultry manure and rice husks, respectively. Lead (Pb) was partitioned to the extent of 12.9 to 18.4%, 23.7 to 32.9%, 11.3 to 22.0% and 12.4 to 29.4% in soils treated with natural clay, calcium phosphate, poultry manure and rice husks, respectively. About 20.9 to 23.4, 16.8 to 18.9, 24.9 to 35.7 and 15.5 to 24.2% of Cu were found to be complexed and/or adsorbed to the organic matter-bound fraction for soils amended with natural clay, calcium phosphate, poultry manure and rice husks, respectively; whereas corresponding yields of Pb in this fraction were 23.2 to 24.2, 17.4 to 27.4, 25.5 to 38.2 and 9.6 to 20.9%. Apparently, Cu appeared to be associated with this fraction to a greater extent than Pb with poultry manure treatment seemingly recording the highest metal associations with the soil organic matter phase. The residual soil fraction retained Cu to the extent of 25.8 to 40.1, 18.4 to 29.2, 9.3 to 19.0 and 11.6 to 19.4% following natural clay, calcium phosphate, poultry manure and rice husks, respectively; while corresponding yields of Pb range was 23.7 to 41.4, 24.6 to 27.5, 13.2 to 16.7 and 17.9 to 20.7%, respectively. More Pb was, thus, in association with the residual soil fraction than Cu. It appeared that natural clay caused the

Table 3. BCR-extracted Cu and Pb (mg/kg) in fractions of soil before and after amendment with different doses of immobilizing agents.

Agent	Metal	Dose (% *w/w*)	M_1	M_2	M_3	M_4
Parent soil	Cu	0	130.5±9.5	10.4±0.5	10.8±2.1	19.1±1.0
	Pb	0	89.5±9.0	15.1±3.0	10.4±1.2	15.7±3.1
Natural clay	Cu	5	70.2±4.5	20.0±4.6	35.5±5.4	43.7±6.7
		10	58.3±4.7	25.7±4.8	40.3±6.6	42.5±6.6
		20	30.1±6.0	30.2±5.0	38.6±5.0	66.3±6.8
	Pb	5	52.3±5.0	16.7±2.4	30.1±3.3	30.7±4.0
		10	38.4±4.5	20.3±2.8	26.4±3.0	40.9±5.9
		20	20.1±3.2	23.2±2.5	30.5±4.0	52.2±6.0
Phosphate salt	Cu	5	60.0±4.6	50.0±5.2	28.5±5.4	31.2±4.0
		10	50.9±3.3	55.7±4.3	36.1±6.6	24.3±2.5
		20	25.4±2.0	59.1±5.9	30.7±4.8	47.5±6.9
	Pb	5	40.5±3.1	30.6±2.4	22.5±3.3	35.5±4.3
		10	28.2±2.3	36.7±3.0	28.5±3.7	34.6±5.0
		20	18.4±2.4	40.3±2.7	33.5±3.0	30.1±3.1
Poultry manure	Cu	5	88.0±5.0	20.3±2.1	41.0±4.0	15.3±3.2
		10	77.6±3.3	27.9±3.0	40.3±4.7	13.3±1.7
		20	40.0±2.0	30.3±2.7	55.5±4.4	29.6±6.9
	Pb	5	60.0±3.1	13.5±2.7	30.5±3.6	15.8±4.0
		10	49.1±2.3	17.6±2.0	37.5±4.2	19.5±2.1
		20	26.3±2.4	25.1±2.3	43.5±5.6	19.0±1.1
Rice husks	Cu	5	100.2±6.5	20.0±2.5	25.5±4.4	19.1±2.0
		10	90.0±5.3	25.5±3.3	31.1±4.7	15.4±2.5
		20	53.3±4.1	35.3±5.0	37.9±3.6	30.4±4.1
	Pb	5	71.5±3.3	15.5±2.1	12.0±3.0	26.1±4.0
		10	50.1±3.5	21.2±3.0	20.3±3.0	28.7±3.3
		20	36.0±3.0	33.3±3.2	23.6±2.3	20.2±3.5

M_1- acetic; M_2- hydroxylamine hydrochloride, pH 1.5; M_3- hydrogen peroxide, then ammonium acetate, pH 2 and M_4- aqua regia.

greatest metal redistributions in the residual fraction, probably due to additional supply of silicates to the soil matrix (Putwattana et al., 2010).

Assessment of changes in Cu and Pb leaching rates

Kinetic leaching experiments were designed to provide further information on the nature of bound metals in the soil and their degree of lability, hence bioavailability before and after amendments. A non-selective complexing agent, EDTA (0.05 M), capable of direct metal extraction from several soil compartments, such as metal bound to organic matter, carbonate and Fe/Mn (hydr)oxides was employed for this purpose. This concentration was previously established as furnishing optimal metal removal at pH 6.6 (Morera et al., 2001). Figure 3 illustrates the cumulative amounts of Cu and Pb leached by EDTA with time from the parent soil and its amended (20% *w/w*) sub-samples.

Apparently all the leaching curves show similar profiles. That is, the leached metal concentrations increased rapidly at short extraction times, and then reduced by the second hour into the experiment furnishing plateaux. The implication is that at short extraction time (1 h), metal extraction rates were very fast, mobilizing high

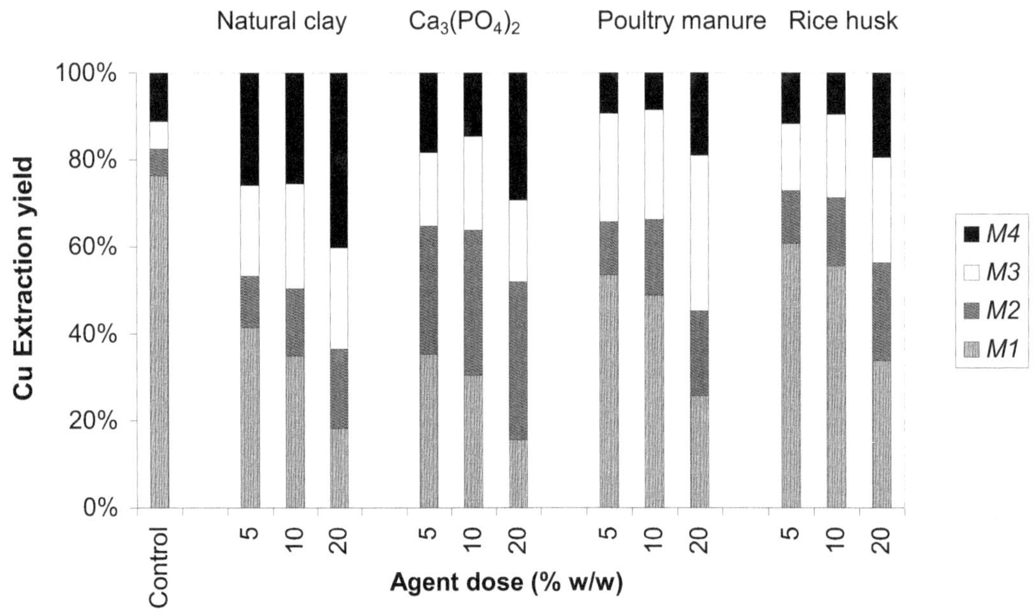

Figure 2a. Extraction yields (%) of Cu in operationally defined pools of contaminated soil amended with different immobilizing agents extracted by the BCR sequential procedure.

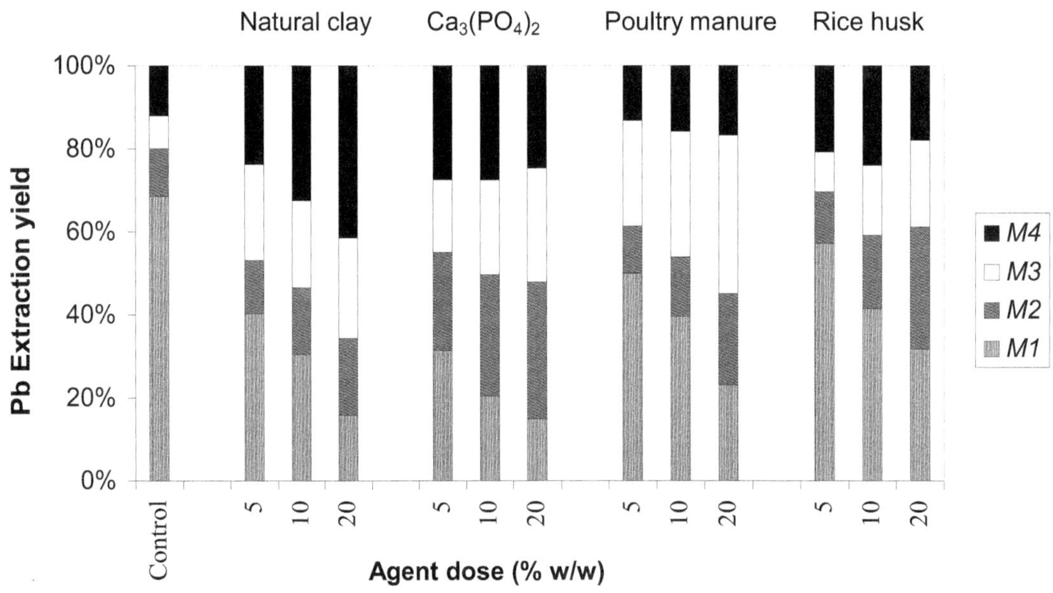

Figure 2b. Extraction yields (%) of Pb in operationally defined pools of contaminated soil amended with different immobilizing agents extracted by the BCR sequential procedure.

amounts of the metals. Plateaux observed for metal leaching implies that the soils had fixed sorption capacities that provided limited sites onto which the metals were originally sorbed (Thompson and Goyne, 2012). EDTA extracted amounts that were higher than those removed by any single step of the BCR sequential extraction method and represented more closely the sum of all the mobilizable metal fractions extracted by the BCR procedure. By the end of the 16 h extraction time, 86.0% Cu and 89.8% Pb were leached from the pre-amended soil relative to their pseudototal concentrations. Following amendments, 53.7, 68.0, 78.3 and 82.9% of Cu were respectively leached from soil treated with natural clay, calcium phos-phate, poultry manure and rice husks, respectively. Approximately 58.0, 57.2, 81.6 and 97.6% of Pb were respectively leached from the soil treated with natural clay, calcium phosphate, poultry manure and rice husks.

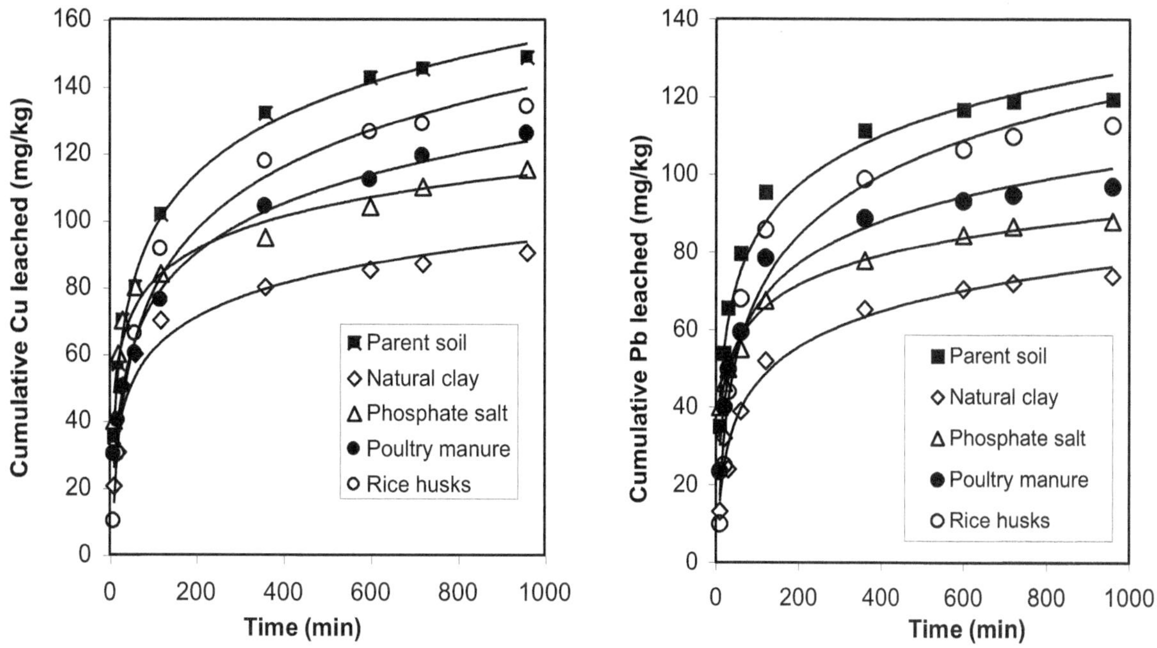

Figure 3. Cumulative Cu and Pb kinetically extracted with 0.05 M EDTA in the parent soil and its sub-samples amended (at 20% w/w) with different immobilizing agents.

Since chemical extractants cannot give information on metal forms in terms of a specific soil compartment unless each compartment presents a different kinetic behaviour, the EDTA kinetic data was fitted into a two-first-order-reaction (TFOR) model by adopting the approach of Labanowski et al. (2008) and Manoucherhi et al. (2011). Average removal rates per unit time, $R_M(t)$ (mg/kg soil/ min) (Equation 3) between t_i and t_{i-1} ($t_{i-1} < t < t_i$), for a given metal were calculated for each lag time and presented as semilogarithmic plots.

$$R_{M(t_i-1<t<t_i)} = \frac{M_{(t_i-1<t<t_i)}}{(t_i - t_{i-1})} \qquad 3)$$

Where, $M_{(t)}$ is amount of metal extracted at a given time. The evolution of $R_M(t)$ as two distinct linear segments of the semilogarithmic kinetic plots with different slopes and intercepts on the ordinate implied the simultaneous existence of two distinct, labile, M_1 and less labile, M_2 metal pools, each following a decreasing exponential function.

$$R_M(t) = R_{M1}(0)\,e^{-k_1 t} + R_{M2}(0)\,e^{-k_2 t} \qquad 4)$$

Where, $R_{M1}(0)$ and $R_{M2}(0)$ are the initial removal rates (at t = 0) of the two pools and k_1 and k_2 are their apparent rate constants. The metal leaching rates were apparently high at the start of the extraction, dominated by the $R_1(0)$ parameter characteristic of the first pool, but the contribution of this first pool decreased quickly to become insignificant (exponential decrease governed by k_1). Then, after a while, the contribution of the second pool became preponderant, with a clear slower extraction rate which still decreased with time (following k_2).

From Table 4, k_1 and k_2 for the leaching of Cu from the unamended (parent) soil were 4.71 x 10^{-2} and 4.00 x 10^{-3} min^{-1}, respectively; while corresponding values recorded for the amended soil were natural clay (k_1 = 4.00 x 10^{-2}/min; k_2 = 3.00 x 10^{-3}/min), phosphate salt (k_1 = 4.83 x 10^{-2}/min; k_2 = 1.20 x 10^{-3}/min), poultry manure (k_1 = 3.93 x 10^{-2}/min; k_2 = 2.60 x 10^{-3}/min) and rice husks (k_2 = 1.23 x 10^{-2}/min; k_2 = 3.30 x 10^{-3}/min). The leaching of Pb from the parent soil gave k_1 and k_2 as 3.84 x 10^{-2}/min and 5.10 x 10^{-2}/min, respectively. Following amendment with the immobilizing agents, these values were natural clay (k_1 = 1.92 x 10^{-2}/min; k_2 = 4.00 x 10^{-3}/min), phosphate salt (k_1 = 5.44 x 10^{-2}/min; k_2 = 3.30 x 10^{-3}/min), poultry manure (k_1 = 4.04 x 10^{-2}/min; k_2 = 3.30 x 10^{-3}/min) and rice husks (k_1 = 4.50 x 10^{-2}/min; k_2 = 3.70 x 10^{-3}/min). These observations indicate that the leaching rates for both the labile and the less labile Cu pools from the soil were apparently retarded for all the agents relative to the unamended soil. In the case of Pb, however, only the leaching rate for the less labile pool appeared to be suppressed upon amendment.

Conclusion

Single, sequential and kinetic leaching tests performed to assess the effectiveness of some cheap and non-toxic

Table 4. Rate constants, k_1 and k_2 for leaching of labile and less labile pools* of Cu and Pb from soil with 0.05 M EDTA before and after amendment with 20% w/w of immobilizing agents.

Agent	Metal	$k_1 \times 10^{-2}$/min	$k_2 \times 10^{-3}$/min
Parent soil	Cu	4.71	4.00
	Pb	3.84	5.10
Natural clay	Cu	4.00	3.00
	Pb	1.92	4.00
Phosphate salt	Cu	4.83	1.20
	Pb	5.44	3.30
Poultry manure	Cu	3.93	2.60
	Pb	4.04	3.30
Rice husks	Cu	1.23	3.30
	Pb	4.50	3.70

*Defined by two-first order reaction model.

agents at reducing the bioavailability of Cu and Pb in contaminated soil revealed that the immobilization efficiencies of the agents were high enough, increased with the level of amendment and varied in the sequence: calcium phosphate > natural clay > poultry manure > rice husks. The immobilizing agents retarded leaching rates for both the labile and the less labile pools of Cu but suppressed only the leaching rate for the less labile pool in the case of Pb. These agents may reduce risks resulting from Cu and Pb contamination in soil since metal mobility is related to its bioavailability.

REFERENCES

Alvarenga PMLF, Goncalves AP, Fernandes RMCSC, de Varennes APA, Duarte ECNFA, Vallini G, Cunha-Queda CF (2008). Effect of organic residues and liming materials on metal extraction from a mining-contaminated soil, Bioremediation J. 12(2):58-69.

Awodun MA (2007). Influence of sawdust ash on soil chemical properties and cowpea performance in Southwest Nigeria. Int. J. Soil Sci. 2(1):78-81.

Belviso C, Cavalcante F, Ragone P, Fiore S (2010). Immobilization of Ni by synthesising zeolite at low temperatures in a polluted soil. Chemosphere 78(9):1172-1176.

Bhat UN, Khan AB (2011). Heavy metals: An ambiguous category of inorganic contaminants, nutrients and toxins. Res. J. Env. Sci. 5(8):682-690.

Black CA (1965). Methods of Soil Analysis, Agronomy 9(2):63-67.

Bouyoucos GJ (1962). Improved hydrometer method for making particle size analysis of soils. Agron. J. 54:464-465.

California Department of Pesticide Regulation Environmental Hazards Assessment Program (CDPRHAP), (1998). Standard Operating Procedure, Sacramento, California. http://www.cdpr.ca.gov/docs/emon/pubs/sops/eqwa001.pdf

Cao RX, Ma LQ, Chenb M, Singh SP, Harris WG (2003). Phosphate-induced metal immobilization in a contaminated site. Environ. Pollut. 122(1):19-28.

Chukwuka KS, Omotayo OE (2008). Effects of Tithona green manure and water hyacinth compost application on nutrient depleted soil in South-Western Nigeria. Int. J. Soil Sci. 3(2):69-74.

Corrochano BG, Azcárate JA, Gonzalez MR (2011). Heavy metal chemical fractionation and immobilization in lightweight aggregates produced from mining and industrial waste. Int. J. Environ. Sci. Tech. 8(4):667-676.

Cui Y, Du X, Weng L, Van Riemsdijk WH (2010). Assessment of in situ immobilization of lead (Pb) and arsenic (As) in contaminated soils with phosphate and iron: solubility and bioaccessibility. Water Air Soil Pollut. 213:95-104.

D'amore JJ, Al-abed SR, Scheckel KG, Ryan JA (2005). Methods of speciation of metals in soils. J. Environ. Qual. 34(5):1707-1745.

Fawzy EM (2008). Soil remediation using in-situ immobilization techniques. Chem. Ecol. 24(2):147-156.

Finžgar N, Kos B, Leštan D (2006). Bioavailability and mobility of Pb after soil treatment with different remediation methods. Plant Soil Environ. 52(1):25-34.

Florido MC, Madrid F, Madrid L (2011). Effect of an organic amendment on availability and bio-accessibility of some metals in soils of urban recreational areas. Environ. Pollut. 159(2):383-390.

Gatimu E, Mwinyihija M, Killham K (2010a). Determination of the effect of red earth as an ameliorant for lead contaminated soils. Res. J. Environ. Toxicol. 4(3):151-161.

Gatimu E, Mwinyihija M, Killham K (2010b). Determination of the effect of bone meal as an ameliorant on lead contaminated soils. Res. J. Environ. Toxicol. 4(4):237-247.

International Organization for Standardization, ISO (1995). Soil quality: Extraction of trace elements soluble in aqua regia, ISO 11466:1995(E).

Irmak S (2010). Copper correlation of irrigation water, soils and plants in the Cukurova Region of Turkey. Int. J. Soil Sci. 5(3):185-195.

Jackson ML (1962). *Soil Chemical Analysis*, Prentice-Hall, New York.

Labanowski J, Monna F, Bermond A, Cambier P, Fernandez C, Lamy I, van Oort F (2008). Kinetic extractions to assess mobilization of Zn, Pb, Cu, and Cd in a metal-contaminated soil: EDTA vs. citrate, Environmental Pollut. 152(3):693-701.

Manoucherhi N, Besancon S, Bermond A (2011). Kinetic characterizing of soil trace metal availability using soil/EDTA/chelex mixture. Chemosphere 83(7):997-1004.

Matini L, Ongoka PR, Tathy JP (2011). Heavy metals in soil on spoil heap of an abandoned lead ore treatment plant, SE Congo-Brazzaville. Afr. J. Environ. Sci. Technol. 5(2):89-97.

Maxted AP, Black CR, West HM, Crout NMJ, McGrath SP, Young SD (2007). Phytoextraction of cadmium and zinc from arable soils amended with sewage sludge using *Thlaspi caerulescens*:Development of a predictive model. Environ. Pollut. 150(3):363-372.

McLaughlin MJ, Hamon RE, McLaren RG, Speir TW, Rogers SL (2000a). Review: A bioavailability-based rationale for controlling metal and metalloid contamination of agricultural land in Australia and New Zealand. Aust. J. Soil Res. 38(6):1037-1086.

McLaughlin MJ, Zarcinas BA, Stevens DP, Cook N (2000b). Soil testing for heavy metals. Commun. Soil. Sci. Plant Anal. 31(11-14):1661-1700.

Mench MJ, Didier VL, Löffler M, Gomez A, Masson P (1994). A mimicked *in situ* remediation study of metal-contaminated soils with emphasis on cadmium and lead. J. Environ. Qual. 23:58-63.

Morera MT, Echeverría JC, Mazkiarán C, Garrido JJ (2001). Isotherms and sequential extraction procedures for evaluating sorption and distribution of heavy metals in soils. Environ. Pollut. 113(2):135-144.

Nara S (2010). Structural development of clay soil with the amendment of coal combustion by-product FGD Gypsum in Lusatia Germany. Int. J. Soil Sci. 5(2):62-72.

Ogbo EM, Okhuoya JA (2011). Bio-absorption of some heavy metals by *Pleurotus tuber-regium* Fr. Singer (an edible mushroom) from crude oil polluted soils amended with fertilizers and cellulosic wastes. Int. J. Soil Sci. 6(1):34-48.

Okieimen FE, Okieimen CO, Wuana RA (2007). Preparation and characterization of activated carbon from rice husks. J. Chem. Soc. Niger. 32(1):126-136.

Pueyo M, Mateu J, Rigol A, Vidal M, López-Sánchez JF, Rauret J

(2008). Use of modified BCR three-step sequential extraction procedure for the study of trace element dynamics in contaminated soils. Environ. Pollut. 152:330-341.

Putwattana N, Kruatrachue M, Pokethitiyook P, Chaiyarat R (2010). Immobilization of cadmium in soil by cow manure and silicate fertilizer, and reduced accumulation of cadmium in sweet basil (*Ocimum basilicum*). ScienceAsia 36(4): 349-354.

Rajaganapathy N, Xavier F, Sreekumar D, Mandal PK (2011). Heavy metal contamination in soil, water and fodder and their presence in livestock and products: A review. J. Environ. Sci. Technol. 4(3):234-249.

Santos S, Costa CA, Duarte AC, Scherer HW, Schneider RJ, Esteves VI, Santos EB (2010). Influence of different organic amendments on the potential availability of metals from soil: a study on metal fractionation and extraction kinetics by EDTA. Chemosphere 78(4):389-396.

Sarma H (2011). Metal hyperaccumulation in plants: A review focusing on phytoremediation technology. J. Environ Sci. Technol. 4(2):118-138.

Schumacher BA (2002). Methods for the determination of total organic carbon, TOC in soils and sediments. USEPA-NCEA-C-1282, EMASC-001, pp.1-21.

Suruchi Khanna P (2011). Assessment of heavy metal contamination in different vegetables grown in and around urban areas. Res. J. Environ. Toxicol. 5(3):162-179.

Thompson A, Goyne KW (2012). Introduction to the sorption of chemical constituents in soils. Nature Education Knowledge 4(4):7

Ure AM, Quevauviller PH, Muntau H, Griepink B (1993). Speciation of heavy metals in soils and sediments. An account of the improvement and harmonization of extraction techniques undertaken under the auspices of the BCR of Commission of the European Communities. Intern. J. Environ. Anal. Chem. 51(1):5-151.

Wuana RA, Okieimen FE (2011). Heavy metals in contaminated soils: A review of sources, chemistry, risks and best available strategies for remediation. ISRN Ecology. pp.1-20.

Yap CK, Edward FB, Emila RAA, Ainey FI, Ismail A, Tan SG, Sharizat Y (2008). Determination of contamination and bioavailability of heavy metals (Cu, Cd, Zn, Pb, and Ni) in the Serdang urban Lake by using Guppy fish *Poecilia reticulate*. Trends in Applied Sci. Res. 3(1):69-75.

7

Evaluation of soil and water salinity for irrigation in North-eastern Ethiopia: Case study of Fursa small scale irrigation system in Awash River Basin

Mohamed Seid and Tessema Genanew

Department of Soil and Water Resources Management, Wollo University, Dessie, Ethiopia

For sound land use and water management in irrigated area, knowledge of the chemical composition of soils, water, climate, drainage condition and irrigation methods before action are crucial for sustainability of irrigation projects. The study aimed to evaluate the physicochemical properties of soils and water for intended irrigation scheme with reference to standard suitability classes. With regard to this, a study was conducted in Awash River Basin at *Fursa* small scale irrigation scheme in Northeastern Ethiopia. Soil samples were collected from bore holes of three soil mapping units of the study site across depth and water samples were taken from irrigation water with plastic bottles for analysis of range of physical and chemical properties. The results of the analysis reveal the existence of potential sodicity not only in the soil but also in the irrigation water. The study underlines the need for selection of salt tolerant crops and good water management by using appropriate irrigation methods to sustain productivity of soil in the proposed irrigation site. This has significant contribution to deciding the type of crop to be produced and appropriate irrigation methods for sustainability of soil productivity.

Key words: Fursa irrigation scheme, salinity, sodicity, soil analysis, water analysis.

INTRODUCTION

Sound irrigation agriculture contributes towards achieving food security and livelihood improvements for the increasing population through enhancement of agricultural productivity. Lessons from the past indicate that the development of a sound irrigated agriculture depends on a catena, or chain of related factors, involving soils, waters, crops and man. Failure of any one of these links can bring hardship or even disaster to an irrigation enterprise (Tessema, 2011). Poor irrigation agriculture in arid and semiarid regions results in land degradation through soil salinity and sodic soil developments in different parts of the world. Hence, the study of arid lands and salt affected soils has been an important topic for modern agricultural management and particularly for poor countries like Ethiopia where agriculture is the backbone of its economy while arid and semi arid climatic zones occupy over 60% of the total land area (Awulachew et al., 2007).

The total land area affected by salinity and sodicity in Ethiopia estimated at about 11 thousand ha and soils have been reported to occur for the most part of the rift valley zone (FAO, 1985a; Tadesse and Bekele, 1996). Nowadays, soil salinity has become important problem in irrigated soils of Awash River basin in central and Eastern Ethiopia. The effect of the quality of irrigation water on soil properties has been discussed by many researchers (Richards, 1954; Westcott and Ayers, 1985; Kinfe, 1999). Water quality related problems in irrigated agriculture are identified as salinity, sodicity, specific-ion toxicity and impeded infiltration rate as well as hydraulic conductivity (Ayers and Westcott, 1985; Frenkel, 1984).

For appropriate land use and water management in irrigated area, knowledge of the chemical composition of the soil characteristics, water, climate, drainage condition and irrigation methods should be evaluated before implementation of irrigation projects (Al-Ghobari, 2011). With regards to soil studies, a number of surveys have been carried out for different purposes at different times by different institutions. However, the scale and purpose of the studies allow only planning for development undertakings. A very detailed survey is necessary to characterize soils as well as water to identify the salinity hazard and level of nutrients (major or minor) at each irrigation sites for the proper understanding of the hazard and appropriate mitigation measures. Salt affected soils and the associated poor soil drainage conditions results from poor management of soils and irrigation systems (Tessema, 2011). The Middle and Lower Awash River Basins in Ethiopia appear to be the most recent examples of such a situation (Tadesse and Bekele, 1996). In addition to these, salt affected soils are not only the result of the saline soils but also attributed to application of low quality irrigation water. All waters used for irrigation caries varying amounts of dissolved salts and other constituents. Some dissolved constituents can improve crop growth if present in small to moderate amounts, otherwise can harm soils and restrict plant growth if they are present in excessive amount.

Due to the above facts and indication of salt in the immediate upstream Woreda (district) along the gullies, intermittent streams and deep wells dug for domestic use made us to assess the salinity hazard for soils and water of Fursa irrigation project. Therefore, the objective of this study was to evaluate the soil physicochemical properties of intended irrigation areas and to assess chemical composition (quality) of irrigation water for the project understudy based on the standard suitability classes. This has significant contribution to decide the type of crop to be produced and appropriate irrigation methods for sustainability of soil productivity.

MATERIALS AND METHODS

Study area

The study was conducted in Afar Regional State, Awsi-Resu administration, Ada'Ar Woreda (district) at Fursa River diversion irrigation project in Northeastern Ethiopia. It lies with UTM coordinate of 645,886E and 1,237,969N with altitude of 706 m above sea level (m.a.s.l) where the slope rangefrom 0 to 2%(Figure 1).

Climate

According to the traditional agro-climatic zone classification of Ethiopia and considering the temperature, length growing period (LGP) and elevation (500 to 1500m.a.s.l), the study area lies under *Dry kolla* climatic zone. Records from Elliwuha metrological station showed that the mean annual rainfall was about 458 mm. It receives a bimodal type rainfall pattern with the first peak from February to May and the second from August to September. However, both seasons were not sufficient to support reliable crop growth. The mean annual maximum and minimum temperature was 38and 10.3°C, respectively. December is the coldest month and June is the hottest month of the year. The LGP is less than 90 days which shows that there is no reliable growing period.

Survey methods

Office work

Prior to commencement of the field soil investigation, necessary field materials like GPS, topographic map of scale 1:50,000, knife, hoe, shovel, plastic bags, hard paper or labeling, markers, rope, etc, were collected.

Field work procedure

Profile sampling

During site mapping, altitude and geographic locations of block boundaries and important landmarks were recorded using total station with association of GPS to locate local bench marks. Profile locations were taken by total station along with topographic survey. For further soil characterization, soil profile pits were dug on three representative sites for this irrigation project. The soil profile descriptions made was according to FAO (1990) guidelines for soil profile description. Soil samples were collected from the flat face of each natural soil horizons. The pits were dug with size of one meter width, two meter length and two meter depth of bore hole (1 x 2 x 2 m).

Laboratory analyses

Soil analysis

Soil samples, taken from each pit across depth, were air dried, weighed, and grinded and passed through a 2 mm sieve. Then soil pH was measured by pH meter using suspension of distilled water to soil solution ratio (1:2.5) and EC measurement was performed using saturated paste extracts. Exchangeable bases and cation exchange capacity (CEC) of the soils were determined by the 1 M ammonium acetate (pH 7) method according to the percolation tube procedure (Van Reeuwijk, 1993) while calcium carbonate ($CaCO_3$) was determined by acid neutralization (titration) method using HCl. Available phosphorous content of the soils was determined by 0.5 M sodium bicarbonate extraction solution (pH 8.5) using the method of Olsen et al., (1954) as outlined by Van Reeuwijk (1993). The total nitrogen content of the soil was determined by wet-oxidation procedure of the Kjeldahl method (Bremner and Mulvaney, 1982) and organic carbon content by the wet combustion procedure of Walkley and black (1934). Soil texture was determined by the hydrometer method (Bouyoucos, 1951) and soil color was identified using munsell color chart.

Water quality analysis

As long as the natural replenishment of the aquifer equals the withdrawal of irrigation water, there will be little change with time in the chemical characteristics of water due to sampling problem (FAO, 2000). Thus, samples for irrigation water analysis can be collected in 1- or 2-litre clean glass or polyethylene (plastic) bottles. Based on this principle one liter water sample was taken at the middle of *Fursa River* for analysis of pH, EC, Ca, Mg, Na, K and SAR.

$$\mathrm{SAR\ (mmoles_c/L = meq/L)} = \frac{Na^{+}}{[(Ca^{2+}+Mg^{2+})/2]^{0.5}} \quad (1)$$

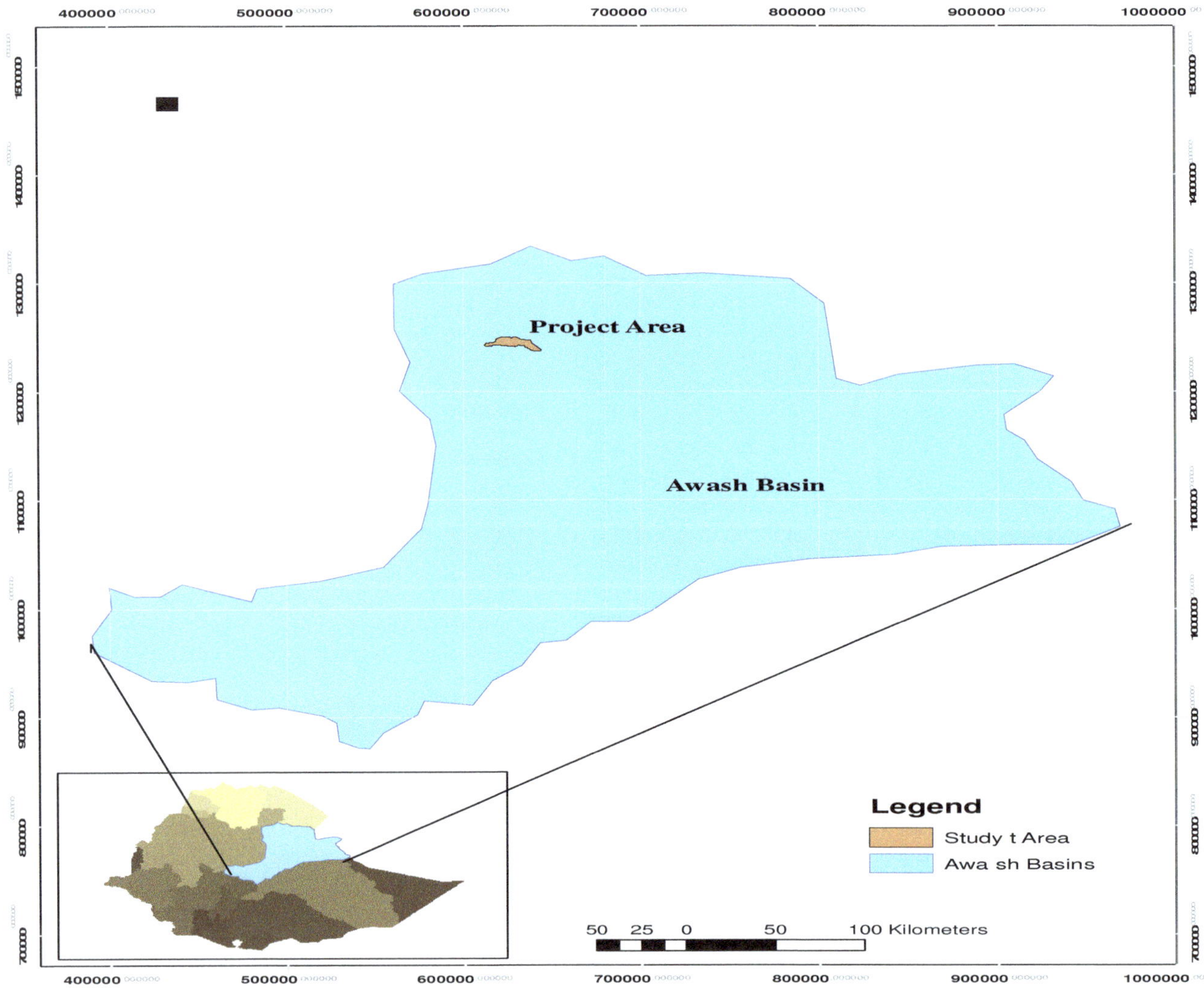

Figure 1. Location map of the study area.

The SAR of a water adjusted for the precipitation or dissolution of Ca^{2+} and Mg^{2+} is expected to occur where water reacts with alkaline earth carbonates within the soil. This was calculated by using FAO Irrigation and Drainage paper No. 29.

Data analysis

The data generated from laboratory analyzed descriptive statistics. One way analysis of variance was used to compare the physical and chemical properties of soil between and within soil mapping units. All the data were edited, coded and analyzed using SPSS software version 16.0.

RESULTS AND DISCUSSION

Soil quality evaluation

For characterizing the physico-chemical properties of soils of intended irrigation area, the catchment were partitioned into three soil mapping units based on uniformity of land attributes. Characteristics of the mapping units and detailed analytical data of the soil profile are given in the subsequent sections.

Soil physical properties

In the soil profile description, the soils of mapping unit one and two were found to be very deep (>150 cm). The texture is clay loam for the first top layer (0 to 35 cm depth) and clay for the next layer (35 to 150 cm depth) (Table 1), whereas clay for the first two top layers (0 to 50 cm depth) and silt-loam for the next layer (50 to 90 cm depth) for site two (Table 1). No sample was taken for the last layer (90 to 200+) due to its gravel layer. A medium soil depth (0 to 87cm) was observed in mapping unit three (site three). The soil texture third site was found to be sandy-clay for the first top layer (0 to 35 cm depth), clay for the next layer (35 to 47 cm depth) and clay loam for the last layer (47 to 87) (Table 1). With regards to soil

Table 1. Some physical characteristics of the studied soils in three mapping units (sites).

Depth (cm)	Clay (%)	Silt (%)	Sand (%)	Soil textural class
Site one				
0-35	28	30	42	Clay loam
35-80	40	28	32	Clay
80-150	48	40	12	Clay
Site two				
0-30	52	38	10	clay
30-50	44	20	36	clay
50-90	18	58	24	Silt Loam
Site three				
0-35	50	2	48	Sandy clay
35-47	48	36	16	clay
47-87	38	38	24	Clay loam

depth, the texture of the soil in the catchment generally lay clay-loam for the upper layer, clay for the middle and lower layer. The top soil texture is dominantly clay loam for mapping unit one, clay for site two and sandy clay for mapping unit three (site three). All sites have soil color of light brown when moist. This physical property has significant influence on infiltrate rate and hydraulic conductivity of the soil as well as the retention and movement of irrigation water around root zone (FAO, 2000).

Generally, results of the particle size analysis indicate that the majority of the soils are heavily textured. Moreover, mass of surface soil layers have weakly developed platy structure while the subsurface soil layers have developed platy structure with good porosity. These properties have impact on movement of air and water within the soil (Brady and Weil, 2002). Medium infiltration rate was registered in all the sampled sites. However, infiltration rate at the beginning of the test was lower when compared with the theoretical rate that is attributed to platy structure of the soil, that is, it could impede the down ward movement of water. This is in agreement with the findings of Mass and Hoffman (1977). As well, availability of deep soil depths in the study site imply that soil depth is not a limiting factor for the production of most field crops.

Soil chemical properties

Soil reaction

Soil pH is a good indicator of intensity of acidity or alkalinity of the soil. In site one, the pH of surface soil is 8.3 (strongly alkaline); decreasing to 8.0 (strongly alkaline) in surface horizons and it comes back again to 8.3 in the subsurface layer. Similar pattern was observed in the second site, it was 8.3 (strongly alkaline) in surface horizons; decreasing to 7.8 (moderately alkaline) and it comes back again to 8.4 (strongly alkaline) in the last subsurface layer. On the third site, pH of 7.9 (moderately alkaline) was recorded in surface horizons and it decreased from 7.8 to 7.7 (moderately alkaline) subsequently in two subsurface layers (Table 2).

The electrical conductivity (EC) is generally non-saline that ranges between 0.083 dS/m on the surface soil (0 to 35) to 0.319 dS/m in the sub soils of the lower layers (35 to 180 cm) soil depth in the first site (Table 2). EC of the second site ranged from non-saline to moderately saline (0.831 to 7.52 dS/m) on the surface soil (0 to 50) and (0.482 and 5.08 dS/m) subsequent sub surface layer (50 to 200 cm soil depth). Similarly results of EC obtained in the third site that ranges from non-saline to moderately saline (0.63 to 7.75 dS/m) on the surface soil (0 to 50cm) and it is slightly-saline in the next layer (2.76 dS/m).

CEC and exchangeable bases

The cation exchange capacity (CEC) of the soils is high [(31.58 cmol (+)/kg soil), (7.24 to 45.40 cmol (+)/kg soil)] low to high in the surface and sub surface layers of site one respectively and generally very high [(71.08 to 73.12 cmol (+)/kg soil), surface, (32.98 cmol (+)/kg soil) last layer] at site two (Table 2). Medium to high CEC, (24.76, 40.38, 37.18 cmol (+)/kg soil) surface to subsurface, was registered at site three (Table 2).

The exchangeable sodium percentage (ESP) generally ranges from low to medium (0.8 to 2.3%, 1.5 to 2.1%) in site one and two, respectively. The exchangeable sodium percentage generally ranges from low to medium (1.5 to 2.1%). However, ESP is medium in the first and last layers (2.3 and 2.74%) but it is extremely high in the middle layer (40.1%) of site three (Table 2).

Organic matter and nitrogen

Organic matter was very low (0.52, 0.62 and 0.29) in site one and (0.72, 1.00, 0.86) in site two, organic matter content (%) obtained from surface layer to subsurface layers across soil depth is shown in Table 2. Similarly, it was low (0.52 to 0.55%) in the surface layers with the

Table 2. Chemical composition of soils irrigated at three sites across soil depth.

Depth (cm)	H_2O pH	EC_e (dS/m)	TN (%)	OC (%)	OM (%)	Available P (ppm)	CEC (cmol (+)/kg)	K (cmol (+)/kg)	Na (cmol (+)/kg)	ESP (%)	$CaCO_3$ (%)
Site One											
0-35	8.3	0.083	0.01	0.3	0.52	5.97	31.58	0.58	0.28	0.89	8.02
35-80	8	0.218	0.07	0.36	0.62	7.26	7.24	0.54	0.37	5.11	9.25
80-180	8.3	0.319	0.06	0.17	0.29	3.42	45.4	1.27	0.46	1.01	8.03
Site Two											
0-30	8.3	0.831	0.09	0.42	0.72	8.4	73.12	0.72	1.07	1.46	8.16
30-50	7.8	7.52	0.07	0.58	1.00	11.51	71.08	0.82	1.13	1.58	13.77
50-90	8.4	0.482	0.06	0.5	0.86	9.92	32.98	0.49	0.7	2.12	7.5
Site Three											
0-35	7.9	0.63	0.05	0.3	0.52	6.04	24.76	0.44	0.57	2.3	7.56
35-47	7.8	7.75	0.04	0.32	0.55	6.42	40.38	0.59	16.2	40.11	9.25
47-87	7.7	2.76	0.04	0.29	0.50	5.89	37.18	0.64	1.02	2.74	14.68

Source: Own soil analysis result.

same range in the sub soils (0.50%), organic matter content registered on site three is shown in Table 2. Likewise, the observed total nitrogen (%) in the three sampling sites was generally low (0.01, 0.07, 0.06) in site one, (0.07, 0.09, 0.06%), site two (0.05, 0.04, 0.04) and site three across surface to subsurface soil depth, respectively (Table 2).

Available phosphorus and carbonates

Available phosphorous is low (5.97 ppm) for the surface layer and for the next subsurface layer (7.26 ppm) but it decreases to 3.42 ppm for the last sub surface in site one. Whereas it ranges low to medium (8.40 to 11.51 ppm) increasing down ward for the surface layer and being in the medium range for the next subsurface layer (9.92 ppm) on site two. Likewise, available phosphorous is low in all the layers ranging 6.04 to 6.42 ppm for the surface layer and 5.89ppm for the next subsurface layer of site three (Table 2).

Generally the surface soils have strongly alkaline reactions on the surface (pH ranges from 7.9 to 8.3) and the subsurface soils do have lower but similar alkaline reactions in the lower layers (pH 7.7 to 8.30). Electrical conductivity (EC) measurements are used as indications of total quantities of soluble salts in the soil. The increase in EC with soil depth is generally registered but it is non-saline and ranges between 0.063 and 0.0831 dS/m on the surface soil, 0.218 and 2.76 ds/m in the sub soils of the lower layers. On the other hand, the organic carbon (OC) content of the soils ranges from 0.30 to 0.42% in the surface layer and its content decreases regularly (0.17 to 0.58%) with soil depth. In almost all the soils, the OC content of the soil layer is <1% for all layers. Such a low OC content of the soils could be due to the association of low humic substances (Brady and Weil, 2002). Similarly, total nitrogen (TN) content of the soils generally ranges from very low to low (0.01 to 0.09%). Subsurface soils have higher values than surface soils. The available phosphorous content of the soils is low on surface layer (5.97 ppm) and subsurface (7.26 to 3.42 ppm).

A widely used measure of the deleterious effects of high sodium level is the exchangeable sodium percenttage, which is defined as [exchangeable Na/CEC] x 100 (Brady and Weil, 2002). An ESP value of 15 is often regarded as the boundary between sodic and non-sodic soils (Brady and Weil, 2002), although, it has been realized that this is an arbitrary figure, since the properties of soils often exhibit no sharp change as the content of exchangeable Na increases. In some soils, exchangeable Na content of 2 to 3 cmol(+)/kg soil may be a more suitable criterion for distinguishing sodic samples. In general, soils with exchangeable Na >1 cmol(+)/kg should be regarded as potentially sodic (Frenkel, 1984; Brady and Weil, 2002). The exchangeable sodium percentage was generally classified as medium (0.8 to 2.1%) and commonly increases with depth, except in the mapping unit three which is extremely a high value (40.1%). This indicates that the soils are potentially sodic. The CEC varies between 24.76 and 73.12 cmol (+) kg^{-1} soil on the surface layers, while it ranges from 7.26 to 71.08 cmol (+) kg^{-1} on the sub surface soils. The CEC of the last layer is higher than the middle layer that ranges from 32.98 to 45.40 cmol (+) kg^{-1}. This indicates the availability of cation saturation in the study site.

Soil chemical characters among mapping units

The result of the analysis indicated that there is no significant difference for pH, electrical conductivity, total nitrogen, CEC, Ca, Mg, K, $CaCO_3$ and exchangeable sodium

Table 3. Physico-chemical properties of soils at three sites (mapping units) (Mean+SE).

Soil parameter	Mapping Unit			Level of Significance		
	Site one	Site Two	Site Three	F-value	(P-value)	Rating
pH	8.2±0.10	8.17±0.19	7.8±0.06	3.093	0.119	High
EC (dS/m)	0.21±0.07	2.94±2.29	3.71±2.11	1.051	0.406	Low
TN (%)	0.047±0.019	0.073±0.008	0.043±0.003	1.872	0.234	Low
OC (%)	0.28±0.06[a]	0.50±0.046[b**]	0.30±0.016[a]	8.33	0.019	Low
OM (%)	0.48±0.10[a]	0.86±0.08[b**]	0.52±0.014[a]	8.057	0.020	Low
Avail. P(ppm)	5.55±1.13[a]	9.94±0.89[b**]	6.11±0.16[a]	8.14	0.020	Low to medium
CEC(cmol(+)/kg)	28.07±11.15	59.06±13.05	34.11±4.76	2.55	0.158	High
K	0.79±0.236	0.68±0.097	0.56±0.06	0.623	0.568	High
Na	0.37±0.057	0.97±0.13	5.93±5.13	1.059	0.404	Medium to very high
ESP (%)	2.34±1.39	1.72±0.20	15.05±12.53	1.068	0.401	Low to medium
$CaCO_3$	8.43±0.41	9.81±1.98	10.49±2.14	0.379	0.700	*

Similar letters or no letters with rows indicate that there is no significant difference among parameters, α = 0.05.

Table 4. Chemical composition of water used for irrigation of the three sites.

Water parameter	Units	Degree of Restriction (Biswas, 1998; FAO soil Bulletin 55, 1985b; Wesstcott and Ayers,1985)			Values for Fursa River	Severity status
		None	Slight to moderate	Severe		
Electrical Conductivity(EC)	ds/m	<0.70	0.70-3.00	>3.00	1.31	Slight to moderate
Total dissolved solids (TDS)	ml/l	<450	450-2000	>2000	838.40	Slight to moderate
Sodium (Na^+)	meq/l		-		70.00	Severe
SAR	$meql^{-1/2}$	<3.00	3.00-9.00	>9.00	10.00	Severe
Adjusted SAR			-		20.40	
Calcium (Ca^{2+})	meq/l	0 to 800: normal range			88.80	Normal
Magnesium (Mg^{2+})	meq/L	0 -120: normal range			8.51	Normal
Potassium (K^+)		-	-		30.00	Normal
pH	pH scale	6.5 -8.4: normal range			7.36	Normal

percentage among the three mapping units. However, significantly higher (p≤0.05) OC, OM and available phosphorus were registered on mapping unit two (site two) as compared to mapping unit one and three (Table 3). Most soil properties (EC, TN, OM and OC) was found to be low and others such as pH, CEC, K and Na were available at high concentration when compared with standards set by FAO guideline (Table 3).

Irrigation water quality

Knowledge of the chemical composition and quality issues related to irrigation water sources in the study area are of particular significance for the proper management and wise utilization. The quality of irrigation water needs to be determined through laboratory analysis in order to maintain a safe salt balance in the soil (Brady and Weil, 2002). In an effort to this end, the water sources used for irrigation water resources of the area was surveyed and identified. Irrigation water quality was analyzed for SAR, pH and EC of the water.

Sodium absorption ratio (SAR)

Irrigation water containing large amounts of sodium is of special concern due to sodium effects on the soil and poses a sodium hazard usually expressed in terms of sodium adsorption ratio (SAR). SAR is calculated from the ratio of sodium to calcium and magnesium as they tend to counter effects of sodium. To determine the SAR value, water sample was taken at edge and middle of Fursa River for analysis of Na, Ca and Mg. And then SAR of the water solution was calculated from the concentrations of soluble Na, Ca and Mg. At the end of the analysis, the SAR value of Fursa River was found to be 10.0, while the adjusted SAR approaches 20.40 (Table 4). However, SAR value of water greater than nine (>9) was severely restricted to use for irrigation due to risk of sodium hazard (Biswas, 1998; FAO soil Bulletin 55, 1985; Wesstcott and Ayers, 1985). This implies early warning for potential hazard of sodic soil in the area.

pH and EC of the water

The pH of the water samples could be measured either in the field or in the laboratory using digital pH meter. From the same sample (with the same method) used for SAR analysis, the pH data was generated. The result of the analysis indicate that the pH of the Fursa River is slightly saline (7.36) or nearly neutral (Table 4). Electric conductivity (EC) of the water samples have been done like the pH analysis (Richards, 1954). The values of EC obtained from the analysis at room temperature needs to be corrected to 25°C using a temperature coefficient of 2.3%. By doing so, the laboratory result showed that the EC of the water is classified as medium (1.31 EC (dS/m) (Table 4) as outlined by different researchers (Biswas, 1998; FAO soil Bulletin 55, 1985; Wesstcott and Ayers, 1985). Moreover, the total dissolved solids were found to be slight to medium range (Table 4).

Soil salinity and sodicity

As a result of the strong relationship between the electrical conductivity of soil extract and the soil salt concentration, the salt content of a soil is commonly expressed by as EC. Measured at a reference temperature of 25°C, the EC is nowadays expressed in decisiemens per meter (dS/m). To evaluate soil salinity, we can measure the EC or the salt concentration in several soil water extracts. The most reliable evaluation is obtained by measuring the salt concentration in soil water at field capacity. The laboratory result shows that the ESP ranges from 0.8 to 2.3% in the surface layer (low to medium) and this indicates that the soils are classified as none saline to slightly saline on the surface but there is an increase in salinity in lower layer (Table 3). The salinity in the mapping unit three is however extremely sodic in lowest layer. Hence coupled with water and soil analysis results, there will be a potential danger of sodicity development in the intended irrigation scheme. Thus, selection of crop type and proper irrigation methods should be designed for sustainability of soil productivity in the study area.

Conclusion

The study showed that the distribution of silt, clay, sand, EC, pH, CEC, available phosphorus, exchangeable bases, carbonates, total nitrogen and organic matter was not uniform with profile/depth wise trend in the three mapping units. The soil salinity and sodicity level of the study area is classified as non-saline and non-sodic for mapping unit one, non-saline and non-sodic for the surface layer of mapping unit two and three, while it is saline for the subsurface layer of site three. There is indication of potential sodicity in the subsurface layer. Moreover, the SAR of the water is medium while the EC of the water is classified as medium. Values of sodicity and combining effect of Salinity and sodicity levels were above the FAO guidelines for water quality restriction limit. Thus, sodicity problem could be expected in the long run. There is indication of potential sodicity not only in the soil but also in the irrigation water. Hence, the study underscores the need for selection of salt tolerant crops and good water management by using appropriate irrigation methods to sustain productivity of soil in the intended irrigation scheme.

ACKNOWLEDGEMENTS

The authors would like to extend sincere thanks to Ada'Ar Woreda Pastoral, Agricultural and Rural Development Office for the financial support. National Soil Testing center professionals are greatly acknowledged for their assistance in analyzing the soil samples in the laboratory.

The kind collaboration from Wollo University Soil and Water Resources Management is also highly appreciated.

REFERENCES

Al-Ghobari HM(2011). Effect of Irrigation Water Quality on Soil Salinity and Application Uniformity under Center Pivot Systems in Arid Region. Aust. J. Basic Appl. Sci. 5:72-80

Awulachew SB, Yilma AD, Loulseged M, Loiskandl W, Ayana M, Alamirew T (2007). Water Resources and Irrigation Development in Ethiopia. Working Paper 123. Colombo, Sri Lanka: International Water Management Institute. pp.78.

Ayers RS, Westcott DW(1985).Water Quality for Agriculture. Irrigation and Drainage Paper No. 29.Rev.1FAO.Rome.

Biswas A (1998). Environmental Planning Management and Development; Tata McGraw-Hill Publishing Company, New Delhi. pp. 208–402.

Bouyoucos GH(1951). A Recalibration of the hydrometer for making mechanical analysis of soils. Agron. J. 43: 434-438.

Brady NC, Weil, RR(2002). The nature and properties of soils. 10th ed. Prentice Hall.

Bremner JM, Mulvaney CJ (1982). Nitrogen-total. In: AL. Page RH Miller and DR Keeney (eds.) Methods of Soil Analysis. Part II. Chemical and microbiological properties. Agronomy 9. American Society of Agronomy Inc. Madison, Wisconsin. pp. 595-624

FAO (1985b). Guidelines: Land Evaluation for Irrigated Agriculture. Soil Bulletin 55. Food and Agriculture Organization of the United Nations, Rome, Italy.

FAO(1985a). Water quality for agriculture. Irrigation and Drainage Paper 29, Rev. 1, FAO Rome.

FAO(1990). Guidelines for soil profile description. 3rd ed. (revised). Soil Resources, Management and Conservation Service, Land and Water Development Division, FAO, Rome.

FAO(2000). Simple Soil, Water and Plant Testing Technique for Soil Resource Management. Land and Water development Division, FAO, Rome.

Frenkel H(1984). Reassessment of water quality criteria for Irrigation. In: I. Shainberg and J.Shalhevet(eds). Soil Salinity under irrigation; processes and management. Springler Verlag, NY. Inc, New Jersey. pp.143-172.

Kinfe H(1999). Impact of climate change on the water resources of Awash River Basin, Ethiopia. Clim. Res. 12: 91-96.

Mass EV, Hoffman GJ (1977). Crop salt tolerancecurrent assessment. J. Irrigation Drainage Div. (Proceedings of American Society of Civil Engineers). 103:115-130.

Olsen SR, Cole CV, Watanabe FS, Dean LA(1954). Estimation of available Phosphorus in soil by extraction with sodium bicarbonate. USDA Circular 939:15-19

Richards LA (1954). Diagnosis and improvement of saline and alkali soils. USDA Agric. Handbook, No.60, Washington.

Tadesse G, Bekele E(1996). Saline and Saline Sodic soils of the Middle Awash Valley Ethiopia. In: Teshome Yizengaw, Eyasu Mokonnen, and Mintesinot Behailu (eds.). 1996. Proceedings of the third conference of the Ethiopian Society of soil Science, February 28 -29, 1996. Addis Ababa, Ethiopia. pp.194

Tessema G (2011). Management of Salt Affected Soils. Soil Analysis for Soil Salinity and Fertility Status. VDM Publishing House Ltd.Verlag Dr.Muller GmbH and Co.KG. Germany. pp. 84.

Van Reeuwijk L (1993). Procedures for soil analysis. 4th ed. International Soil Reference and Information Center, The Netherlands.

Westcott DW, Ayers RS(1985). Irrigation water quality criteria. In Irrigation with Reclaimed Municipal Wastewater: a Guidance Manual, Pettygrove GS, T. Asano T (eds). Lewis Publishers Inc.,Chelsea, Mississippi.

Geo-environmental/ecological management study and audit scheme of Alakiri Field, Niger Delta

P.O Youdeowei[1] and H.O Nwankwoala[2*]

[1]Department of Environmental Management, Institute of Geosciences and Space Technology, Rivers State University of Science and Technology, P.M.B 5080, Nkpolu-Oroworukwo, Port Harcourt, Nigeria.
[2]Department of Geology, University of Port Harcourt, P.M.B 5323, Choba, Port Harcourt, Nigeria.

Ten (10) water boreholes for water level monitoring and sub-soil analysis were carried out in the Alakiri Field, Niger Delta, Nigeria. The positions of the boreholes were chosen on the basis of grid sampling to reflect a good representation of the entire Alakiri field. Boring of the drill holes to 20 m each was carried out with the use of a light cable percussion rig. Stratigraphic synopses of the boreholes display considerable variations and heterogeneity in the soil types, indicating that the soils were deposited under varying energy environmental conditions of river (fluviatile) and tidal channel (marine) activities. Grain size analysis results show a range of well graded (poorly sorted) aquiferous sands with very little fines (1.60 to 3.10% finer than No.200 sieve) to uniformly graded (well sorted) clayey sands, sandy and silty clays with moderate to very high fines or silt/clay fractions (17.41 to 94.60% passing 0.075 mm sieve). Permeability tests of the samples vary from low of 6.40 x 10^{-3} to 7.82 x 10^{-3} cm/s for the silty clayey mud to sandy clay, to fairly high of 7.20 x 10^{-1} to 1.01 cm/s for the clayey sand to sand. Coefficient of permeability increases with increasing size of voids, which in turn increases with increasing grain size. Soil consistency results show a low plasticity range (2.35 to 13.40%), with the predominantly sandy soils being non-plastic. The low value of plasticity indices of the plastic soils is an indication of low water retaining capacity. The soils fall within the soil classification groups of SC, SW, CL, OL and PT under the Unified Soil Classification System (U.S.C) scheme. Resulting from almost saturated state of the soils, water boreholes sited in the area, for domestic and industrial utilization will have to go considerable depths (over 200 m) to obtain uncontaminated water in case of pollution and because of the prevalent saline water.

Key words: Environmental assessment, ecological management, soils, geotechnical engineering, Niger Delta.

INTRODUCTION

To date, the assessment of the qualitative and quantitative status of groundwater is based exclusively on hydrogeological parameters (Steube et al., 2009). Reliable ecological criteria and the design of appropriate indices/ development of applicable assessment schemes are very necessary (Danielopol et al., 2003). Evaluation of the hydrological and subsoil analysis of the study area is very desirable, especially with the intensive oil and gas exploration activities that have been going on for the past three decades.

Alakiri Field which lies on the mangrove swamp environment of the Nigerian coastline in the Niger Delta area is ecologically sensitive. The dense network of rivers and creeks in the area ensures that within the saltwater mangrove swamps, there are small islands with limited area with low ground elevation and shallow depth to groundwater. The water table is also permanently at or near the ground surface. This is mainly influenced in the same way as surface waters hence the prevailing hydrostatic relationship between the ground and surface water in the area which tends to be hydrodynamic. Therefore, assessment of subsoil characteristics, water level determination and hydro-geological investigations become imperative. The impact of recharge and discharge as a

*Corresponding author. E-mail: nwankwoala_ho@yahoo.com.

measure of the aquifer vulnerability and its temporal dynamics, the contaminant degradation and *in-situ* biodegradation rates and detection of effects by short-term contaminant impacts (if any) can only be corrected through regular geo-environmental monitoring.

In this study therefore, the saturated state of the soils and water boreholes was assessed for domestic and industrial utilization to obtain uncontaminated water in case of pollution and the prevalent of saline water encroachment.

The study area

The study area (Alakiri Field) lies on the mangrove swamp environment running parallel to the Nigerian coastline in the Niger Delta area. The area is sandwiched between the outer barrier island complex and the older sands of the Benin Formation, in a belt which reaches widths of between 30 and 40 km at the delta flanks and across the delta. This mangrove swamp zone crisscrossed by a network of dynamic rivers and creeks and has evolved primarily through the process of repeated bifurcation of the River Niger and its distributaries. The frequent flooding with freshwater from rainfall and/or saltwater at high tide has resulted in dense vegetation typical of a brackish environment. The decaying vegetable remains, mixing with deposit of silts and clays to form a swampy landmass.

Intra-swamp deltas form as complexes of islands and mudflats in the widest parts of the larger channels. Mangrove tidal flats soils are polygenetic, consisting of cross-stratified fine to coarse sand and organic - rich clays with abundant drifted debris, which were deposited during the Pliestocene epoch. The predominant vegetation in the area is the hylophytic red mangrove species, *Rhizophora racemosa.*

METHODOLOGY

Drilling technique and soil sampling

Drilling of ten (10) water boreholes for water level monitoring and sub-soil analysis was carried out onsite (positions of water wells are shown on the location map and the groundwater level contour map). Their positions were chosen on the basis of grid sampling to reflect a good representation of the entire Alakiri Field. Boring of the drill holes to 20 m each was carried out with the use of a light cable percussion rig of the Pilcon Wayfarer type. Percussion rig drilling method permits more accurate determination of groundwater levels, more accurate sampling of groundwater for quality analysis, and more accurate sampling of the cuttings as brought up with the bailer. Furthermore, the equilibrium water level in the well is an accurate measure of the pressure head in the aquifer at the bottom of the hole.

This drilling technique entails the boring of a vertical hole by repeatedly dropping a heavy, cylindrical chisel-type drill bit combined with a bailer. This loosened up the soil material and drilling is periodically interrupted to remove the soil samples. The length of the combined tool is about 1.5 m and was suspended from a wire cable, which was strung over a pulley at the top of a near-vertical mast erected over the hole. The bottom of the bailer consisted of a flat valve which was open when the bailer was lowered into the hole and closed when the bailer was pulled up, bringing out drill cuttings with it. The boreholes were logged on site and soil samples recovered at intervals where lithologic changes occurred for laboratory analysis of textural characteristics, classification and infiltration properties.

The wells were screened (0.5 mm slot) at the aquifer depth, gravel packed (3/8" gravel size), cased, and capped with PVC plastic pipes (3" diameter). Their bases were also cemented. These were all done to prevent well collapse and groundwater contamination within the borehole sites as per casing and/or sealing of the wells may result in the direct ingress of contaminated surface water into groundwater. Soil samples recovered from the boreholes were subjected to both visual examinations and laboratory tests. Laboratory tests carried out on representative soil samples in accordance with B.S 1377 (1990) Standard.

RESULTS AND DISCUSSION

Soil stratigraphy

Stratigraphic sections of the ten boreholes drilled are detailed in the log sheets (Figure 1). The stratigraphic synopses of the boreholes display considerable variations and heterogeneity in the soil types. This is an indication that the soils were deposited under varying energy environmental conditions of river (fluviatile) and tidal channel (marine) activities. The sandy lenticular sediments represent the high energy environmental deposits of river and tidal channel processes. The interbedded silty, clayey sediments were deposited in the low energy environment.

The most commonly occurring soil type as revealed by the loss of the drill holes are the light to dark grey, very fine grained, organic silty clays. They are prominent in the fossiliferous (woody plant remains) muddy form as peaty silty clayey mud (clayey shale) also locally known as *Chikoko mud.* This latter soil type represents a marker bed, as it occurs consistently across the area. In some places, it is the superficial, overlying soil with thickness up to 5 m, in other places, it lies about 1 m below the dredge spoils to depths of about 8 m. It is however absent in BH4 and BH10. Clastic sequences of coarse sand (2 to 6 m thick) interpose the various strata and occur at end of some of the boreholes. These are the aquiferous layers. The boreholes are also ineter-stratified with other detrital deposits such as sandy clay, clayey sand, gravelly silty clay, and sandy silty clay which vary in thickness, grain size, relative density and consistency, but with minor colour alterations (light to dark grey and light to dark brown).

Soil textural analysis and classification

Grain size distribution analysis, permeability (using both falling and constant head permeameters) and Atterberg (Consistency) limits were carried out. The tests were carried

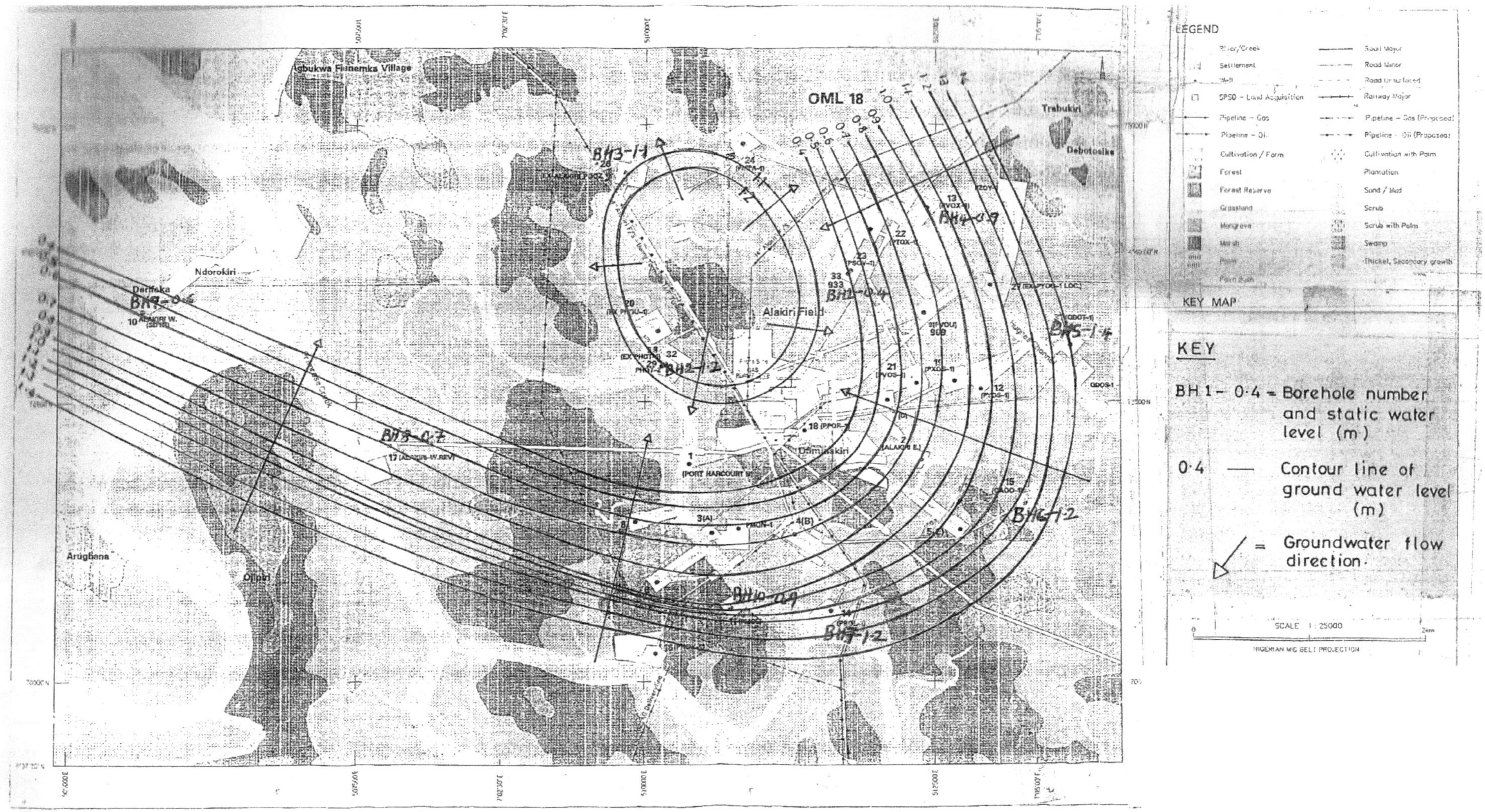

Figure 2. Groundwater level contour map of Alakiri Field.

Table 1. Summary of laboratory soil test results in the study area.

Borehole number	Analyzed soil sample	Sample depth (m)	Soil group nomenclature	Grain size distribution (Percent passing sieves)				Atterberg limits			Permeability (cm/s)	Classification
				No. 4 (4.75 mm)	No. 10 (2.00 mm)	No. 40 (0.42 mm)	No. 200 (0.075 mm)	LL (%)	PL (%)	PI (%)		Unified soil classification system (U.S.C.S)
BH1	BH (0.00)	0.00	Clayey sand	99.50	92.70	41.300	30.00	18.50	12.60	5.10	1.01×10^{0}	SC.
	BH (15.00)	15.00	Sandy clay	100	98.80	77.60	43.20	22.50	11.90	10.60	7.34×10^{-3}	CL.
BH2	BH (0.00)	0.00	Peaty mud	-	-	-	-	34.20	22.30	11.90	6.51×10^{-3}	PT.
BH3	BH (11.00)	11.00	Silty clayey mud	-	-	-	-	39.20	25.80	13.40	6.40×10^{-3}	PT.
BH4	BH (6.00)	6.00	Sandy silty clay	99.00	97.00	74.30	66.60	26.50	20.37	6.13	7.81×10^{-1}	OL.
BH5	BH (0.00)	0.00	Clayey sand	96.77	55.41	33.07	17.41	-	-	Non plastic	9.31×10^{-1}	SC.
	BH (16.00)	16.00	Sandy silty clay	99.00	97.50	34.50	20.00	20.80	17.55	3.25	8.60×10^{-1}	OL.
BH6	BH (17.00)	17.00	Clayey sand	97.20	92.40	58.40	29.10	15.00	12.65	2.35	7.20×10^{-1}	SC.
BH7	BH (20.00)	20.00	Silty clay	100	100	96.40	94.60	39.60	30.05	9.55	6.61×10^{-3}	OL.
BH8	BH (20.00)	20.00	Sand	94.40	80.60	26.70	1.60	-	-	-	8.73×10^{-1}	SW
BH9	BH (0.00)	0.00	Silty clay	100	99.60	95.30	83.80	34.00	27.50	6.50	6.80×10^{-3}	OL
	BH (3.00)	3.00	Sandy clay	99.70	96.40	70.70	46.00	18.00	14.10	3.90	7.82×10^{-3}	CL
BH10	BH (0.00)	0.00	Sand	98.70	90.70	21.40	3.10	-	-	-	7.50×10^{-1}	SW

out to enable the evaluation of the gradation, hydraulic conductivity (coefficient of permeability) and consistency (water absorbing and adsorbing ability) properties of the soil samples, as well as their classification. These para-meters were also used as indices of the infiltration capa-city of the soils at the site. The results are summarized in Table 1.

Grain size patterns

Grain size analysis results obtained from the test showed a range of well graded (poorly sorted) aquiferous sands with very little fines (1.60 to 3.10% finer than No. 200 sieve) to uniformly graded (well sorted) clayey sands, sandy and silty clays with moderate to very high fines or silt/clay fractions (17.41 to 94.60% passing 0.075 mm sieve). Uni-form graded soils have relatively low densities since the number of particles per unit area is small. Infiltration of fluids through the uniformly graded fine grained soils such as the silty clay and the peaty mud will be greatly reduced due to the reduction of void spaces in them, so that they are characterized by poor drainage capacity.

Permeability

The values obtained from permeability tests of the sam-ples varied from low of 6.40×10^{-3} to 7.82×10^{-3} cm/s for the silty clayey mud to sandy clay, to fairly high of 7.20×10^{-1} to 1.01 cm/s for the clayey sand to sand. Coefficient of permeability increa-sed with increasing size of voids, which in turn increased with increasing grain size. The larger grains will permit more fluid flow as a result of high

Table 2. Static water levels for the wells.

Bore holes (wells)	Static water levels (m)
BH1 (Oil Well 33)	0.47
BH2 (Oil Well 32)	1.22
BH3 (Oil Well 26)	1.13
BH4 (Oil Well 13)	0.97
BH5 (Oil Well 31)	1.40
BH6 (Oil Well 15)	1.24
BH7 (Oil Well 14)	1.26
BH8 (Oil Well 17)	0.78
BH9 (Oil Well 10)	0.65
BH10 (Oil Well 11)	0.97

permeability than the finer ones.

Soil consistency

The consistency of the soils at the study site was studied by measuring the natural moisture contents, determining both the liquid and plastic limits and the plasticity indices. The results show a low plasticity range (2.35 to 13.40%), with the predominantly sandy soils being non-plastic. The low values of plasticity indices of the plastic soils is an indication of that they have a low water retaining capacity. The soils fall within the soil classification groups of SC, SW, CL, OL and PT under the Unified Soil Classification System (U.S.C) scheme. Soils with the same classification tend to have the same engineering behaviour (Krynine and Judd, 1957).

Hydrological conditions, pollution susceptibility and pathways

Several factors govern the occurrence and distribution of groundwater in the area. This includes the overburden geologic formation, the low-lying topography, the high – rainfall regime (annual average over 3000 mm), its surrounding rivers and creeks, and even the occasional canalization activities. Surface water comes from either rivers and primary freshwater flows or creeks which receive freshwater flows but are also subject to tidal influences, twice daily, pushing saltwater into the area. The area is almost totally submerged during high tides and seasonal floods giving rise to very severe drainage problems. The insufficient drainage capacity is caused by the generally flat topography and superficial clayey deposits, giving rise to a wide distribution of swamps within the area.

The dense network of rivers and creeks ensures that within the saltwater mangrove swamps, there are small islands with limited area and normally low ground elevation as well as shallow depth to groundwater. The water table is permanently at or near the ground surface and is mainly influenced in the same way as surface waters, so that the prevailing hydrostatic relationship between the ground and surface water tends to be hydrodynamic. Because of the fairly uniform nature of the terrain, gravity drainage is low and the hydraulic gradient between two points will change very little from point to point. Subsequently, both groundwater movement and surface discharge will vary only little from hour to hour. Topography is the primary control over flowing wells and there is head loss, the hydraulic gradient will cause the flow of groundwater.

From field measurements of static water levels in the boreholes (Table 2), a ground water level contour map was constructed for the study site (Figure 1). Flow lines, sketched perpendicular to contours show directions of movement (Todd, 1980; Davies and DeWeist, 1966). The velocity of groundwater, other things being equal, is greatest where the contour lines are closest together such as at the groundwater cascade between BH7 and BH10. Leachate or any pollutants introduced at a given point will migrate as pollution plumes along these flow paths. Effluents may fail to be mineralized or metabolized at significant rates in such aquatic ecosystems. Some chemicals readily dissolve in water while others have low chemical solubility. Solvents with lower density tend to float while those with heavier density than water tend to sink or migrate vertically down the aquifer. Except for small amounts of hydrocarbons that go into solution, oil does not penetrate to below the water table (oil is immiscible in water and is less dense). As oil accumulates on the water table, the oil zone spreads laterally, initially under the influence of gradients caused by gravity and latter in response mainly to capillary forces (Freeze and Cherry, 1979).

Conclusion

This study has revealed, from lithostratigraphic setting the intercalations of predominantly sand and clay formations which constitute a system of aquifers separated by aquitards. The peaty silty clay mud which overlies most parts of the terrain will exhibit a measure of transmissivity due to the presence of the thickly fibrous materials in them. However, the near-saturation state of the soil will restrict infiltration (Davies and DeWiest, 1966). Parameters that influence infiltration include soil type and cover, surface texture, limiting slope value, initial moisture content of soil, and intensity of spill. Infiltration decreases exponentially with time, and as the soil becomes saturated and as its clay particles swell. This is caused mainly by the filling of the soil pore by water.

The decline is more rapid and the final constant rate is lower for fine grained clay soils than for loose sandy soils (Freeze and Cherry, 1979). Resulting from almost saturated state of these soils and water boreholes sitting in the area, for domestic and industrial utilization, one will have to go considerable depths (over 200 m) to obtain uncontaminated water in case of pollution and because of the prevalent saltwater.

REFERENCES

BS 1377 (1990). Methods of test for soils for civil engineering purposes. British Standards Institution (BSI), London.

Danielopol DL, Griebler C, Gunatilaka A, Notenboom J (2003). Present state and future prospects for groundwater ecosystems. Environ. Conserv., 30: 104 – 130

Davies SN, DeWiest RJM (1966). Hydrogeology. John Willey and Sons. Inc.

Freeze RA, Cherry JA (1979). Groundwater. Prentice-Hall, Inc. Englewood Cliff, N.J.

Krynine DP, Judd WR (1957). Principles of Engineering Geology and Geotechnics. McGraw-Hill Book Company.

Steube C, Richter S, Griebler C (2009). First attempts towards an integrative concept for the ecological assessment of groundwater ecosystems. Hydrogeol. J., 17:23 – 35.

Todd DK (1980). Groundwater Hydrology. 2nd Edition. John Wiley and Sons. Inc. N.J.

Some crustacean zooplankton of Wular lake in Kashmir Himalaya

Javaid Ahmad Shah and Ashok K. Pandit

Centre of Research for Development (CORD), University of Kashmir, Srinagar- 190006, Kashmir, India.

In a taxonomic survey of crustacean zooplankton collected from Wular lake of Kashmir, a pictorial key was developed. Thirty-six (36) pictures of 25 species of crustacean zooplankton, out of which 21 represented 16 Cladocera taxa belonging to Chydoridae (*Alona affinis, A. rectangula* and *A. monacantha, Chydorus sphaericus* and *C. ovalis*), Daphnidae (*Daphnia magna, D. catawba, D. magna, D. pulex*; *D. rosea, D. galeata, D. retrocurva* and *Moinodaphnia*), Polyphemidae (*Polyphemus pediculus*) and Sididae (*Sida crystallina*) and 15 pictures of Copepoda belonged to 9 species being represented by three families viz Cyclopidae (*Cyclops scutifer, C. bicuspidatus, C. vernalis, C. panamensis, Eucyclops agilis, Megacyclops viridis*), Canthocamptidae (*Bryocamptus hiemalis*) and Diaptomidae (*Diaptomus* sp.).

Key words: Crustacean, Cladocera, Copepoda, taxonomy, Wular lake.

INTRODUCTION

An identification key is a tool to assist a user to identify an organism. It guides the user along a path towards identification. An identification is performed by selecting from among the listed features those that match your specimen; essentially you describe the features of your specimen to the key. The tool then compares your described features against a database of the features of all the organisms in the key, and returns a list of those organisms that match your choices. As you choose answers to the questions asked by the key, the list of potential matches is progressively reduced. Crustaceans a globally-distributed faunal group found across all habitats from the equator to the poles and form important link as food items of higher trophic levels, besides being used as bioindicators in the aquatic systems (Pedrozo and Rocha, 2005; Burns and Galbraith, 2007; Rajashekhar et al., 2009; Joshi, 2011; Shah et al., 2013; Shah and Pandit, 2013). It is in view of their ecological importance in understanding the health of aquatic ecosystem, a taxonomic survey based on key features of crustacean zooplankton, was carried out in the largest freshwater body of Indian subcontinent, the Wular lake of Kashmir Himalaya. Till date there is no photographic key available for crustacean zooplankton from Kashmir Himalaya and therefore the study aimed to develop the same.

MATERIALS AND METHODS

Wular lake, an ox-bow type lake located in the north-west of Kashmir about 55 km from Srinagar city, is situated at an altitude of 1,580 m (a.m.s.l), and lies between 34°16´-34°20´N and 74°33´-74°44´E geographical co-ordinates. The lake is drained in the north-east by the only single outlet in the form of River Jhelum (Figure 1). The lake has been declared as the wetland of national importance (1986) under wetland programme of Ministry of Environment and Forests (MoEF), Government of India and is designated as Ramsar Site (a Wetland of International Importance under Ramsar Convention, 1990).

Samples for crustacean zooplankton were collected by filtering 100 litre sub-surface water from five study sites through 80 µm nylon Bolt mesh and fixed in 4% formalin to which 4-5 drops of

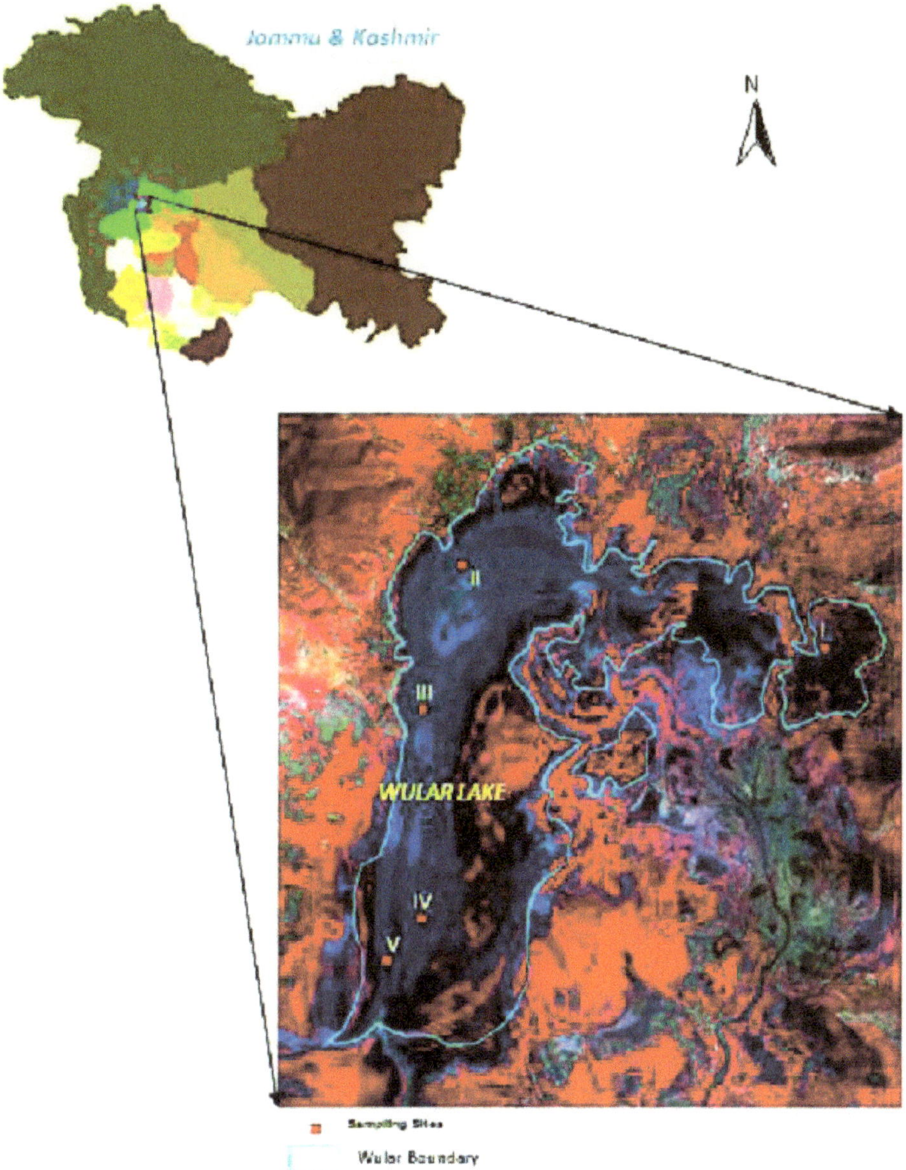

Figure 1. Location of study sites within Wular lake.

glycerine and 5% sucrose were added. Identification was made to the lowest possible taxon using standard taxonomic keys (Smith and Fernando, 1978; Bacler et al., 1984; Pennak, 1989; Edmondson, 1992). All pictures are original taken with digital camera (14.2MP) under Magnus binocular microscope (MLX-DX, no.4 B 523861).

RESULTS AND DISCUSSION

Identifying features of Cladocera (Figure 2)

Alona affinis (Leydig, 1860)*:* The rostrum is more or less pointed; antennule is rod like in appearance. Wider distribution and are considered lacustrine zooplankton (plate 1). Postabdomen of *Alona Affinis:* 10-14 marginal denticles basal spine long (2-3 spinules) (plate 2).

Alona rectangula (Sars, 1861): Littoral species, Postabdomen of *Alona rectangula* with 8-9 denticles (plate 3). Postabdomen with 8-9 denticles *Alona rectangula* (plate 3)

A. monacantha (Sars, 1901): Valves with distinct longitudinal striae. Inferoposteal angle with 1-3 small teeth, Postabdomen with 9-10 denticles (plate 4). Claws with very long basal spine *A. monacantha* (plate 5)

Bosmina longirostris (Müller, 1776): World-wide distribution. Postabdomen long, proximal teeth with 2-3 pectens (plate 6).

Ceriodaphnia sp. Wider distribution. Head rounded with no rostrum, have a little, inflexible antennules. Upper rim of value is arched but lower rim of values delicately thorned branch, single compound eye present. Rostrum

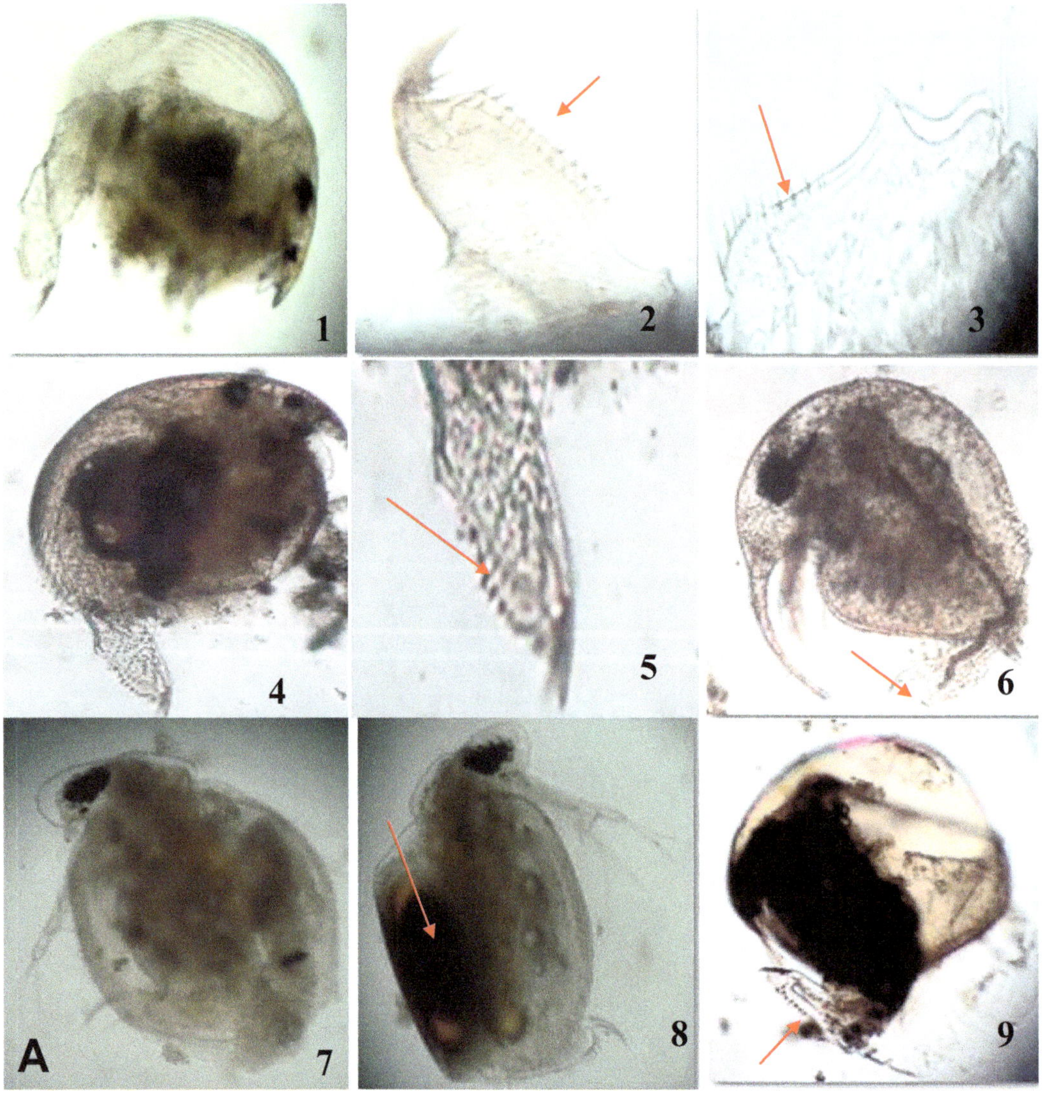

Figure 2a. Plates 1 - 9 are photographic pictures of Cladocera. Plate 1 = *Alona affinis;* plate 2=Postabdomen of *Alona affinis*; plate 3= Postabdomen of *Alona rectangula* with 8-9 denticles; plate 4 = *A. monacantha*; plate 5= Postabdomen of *A. monacantha*; plate 6= *Bosmina longirostris*; plate 7 = *Ceriodaphnia* sp.; plate 8= *Ceriodaphnia* sp. with ephippium; plate 9 = *Chydorus ovalis*.

(beak) absent, head is small and compacted, ventral side is rounded antennae fixed and branched; 3 segments on each Setae at the base of the second segment of the dorsal ramus of the antennae is longer than the ventral ramus (plate 7)

Ceriodaphnia sp. with ephippium containing the winter eggs (plate 8)

Chydorus ovalis (Kurz, 1874): Rounded to oval Post-abdomen with rounded apex 12-14 marginal denticles. Claws with 2 basal spines, the proximal one minute (plate 9)

Chydorus sphaericus (Mueller, 1785): Species of cosmopolitan distribution and a wide range tolerance to a multiplicity of environmental factors Nearshore and littoral, usually found clinging to filamentous algae. Usually more abundant in eutrophic waters. Spherical or broadly elliptical, shell usually reticulate, sometimes smooth. Postabdomen with 8-9 marginal denticles. Claw small proximal basal spine very minute (plate 10).

Daphnia magna (Straus, 1820): Is found in marginal habitats such as small ponds with abundant organic matter and saline ponds. Occasionally found along the margins of lakes (plate 11)

Daphnia catawba (Coker, 1926): Postabdominal process well developed with upper one well developed than the middle one (plate 12).

Posterior margin of post-abdomen deeply sinuate in lateral view; carapace spine usually short with alternate spines on both the sides of carapace spine. *Daphnia magna* (plate 13).

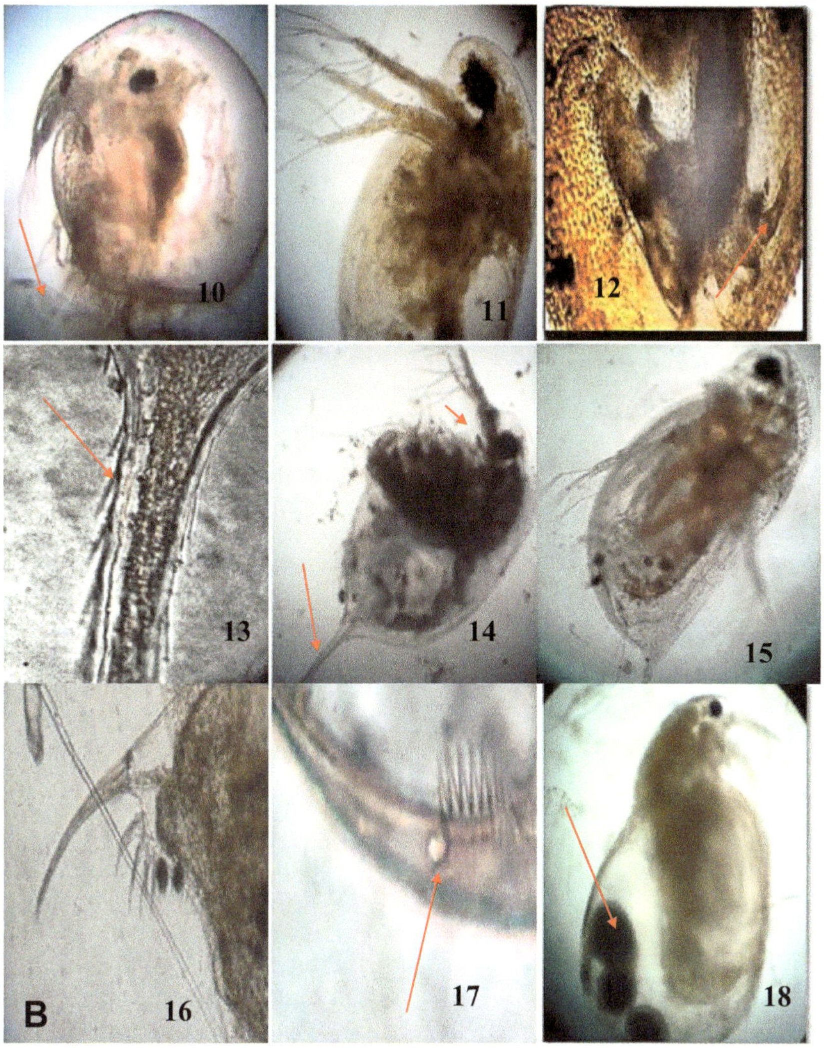

Figure 2b. Plates 10 -18 are photographic pictures of Cladocera. Plate 10=C. *sphaericus;* plate 11= *Daphnia magna;* plate 12=Postabdomen of *D. catawba*; plate 13= Basal spine with setae of *D. magna*; plate 14= *D. pulex*; plate 15= *D. rosea;* plate 16=*D. galeata*, plate 17= *D. retrocurva=;* plate 18= *Moinodaphnia* sp. with eggs.

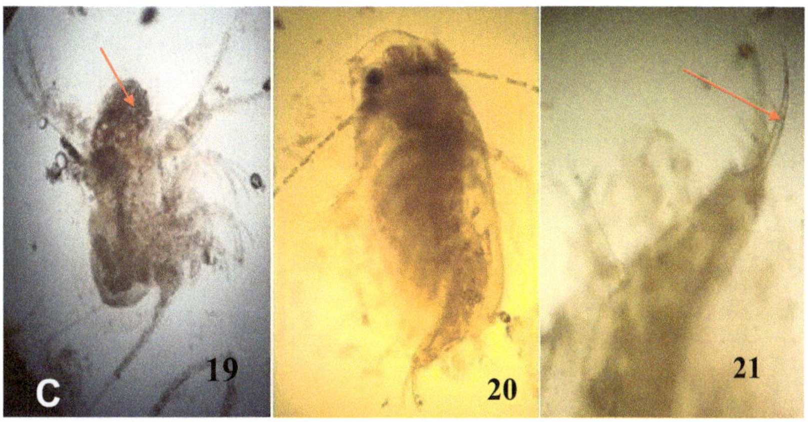

Figure 2c. Plates 19 - 21 are photographic pictures of Cladocera. Plate 19 = *Polyphemus pediculus;* plate 20= *Sida crystallina;* plate 21= basal spine of *S. crystallina.*

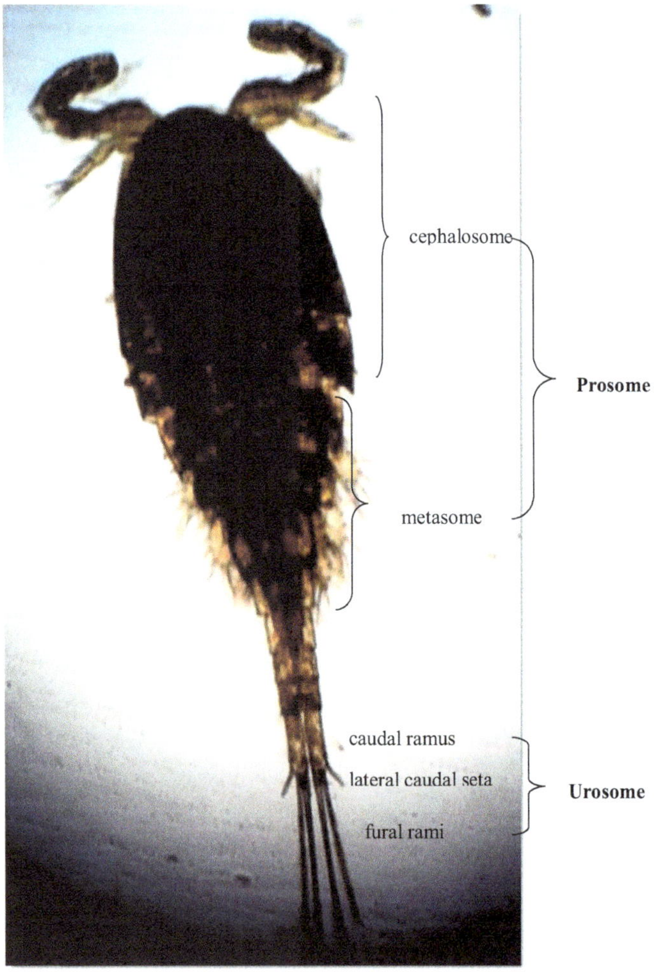

Figure 3. Structure of a typical copepod.

Daphnia pulex (Leydig, 1860): Occasionally in shallow water along the edge of lakes, carapace spine 1/5 to 1/3 of carapace length in mature specimens, ventral margin of head concave, tip of rostrum directed postero-ventral (plate 14)

Daphnia rosea (Sars, 1862): Species bears toothed crust on antero-dorsal margin of the head, although usually crested, is never produced into a helmet. The optic vesicle is very close to antero-ventral margin of the head (plate 15)

D. galeata (Sars, 1864)*:* Postabdomen with well developed setae with plain postabdominal claw (plate 16).

Daphnia retrocurva (Forbes, 1882): Middle pecten of the Postabdominal claw well differentiating from the rest (plate 17).

Moinodaphnia sp.: First antennae attached posterior to compound eye; post-abdomen with a row of lateral feathered setae, and a distal bident tooth on post-abdominal margin near base of post-abdominal claw (plate 18).

Polyphemus pediculus (Linnaeus, 1761): Carapace much reduced, not covering the entire body, forming the brood chamber, body very rounded, short with elongate "tail" huge compound eye - specimens very uniform in shape, immediately recognizable (plate 19).

Sida crystallina (Müller, 1776): Head with dorsal gland, pointed rostrum, dorsal ramus of antennae with two segments (plate 20).

Postabdominal claw long cylindrical tapering toward the end with 3-4 setae.

Identifying features of copepods

Copepods: The major articulation divides the body of Copepoda into an anterior prosome and a posterior urosome. Anteriorly is the cephalosome bearing the five head appendages and the maxillipeds and posterior to this are three free prosomites bearing the second to fourth swimming legs. The urosome comprises an anterior somite bearing the fifth pair of legs and five other somites (referred to as abdomen) with caudal rami elongated into fural rami with or without setae the distinguishing feature which help in identification of majority of the copepod species (Figures 3 and 4)

Cyclops scutifer (Sars, 1863): Inner margin of caudal rami with fine hairs (using high magnification), caudal rami 4 times as long as broad, Fourth and fifth metasomal segments expand into pointed wings, and the positions of the lateral setae (65 to 73% from base) distinguish it from (plates 22 and 23).

Eucyclops agilis (Koch, 1838): Caudal ramus usually not more than 5 times as long as broad, lateral spinules conspicuous on outer margins of caudal ramus and lateral hairs (plates 24 and 25).

Cyclops bicuspidatus thomasi (Forbes, 1882): Species associated with eutrophy (plate 26). The interior of the caudal rami is covered with hairs. *Cyclops bicuspidatus thomasi* (plate 27).

Megacyclops viridis (Jurine, 1820): Body is somewhat robust and leg 5th is very characteristic. Length inner seta of the fural rami is more than twice as long as outer seta (plates 28 and 29).

Cyclops vernalis (Fischer, 1853): *Fural rami* bear prominent hairs on both the sides, inner margin of caudal ramus without hairs (plates 30 and 31).

C. panamensis (Marsh, 1913): Caudal rami with hairs (lateral seta without hairs) (plate 32).

Diaptomus minutus (Lilljeborg1889): Leg 5th with exopod flattened, sharply pointed inner process (plates 33 and 34). *Diaptomus* sp. First antenna as long as body, caudal ramus of equal length, the forth leg not differentiating (plate 35).

Bryocamptus hiemalis (Pearse, 1905) : With egg sac on genital segment (plate 36).

ACKNOWLEDGEMENT

Thanks are due to the Director, Centre of Research for Development and Head, Environmental Science, University of Kashmir for providing necessary laboratory facilities.

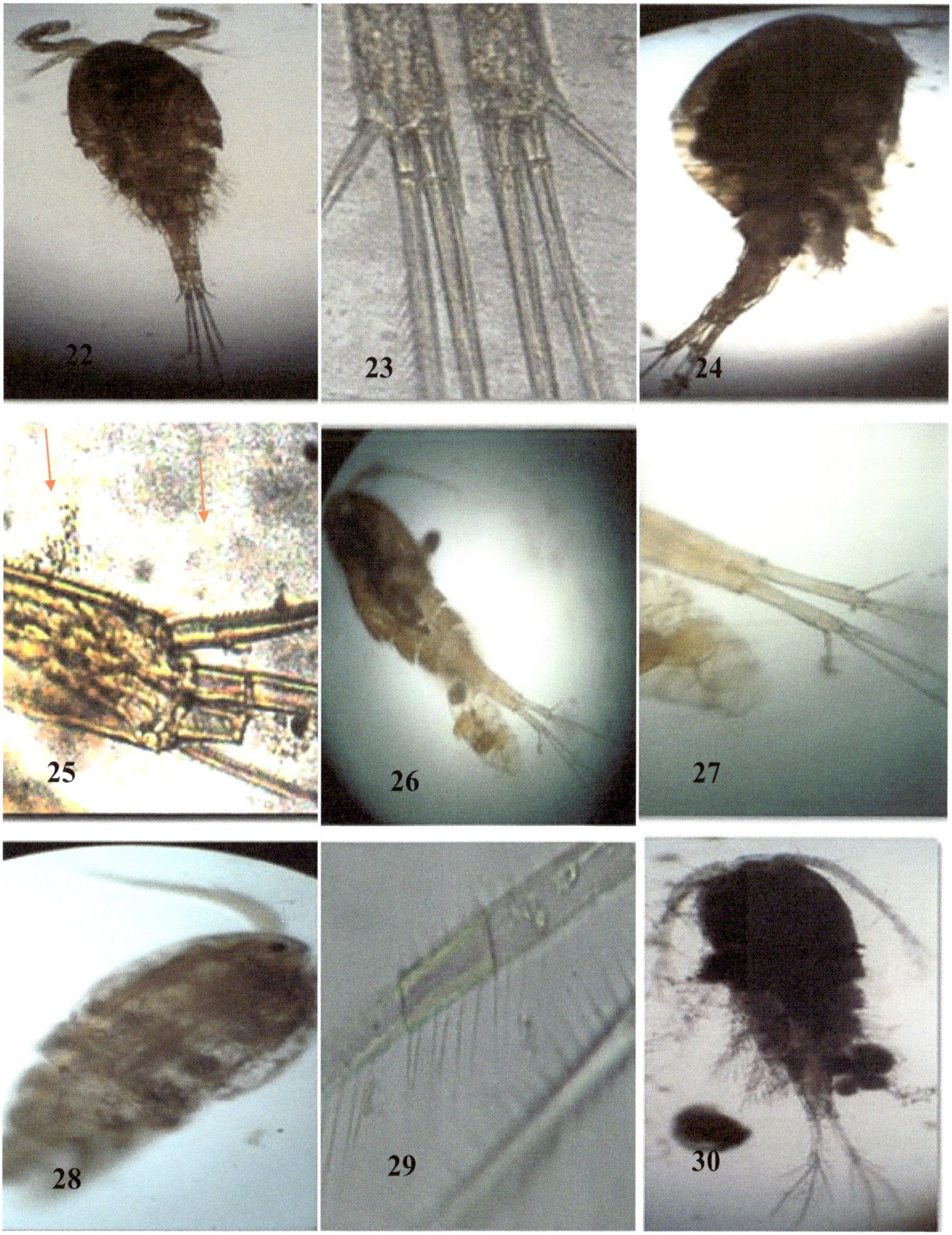

Figure 4. Plate 22-30 are pictures of copepods. Plate 22 = *Cyclops scutifer*; plate 23 = *C. scutifer* caudal ramus; plate 24 = *Eucyclops agilis*; plate 25 = Caudal ramus with minute setae on outer margin of *Eucyclops agilis;* plate 26 = *Cyclops bicuspidatus;* plate 27 = Urosome and caudal ramus of *C. bicuspidatus;* plate 28 = *Megacyclops viridis;* plate 29 = Fural rami of *Megacyclops viridis;* plate 30 = *Cyclops vernalis.*

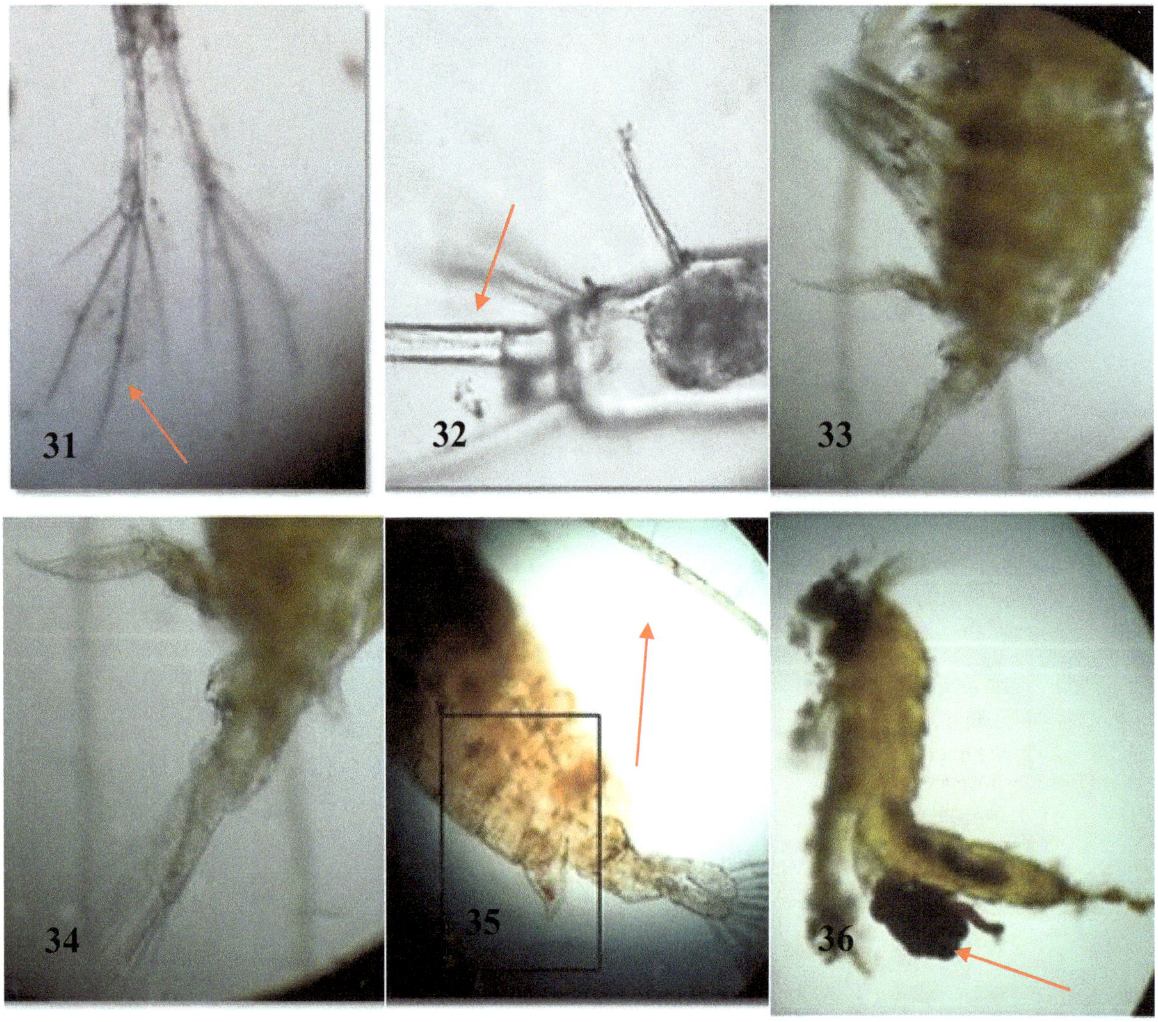

Figure 4. Plate 31-36 are pictures of copepods. Plate 31 = Fural ramus of *C. vernalis;* plate 32 = Caudal rami with hairs (lateral seta without hairs) of *C. panamensis;* plate 33 = *Diaptomus minutus* ; plate 34 = *Diaptomus minutus* with 5th leg*;* plate 35 = *Diaptomus* sp.*;* plate 36 = *Bryocamptus hiemalis* with egg sac.

REFERENCES

Bacler M, Korda N, Dodson SA (1984). Zooplankton of the Great Lakes: A Guide to the Identification and Ecology of the Common Crustacean Species. University of Wisconsin Press. pp.74-101.

Burns CW, Galbraith LM (2007). Relating planktonic microbial food web structure in lentic freshwater ecosystems to water quality and land use. J. Plankton Res. 29:127-139.

Edmondson WT (1992). Ward and Whiple Freshwater Biology. 2nd ed. Intern. Books and Periodicals Supply Service, New Delhi.

Joshi PS (2011). Studies on zooplanktons of Rajura Lake of Buldhana district, Maharashtra India. Sci. Res. Rep. 1(3):132-137.

Pedrozo CS, Rocha O (2005). Zooplankton and water quality of lakes of the Northern Coast of Rio Grande do Sul State, Brazil. Acta Limnol. Brasiliensia, 17:445-464.

Pennak RW (1989). Fresh-water Invertebrates of the United States: Protozoa to Mollusca 3rd ed. Wiley-Interscience, New York. p. 769.

Rajashekhar M, Vijaykumar, K Parveen Z (2009). Zooplankton diversity of three freshwater lakes with relation to trophic status, Gulbarga district, North-East Karnataka, South India. Int. J. Syst. Biol. 1(2):32-37.

Shah JA, Pandit A K Shah G M (2013). Distribution, diversity and abundance of copepod zooplankton of Wular Lake, Kashmir Himalaya, J. Ecol. Nat. Environ. 5(2):24-29

Shah JA. Pandit AK (2013). Relation between physico–chemical limnology and crustacean community in Wular lake of Kashmir Himalaya, Pak. J. Biol. Sci. 16(19):976-983.

Smith K, Fernando CH (1978). A guide to the Freshwater Calanoid and Cyclopoid Copepod Crustacea of Ontario. Unviersity of Waterloo. Waterloo, Ontario, Canada.

10

Public participation in Malawi's environmental impact assessment (EIA) process

I. B. M. Kosamu, A. A. Mkandawire, W. Utembe and H. W. T. Mapoma

University of Malawi Polytechnic, Private Bag 303, Blantyre 3, Malawi.

Since the adoption of the Environmental Management Act of 1996, Malawi has been using environmental impact assessment (EIA) as a tool for predicting and assessing the impact of development projects on the environment. This study assessed the extent of public participation in Malawi's EIA process. Desktop study of relevant documents such as EIA and environmental audit reports, policies and laws, books, journals and internet articles was done. Further information was obtained from relevant stakeholders through interviews and questionnaires. This involved fifteen respondents divided into three categories: five NCE members, five TCE members and five EIA practitioners. This study reveals that public participation is not adequate at most of the key stages of the EIA process in Malawi which puts the human and ecosystem health at risk. There is an urgent need to enforce the EIA provision as stipulated Malawi's Environmental Management Act. Effort should focus on public awareness and human resource capacity development so that the EIA process is diligently executed and that enforcement and follow-ups are properly done by the Department of Environmental Affairs which is a lead agency on environmental issues in Malawi.

Key words: Public participation, environmental impact assessment, Malawi, enforcement.

INTRODUCTION

Most African countries, including Kenya, South Africa, Tanzania, and Nigeria, have done extensive research on public participation in their environmental impact assessment (EIA) process (Murombo, 2008). However, studies to look at the extent of public participation in EIA activities in Malawi are not existent. Since the implementation of the EMA in 1996, Malawi has had a number of development projects including Kayelekera Uranium Mining Project and the construction of Nsanje World Inland Port. The latter project is meant to connect the Shire River in southern Malawi to the Indian Ocean. Besides that, Malawi parliament recently passed a bill to allow for an extensive feasibility study on the possibility of starting oil extraction activities on Lake Malawi.

For such projects EIA procedures are mandatory and public participation plays a significant role in these processes. This is so because people may be affected, for example, re-allocation, exposure to radiation, contamination of potable water sources and creation of employment opportunities (Carroll and Turpin, 2009). The EMA stipulates the procedure for EIA and makes provisions for its implementation and enforcement. EIA provisions in the EMA are found in Sections 24 (projects for which an EIA may be required), 25 (EIA reports), 26 (review of EIA reports), 27 (environmental audits), 28(monitoring existing projects), 29 (fees), 63 (penalties), 76 (closures) and 77 responsibilities. As such Malawi has its own EIA guidelines published in 1997 which provide the code of conduct for public consultation in line with the section 52 of EMA that stipulates public access to information and participation. The level of public participation on EIA issues in such development projects in Malawi is far from clear. It is not surprising therefore that in several projects including the Kayelekera uranium mining project in northern Malawi and the Kapani meat processing facility in the southern city of Blantyre, the lack of public information on potential costs and benefits of development projects has resulted in popular disapproval of otherwise helpful initiatives.

EIA as a tool for sustainable development was adopted much later in developing countries (IAIA, 2006; Glucker, 2012) resulting in most development projects taking place without any EIA (Appiah-Opuku, 2001). In most projects, EIAs were conducted as a precondition from donor and multilateral development agencies (World Bank, 1993). This happened until EIA as a procedure was adopted by most countries during the "post Agenda 21" era in the 1990s (Kakonge, 1996; Tarr, 2003). 'Public participation' has been defined in this study as a process of engagement, where people are enlisted into the decision making process; requiring that those initiating it are open to the potential need for information exchange and are prepared to work with different interests to develop plans or to amend or even drop existing proposals (Petts, 1999). Glasson et al. (2005) emphasize the need for consultation and participation at all stages of the EIA process. This study therefore assessed the extent of public participation in Malawi's EIA process.

MATERIALS AND METHODS

This study used both qualitative and quantitative data collection methods based on multi-stage (two-stage) sampling technique. From the EIA and Environmental Audit reports produced per year from 2001- 2012, a total of 60 reports were randomly selected and analysed. An average number of reports were produced per year. At least fifteen members from the TCE, the NCE, and EIA officers/experts (five interviewees from each category) were also interviewed. Such triangulation is important for ensuring objectivity and representativeness of data.

Description of sample

Forty-five EIA and fifteen Environmental Audit reports for mining, waste disposal, road construction, and water resources and irrigation scheme and residential area development projects were analysed using criteria by Wood (1995). Further information was obtained using both structured and unstructured interviews as well as through a questionnaire for a range of highly knowledgeable experts in the field who could identify key issues in EIA policies, legislation and practice that are related to public participation, health impacts and their assessment and sustainable development. Some of the issues covered include: Is the public consulted? At which stage(s) of the EIA process is the public consulted/allowed to participate in Malawi's EIA system? What are the major challenges facing public participation process in Malawi EIA system? What are the opportunities for the success of public participation in Malawi's EIA System?

Every effort was made to involve respondents with the most extensive experience, that is, those recognized by their peers as having the extensive expertise in the field. Altogether fifteen respondents were interviewed.

Data analysis

The qualitative data was analyzed by organizing the information and identifying the pattern, developing relevant themes and drawing conclusions. The quantitative data was analyzed using Microsoft Excel (2007) which was presented in the form of charts and tables.

RESULTS AND DISCUSSION

Categories of respondents

Data was captured at five levels of respondents, each of which was made up of various categories or groups which took part in the public participation process. These levels are outlined below.

Project area's information/projects' surrounding area

This level looked at the number of people involved in the process of public consultation from affected and surrounding communities and/or villages as well as people involved from the administrative offices of the central government.

Institutional libraries

This level looked at public involvement from surrounding primary, secondary and tertiary academic institutions across the country.

National/local libraries

This category comprised people who forwarded comments, questions and concerns pertaining to the consequences (both positive and negative) of a particular development using public consultation forms placed at local and/or national libraries present in Malawi.

Local institutions and NGO's

This category was concerned with participants from environmental and human rights NGOs, health institutions (example Malawi's ministry of health),water and sanitation NGOs, trade unions, religious groups, legal institutions, companies, security institutions, media houses, political parties, welfare organizations, statutory

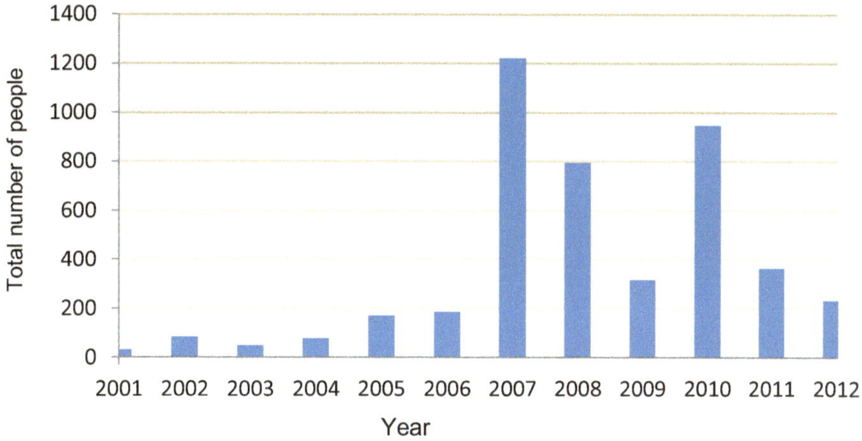

Figure 1. Total number of people consulted each year from 2001-2012.

co-operations and key government departments.

Other sources

This level concentrated on participants from International NGOs and/ or agencies, government ministries, local and/ or international EIA experts, international or outside universities.

Figure 1 presents the number of people that were consulted as indicated in the 60 reports that were submitted to the Department of Environmental Affairs between the years 2001 and 2012.

Figure 1 indicates that public participation was highest in 2007 but the trend decreased with fewer people being consulted in 2012. This is a worrisome scenario as it reflects less incorporation of people's views in project planning and implementation.

Results for the key issues identified

Importance of public participation in EIA

Respondents were asked on how they would rate Public participation in Malawi's EIA System. 80% ($n = 12$) showed that it is very important while 20% ($n = 3$) rated it as just being important as shown in Figure 2.

Success of public participation in Malawi's EIA System

When asked about the success of public participation in Malawi's EIA process, 40% ($n = 6$) answered that it is successful, another 40% ($n = 6$) said it is fairly successful, 13% ($n = 2$) said it is not successful and 7% ($n = 1$) of the respondents did not respond to this question (hence, N/A= not applicable) and this information is shown in Figure 3.

An in depth analysis showed that the responses were based on the following issues:

Successful: The respondents said public participation in EIA is successful because of such reasons as: Department of Environmental Affairs' co-ordination with stakeholders, key stakeholders are normally consulted in the EIA studies, key stakeholders participate in implementation of mitigation measures and some projects' design have been modified based on the recommendation from public participation.

Fairly Successful: Under this rating, respondents attributed the reasons to less time being spent on consultation, that is limited time is allocated to communities to participate; inability of the consultants to consult the right people; unwillingness of the public members to participate since they do not know the importance of public participation in EIA (as a result of low public awareness), low priority accorded to environmental issues by the government, and that most of the communities' views are not taken aboard in the preparation of EIA and Environmental Audit reports.

Not Successful: The reasons alluded to this rating included: participation is only done to fulfill legal requirements, that is, it is pseudo-participation and is closely related to what is called "pseudo- alternatives" (Murombo, 2008); monitoring of public participation process in not effectively done by the government agent (EAD) and also due to inadequacy of human and financial resources. This is consistent with the findings of Kosamu (2011).

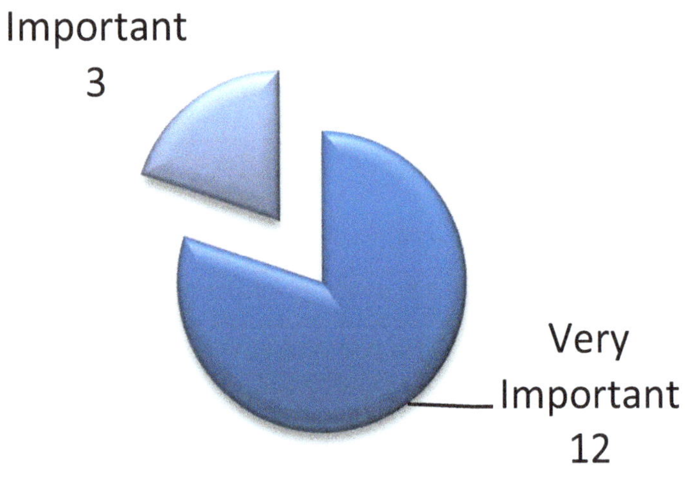

Figure 2. Rating of Public Participation in Malawi by respondents.

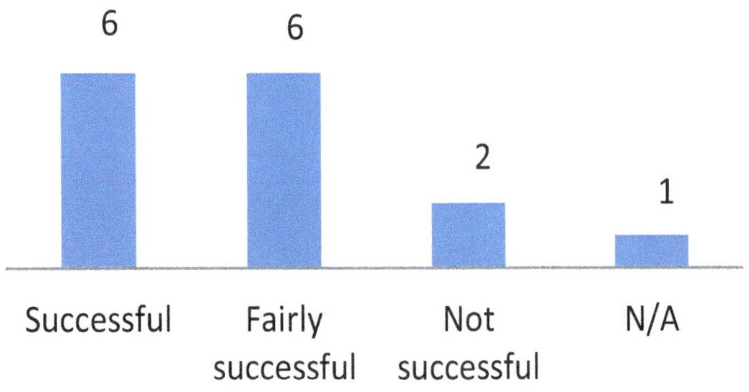

Figure 3. Success of Public Participation in EIA in Malawi.

Stages of the EIA process at which the public is consulted in Malawi's EIA system

The respondents indicated that public participation is done during various stages including: project screening; consideration of alternatives; identification of the main impacts, post-decision monitoring and follow-up; prediction of impacts, mitigation measures; impact analysis; EIA/EA report presentation, decision making and auditing. It was also established that public consultation and participation methods commonly used in Malawi' EIA system are: information notices, community meetings, public hearings, radio stations, manipulative participation, passive participation and participation by use of material incentives.

Challenges facing Public Participation Process in Malawi EIA system

Upon being asked to provide the challenges and/or limitations faced during public participation in Malawi's EIA system, stakeholders raised many issues and the following is a summary of the major ones: tendency for local chiefs to dominate the discussions thereby barring the views of most ordinary people; inadequate briefing of the projects; high illiteracy of the community members makes it difficult to communicate EIA issues since environmental issues are considered complex; there are no independent monitoring of public participation; sometimes it is difficult to meet relevant people in the field, sometimes projects are required to be implemented quickly due to pressure from donors or politicians hence inadequate time for public participation, technocrats and administrators in the government ministries are not always available for consultations due to their busy schedules; environmental issues are given low priority by the government; there is a shortage of number of expertise who are equipped with the skills of public participation; limited resources which also limits the time and area of coverage during the EIA and lastly, poor communication (example absence of road networks) in some areas such that it is difficult to reach some stakeholders. Most of these challenges have previously been reported by Kosamu (2011).

Opportunities for success of public participation in Malawi's EIA system

The respondents were also asked to highlight the opportunities present in Malawi for the success of the public participation and consultation. The following is a summary of the responses provided: the EMA is present and needs to be strictly enforced; a regulatory body (EAD) is present and there is a need to boost its functionality by increasing its workforce an financial allocation; government through its community services may assist in conducting awareness campaigns; traditional and political leaders may be used to influence their subjects to participate in EIA activities; and the revision of the EIA legal framework and preparations of clear guidelines for public consultations.

Conclusion

This study established that public participation in Malawi's EIA process is still very low. However, a number of opportunities and interventions exist which can make public participation a successful approach. Planners of development projects should ensure that incentives for public participation are established. This can be fostered through oral and written briefings and discussions with local people to obtain advice and consent to foster local project ownership and participation.

As a lead agency, the EAD has to make sure that EIA procedures are critically followed by the project developers. The workforce at the EAD also needs to be blended with expertise from different fields so that there is adequate consensus on issues raised by the public.

ACKNOWLEDGEMENTS

Authors would like to thank the University of Malawi for financially supporting this study.

REFERENCES

Appiah-Opoku S (2001). Environmental Impact Assessment in Developing Countries: The case of Ghana. Environ. Impact Assess. Rev. 21:59-71.

Carroll B, Turpin T (2009). Environmentallmpact Assessment Handbook: A Practical Guide for Planners, Developers, and Communities. 2nd ed. Thomas Telford Limited, USA.

Glasson J, Therivel R, Chadwick A (2005). Introduction to Environmental Impact Assessment. 3rd ed. Routledge Taylor and Francis Group, London.

Glucker A N (2012). Public participation in Environmental Impact Assessment (EIA) - An investigation into theory and practice in Costa Rica and Nicaragua. MSc. Thesis, University of Utrecht, The Netherlands.

IAIA (2006). Public participation. International Best Practice Principles. Special Publication Series No. 4. Available at: http://www.iaia.org/publicdocuments/special-publications/SP4%20web.pdf.

Kakonge JO (1996). Problems with Public Participation in EIA Process: Examples from Sub-Saharan Africa. Impact Assessment, Volume, 14. Available at: http://www.ttl.fi/Internet/English/Information/Electronic+journals/Africa+Newsletter/2000-03/03.htm.

Kosamu IBM (2011). Environmental impact assessment application in infrastructural projects in Malawi. Sustain. Sci. 6(1):51–57.

Lee N, George C (2000). Environmental Assessment in Developing and Transitional countries. New York: Wiley & Sons.

Leu WS, Williams WP, Bark AW (1997). Evaluation of environmental impact assessment in three Southeast Asian Nations. Project Appraisal 12(2):89-100.

Murombo T (2008). Beyond Public Participation: The Disjuncture between South Africa's Environmental Impact Assessment (EIA) Law and Sustainable Development. Potchefstroom Electronic Law J. (PER), 3:1-31. ISSN 1727-3781.

Petts J (1999). Public Participation and Environmental Impact Assessment. In Petts, J. (ed.): Handbook of Environmental Impact Assessment Practice: Impact and Limitations (Vol. 1). Blackwell Science Limited, USA.

Sankoh OA (1996). Making Environmental Impact Assessment Convincible to Developing Countries. J. Environ. Manage. 47:185–189.

Tarr P (2003). EIA in southern Africa: Summary and future focus. In Southern African Institute for Environmental Assessment: Environmental Impact Assessment in Southern Africa. Windhoek, Southern African Institute for Environmental Assessment, pp. 329-337.

Wood C (1995). Environmental Impact Assessment: A Comparative Review. Longman Group Ltd., Harlow, UK.

World Bank (1993). Public Involvement in Environmental Assessment: Requirements, Opportunities and Issues. Environmental Assessment Sourcebook Update 5. Environmental Department, World Bank, Washington DC.

11

"Mapping the regional variation in potential vulnerability in Indian Agriculture to climate change"- An exercise through constructing vulnerability index

Arijit Das

Centre for the Study of Regional Development, School of Social Science, Jawaharlal Nehru University, New Delhi-110067, India. E-mail: arijit3333@gmail.com.

India is potentially vulnerable to adverse impacts in agriculture on account of climate change. The Anticipated changes in water availability (surface, freshwater and ground water), temperature rise, soil degradation and the suggested increase in extreme events such as the drought, floods and cyclones will affect agricultural severely. The present paper aims to develop the vulnerability profile of agricultural systems of the Indian states to the changing climate scenarios. It develops two sub indices, namely, Bio-Physical vulnerability index and Socio-Economic vulnerability index to develop the final overall vulnerability index. Bio-Physical vulnerability index has been constructed by considering five indicators, namely: (1) Cropping intensity, (2) Percentage of area under leguminous crops, (3) Percentage of degraded area to total geographical area, (4) Unexploited surface water available for future and (5) Unexploited surface water available for future. The Socio-Economic Vulnerability index has been computed by considering four indicators like: Irrigation intensity, Percentage of people below poverty line, Percentage of agricultural workers to total workers, Literacy rate. Modified UNDP method has been used in developing all the indicators. It has been found that most of the states which are having very high or high vulnerability are centered on the central and north-central part of India.

Key words: Climate change, vulnerability, environmental sustainability.

INTRODUCTION

Agriculture is the backbone of Indian economy more specifically of the rural livelihood security system. However, the contribution of agriculture to GDP has been steadily declining over the years (nearly 25% to the total at present). But the share of agriculture in providing employment and livelihoods to a majority of population continues to remain intact (agriculture provides employment about two-thirds of the total workforce) in spite of the diversification of economic activity. It is also an important instrument of poverty eradication and employment generation (Planning commission 1997; 2002). Among India's population, more than one billion people (about 68%) are directly or indirectly involved in the agricultural sector. Indian agriculture is now facing the several problems like regional variation in productivity is still persisting, the yield has been stagnated or decline over most of the regions, the yield of food grains in major producing, states like Punjab and Haryana has reached a plateau, more than 60% net sown area is still under rained agriculture etc. Above all, the projected climate change is a single largest threat for sustainable agriculture in India. This sector due to its heavy reliance on climatic factors is particularly vulnerable to present-day climate variability, including erratic rainfall and increase in average surface temperature. Predictions made by the scientists by using global circulation model shows that India could experience warmer and wetter conditions as a result of climate change, particularly if the summer monsoon becomes more intense (Mitra et al., 2002). Contrary to this some scientists using the same model assessed that increased rates of evapotranspiration due to the higher temperatures may offset the increased precipitation, leading to negative impacts on soil moisture. There are also considerable uncertainties associated with climate model projections of tropical monsoon behavior, and simulations that include sulfate aerosol forcing indicate decreasing summer monsoon rainfall (Lal et al., 1998). The picture of climate change in India is not crystal clear because it suffers from good esti-

estimates. Scientists at the Centre for science and environment predict that the semi-arid regions of Western India are expected to receive higher than normal rainfall as the temperature soar, while central India will experience a decrease of between 10 to 20% in winter rainfall by 2050, Average surface temperature is likely to increase about 2-4°C by 2050, rainfall intensity is likely to increase by 1 to 4 mm/day, frequency and intensity of the weather extreme events such as cyclone, drought, flood will increase. But still there is a lack of accuracy of measurement of change in climate at regional level.

Indian agriculture is still heavily dependent on climate as indicated from the fact that nearly 60% of the net sown area is under rain fed agriculture. Broadly, agricultural productivity is sensitive to climate-induced effect both directly and indirectly- direct effect from changes in temperature, rainfall amount, pattern and intensity, increased frequency of extreme climate event like drought, flood, cyclones etc., and indirect effect through changes in soil moisture, acceleration of various land degradation effects etc., (IPCC, 1997, 2001). Although the direct temperature and CO_2 effects of climate change may lead to productivity increases for some irrigated crops there is general consensus that major agricultural production areas are likely to be adversely affected by climate change, particularly in areas that become increasingly water-stressed (Dinar et al., 1998; Kumar and Parikh, 2001; Lal et al., 1998; Gadgil, 1995).

The study on impact of predicted climate change on particular crop and over particular agricultural region is few. A 2°C increase in temperature will lead to a reduction in rice yield by 0.75 ton/hectare in high yield areas and coastal regions. On the other hand a 0.5°C increase in winter temperature will lead to a reduction in wheat crop duration by seven days, which will intern reduce yield by 0.45 ton/hectare (Swaminathan and Sinha, 1991). An increase in temperature will lead to 10% reduction in wheat yield in high yielding states of Punjab, Haryana and Uttar Pradesh. World Bank report (1998) analyzed climate change effects on Indian agriculture, through annual net revenues, by using Ricardian method. The three methodologies, as adopted in the study, found Indian Agriculture sensitive to warming. The analyses further showed year-to-year climate sensitivity to the system's response. The studies revealed that net revenues fall precipitously with warmer April's, but also sensitive to warmer January and July. Crop revenues increased with October temperatures. Net revenues were also sensitive to precipitation, but the effects were smaller and offsetting. A warming scenario of +2.0°C rises in mean temperature and a +7% increase in mean precipitation levels will create reduction in the netrevenues, as revealed from the three approaches. The impact is differential on spatial and temporal scales. But the study seemed to be weak for linking with the biophysical aspects.

However, the vulnerability of agricultural production to climate change depends not only on the physiological response of the affected plant, but also on the ability of the affected socio-economic systems of production to cope with changes in agricultural production or the factors like changes in the frequency of droughts or floods which effects the agricultural production. The adaptability of farmers in India is severely restricted by the heavy reliance on natural factors and the lack of complementary inputs and institutional support systems. The central challenge of sustainable agriculture in India today is to meet the food demand of the present generation without sacrificing the needs of future generations. This cannot be achieved without the systemic integration of the social, economic pillars of agriculture to combat with the climatic variability such as changes in water availability (surface, freshwater and ground water), temperature rise, soil degradation and the suggested increase in extreme events such as the drought, floods and cyclones. Recognizing the perceived adverse impact of climate change on Indian Agriculture to maintain the sustainability of the food production India is left with two option either increase adoptability of the agricultural system at the various states or mitigate the climate change. The agricultural system which has high adaptive capacity is less vulnerable to be effected adversely to climate change and vice-versa.

In this regard a brief note on theoretical understanding of Agricultural vulnerability has put forwarded in this section. Vulnerability is the degree to which a system is susceptible to, or unable to cope with adverse effects of climate change, including climate variability and extremes. Vulnerability is a function of the character, magnitude, and rate of climate change and variation to which a system is exposed as well as the system's sensitivity and adaptive capacity. Vulnerability to climate change varies across regions, sectors, and social groups. Understanding the regional and local dimensions of vulnerability is essential to develop appropriate and targeted adaptation efforts (McCarthy et al., 2001).

Vulnerability of Agricultural systems to climate change is generally understood to be a function of a range of biophysical and socio-economic factors. The most recent report of the Intergovernmental Panel on Climate Change (IPCC) provides a useful typology suggesting that vulnerability may be characterized as a function of three components: adaptive capacity, sensitivity, and exposure.

Adaptive capacity describes the ability of a system to adjust to actual or expected climate stresses, or to cope with the consequences. It is considered "a function of wealth, technology, education, information, skills, infrastructure, access to resources, and stability and management capabilities" (McCarthy et al., 2001: 8). Sensitivity refers to the degree to which a system will respond to a change in climate, either positively or negatively. Exposure relates to the degree of climate stress upon a particular unit of analysis, it may be represented as either long-term change in climate conditions, or by changes in climate variability, including the magnitude and frequency of extreme event.

In developing Agricultural vulnerability profile, it has been assumed that climate change exposure (that is, on-going and future exposure) will affect current sensitivity, either positively or negatively, and that farmers will respond to these changes in climate sensitivity if they have sufficient adaptive capacity. Thus the vulnerability profile is constructed by combining indices for adaptive capacity with sensitivity indices that take into account exposure to climate change. Study on agricultural vulnerability in India is very much limited most of study confines to impact of predicted changes in temperature and rainfall on individual crops and agricultural regions. Concentrated efforts to understand vulnerability at regional or at local level only made by TERI and individual study done by Karen O'Brien et al. (2004).

In developing vulnerability profiles TERI considered three dimensions such as Biophysical factors, social factors and technological factors. Biophysical factors that influence agricultural production includes soil conditions and groundwater availability. The social factors that influence adaptive capacity comprise indicators representing the percentage of workers employed in agriculture, the percentage of landless labourers in the agricultural workforce, human capital (as represented by literacy levels), gender discrimination, (as measured by excess girl child mortality), and child mortality and fertility (as measured by female literacy rates). Technological factors that influence adaptive capacity include irrigation and infrastructure. Irrigation was measured by net irrigated area as a percentage of net sown area, while infra-structure was measured by the existing Infrastructure development index (CMIE, 2000).

While Karen et al. (2004) measure the vulnerability of Indian agriculture have identified significant biophysical, socioeconomic, and technological factors that influence agricultural production. The bio-physical indicators used in the profile consisted of soil conditions (quality and depth) and ground water availability. Indicators for soil quality include the depth of the soil cover and severity of soil degradation, while indicators of groundwater availability are based on estimates of the total amount of replenishable groundwater available annually. Socioeconomic factors consisted of levels of human and social capital, and the presence or lack of alternative economic activeties Human capital was represented by adult literacy rates, while social capital was measured by degree of gender equity in a district Presence of alternative economic activities is measured by the percentage of the district workforce that is employed in agriculture and by the percentage of landless laborers in the agricultural workforce.

Both of the study considers very few representative indicators and hence, suffers from the problem of coverage and able to capture very limited information about vulnerability of Indian Agriculture to climate change. Therefore, the present paper aimed at development of vulnerability Profile for agriculture at the state level of India considering of indicators covers wide range of information about vulnerability profile of Indian Agriculture to climate change.

METHODS

The Agricultural vulnerability index has been created very carefully by considering the best representative indicators. The present study tried to cover almost all the dimension of bio-physical vulnerability and socio-economic vulnerability by considering six indicators and four indicators respectively.

Bio-physical vulnerability

Bio-physical vulnerability is the physiological response of the affected plants to changing climatic conditions. It is defined by the following indicators (Table 1).

Cropping intensity

It is the ratio of the gross cropped area to net sown area and is expressed as percentage. It is indicative of the good health and better moisture retention capacity of the soil. Better soil has higher cropping intensity. Hence, high crop intensity indicates low bio-physical vulnerability and vice-versa. Information regarding this is taken from Agriculture Census (2001), Ministry of Agriculture, Government of India.

Percentage of area under leguminous crops

It ensures the sustainable crop patterns because most of the legumes can convert the nitrogen gas from air into soluble form of nitrogen that is, ammonia. They can adopt the harsh climatic conditions. Hence the higher percentage of which represents the low bio-physical vulnerability and vice versa. Data on this is taken from Agriculture Census (2001), Ministry of Agriculture, Government of India.

Percentage of degraded area to total geographical area

Land degradation due to salinisation, erosion, water logging etc. refers to degradation of quantity and quality of soil that result in loss of biological productivity of land and enhance the bio-physical vulnerability. The source of information of this is FSI, State of Forest Report (2001). "Forest Survey of India, Government of India.

Unexploited surface water available for future

It represents the future availability of the surface water which indicates potential availability of surface water for irrigation which can reduce the dependency on climate for water and lessen the bio-physical vulnerability. Data on this is taken from CWC, Surface Water Statistics (2000), Central Water Commission, Government of India.

Unexploited ground available for future

It also represents the future availability of the ground water for irrigation which can reduce the dependency on climate for water and lessen the bio-physical vulnerability. Informa-tion regarding this is taken from CGWB (1996), Ground-water statistics (2000). Ministry of Water Resources, Government of India.

Table 1. Bio-Physical vulnerability index.

Source	Cropping intensity	(%) of degraded area to TGA	(%) of leguminous crops to total crops	Future availability of surface water	Future availability of ground water	Bio-Physical vulnerability index	Modified index (1- original index)	Rank
Andhra Pradesh	0.21	0.75	0.69	0.40	0.74	0.557	0.443	20
Arunachal Pradesh	0.34	0.84	0.00	0.47	1.00	0.530	0.470	16
Assam	0.47	0.68	0.08	0.78	0.93	0.587	0.413	24
Bihar	0.34	0.85	0.24	0.56	0.67	0.530	0.470	17
Goa	0.16	0.84	0.00	0.75	0.92	0.534	0.466	18
Gujarat	0.06	0.69	0.70	0.59	0.51	0.508	0.492	12
Haryana	0.79	0.92	0.18	0.28	0.24	0.484	0.516	9
Himachal Pradesh	0.83	0.61	0.10	0.52	0.83	0.578	0.422	23
Jammu and Kashmir	0.48	0.91	0.08	0.09	0.99	0.508	0.492	11
Karnataka	0.13	0.88	0.67	0.24	0.67	0.518	0.482	14
Kerala	0.27	1.00	0.04	0.46	0.81	0.515	0.485	13
Madhya Pradesh	0.29	0.78	1.00	0.65	0.81	0.706	0.294	25
Maharashtra	0.22	0.76	0.48	0.34	0.65	0.493	0.507	10
Manipur	0.56	0.00	0.00	0.51	1.00	0.414	0.586	6
Meghalaya	0.16	0.25	0.00	0.69	0.96	0.413	0.587	5
Mizoram	0.00	0.70	0.00	1.00	0.00	0.340	0.660	3
Nagaland	0.05	0.13	0.00	0.15	0.00	0.065	0.935	1
Orissa	0.38	0.83	0.25	0.52	0.85	0.566	0.434	22
Punjab	1.00	1.00	0.03	0.07	0.02	0.423	0.577	8
Rajasthan	0.31	0.74	0.70	0.07	0.27	0.419	0.581	7
Sikkim	0.33	0.51	0.00	0.66	0.00	0.299	0.701	2
Tamil Nadu	0.14	0.76	0.69	0.00	0.37	0.394	0.606	4
Tripura	0.63	0.83	0.00	0.70	0.67	0.565	0.435	21
Uttar Pradesh	0.52	0.91	0.30	0.41	0.58	0.545	0.455	19
West Bengal	0.76	0.96	0.07	0.14	0.68	0.520	0.480	15

Source: Authors own calculation from various sources.

.Socio-economic vulnerability

It represents the lack of adoptability of the farmers and the lack of complementary inputs and institutional support systems. It is defined by the following indicators.

Irrigation intensity

This represents the efficient utilization of water and less dependency on rainfall and reduces the socio-economic vulnerability. Data on this is taken from Agriculture Census-(2001), Ministry of Agriculture, Government of India.

Percentage of people below poverty line

This group of population is expected to be worst victim of the decrease in food production and higher proportion of which will enhances the socio-economic vulnerability. This information is taken from State of Poverty in India, Planning Commission report, Government of India.

Percentage of agricultural workers to total workers

This is an indicator of relative economic deficiency and reflects heavily dependence on agriculture and enhances the socio-economic vulnerability. This information is taken from Census of India, 2001, Government of India.

Literacy rate

Literate population indicates real human resource stock and literate population has the higher capability of adaptation of new technology and reduces the socio-economic vulnerability. This data is taken from Census of India (2001), Government of India.

Modified UNDP method has been used in constructing both bio-physical vulnerability index and socio-economic vulnerability index. The following procedure has been adopted in converting the indicators into normalized form. The indicators that have been used in the construction of the food availability index are both negative and positive in nature. First the best and the worst values in an indicator are identified. The best and the worst values are depends on the nature of the indicators. In case of a positive indicator, the highest value will be treated as the best value and the lowest, will be considered as the worst value. Similarly, if the indicator is negative in nature, then the lowest value will be considered as the best value and the highest, the worst value. Once the best and worst values are identified, the following formula is used to obtain normalize values:

$$NVij = 1 - \left(\frac{\{\text{Best Xi- Observed Xij}\}}{\{\text{Best Xi- Worst Xi}\}} \right)$$

Normalized values always lies between 0 and 1. As the value of a particular indicator inclined towards 1 indicates better performance and vice-versa.

Two dimension index that is, bio-physical vulnerability index and socio-economic vulnerability index has been computed by using the following formula:

$$D_i = \frac{\sum NVij}{n}$$

NVij = Sum of all normalized indicator. I=1...9 denotes specific indicators (X_1=Cropping intensity, X_2=Percentage of area under leguminous crops, X_3=Percentage of degraded area to total geographical area, X_4=Unexploited surface water available for future, X_5=Unexploited surface water available for future, X_6=Irrigation intensity X_7=Percentage of people below poverty line, X_8=Percentage of agricultural workers to total workers, X_9=Literacy rate)

j=1...29 means State

n= Total number of indicator used in calculating the indicator

D_i= Dimension index

The overall vulnerability index has been computed on the basis of following formula: $\sum D_i$/ Total number of dimension index

Since all the representative indicators of Agricultural vulnerability are negative in nature to vulnerability that is, higher value of which indicates lower vulnerability and vice versa, the final index is modified by the following equations so that now higher value is indicating higher vulnerability and vice versa. Modified index= 1- Original index value.

RESULTS

Indian agriculture faces the challenge of feeding a billion people in a changing climatic scenario. Agriculture is the predominant means of livelihood for a large number of peasant cultivators and agricultural laborers, for whom it is not easy to shift to other occupations. Due to their low financial and technological adaptability, such groups are potentially vulnerable to both climatic changes. Therefore, the present study made a modest attempt to develop the vulnerability profiles of agriculture to climate change in various Indian states. Vulnerability was seen as a composite of adaptive capacity and climate sensitivity. However, the vulnerability of agricultural system to climate change depends not only on the physiological response of the affected plant, but also on the ability of the affected socio-economic systems of production to cope with changes in climatic factors like changes in the frequency of droughts or floods which effects the agricultural production. Hence, the Agricultural Vulnerability Profile is composite of Bio-Physical vulnerability and Socio-Economic vulnerability (Tables 1 to 5).

The bio-physical vulnerability profile in India clearly indicates that three states (about 12%) namely, Nagaland, Sikkim, Mizoram have very high bio-physical vulnerability. The very high bio-physical mainly influenced by the over exploitation of the surface water and land degradation due to slope erosion. There are five states (about 20%), namely, Tamil Nadu, Meghalaya, Manipur, Rajasthan, and Punjab have high levels of bio-physical vulnerabilities. It is to be noted that major cereal producing state like Tamil Nadu and Punjab belong to High bio-physical vulnerability class because of the over exploitation of the ground water and surface water which makes it high vulnerable. Eleven states (about 44%) belong to the medium vulnerability category these include Haryana, Maharashtra, Jammu and Kashmir, Gujarat, Kerala, Karnataka, West Bengal, Arunachal Pradesh, Bihar, Goa, Uttar Pradesh. There are five states (about 20%) like Andhra Pradesh, Tripura, Orissa, Himachal Pradesh and Assam belongs to the low vulnerability category. Only one state that is, Madhya Pradesh has very low bio-physical vulnerability. The states which are having very low and low vulnerability are very wisely utilized the ground water and surface water and the areaunder leguminous crops are relatively higher which can adopt the harsh climate (Figure 1).

The socio-economic vulnerability profile in India clearly indicates that nine states (about 36%) namely Bihar, Madhya Pradesh, Arunachal Pradesh, Orissa, Meghalaya, Assam, Nagaland, Sikkim, and Uttar Pradesh. Very high socio-economic vulnerability in these states is mainly influenced by the high poverty rate and literacy rate which will reduce the adoptability of the farmers to climate change. There are nine states (about 20%), namely, Manipur, Rajasthan, Maharashtra, Karnataka, Andhra Pradesh, Tamil Nadu, Mizoram, West Bengal, Gujarat have high levels of socio-economic vulnerabilities. Two states (about 8%) belong to the Medium vulnerability category these include Tripura, Jammu and Kashmir. There are no states in low vulnerability category. While five states (about 20%), namely, Haryana, Kerala, Punjab, Himachal Pradesh, Goa has very low socio-economic vulnerability. The states which are having very low and

Table 2. Regional variation bio-physical vulnerability.

Vulnerability	Class range (index value)	Name of the State	Number of State	Percentage (%) of State
Very low	<0.350	Madhya Pradesh.	1	4
Low	0.351-0.450	Andhra Pradesh, Tripura, Orissa, Himachal Pradesh, Assam.	5	20
Medium	0.451-0.550	Haryana, Maharashtra ,Jammu and Kashmir, Gujarat, Kerala, Karnataka, West Bengal, Arunachal Pradesh, Bihar, Goa, Uttar Pradesh.	11	44
High	0.551-0.650	Tamil Nadu, Meghalaya, Manipur, Rajasthan, Punjab.	5	20
Very high	>0.650	Nagaland, Sikkim, Mizoram.	3	12

Source: Authors own calculation from various sources.

Table 3. Regional variation socio-economic vulnerability.

Vulnerability	Class range (index value)	Name of the States	Number of States	Percentage (%) of States
Very Low	<0.350	Haryana, Kerala, Punjab, Himachal Pradesh, Goa.	5	20
Low	0.351-0.450	-	0	0
Medium	0.451-0.550	Tripura, Jammu and Kashmir.	2	8
High	0.551-0.650	Manipur, Rajasthan, Maharashtra, Karnataka, Andhra Pradesh, Tamil Nadu, Mizoram, West Bengal, Gujarat.	9	36
Very High	>0.650	Bihar, Madhya Pradesh, Arunachal Pradesh, Orissa, Meghalaya, Assam, Nagaland, Sikkim, Uttar Pradesh.	9	36

Source: Authors own calculation from various sources.

Table 4. Socio-Economic vulnerability index.

Source	Irrigation intensity	(%) of agricultural worker to total workers, 2001	Literacy rate, 2001	Poverty rate, 1999 to 2000	Socio-Economic vulnerability index	Modified index (1-original index)	Rank
Andhra Pradesh	0.39	0.18	0.31	0.72	0.401	0.599	16
Arunachal Pradesh	0.00	0.23	0.17	0.31	0.176	0.824	24
Assam	0.00	0.46	0.39	0.25	0.275	0.725	17
Bihar	0.33	0.05	0.00	0.10	0.123	0.877	18
Goa	0.63	1.00	0.80	0.98	0.853	0.147	12
Gujarat	0.28	0.24	0.52	0.76	0.447	0.553	9
Haryana	0.89	0.36	0.49	0.88	0.652	0.348	23

Table 4. Contd.

Himachal Pradesh	0.90	0.20	0.68	0.90	0.672	0.328	11
Jammu and Kashmir	0.52	0.40	0.16	1.00	0.520	0.480	14
Karnataka	0.29	0.21	0.45	0.62	0.391	0.609	13
Kerala	0.14	0.99	1.00	0.80	0.731	0.269	25
Madhya Pradesh	0.05	0.00	0.38	0.22	0.163	0.837	10
Maharashtra	0.18	0.10	0.69	0.51	0.369	0.631	6
Manipur	0.17	0.35	0.49	0.43	0.361	0.639	5
Meghalaya	0.17	0.19	0.36	0.30	0.257	0.743	3
Mizoram	0.13	0.07	0.94	0.63	0.443	0.557	1
Nagaland	0.18	0.14	0.45	0.33	0.277	0.723	22
Orissa	0.15	0.23	0.37	0.00	0.187	0.813	8
Punjab	1.00	0.56	0.52	0.94	0.753	0.247	7
Rajasthan	0.28	0.14	0.31	0.73	0.365	0.635	2
Sikkim	0.00	0.41	0.51	0.24	0.290	0.710	4
Tamil Nadu	0.23	0.27	0.60	0.60	0.424	0.576	21
Tripura	0.82	0.45	0.60	0.29	0.541	0.459	19
Uttar Pradesh	0.45	0.14	0.23	0.37	0.295	0.705	15
West Bengal	0.34	0.47	0.50	0.46	0.444	0.556	16

Source: Authors own calculation from various sources.

low vulnerability are very able to translate the economic prosperity with social development through increase in literacy rate, diversion of economy and reduction of poverty rate. The location of the states having very high socio-economic vulnerability is showing a characteristic location pattern, most of them belong to the Central and Northern part together they are forming a contiguous belt (Figure 2).

Overall regional variation (Figure 3) in vulnerability of agricultural system to climate change has been computed simply by averaging the bio-physical vulnerability index value and the socio-economic vulnerability index value (Tables 2 and 3). It is found that there are four states like Bihar, Nagaland, Sikkim, and Meghalaya have very high levels of overall vulnerability out of these four in Nagaland, Sikkim, and Meghalaya bio-physical factors are mainly responsible for the very high vulnerability while in Bihar it is mainly contributed by the social factors. There are ten states (about 40%), namely, Arunachal Pradesh, Orissa, Manipur, Mizoram, Rajasthan, Tamil Nadu, Uttar Pradesh, Maharashtra, Assam, Madhya Pradesh have high levels of Overall vulnerabilities. Five states (about 20%) belong to the medium vulnerability category these include Karnataka, Gujarat, Andhra Pradesh, West Bengal, Jammu and Kashmir. There are five states (about 20%), Haryana, Punjab, Tripura, Kerala, Himachal Pradesh in low vulnerability category. While one states (about 4%), namely, Goa have very low vulnerability. The vulnerability profiles in India clearly indicates that most of the states which are having very high or high levels of vulnerability are clustered around central and the North-Central part here the states are performing very bad in the human resource development and hence, are very much vulnerable to be effect by the changing climate scenarios (Table 6 and Figure 3).

The aforementioned discussions can be summarized in the following manner. Firstly, the bio-physical vulnerability and the social vulnerability are not truly overlapped some regions are bio-physically not much vulnerable but are lack of social development which makes them very low adoptability to climate change. Secondly, most of the states which are having very high or high vulnerability are clustered around Central and North-Central India. Thirdly, most of the states belong to North and West part of India are Bio-Physically vulnerable but are socio-economically developed

Table 5. Regional variation in the vulnerability of the agricultural system.

States name	Socio-Economic vulnerability index	Bio-Physical vulnerability index	Overall vulnerability index	Rank in overall vulnerability index
Andhra Pradesh	0.599	0.443	0.521	17
Arunachal Pradesh	0.824	0.47	0.647	5
Assam	0.725	0.413	0.569	12
Bihar	0.877	0.47	0.674	3
Goa	0.147	0.466	0.307	25
Gujarat	0.553	0.492	0.522	16
Haryana	0.348	0.516	0.432	21
Himachal Pradesh	0.328	0.422	0.375	24
Jammu and Kashmir	0.48	0.492	0.486	19
Karnataka	0.609	0.482	0.545	15
Kerala	0.269	0.485	0.377	23
Madhya Pradesh	0.837	0.294	0.565	14
Maharashtra	0.631	0.507	0.569	13
Manipur	0.639	0.586	0.613	7
Meghalaya	0.743	0.587	0.665	4
Mizoram	0.557	0.66	0.608	8
Nagaland	0.723	0.935	0.829	1
Orissa	0.813	0.434	0.623	6
Punjab	0.247	0.577	0.412	22
Rajasthan	0.635	0.581	0.608	9
Sikkim	0.71	0.701	0.705	2
Tamil Nadu	0.576	0.606	0.591	10
Tripura	0.459	0.435	0.447	20
Uttar Pradesh	0.705	0.455	0.58	11
West Bengal	0.556	0.48	0.518	18

Source: Authors own calculation.

Table 6. Regional variation overall vulnerability.

Vulnerability	Class range (index value)	Name of the State	Number of State	Percentage (%) of State
Very low	<0.350	Goa.	1	4
Low	0.351-0.450	Haryana, Punjab, Tripura, Kerala, Himachal Pradesh.	5	20
Medium	0.451-0.550	Karnataka, Gujarat, Andhra Pradesh, West Bengal, Jammu and Kashmir.	5	20
High	0.551-0.650	Arunachal Pradesh, Orissa, Manipur, Mizoram, Rajasthan, Tamil Nadu, Uttar Pradesh, Maharashtra, Assam, Madhya Pradesh.	10	40
Very high	>0.650	Bihar, Nagaland, Sikkim, Meghalaya.	4	16

Source: Authors own calculation from various sources.

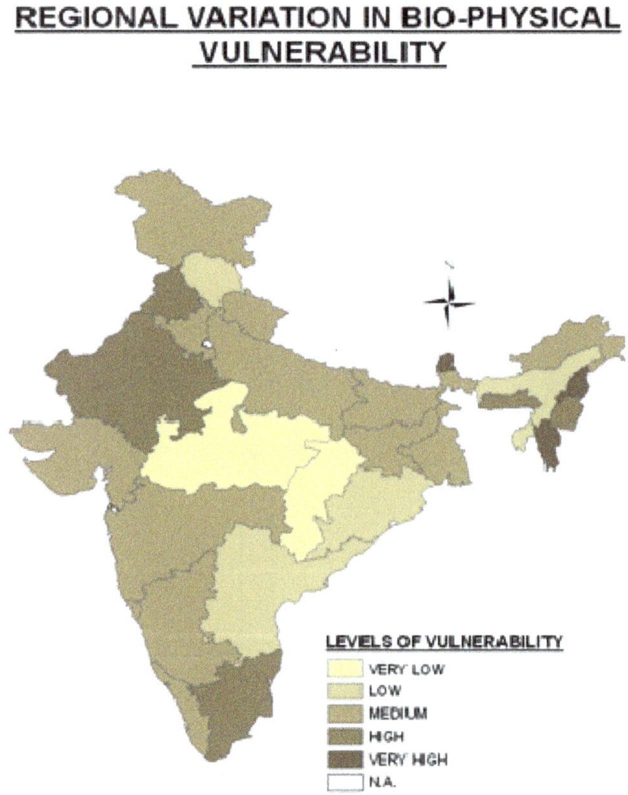

Figure 1. Regional variation bi-physical vulnerability.

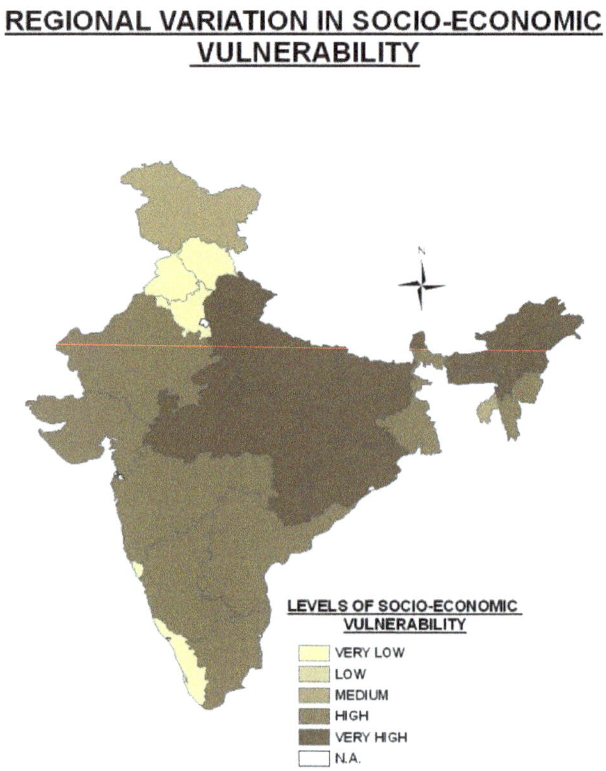

Figure 2. Regional variation socio-economic vulnerability.

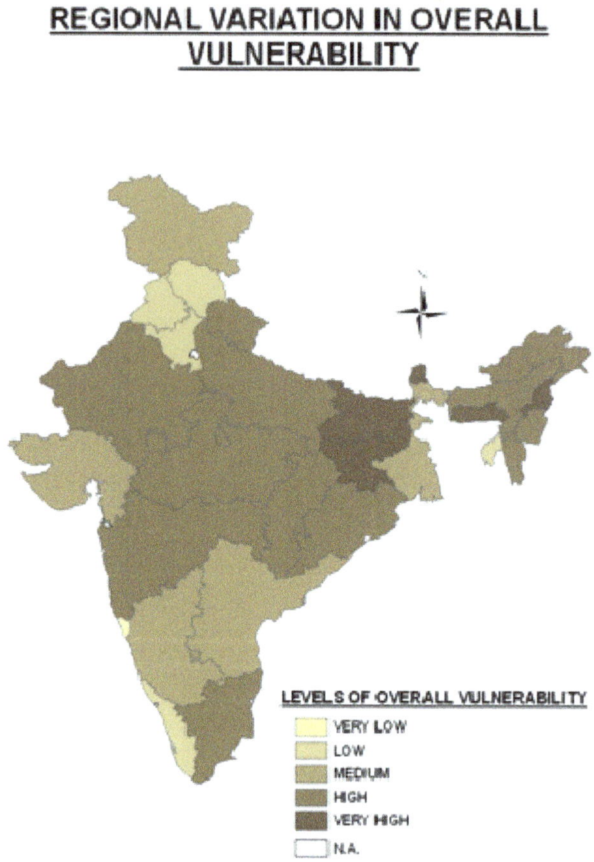

Figure 3. Regional variation overall vulnerability.

and are capable of high adaptation. Fourthly, most of the states of central India and North Central India are socio-economically vulnerable.

DISCUSSIONS

The likely climate change scenarios have very strong adverse impact on the agricultural system. Like the regional variation in agricultural development there is vast regional disparity exists in the levels of adoptive capacity of the Indian states so far as the vulnerability of the agricultural system is concerned. Numerous physical and socio-economic factors come into play in enhancing or constraining the current capacity of farmers to cope with adverse changes. Prominent among the physical factors are cropping patterns, crop diversification, and shifts to drought /salt resistant varieties. The most important socioeconomic factors include ownership of assets (like land, cattle, pump-sets, and agricultural implements), access to services (like banking, health, and education), and infrastructural support (like irrigation, markets, and transport /communication networks). Policies that are designed to fortify current coping capacity also have the power to strengthen long-term adaptive capacity. Some of the adoptive measures which could be beneficial regardless of how or whether climate changes – include, identification of the present vulnerabilities of agricultural systems, agricultural research to develop new crop varieties, improved training and general education of populations dependent on agriculture, food programs and other social security programs to provide insurance against supply changes.

ACKNOWLEDGEMENTS

I would like to express my gratitude to my supervisor Prof. Sachidanand Sinha and my friends Nilanjan Patra, Sambuddha Goswami, Safayat Karim for their insightful comment, which contributed much to this paper.

REFERENCES

Agriculture Census (2001). Ministry of Agriculture, Government of India.

Census of India (2001). Government of India.

CGWB (1996). Groundwater statistics. Ministry of Water Resources, Government of India.

CWC (2000). Surface Water Statistics, Central Water Commission, Government of India.

Dinar A, Mendelsohn R, Songhi A (1998). Measuring the impact of climate change on Indian Agriculture. World Bank Technical Paper 402, Washington.

FSI (2001). State of Forest Report-2001. Forest Survey of India, Government of India.

Gadgil S (1995). Climate change and agriculture: an Indian perspective. Current Science 9:649–659.

IPCC (1997). Climate Change: Impacts, adaptations, and mitigation of climate change – scientific-technical analyses. Cambridge University Press, Cambridge.

IPCC (2001). Climate Change: Impacts, Adaptation, and Vulnerability: Summary for Policy Makers and Technical Summary of the Working Group II Report. Cambridge University Press, Cambridge.

Lal M, Singh KK, Rathore LS, Srinivasan G, Saseendran SA (1998). Vulnerability of rice and wheat yields in north-west India to future changes in climate. Agric. For. Meteorol.89:101–114.

McCarthy JJ, Canziani OF, Leary NA, Dokken DJ, White KS (Eds.) (2001). Climate Change 2001: Impacts, Adaptation, and Vulnerability. Cambridge University Press, Cambridge.

Ravindranath NH, Sathaye J (2002). Climate Change and Developing Countries. Kluwer Academic Publishers, Dordrecht, Netherlands.

Sinha SK, Swaminathan MS (1991). Deforestation, Climate Change and Sustainable Nutrition Security: A case study of India. Climatic Change 19:201-209.

PCI (2001). State of Poverty in India, 2001.Planning Commission Report, Government of India.

Ambient air pollution and assessment of ozone creation potential for reactive volatile organic compounds in urban atmosphere of Southwestern, Nigeria

Olumayede E. G.[1] **and Okuo J. M.**[2]

[1]Department of Chemical Sciences, Faculty of Science, Ondo State University of Science and Technology, Okitipupa. Ondo state, Nigeria.
[2]Chemistry Department, University of Benin, Benin City. Edo State, Nigeria.

Volatile organic compound (VOC) species react at different rate and exhibit differences in reactivity with respect to ozone formation in polluted urban atmosphere. To assess this, the variations pattern, reactivity relative to OH radical and ozone creation potential of ambient VOCs were investigated in field studies at designated sites in urban centers of Akure and Ado-Ekiti, Southwestern, Nigeria, between June, 2009 and May, 2010. The ambient air samples were drawn into a tube using active sampling method with the low volume sampling pump (Acuro, Drager, Lubeck, Germany). The tube contains Chromosorb 106 as adsorbent. Air samples were collected four times per month at the height of 1.5 m from the ground surface so as to reflect the human breathing zone. On each sampling day, air samples were collected three times between 6.00 a.m and 7.00 p.m to reflect morning, afternoon and evening period. After sampling, the adsorbed VOCs were desorbed with carbon disulphide (CS_2) and the solution analyzed using gas chromatography coupled with mass spectrophotometer. Analysis of the VOCs data revealed that four major classes of abundant VOC groups were captured in the ambient air of the two studied centers. Among the four classes of abundant VOC groups, aromatic dominated with 61 to 77 %, followed by aliphatics (7 to 20%), halogenated (5 to 21%). The results reveal that the total VOCs at various sites showed significant ($p<0.05$) diurnal pattern with bimodal peaks that followed the cycles of economic activities and traffic loads in the areas. The total ozone formation by abundant non-methane hydrocarbons in this study were 52.54 and 71.53µg/m^3 at Akure and Ado-Ekiti, respectively with m,p-xylene, ethylbenzene being the lead contributors at both centres. The abundant alkane in the atmosphere calls for monitoring as they can be move farther distance from their sources and produce a higher concentration of ozone even though the local concentrations are very low.

Key words: Volatile organic compounds, photochemical reactive hydrocarbons, ozone formation potential.

INTRODUCTION

In recent years, it has become apparent that there are close links between climate change and tropospheric ozone. Ozone is the most important photochemical oxidant in the troposphere and holds a special position among the criteria pollutants and radiative trace gases found in the atmosphere. Ozone has been demonstrated equivocally to cause serious injury to sensitive species of vegetation (Ashmore, 1984; Skarby and Sellden, 1984). Clinical and epidemiological researches have also established that exposures to levels of ozone at, or even below the 120 ppb level can produce significant changes in respiratory function in humans (McDonnell et al., 1983;

Linn et al., 1986; Avol et al., 1984; Lippmann, 1989), therefore elevated concentration of ozone in the surface boundary layer will have lots of detrimental effects on human race and ecological environment.

Other peculiarity of ozone is not only because it is not released from either natural or anthropogenic sources but its tropospheric concentration has generally been rising. In the troposphere, ozone is transported downward from the stratosphere (Seinfeld and Pandis, 1998; Singh and Zimmermon, 1992). The mean annual ozone transport from stratosphere to troposphere is estimated at between 720 and 1220 million tones per annum (Vaughan, 1988). Another major source of ozone in the troposphere is that it is produced by photochemical reactions involving nitrogen oxides (NOx) and volatile organic compounds (VOCs), facilitated by sunlight (as shown in Equation 1) (Crutzen, 1974; Atkinson, 2000).

$$VOC + NO_X + \text{sunlight} \rightarrow O_3 + \text{other products} \quad (1)$$

Other products in this reaction include gaseous peroxyl acetyl nitrate (PAN), nitric acid and oxygenated hydrocarbons.

Air pollution has become a problem in the main urban cities of the world, as energy demanded is increasing. Elevated levels of volatile organic compounds are known to be prevalent in the atmosphere of most world cities (Lee et al., 2002; Barletta et al., 2005; Srivastava et al, 2005; Srivastava, 2004, Na et al., 2001, 2003). In most of these studies, anthropogenic sources including motor vehicle exhaust and industrial emissions, power plants, gasoline vapors, chemical solvents, and vegetation are some of the major sources of VOCs (Srivastava et al, 2005; Lenicek et al., 1993; Sims and Overcash, 1983). High concentrations of photochemical reactive VOCs, such as olefins and aromatics, have been reported as well. These two groups have been identified as the main species responsible for ozone formation (Gasca et al., 2004; Mugica et al., 2002). Meanwhile, Sillman (1994) reported that ozone formation responds differently to different VOCs reactivity in different locations. In Nigeria, many researchers have addressed air pollution problem in the country cites (Baumback et al., 1995; Okuo and Ndiowokere, 2006; Ukepbor and Ahonkhia, 2000; Ukepbor et al., 2010; Obioh et al., 1994, Abdul-Raheem et al., 2008). Most of these researchers reported that emissions of large number of anthropogenic are known to be prevalent in the cities with vehicular emissions account up to 70% (Obioh et al., 1994) while some also showed significant temporal and spatial variations. Until the recent time, when we began the study of ambient volatile organic compounds, there has not been existing report on hydrocarbons concentrations and their variations characteristic in urban atmosphere of Nigerian cities (Olumayede, 2011; Olumayede and Okuo, 2012). The results from most of our studies showed atmospheric hydrocarbon levels of 30.64 ± 8.40; 26.42 ± 6.83 and 24.16 ± 3.62 $\mu g \cdot cm^{-3}$ were observed in Benin City, Akure and Ado-Ekiti, respectively, with significant spatial and temporal patterns of distribution. The results also revealed the presence of photochemical reactive VOCs, including olefins and aromatics in the abundant VOCs composition. Unfortunately, the relative reactivity and contribution to production of ozone by these anthropogenic pollutants in Nigerian Cities has not been investigated.

The Southwestern Nigeria falls within the tropical rainforest region. The region has two distinct seasons (the long wet season and short dry season), the raining season between April and October, and the dry season between November and March. The temperature throughout the year ranges between 21 and 29°C except when it is extreme. Humidity is relatively high ranging between 50 and 90% depending on the season, while the rainfall ranges between 150 and 200 mm (NMA, 2009). Recently, there has been tremendous increase in the number of small-scale industries in most urban centers of this region. The small-scale industries are scattered all over the cities. Some of them are located within the premises of residential buildings. These small-scale indus-tries have their own characteristic pollutants, mostly VOCs, emitted to the atmosphere. It is therefore pertinent to quantify the relative contribution of the different atmospheric VOC species on photochemical cycle of ozone in the region.

The work presented in this paper is intended to provide data for better understanding of variability characteristics and relative reactivity of different abundant atmospheric VOC species within Akure and Ado-Ekiti, southwestern, Nigeria. The two centers used in this work are fast growing urban settlements in the Southwestern Nigeria. It is envisaged that the data generated would help to provide information on ozone formation potential of different captured VOC species in the areas and identify the potential threats to human health and natural ecosystem.

METHODOLOGY

Sampling locations

The present study was focused on Akure, Ondo State (50°.10' N and 7°. 15' E) and Ado Ekiti, Ekiti state (5°13' N and 7° 41' E). The population of Akure and Ado-Ekiti during 2006 national population census was estimated at about 360,268 and 313,690 people, respectively (NPC, 2010). They are both state administrative headquarters' and transitory towns, linked with other state capitals, both in the northern and southern parts of the country. In both centers, many small scales industrial plants in the town are located close to the residential areas such as liquid petroleum gas storage plants and most buildings have many ground level shops. Analysis of traffic in both centers revealed that the most predominant mode of transport in the metropolis were cars which constituted 71% of traffic.

Ambient VOCs sampling and sampling routine

The coordinate and description of the sampling locations for the collection of our data, sampling device, chemical extraction and analysis have been discussed in our previous work (Olumayede, 2008; Olumayede and Okuo, 2011).

Meteorological data, average monthly air temperature (°C), relative humidity (%), and wind speed (ms^{-1}) for the sampling period were measured simultaneously before and after VOCs measurements using a humidity/temperature meter, with resolutions of 0.1% RH and 0.1°C (model RS 1364, RS components Ltd, UK). At the same time, wind speeds were measured using an LM -8000 anemometer with a resolution of 0.1 ms^{-1}(Heatmiser UK Ltd). The sampling period covers wet and dry seasons at different sites to highlight the temporal variability.

Traffic survey

A traffic count was performed during the sampling period on avenue close to each sampling site. The traffic data collected were analyzed for traffic composition, traffic volume and variation along the route. The hourly traffic volume was then determined.

Chemical extraction and analysis

After sampling, adsorption tubes were labeled and closed with special caps to avoid contamination and desorption (ASTM, 1988). Chemical extraction and analysis have been discussed in our previous work (Olumayede, 2008; Olumayede and Okuo, 2011).

Quality control

A quality assurance program was implemented in the framework of which all analytical systems related to the analysis of VOCs have been checked for their performance. Various quality control tools were used in order to ensure that adequate laboratory performance was maintained. These included control charts for standard solutions, analysis of control standards as unknown and analysis of the blank activated carbon tubes as process blank to determine if there was any contamination in the activated carbon tubes. None of the compounds included in this study were detected in CS_2 and in process blanks. The calibration curve was found to show good linearity, with determination coefficients (r^2) greater than 0.999 for all the compounds. The limits of detection for BTEX were determined for most VOCs using sample volume of 10 ml and the values calculated in absolute mass. The limits of detection for carbonyls were found to be 0.1 mgm^{-3} for acetone and 0.1 $mg\ m^{-3}$ for benzene, toluene, ethyl benzene, m,p-xylene and 0.2 mgm^{-3} for o-xylene.

Statistical analysis

In order to test the significance of the interaction between the mean total VOC obtained at different locations, two- way analysis of variance (ANOVA) statistical test was used.

Estimation of ozone formation potential

Recently, different methods are used to define the reactivity for ozone forming potential from non-methane hydrocarbon. In this study, the contribution individual VOC to ozone formation were also studied according to the propylene-equivalent concentration (Chameides et al., 1992) and maximum incremental reactivity (MIR) (Carter and Atkinson, 1989) methods.

Propylene-equivalent is a measure of the concentration of hydrocarbon on an OH- reactivity based scale normalized to the reactivity of propene. The propylene-equivalent is defined as:

$$\text{Propylene-equivalent} = C_i \times K_{OH}\ (i) / K_{OH\ (C3H6)} \quad (1)$$

Where C_i is defined as a concentration of individual VOC species; $K_{OH\ (i)}$ is the rate constant for the reactivity of individual VOC; and $K_{OH\ (C3H6)}$ is the rate constant for reaction of C_3H_6 with OH radical. The ozone formation potential is evaluated as product of the concentration of each VOC and the maximum incremental reactivity coefficient (MIR). The MIR is defined as:

$$OFP_{(i)} = \text{concentration}_{(i)} \times \text{MIR coefficient}_{(i)} \quad (2)$$

Where, $OFP_{(i)}$ is defined as ozone formation potential of individual hydrocarbon i and MIR coefficient; (i) the maximum incremental reactivity coefficient of compound i, (Carter, 1994).

RESULTS AND DISCUSSION

Distribution and mixing ratio of ambient VOC species in atmosphere of the sampling sites

Ambient VOCs data for this study were the measured data generated between May, 2009 and June, 2010. In our previous report (Olumayede and Okuo, 2011), a total of 15 volatile organic compounds species were successfully identified and quantified in ambient air of Akure and Ado-Ekiti, among which were four alkanes, six aromatic compounds, three chlorinated hydrocarbons and two carbonyls. These compounds were used in this study.

The total concentration of VOCs was obtained by summing the concentrations of individual species detected. Figures 1 and 2 present the TVOCs distribution by compound type in the sampling sites of Akure and Ado-Ekiti, respectively.

In term of composition, the most abundant compound types in atmosphere of the two studied centers are: aromatic hydrocarbons, aliphatics, carbonyl and halogenated compounds. Aromatic accounted for between 61 and 77% in most sites; this was followed by aliphatics 20-7%, while 5-21% halogenated compounds and carbonyl followed. A comparative study of TVOC in this study with similar urban centers earlier studied in Foshan, China (Tan et al., 2012), shows that the compositions of VOCs in Akure and Ado-Ekiti showed similarities with those of anthropogenic traffic impacted urban centers.

According to Sharif et al. (2008), factors responsible for diurnal variations of pollutants include (i) difference in local traffic volume at different times of the day (ii) traffic flow (iii) meteorological factors. In attempt to further identify how these factors affect pollution patterns in the study centers, we examined traffic and meteorological pattern in the centers.

The hourly traffic volume measured at the various sites (Tables 1 and 2) revealed significant spatial variation dependence of the economic activities around the sites. The market areas at the two studied centers have the highest traffic volume. This is expected as it is common in

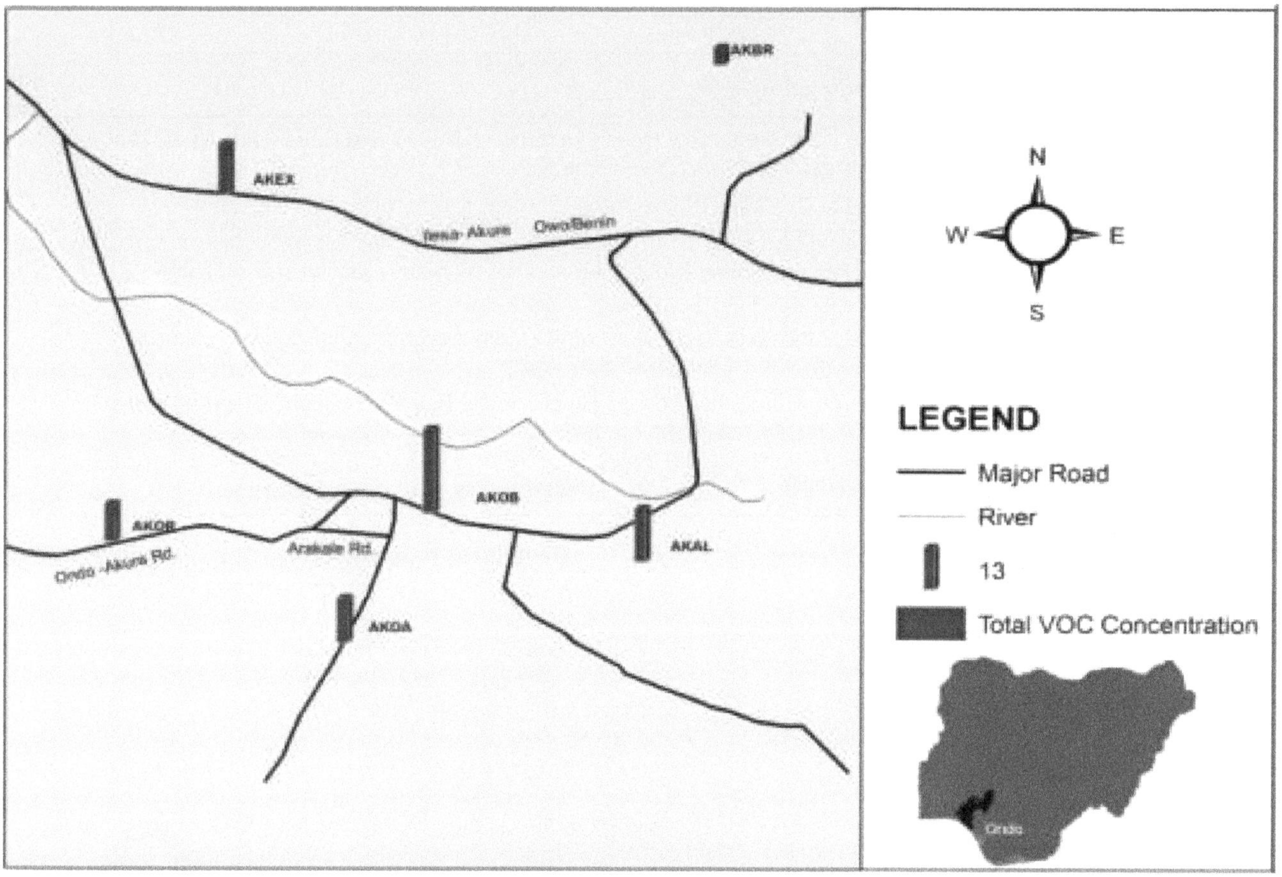

Figure 1. Map of Akure showing the TVOCs at the various sampling sites.

Table 1. Monitoring locations, their characteristics and co-ordinates at Akure.

Site	Code	Coordinate	Site description	Hourly Traffic volume (Vph)
1	AKLR	N°07 17’ 06.2’ E005° 10’ 34.0’	Created along Ilesa- Akure- Owo/Benin road, an express way and close to the NNPC station.	682
2	AKOR	N07° 14’ 25.3’ N005° 11’ 15.0’	Oke - Aro monitoring site, a residential area with many local buildings and close to furniture workshops.	986
3	AKOB	N07° 15’ 12.7’ E005° 11’ 43.8”=	Created at the Oja Oba, a market square with heavy traffic.	1758
4	AKAL	N07° 14’ 55.1’ E005° 12’ 54.4’	Alagbaka monitoring site, the site of many government offices.	986
5	AKOD	N07° 15’ 0.32’ E005° 09’ 96.6’	Ondo road park monitoring site, a bus terminus with a cluster of petrol dispensing stations.	563
6	AKBR	N07° 16’ 10.8’ E005° 19’ 22.4’	Aponmu monitoring site, a small community and a background site for the study.	1

Nigerian cities to have the market located around the bus stops.

In Nigeria, it is customary to classify the day into three time zones- morning (from dawn to noon), afternoon (noon to 4:00 pm) and evening (4:00 pm to 7:00 pm). Figure 3a and b present the diurnal distribution of the TVOCs in each sampling site. A cursory look at these figures revealed that the distribution showed bimodal pattern with morning and evening peaks in most sampling sites of both centers. Emissions from vehicle traffic begin around 6.00 a.m and stay till 9.00 a.m. This is followed by various degrees of decrease in the afternoon and

Table 2. Monitoring locations, their characteristics and co-ordinates at Ado-Ekiti.

Site	Code	Coordinate	Site description	Hourly traffic volume (Vph)
1	ADOM	N07° 35' 11.7" E005° 13' 05.3"	Created along Ikere road, close to a furniture factory and Coca-cola Depot. The roadway carries large traffic volume daily.	662
2	ADOG	N07° 37' 30.1" N005° 13' 13.2"	Created at Oba Market/old garage. Sources situate within close proximity are dump site, petroleum dispensing stations, wholesaling and large traffic volume.	1183
3	ADFP	N07° 37' 51.3" E005° 13' 52.3"	Fajuyi Park monitoring site is characterized by many roads intersection with large traffic volume.	860
4	ADIL	N07° 38' 16.7" E005° 12' 15.8"	Created at the state secretariat, where many government offices are situated, the roadway carries large traffic volume.	611
5	ADAD	N07° 40' 41.1" E005° 14' 39.2"	Adehun monitoring site close to Hotel and the Petroleum Depot. The road leads to the University; hence the roadway carries large traffic volume daily.	713
6	ADOD	N07° 36' 76.7" E005° 07' 83.7"	Odo camp site, a background/control site	3

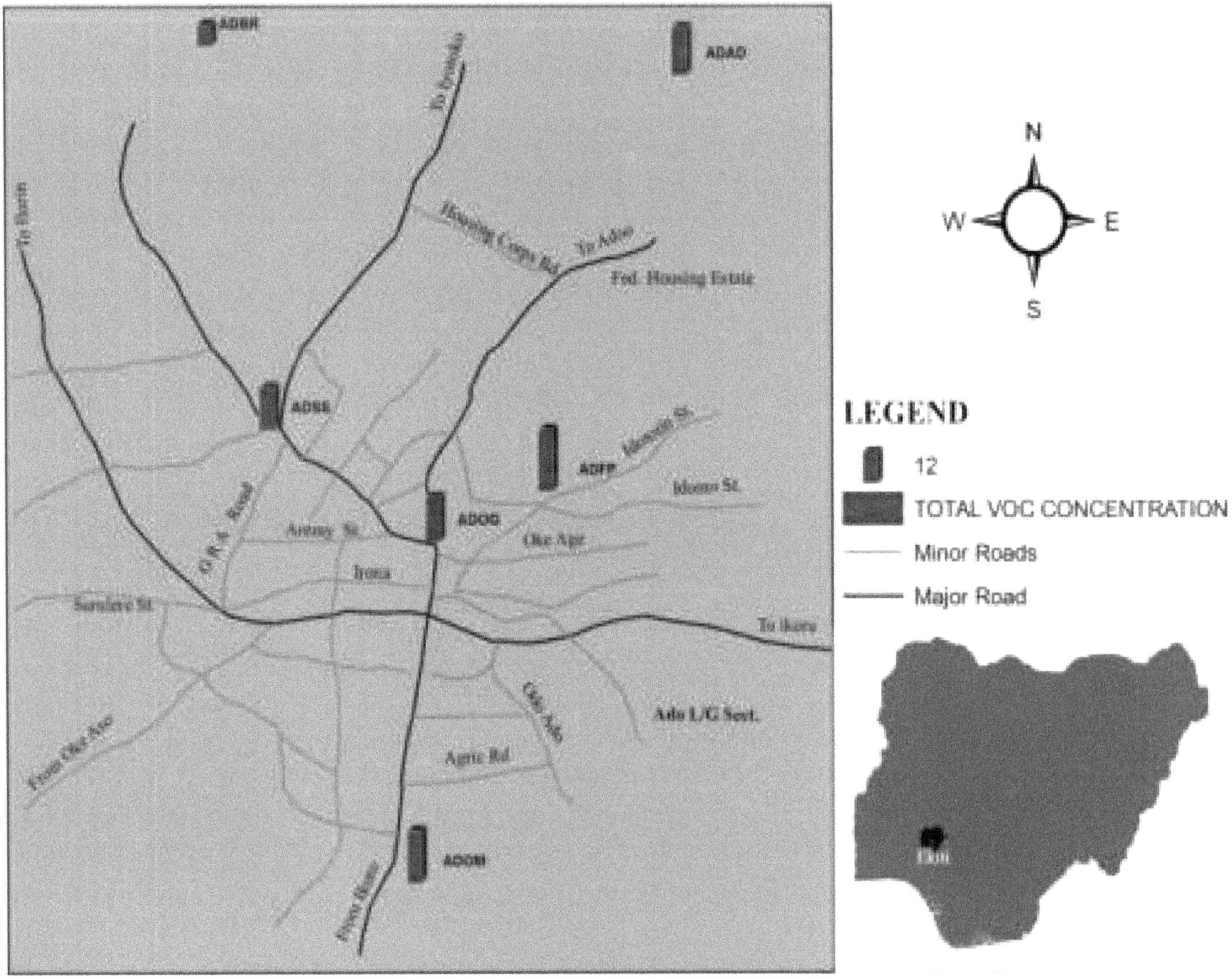

Figure 2. Map of Ado-Ekiti showing the TVOCs at the various sampling sites.

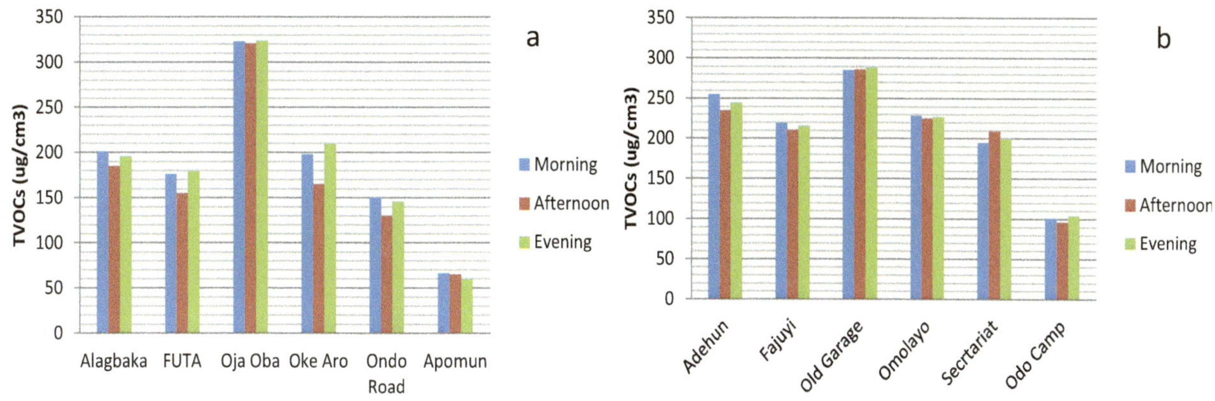

Figure 3a. Diurnal Pattern of TVOCs in Akure sampling site. **b.** Diurnal Pattern of TVOCs in Ado-Ekiti sampling sites.

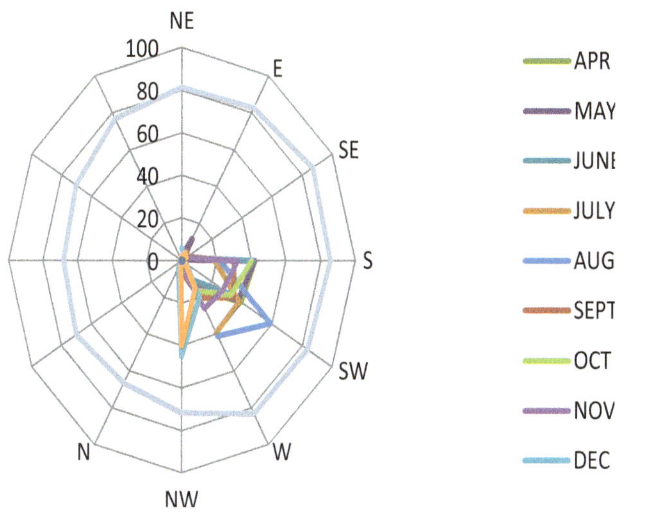

Figure 4a. Wind direction rose in Akure during the sampling period.

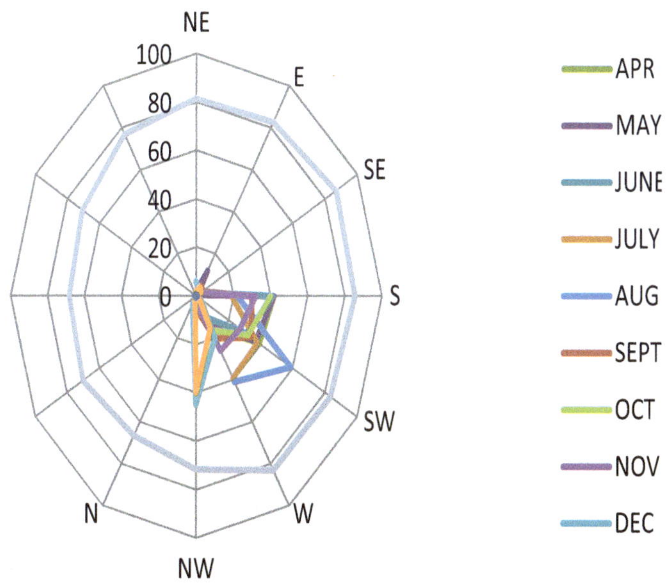

Figure 4b. Wind rose in Ado-Ekiti during the sampling period.

increase in the evening. The early morning period coincide with the hours of maximum traffic volume as a result of the rush to get to the offices, schools and the city markets. The early morning rush can further be attributed to frequent traffic jams resulting from high traffic density, poor road network, poorly maintained road, poor traffic handling and inadequate traffic disciplines. The observed slight uniformity at Oja-Oba and Old garage at Akure and Ado-Ekiti respectively may be attributed to uniformity in traffic volume as the sites are located around market and bus stops.

The same reason has been deduced for similar observation for other pollutants in previous studies (Ukpebor et al., 2010; Baumbach et al., 1995). Meanwhile, the deviation observed at sites close to government offices (Alagbaka site and Secretariat site of Akure and Ado-Ekiti centres, respectively), TVOC concentrations have a distinctive feature with the peak occurring at the afternoon. This suggests the possibility of contributions from sources other than vehicular emission. These sources may include emission from uses of various gargets such as air-conditioning, photocopies, lamination machines, etc. are common. The activities in the area take place mostly during afternoon.

In order to investigate the influence meteorological conditions on the distribution of TVOCs, we determined the monthly weighted-average in each site along with the frequencies of wind speeds and the wind direction rose in the studied centers as presented in Figure 4a and b. We also determine the specific site monthly weighted-average in each sampling sites (Figures 5a and b). In Nigeria, wet season includes June, July, August, September and October while dry season includes November, December, January, February, March and April. The most frequency wind direction was the south-southwest (S-SW) wind during the wet months while the NW wind dominates during the dry months.

In our results, we observed that the TVOC values were lower when the wind speed and relative humidity were high. Figures revealed very calm wind speeds during dry season months. Meanwhile, De Foy et al. (2005) reported

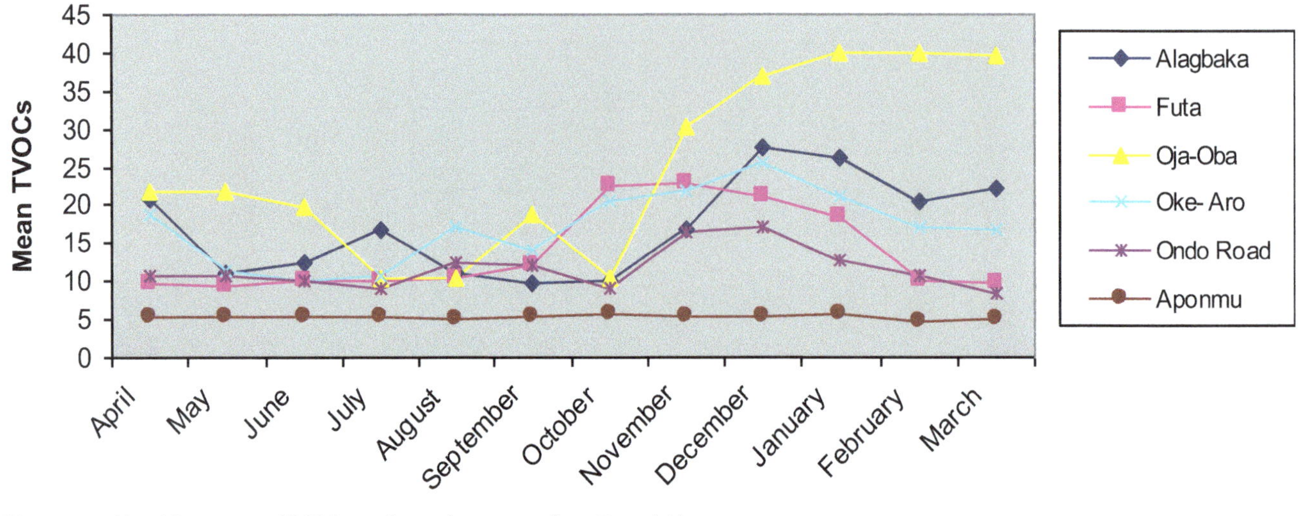

Figure 5a. Monthly average TVOCs at the various sampling sites of Akure.

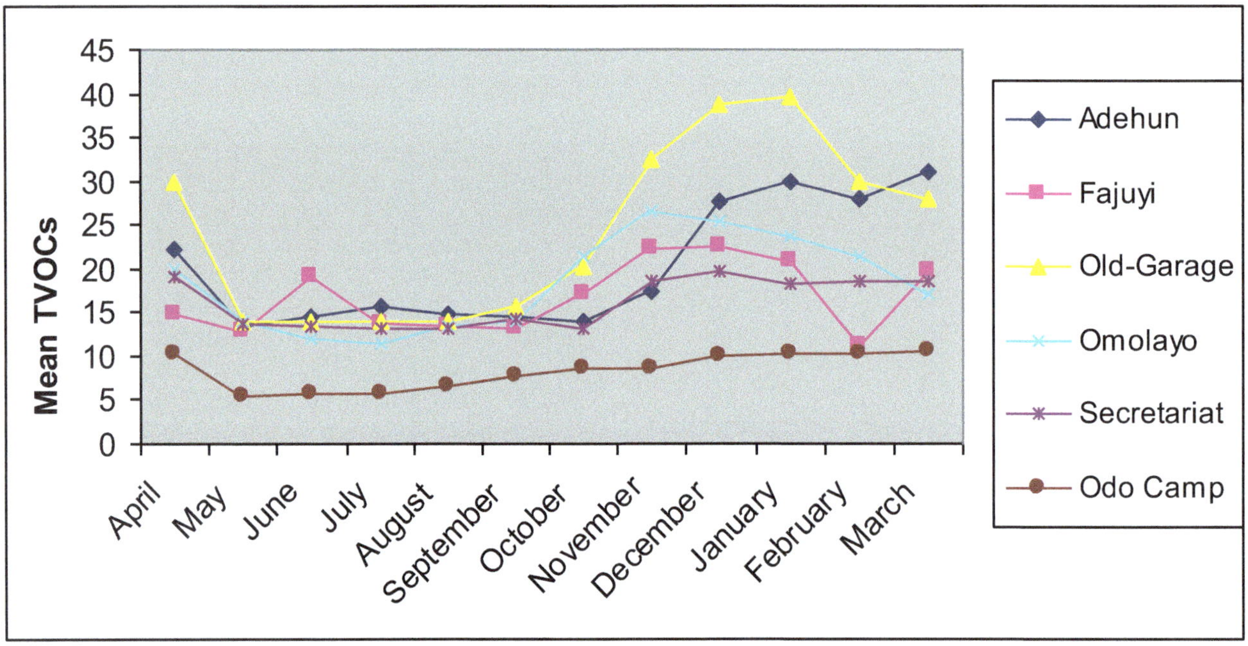

Figure 5b. Monthly average TVOCs at the various sampling sites of Ado-Ekiti.

that winds are always calm at sunset leading to minimal horizontal dispersion. The observations in this study agree well with earlier reports that low wind speed is responsible for poor dispersion and dilution of pollutants in Benin City, Nigeria (Ukpebor et al., 2010; Olumayede et al., 2012).

In general, pollutant emissions follow the cycles of human activity and traffic load with rises near the start of each day's activities, followed by various degrees of decrease or increase around midday, increases in early evening and declines late in night. Such traffic load activity obviously depends on various characteristics of meteorological factors.

Mixing ratio and ozone formation potential

Table 3 summarizes the mean, standard deviation of individual abundant VOC species at Akure and Ado-Ekiti. The species considered for analysis accounted for about 90% of the total VOCs. On the basis of average concentrations, the most abundant VOCs at Akure were in decreasing order; ethylbenzene (3.76 ± 0.21); benzene

Table 3. Concentration, reactivity and ozone formation potential of measured non-methane hydrocarbons at Akure and Ado-Ekiti.

VOC species	Group	k_{OH}	Akure Mean ± SD	OH reactivity	Prop-equiv	OFP	Ado-Ekiti Mean ± SD	OH reactivity	Prop-equiv	OFP
n-Butane	Aliphatic	2.54	0.34±0.03	0.86	0.03	0.37	0.33±0.08	0.84	0.03	0.36
2,2-Dimethylbutane		2.59	0.37±0.05	0.96	0.04	0.41	1.53±0.12	3.96	0.15	1.70
Isopentane		5.6	0.35±0.07	1.96	0.08	0.48	0.69±0.04	3.86	0.15	0.94
3-Methylpentane		3.7	0.35±0.10	1.30	0.05	0.60	0.17±0.01	0.63	0.02	0.29
Undecane		11.2	2.36±0.74	26.43	1.01	1.30	2.06±0.03	23.07	0.88	1.13
Benzene	Aromatic	1.23	2.36±0.22	2.90	0.11	1.63	2.23±0.16	2.74	0.1	1.54
Toluene		5.96	2.06±0.51	12.28	0.47	7.99	2.03±0.18	12.10	0.46	7.88
Ethylbenzene		7.1	3.76±0.21	26.70	1.02	11.02	2.21±0.31	15.69	0.6	6.48
m,p-Xylene		19	1.69±0.09	32.11	1.22	25.7	2.42±0.56	45.98	1.74	36.81
o-xylene		13.7	0.52±0.14	7.12	0.27	3.87	1.44±0.17	19.73	0.75	10.71
Napthalene		NA	0.06±0.01	-	-	0.19	0.05±0.01	-	-	0.16
Methylene Chloride	Halogenated	NA	0.16±0.01	-	-	0.01	0.31±0.11	-	-	0.01
Carbon tetrachloride		NA	0.26±0.05	-	-	0	0.16±0.04	-	-	0
Chloroform		NA	0.23±0.16	-	-	0.01	0.13±0.02	-	-	0.00
Ethylacetate	Carbonyl	NA	0.11±0.02	-	-	0.41	0.71±0.13	-	-	0.42
Acetone		NA	0.42±0.03	-	-	0.18	0.28±0.04	-	-	0.10

*K_{OH} corresponds to the reaction rate coefficient ($cm^3molecule^{-1}s^{-1}$) $x10^{-12}$ published by Atkinson (1994,1997); OH reactivity = Product of mean individual VOC concentration and OH reaction rate coefficient.

(2.36 ± 0.22); undecane (2.36 ± 0.74), toluene (2.06 ± 0.51), m,p-Xylene (1.69 ± 0.09); ethylacetate (0.11 ± 0.02) and the lowest being napthelene (0.05 ± 0.02). At Ado-Ekiti, the order was m,p-Xylene (2.42 ± 0.56); benzene (2.23 ± 0.16); ethylbenzene (2.21 ± 0.31); undecane (2.36 ± 0.74), toluene (2.06 ± 0.51), and the lowest being napthelene (0.06 ± 0.01). Similar trend had been observed in other studied (Barletta et al., 2008; Tang et al., 2007) but with different order. A comparison of the mixing ratios in our study with other cities previously studied revealed that value in Akure and Ado-Ekiti are lower. This may be attributed to difference in population of the cities.

The actual amount of ozone produced by a given hydrocarbon depends on their particular oxidation mechanism and their abundance (Carter, 1994). One of the objectives of this study was to examine the relative importance of the abundant atmospheric hydrocarbon compounds for their potential role in the production of ozone. For this purpose, we estimated the reactivity of the abundant hydrocarbon with hydroxyl radical (OH) in each studied centre, which represents the contribution of each species to the OH lost rate (Atkinson, 1994).

Table 3 lists the major hydrocarbons in the atmosphere of the studied centers by their annual mean concentrations along with their reactivity relative to OH radical. In our study, the result shows that ranking by propyl-equivalent, the alkanes included in this study; butane, isopentane, methylpentane and methylhexane account for 58% of the total propyl- equivalent concentration. It is important to point out that many oxygenated and halogenated VOCs were not considered in our analysis, due to absence of the K_{OH} values; therefore, we considered that it is appropriate to use the term non-methane hydrocarbons.

The total ozone creation potential in the two centres of our study was 52.54 and 71.53 μg/m^3 at Akure and Ado-Ekiti respectively. At Akure, the highest photochemical O_3 formation of 25.70 μg/m^3 comes from m,p-xylene. This is followed in the decreasing order by ethylbenzene (11.02 μg/m^3), toluene (7.99 μg/m^3), o-xylene (3.87 μg/m^3). Similar trend was observed at Ado-Ekiti centre with m,p-xylene (36.81 μg/m^3), toluene (7.88 μg/m^3), ethylbenzene (6.48 μg/m^3). These results indicate that aromatics played the most significant role in O_3 formation in the studied centers. So and Wang (2004) also found that m,p-xylene and toluene were ranked high in ozone formation in urban Hong kong where the reactivity of VOC was dominated by anthropogenic sources. The fact that undecene and 22-dimethylbutane, aliphatic group and heavy hydrocarbon were ranked among the highest contributions to ozone formation in our study suggested that pollutants associated with diesel combustion and exhaust from generators need to be monitored in the area. These alkanes have longer lifetimes and can be transported farther from their sources hence they can produce a higher concentration of ozone even though the local concentrations are very low.

A comparison of total ozone formation potential (OFP) in our study with results from other cities of the world revealed that the total concentration of ozone production in our study is 12 fold lower than the value of 863.4 μg/m^3

reported in Foshan, China (Tan et al., 2012).

Conclusions

The VOC concentrations in ambient air of two Nigerian urban centers of Akure and Ado-Ekiti were studied to understand the distribution patterns. The diurnal trend of TVOCs concentrations showed bimodal peaks with morning and evening peaks which followed the traffic pattern of the study centers and suggest the importance of vehicular emission. The prevalence of low wind speed in the studied centers was also observed to be responsible for the poor dilution and dispersion of the emitted VOCs. Evaluation of ozone formation potentials of the four classes of hydrocarbons detected showed that aromatics play important roles in ozone formation in the centres.

REFERENCES

Abdul-Raheem AMO, Adekola FA, Obioh IO (2008) The seasonal variation of Sulphur dioxide and Nitrogen Oxides in Two Nigerian Cities. Environ. Model. Assess. DOI10.1007/s10666-008-9142-x.

Ashmore MR (1984). Effects of ozone on vegetation in United Kingdom. In: the evaluation and assessment of the effects of photochemical oxidants on Human Health, Agricultural crops, forestry, material and visibility. P, Grennfelt, edition (Swedish Environmental Research Institute, Goteborg). pp. 92 -104.

ASTM (1988). Method D 3686- 84: Standard practice for sampling atmospheres to collect organic compound vapours. Annual book of ASTM standards 11(03): 234-240.

Atkinson R (1997). Gas-phase Tropospheric Chemistry of Volatile Organic Compounds: 1. Alkanes and Alkenes. J. Phys. Chem. Ref. Data 26: 215-290.

Atkinson R (2000).Atmospheric Chemistry of VOCs and NOx. Atmos. Environ. 34: 2063-2101.

Barletta B, Meinardi S, Rowland FS, Chan CY, Wang XM, Zou XC, Chan LY, Blake DR (2005). Volatile Organic Compounds in 43 Chinese Cities. Atmos. Environ. 39: 5979-5990.

Carter WPL (1994), Development of ozone reactivity scales for volatile organic compound. J. Air Waste Manage. Assoc. 44:881-889

Carter WPL, Atkinson R (1989). Computer modeling study of incremental hydrocarbon reactivity. Environ. Sci. Technol. 23:864-880.

Chameides WL, Fehsenfeld F, Rodgers MO, Cardelino C, Matinez J, Parrish D, Lonnemans W, Lawson DR, Rasmussen RA, Zimmerman P, Greenberg J, Middleton P, Wang T(1992). Ozone precursor relationships in the ambient atmosphere. J. Geophys. Res. 97(D5):6037-6055.

Crusten JP (1974). Photochemical reactions initiated by and influencing ozone in unpolluted tropospheric air. Tellus 26:47-57

De Foy B, Caetano E, Magana V, Zitacuaro A, Cardenas B, Retama A, Ramos R, Molina LT, Molina MJ (2005) Mexico City basin wind circulation during MCMA-2003 field campaign. Atmos. Chem. Phys. 5: 2267 - 2288.

Lee SC, Chiu KF, Zou SC, Wang Xinming (2002). Volatile Organic Compounds(VOCS) in Urban atmosphere of Hong Kong. Chemosphere 48:375-382.

Lenicek J, Sekyra M, Panmdey P, Citkova M, Benes I, Novotna J, Koclanova S, Helaskova J, Sinonva M (1993). Polycyclic Aromatic Hydrocarbons at program Teplice Site in the Czech Republic, Toxicol. Environ. Chem. 58:25.

Na K, Kim YP, Moon K (2003). Diurnal Characteristics of Volatile Organic Compounds in the Seoul Atmosphere. Atmos. Environ. 37:733-742.

Na K, Kim YP, Moon KC, Moon I, Fung K (2001). Concentrations of Volatile Organic Compounds in an Industrial Area of Korea. Atmos. Environ. 35: 2747-2756.

NMA (2009). Nigerian Meteorological Agency, Weather Report Bulletin.

Obioh IB, Oluwole AF, Akeredolu FA (1994) Lead Emissions and Source Strength In Nigeria: 1998 Inventory in Heavy Metals in The Environment. Allen and Nriagu (eds). Proceeding of an international conference, Toronto Canada, 12 -17 september1993 CEP Publishers ISBN 0-905941-51-9, 271-274

Okuo JM, Ndiokwere CL (2006) Elemental characterization of total suspended particulate matter in an urban: A case of Benin City, Nigeria. J. Chem. Soc. Niger. 31 (1&2): 6 - 11

Olumayede EG, Okuo JM (2011) Baseline Levels Of Volatile Organic Compounds In Atmosphere Of Two Urban Centres Of Southwestern, Nigeria *Intern.* J. Chem. Sci. 4(1):42-55.

Olumayede EG (2011). Characterizations, source identification and apportionment of volatile organic compounds in atmosphere of three Nigerian urban centres, Unpublished Ph.D thesis, University of Benin, Edo, Nigeria.

Olumayede EG, OkuoJM (2012). Variation Characteristics Of Volatile Organic compounds in an Urban Atmosphere in Nigeria Polish J. Environ. Stud. 21:177-186.

Seinfeld, J.H.,and Pandis S.N. (1998): Atmospheric chemistry and Physics: From Air pollution to climate change. Willey, New York, NY, pp 234-336.

Sharif TA, El-Henshir AK, Treban MM (2008). Variations of Air pollution with related meteorological factors in Tripoli (Case study); Air Pollution XVI; Edited by C.A. Brebbia, J.N.S Longhurst.

Sillman S (1994). The use of NO_X, H_2O_2 and HNO_3 as indicators for ozone-NO_X hydrocarbon sensitivity in urban locations. J. Geophys. Res. 100:14175-14188.

Sims RC, Overcash MR (1983). Polycyclic Aromatic Hydrocarbons (PANs) in soil - plant system. Residue Rev. 88: 1.

Singh, H.B and Zimmermon, P.B. (1992); Edited by O.Nriagu, Wiley Services On Advance In Environment Science And Technology 24, 177.

Skarby L, Sellden LR (1984). The effects of ozone on crops and forests. Ambio 13:68 – 72.

So KL, Wang T (2004). C_3-C_{12} Non methane hydrocarbons in subtropical Hong kong: Spatial- Temporal Variations, Source Receptor Relationship and Photochemical Reactivity. Sci. Total Environ. 328:161-174

Srivastava A (2004). Source Apportionment of Ambient Volatile Organic Compounds in Mumbia City. Atmos. Environ. 38 (39):6829 - 6843.

Srivastava, A.; Joseph; A. E., More, A. and Patil, S., (2005), Emissions of VOCs at Urban Petrol Retail Distribution Centres in India (Delhi and Mumbia), Environmental Monitoring and Assessment 109(1-3):227- 242.

Tan JH, Guo SJ, Ma YL, Yang FM, He KB, Yu YC, Wang JW, Shi ZB, Chen GC (2012). Non-methane Hydrocarbons and their Ozone formation potentials in Foshan. Aerosol Air Qual. Res. 12:387-398.

Ukpebor EE, Ukpebor JE, Eremonene F, Odiase JI, Okoro D (2010). Spatial and Diurnal variation of Carbon monoxide pollution from Motor Vehicles in an Urban Centre. Pol. J. Environ. Stud. 19 (4): 817-823.

Vaughan G (1988). Stratosphere-Troposphere Exchange of ozone: "In troposheric interactions", Iasken, I.S.A (edition), NATO ASI Series C, Volume 227; Reidel Publishing Company, Dordrecht, Holland. pp. 125-135.

13

Sustainable development in a developing economy: Challenges and prospects

P. C. Okonkwo

Chemical Engineering Department, Ahmadu Bello University, Zaria, Kaduna State, Nigeria.

Sustainable development implies development which ensures maximization of human well being for today's generation which does not lead to declines in future well being. Attaining this path requires eliminating those negative externalities that are responsible for natural resource depletion and environmental degradation. All human activities and developmental projects are associated with environmental degradation in one form or the other with the attendant generation of wastes. As a result of these, environmental problems of various types and intensities have emerged to threaten man's well-being and the natural environment which serves as his life support system. In the light of the present global drive towards sustainable development and the concern of the Federal Government of Nigeria, fundamental strategies have been presented here to provide sound basis for comprehensive plans towards environmental management in Nigeria which is a fast developing economy. The basic paradigms of the interaction of man and the environment were used as basis for this study. The socio-economic potentials of effective environmental management were presented. Application of the strategies has the potentials of contributing significantly to the national GDP and will also ensure that development is in harmony with the environment.

Key words: Environment, degradation, sustainable, development, paradigms, pollution, recycling.

INTRODUCTION

The world Commission on Environment and Development, commonly referred to as the Brundtland Commission, defined the concept of sustainable development as 'development that meets the needs of the present without compromising the ability of future generations to meet their own needs, and at same time takes into account the needs of the poor in the developing world'. Sustainable development can be defined in technical terms as a development path along which the maximization of human well being for today's generations does not lead to declines in future well being. Attaining this path requires eliminating those negative externalities that are responsible for natural resource depletion and environmental degradation. It also requires securing those public goods that are essential for economic development to last, such as those provided by well functioning ecosystems, a healthy environment and a cohesive society. Sustainable development also stresses the importance of retaining the flexibility to respond to future shocks, even when their probability, and the size and location of their effects, cannot be assessed with certainty.

The scope and scale of environmental problems has expanded considerably over the past three decades (Colby, 1991). This expansion range from pollution issues at local, regional and then international levels, to deforestation, soil erosion, declining water tables, and other forms of national resource depletion and degradation, to global concerns such as climate change and the ozone layer. This expansion has coincided with cedented growth in the scope and scale of human activities, and in many countries, improvements in human welfare. All human activities take place in the context of certain types of relationships between society and bio-physical world (the rest of nature). 'Development' involves transformations of these relationships. When human activities took place on a scale that was minor as compared to that of nature's own, it did not matter much whether the relationships were of a 'parasitic' or 'mutualistic' type. However, world population has tripled and the world economy has expanded to 20 times of its size in 1900 (Speth, 1989). Vitousek et al. (1986) have estimated that humankind now is responsible for the consumption of some 40% of all terrestrial primary productivity. Matter and energy flows- the physical presence of the economy within the ecosphere now rival in magnitude the flow rates of many natural cycles and fluxes.

RECENT SUSTAINABLE DEVELOPMENT ISSUES

When modeling the impacts of the latest trends in CO_2 emissions, projections show that global average temperatures will increase by about 3.5°C by 2100 (Climate Action Tracker, 2012). This is well above the 2°C of warming considered by many to be threshold for triggering dangerous, runaway climate change (UK Met. Office, 2010). Even with rapid decarbonisation and a green growth revolution, most scientists now consider 2°C to be unobtainable, though this remains a target for political negotiations. Such rapid warming has fundamental implications for development and economic activity. More frequent and severe extreme weather, combined with ever growing numbers of people and assets in exposed coastal areas and floodplains will lead to massive economic losses. This is particularly so in Asia, where 125 millon people are expected to be exposed to tropical cyclones by 2030, double in number in 1990 (IPCC SREX, 2011; Peduzzi et al., 2011). Significant long term shifts and inter-annual variability in agricultural yields will amplify food insecurity through unpredictable supply. In a world of global food supply chains, direct climate impacts will have diverse, distant and indirect effects. For example, based on modeling warming of 4°C, soya bean yield will be halved, at least in almost every developing country in which it is grown (Osborne et al., 2009).

This threat of dramatic climate change hangs over the world in which resources are already scarce in many regions, with global scarcity of key resources a real risk under business as usual scenarios.

By 2030, the world will need at least 50% more food, 45% more energy and 30% more water (United Nations Secretary General High Level Panel on Global Sustainability, 2012). Almost one quarter (23%) of the substantial increase in crop production achieved over the past four decades was due to the expansion of arable land. Agriculture accounts for about 70% of water withdrawals, while water extraction from rivers and lakes has doubled since 1960(Turral et al., 2011). Only 13% of global energy comes at present, from renewable sources, but the imperative of emissions reduction means that renewable energy must increase, with consequences for both land and water resources (Intergovernmental Panel on climate change 2012).

Contemporary globalization presents a paradox of inequality. Inequality between countries (by money - metric measures) is declining as a large cohort of developing countries catches with OECD nations in terms of national income and wealth. There is a corresponding change in the balance of global middle class, which Kenny and Summer (2011) expects to grow massively in developing countries in the next 20 years. Taking a metric of an annual level of per capita house-hold consumption of between \$10 and \$100 dollars of purchasing power parity (PPP) per day, Kenny and Summer (2011) estimates that the global middle class will increase from 1.8 billion people in 2009 to 4.9 billion by 2030 (United Nations, 2011).

The world environmental conference that took place in Stockholm in 1972 drew world attention to the inextricable links between development and the environment. Incidentally, that conference took place at the height of the draught in the West African Sahel that caused so much human misery and death in that part of the African continent. Since 1972, the twin issues of economic development and environmental protection have engaged the attention of scientists and non scientists alike all over the world (Okonkwo, 2000).

Environmental problems are manifestation of disharmony between human activities and the environment. When human population was small and his technological ability limited his activities inflicted little damage on the environment and such damages were repaired by the regenerative powers of nature.

Although, as population increased and as man's technological capabilities increased, man was able to temporarily dominate nature but at an increasing cost to his well-being and survival. Environmental problems of various types and intensities have emerged to threaten his well-being, and the natural environment which serves as his life-support system. Man has thus realized that development cannot be sustained and managed. There are essentially ecological limits to economic growth. Sustainable development and environmental protection and management are now the major issues facing mankind.

The problem of sustainable development also brought the world together in a convention tagged United Nations Conference on Environment and Development (UNCED of Earth Summit) in Riode Janeiro (Brazil) in 1992. The conference took a critical appraisal of the state of global environment as well as proffered strategies for the mitiga-

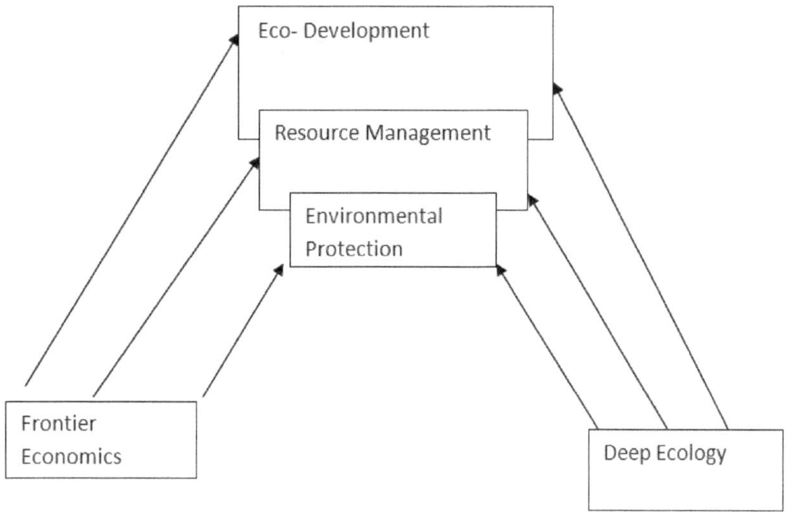

Figure 1. Evolution of environment-development paradigms.

gation of environmental pollution for sustainable development.

To ensure a practicable balance between development and environmental protection conscious efforts must be made as have been highlighted here. These efforts are to be undertaken by all stakeholders (government, community and industries) and should be pursued in a comprehensive manner.

RELATIONSHIP BETWEEN ENVIRONMENTAL MANAGEMENT AND DEVELOPMENT

There are five basic 'paradigms' of the relationship between humans and nature or of 'environmental management in development' (Colby 1991). Figure 1 shows graphically the nature of the evolutionary relationships between the five paradigms. Each paradigm has different assumptions about human nature, about nature itself and their interactions.

The Frontier Economics treats nature as an infinite supply of physical resources (raw materials, energy, water, soil, air) to be used for human benefit, and as an infinite sink for the by-products of the consumption of these benefits, in the form of various types of pollution and ecological degradation. Deep ecology (Naess, 1973; Sessions and Devall, 1985) advocate the merging of scientific aspects of systems ecology with a 'biocentric' (non-anthropocentric) or harmonious view of the relationship between man and nature. Among the basic tenets are intrinsic' biospecies equality, major reductions in human population, bioregional autonomy, promotion of biological and cultural diversity, decentralized planning utilizing multiple value systems, non growth oriented economies, non dominant (simple or low) technology and more use of indigenous management and technological systems. Environmental protection emphasized rational means for assessing the costs and benefits of development activities. This led to the institutionalization of 'environmental impact statements'. Resource management aim at incorporation of all types of capital and resources- biophysical, human, infrastructural and monetary into calculations of natural accounts, productivity, and policies for development and investment planning. Eco-development (Riddell, 1981; Glaeser, 1984) explicitly sets out to restructure the relationship between society and nature into a ' positive sum game' by reorganizing human activities so as to be synergetic with ecosystem processes and services. Eco-development emphasizes biophysical economics model of a thermodynamically open economy embedded within the ecosystem: biophysical resources (energy, materials and ecological processing cycles) flow from the ecosystem into the economy, and degraded (non- useful) energy and other by products (pollution) flow through to the ecosystem as shown in Figure 2.

In addition to the economic justifications for adopting improved environmental governance practice, there are also strategically defensive justifications for doing so. Negligent corporate environmental stewardship cases have raised the ire of environmental interest groups, government regulators and society in general. Accordingly, improved environmental governance practices are viewed as a way to stave off both public protest and regulatory intervention (Reinhardt, 1999). In short, even for critics who view environmental governance as an overall cost of doing business, there is a degree of acceptance that environmental management practices have strategic defensive value (Palmer et al., 1995; Kiernan, 2001). Khanna (2005) framework depicted in Figure 3 summarizes many of the diverse forces that compel firms to adopt improved environmental manage-

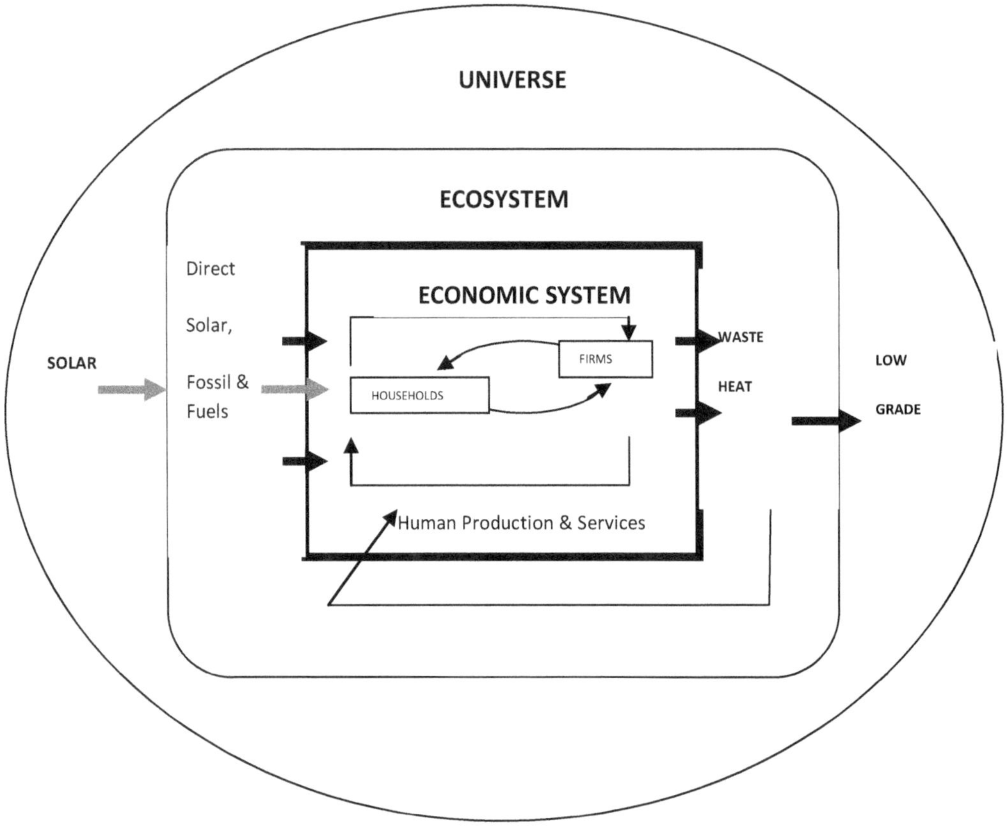

Figure 2. Evolution of environment and development paradigms.

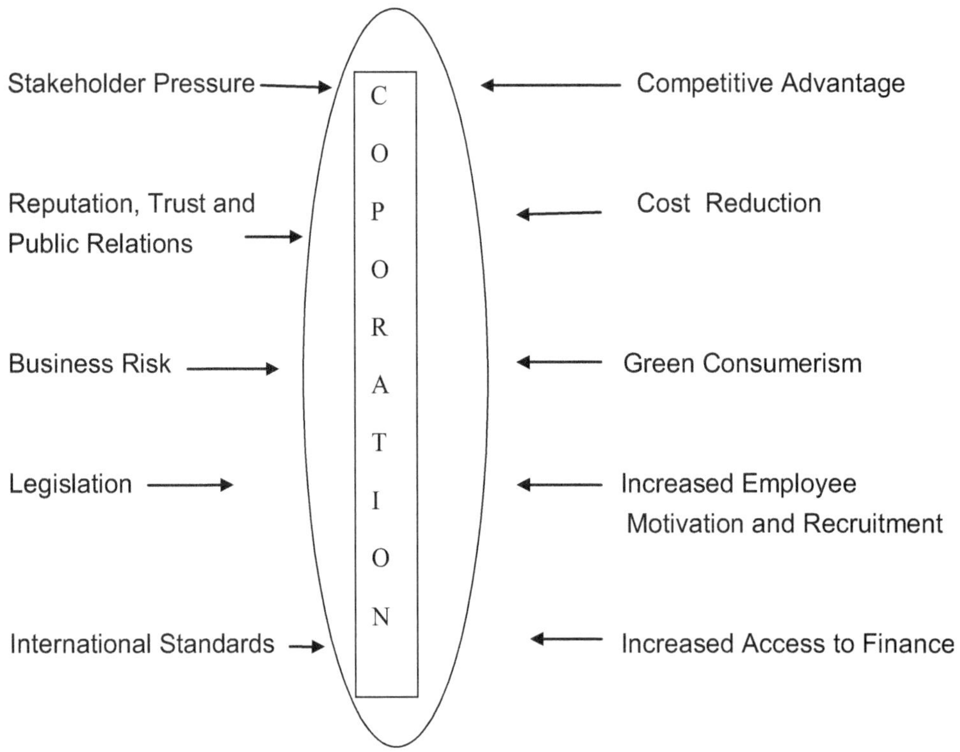

Figure 3. Summary of environmental management catalysts (Khanna, 2005).

Table 1. Some reported environmental degradation cases.

Data/place	Case	Impact
1960-Nigeria	Human Poisoning by gamalin 20 at UCH Ibadan from Cocoa Production Areas	Death of victims
1953-1960	methyl mercury poisoning of people eating fish polluted by mercury	Death of 120 victims
1967 Ghana	Methyl mercury poisoning from fish	Death of 144 victims
1972 Iraq	Methyl mercury poisoning from eating of rice	1000 deaths
1988 Onno-Nigeria	Discharging of water containing high ammonia level into Okinka river by NAFCON a fertilizer company	Massive killing of fish and aquatic, life and huge socio-economic problem

ment techniques. With respect to earlier studies, a great deal is now known about the myriad of ways in which improved environmental governance can benefit firms; yet, the amalgamation of such knowledge into a functional strategic planning framework is still at an evolutionary stage. This has been highlighted previously by Porter and Kramer (2006), who in relation to the broader field of corporate social responsibility (CSR) pointed out that the prevailing approaches to CSR are so fragmented and so disconnected from business and strategy as to obscure many of the greatest opportunities for companies to benefit society.

NATURE OF ENVIRONMENTAL PROBLEMS

The broad divisions of environment comprises of land (terrestrial), water and air (atmosphere). As a result of man's activities on the environment the natural characteristics and features of these are deterioration of water bodies and devastation of aquatic life, defacing of the land and deforestation, denaturization of the atmosphere,global warming etc. Table 1 shows some reported environmental degradation accidents.

TYPES AND CAUSE OF ENVIRONMENTAL POLLUTION AND DEGRADATION

There are three main types of environmental pollution namely, air, water and land population. Air pollution is caused by emission of gaseous pollutants into the atmosphere. These pollutants includes: sulphur dioxide, sulphur trioxide, nitrogen dioxide, nitrous oxide, hydrocarbon vapours, photochemical oxidants, particulates, hydrogen sulphide, asbestos dust, herbicides, pesticides, ammonia, carbon monoxide, radioactive substances and combustion products of fossil fuel, etc.

Water pollution is caused by discharge into water bodies the following pollutants: oil and grease, heavy metals, chemical sludges, dyes, acid, bases, hospital wastes, wastes chemicals, etc.

Land pollution is caused by accumulation of machinery scraps, municipal solids wastes (MSW), used packaging materials and plastics, industrial sludges, etc. The source of these environmental pollutants are industries, households, offices and small scale business centers.

Industrial wastes

Industrialization is the linchpin of development but various disasters which have continuously occurred over the last decades, implicate industries as major contributors to environmental degradation and pollution problems of various magnitudes (Katsina, 2004).

Industrial wastes and emissions contain toxic and hazardous substances most of which can be detrimental to human health. These include heavy metals such as lead, cadmium and mercury, toxic and hazardous substances most of which can be detrimental to human health. These include heavy metals such as lead, cadmium and mercury, toxic organic chemicals like pesticides, polychlorinated bi phenyls (PCB5), petrochemicals and phenolic compounds (FEPA, 1991). In Nigeria, most industries discharge untreated and toxic liquid effluents into open drains, rivers, streams, etc. The solid wastes they generate are often dumped in heaps within the premises, while gaseous emissions and particulates matters are freely discharged into the air (Katsina, 2004). The effect of such uncontrolled pollution as seen in most of the heavily industrialised centers in the country adversely affects the nearby rivers, streams and underground water systems. Table 2 shows some wastes generated in typical manufacturing processes.

Household wastes

As households (a group of people sharing common cooking and housekeeping arrangements) carryout normal family activities, wastes are generated. Various factors influence the rate and modes of wastes generation, these include population growth, urbanization, industrialization, general economic growth, consumption

Table 2. Typical waste materials generated from industrial manufacturing.

Types of industry products	Waste product
Metal and metal product	Cans, ferrous and non ferrous metals rusty scrap iron, metal dust spent oils, metal dust.
Wood and wood fibre	Sawn timber, soft board, hard board, particle board.
Automotive related industry	Motor oil, transmission fluid, engine oil, coolant, batteries. Outdated supplies, paint waste, solvent.
Paint manufacturing	Waste rinse waster, solvent spills, paint sludge inks, mixing bath scrapping, paint soil rags, expired chemicals.
Paper and paper products	Boxes used cartoons waste paper bags.
Plastics	Soft drink bottles, milk bottles, carriers bags, wrapping, plastic sachets.
Photographic laboratories	Waste chemicals from film development and printing papers, contaminated water.
Cosmetics and detergents (lotion, perfumes, powders and liquid detergents)	Oil and grease, soaps, sludge, waste water.
Leather	Waste chemicals, crushers bagaruwa seeds, trimmings, flashings, hair skin residues.
Furniture making	Saw dust, wood pieces
Footwear & leather goods	Pieces of leather, trimmings, lining, leather dust, adhesives finishing chemicals
Oil milling	Sludge and cake
Maize and flour production	Bran
Chalk production	Sediments
Youghurt production	Water and waste water, waste containers.
Packaged waster	Plastics
Juice production	Plastics
Juice production	Plastics, waste water
Bread and bakery products	Ash, waste plastic
Soap/pomade making	Sludge waste chemicals
Polythene making	Plastic wastes
Foam making	Waste foam, waste chemicals
Tourism (hotels)	Food scraps and leftovers, wrapping materials (plastic, paper and packaging, materials)

patterns and practices of individuals and families (Nwaedozie, 2001). Nigeria Environmental Society in 1991 estimated that 20 kg of solid wastes is generated per capital income in Nigeria. This s is equivalent to 3.0 million tonnes in a year given Nigeria's estimated population of 150 million. It is therefore expected that this quantity will continue to grow as the population continue to grow.

Furthermore, using the United Nations estimated figures of 0.54 kg/day of wastes generation in developing countries, to calculate/estimate the quantity of municipal solid wastes (MSW) generated in Kaduna metropolis and using a population figure of about 1.5 m people, the waste generated is about 810 tons/day (Hussaini, 2004). The non-biodegradable wastes that are commonly generated in Nigerian household are various forms of plastic wares-bags, wrappers, containers, etc. There are also plastic kitchen and table-wares, e.g buckets, jerry cans, basins, cups, spoons, clothing articles as shoes, etc. the biodegradable wastes constitutes mainly of kitchen wastes. The generation of liquid wastes is essentially minimized by the adoption of water closet system by households however municipal wastes water that run through gutters is equally large in volume hence the need to handle it properly. This waste water has high bio-oxygen demand content.

Offices and small scale business wastes

The wastes generated by this category of operators include solid, liquid and gaseous wastes. Waste generators in this category include mechanic workshops, restaurants, small scale manufacturers, filling stations, retail and wholesale shops, government offices, etc.

The typical wastes generated include waste oils and grease machinery scraps, kitchen wastes, plastic containers, scrap papers and office equipment, packaging materials both paper and plastics, hydrocarbon vapours and gases, fuel combusting gases, etc.

STRATEGIES FOR ENVIRONMENTAL MANAGEMENT

The current global magnitude and spread of environmental pollution requires a comprehensive approach to the realization of a balance between development and environmental protection to minimize the adverse effects of urbanization, population growth and industrialization which are typical of developing economies on people's lives, the following steps need to be taken on wholesome basis.

The five basic paradigms of man's interaction with nature and his environment provide a sound basis for comprehensive approach to environmental management.

Governmental sectoral policies

The government needs to develop robust and integrated sustainability goals in sectoral policies. These sectors through the goods and services they provide, contribute to meeting human needs but, through their activities, also impinge on the resources available to other sectors and to future generations. The neglect of this interdependence in sectoral policies may jeopardize other policy objectives and reduce total well being. The following sectors need to be critically integrated.

a) Energy is a key requirement for economic and social development, but certain forms of energy can damage environmental quality when they are produced, transported and used. Energy accounts for 85% of total greenhouse gas emissions in OECD countries. It also contributes substantially to emissions of sulphur oxides, volatile organic compounds and particulates. The challenge for energy policy is that of reducing the environmental costs of energy production and use while extending access to basic services in developing countries and preserving energy security.
b) Transportation contributes to economic growth and to meeting social needs for access and mobility. But it also contributes to environmental degradation, depletion of non renewable resources, and damage to and loss of human health. The sectoral policy should address:(i) a better integration of transport and land planning policies, (ii) improvements in the use of transport infrastructure, (iii) shifts of demand for new vehicles towards more fuel efficient ones (through fiscal incentives)
c) Past agricultural growth has been achieved with fewer workers and less land, but using more water, chemicals and machinery. This has led to increased pollution and natural resource use, greater homogenization of landscape and destruction of wildlife habitat. The policy direction should address (i) the strengthening of the agricultural knowledge system, to encourage farmers to adopt sustainable methods, (ii) measures to facilitate the structural adjustment of affected workers and communities, (iii) increased use of pollution charges, to correct environmental damage caused by agriculture.

Public education

The inter-governmental conference on environmental education held in Tblisi (in former USSR) on October 1997 stressed the need for an all - out education programme on environmental problems if nations are to be saved from environmental disasters (Atachia, 1989), it is then necessary to integrate the environmental education into the formal education system, that is, from pre-primary level to tertiary level of education. This will provide the necessary knowledge, understanding, values and skills required by the general public and many occupational groups for their participation in devising solutions to environmental questions. It is also necessary to evolve a vigorous non formal education programme for the "man-in-the street".

Non-governmental organizations (NGOs)

The National Policy on the environment recognize the need to include NGOs and community based organization (CBOs) in the implementation of its policy objectives (FEPA, 1991). The emerging global rapid, complex and often unpredictable political, institutional, environmental, demographic, social and economic changes has brought to fore the need for alternative solutions to the man's problems and this will essentially revolve around NGOs and CBOs (Achi, 2001).

The establishment of specialized NGOs with thrust towards solving environmental problems should be encouraged. Trust funds should also be established by corporate organization and well meaning individuals from which these NGOs can draw funds for their projects.

Effective town planning

Government at all levels should embark on development of functional and effective master plans for all centres of development, towns and villages. This will go a long way towards solving many ecological and environmental degradations being experienced. These master plans must also include adequate provision of central waste management schemes for handling of liquid, solid and gaseous effluents from the various human activities. The central waste handling facilities should be part of the plans for housing estates, markets, shopping complexes, industrial areas and designated sections of the towns and cities.

It has been observed that this approach is grossly neglected in the recent developments across the country.

Enforcement of existing environmental Regulations

Pollution control and waste management objectives can be attained through a variety of policy instruments. These instruments can be categorized into:

a. The command-and-control or direct regulation along with monitoring and enforcement.
b. The economic strategies.

The regulatory approach generally requires government to set health or ecology based ambient environmental objectives and specify the standards or amount of pollutants that can be discharged or the technology by which polluters should meet these objectives. In most cases, the command- and-control approach also specifies schedules or approach to the standards, permitting and enforcement procedure for facilities, liability assignment and penalties for non-compliance. The responsibility for defining and enforcing the standards and other requirements is shared in legislatively specified ways between the national, state and local governments.

The major environmental laws in Nigeria could be said to be contained in the following legislations and their subsidiary legislations:

i. Federal Environmental Protection Agency Decree 1992 (AS Amended).
ii. National guidelines and standards for Environmental Pollutions Control in Nigeria (1991).
iii. National Effluent Limitation Regulation section 1.8 of 1991.
iv. Pollution Abatement in industrial and Facilities Generation Wastes S.1.9, 1991.
v. Waste Management Regulations 5.1.15 of 1991.
vi. Environmental Impact Assessment (E.LA) Decree No. 86 of 1992.
vii. Sectional Guidelines on E.I.A.
viii. State Environmental Protection Edicts.
ix. National Environmental Standards Regulations and Enforcement Agency Act 2007 Etc.

These standards however should be continuously updated to meet international standards and accommodate emerging challenges imposed by technological developments, socio-political and cultural changes. Developing economies face challenges of being dumping grounds of the more developed economies and poor implementation of development plans, which lead to inundation of these economies with goods which sometimes are not environmentally friendly, there thus have a robust regulatory framework that will respond adequately to these challenges.

The economic approach on the other hand, incorporating among other things the polluter pay principle (PPP) is usually adopted to introduce more flexible efficiency and cost effectiveness into pollution control measures. Under the PPP, a polluter pays a financial penalty or receives a financial reward for lower levels of pollution, the strict enforcement of the existing regulations and continuous upgrading of these will assure better environmental protections. Furthermore, the polluter pay principle is expected to contribute to the pool of fund for government interventions in environmental and ecological degradation management.

Waste minimization and recycling waste

Waste minimization is the reduction of waste at source through technological innovation and behavioural change. Waste reduction is considered the top most priority in waste management hierarchy. Recycling involves the basic steps of separating and processing of recyclable waste materials from the waste stream and reselling of the reprocessed items. Recycling materials include glass, paper, cardboard, plastic, metals, etc, recycling is the third preferable option in the waste management hierarchy. The waste management scheme and the potentials for waste recycling activities is shown in Figure 4.

Waste-to-wealth is a product of wastes recycling. In developing economies that face the twin problem of high unemployment and under employment rates, recycling can create a window of opportunities of engaging the population in productive activities of creation of goods and services which can contribute immensely to the GDP of the economy.

Waste minimization an recycling strategy should be practiced at household and industrial sectors to achieve a safe environment.

CONCLUSION

Environmental management is a complex activity and needs a global and comprehensive approach, to achieve sustainable development. Comprehensive strategies for management of the environment have been presented. Nigeria being a fast developing nation should adopt these strategies to ensure that her developmental efforts are achieved in an environmentally friendly manner.

The need for government action to limit environmental degradation is emphasized however it is expedient that the policy framework need to integrate all the sectors.

The five basic paradigms of relationship between nature and environmental management and development was employed to proffer strategies for comprehensive environmental management in developing nations like Nigeria. The socio-economic potentials of environmental management were highlighted. It has been shown that effective environmental management also provide a window of economic emancipation through engagement in activities of waste-to-wealth; which will result in improved GDP of the nation. It has been suggested that environmental management being dynamic in nature will require continuous review so as to encourage best practices.

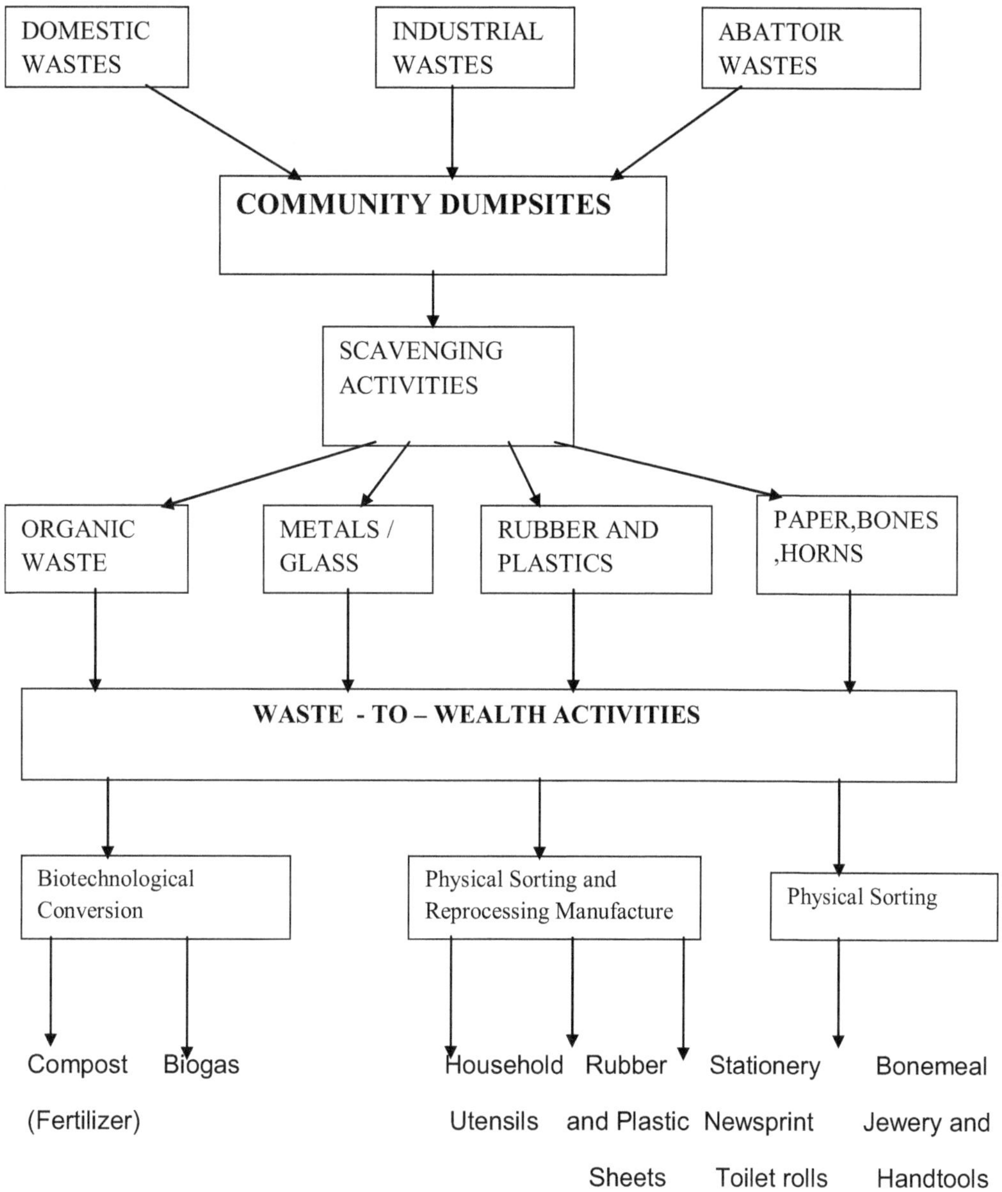

Figure 4. Waste-to-wealth activity and processing profiles.

REFERENCES

Achi LB (2001). The roles of non-governmental organizations in environmental protection. Nig. J. Ind. Pollut. 1:13-17.

Atachia M (1989) UNEP Environmental Training Policy in the perspective of Development in Formal and Integrated Management of Resources is Africa UNEP-1989.

Climate Action Tracker(2012). http://www.climatectiontracker.org.

Colby ME (1991). Environmental Management in development : the evolution of paradigms, Ecol. Econ. 3:193-213.

FEPA (1991), Guidelines and Standard for Environmental Pollution control in Nigeria. Federal Environmental Protection Agency.

Glaeser B (Editor) (1984). Ecodevelopment: Concepts, Policies, Strategies. Pergamon New York.

Hussaini IG (2004), Technologies and Procedures for sustainable Waste management in Industries, paper presented at workshop on Practical Approach to Waste Management organized by Kaduna Environmental Protection Authority July 4-6 2004.

Intergovernmental Panel on Climate Change (2012). Special Report, Managing the Risks of Extreme Events and Disasters to Advance Climate Change Adaptation.

Katsina MS (2004). Policy issues on Industrial Development and waste Management,paper presented at Workshop on Practical Approach to Waste Management organized by Kaduna Environmental Protection Authority. July 4-6 2004.

Kenny C, Summer A (2011). More money or More Development:What have MDGs achieved? Forthcoming working paper. Washington DC; Center for Global Development

Khanna M (2005). Measuring corporate environmental governance: Adelphi study on ranking corporate environmental governance of companies in Singapore:MSc. Environmental Management Dissertation, Graduate School of Design and Environment. National University Singapore.

Kiernan MJ (2001). Eco- value, sustainability and shareholder value: Driving environmental performance to the bottom line. Environ. Qual. Manage. 10:1-12.

Naess A (1973) The shallow and deep, long range ecology movements a summary Inquiry 16. pp. 95-100.

Nwaedozie JM (2001). Industrial Waste Water Management Technology in the Tropics. J. Ind. Pollut. 1: 46-53

Okonkwo PC (2000). Unpublished Environmental Audit Report on UNIPETROL PLC Lbricating Oil Blending Plant submitted to Federal Ministry of Environment Nigeria : 2.

Osborne T, Slingo J, Lawrence D, Wheeler T (2009). Examining the interaction of growing crops with local climate using a coupled clop climate model. J. Clim. 22:6.

Palmer K, Oates WE, Portney PR (1995). Tigthening environment standards: The benefit - cost or non-cost paradigm. J. Econ. Perspect. (1986 - 1998) 9:132.

Peduzzi P, Chatenoux B, Dao H, De Bono A, Herold C, Kossin J, Mouton F, Nordbeck O(2011). Global Trends in Tropical cyclone Risk, Nature, Climate, Change 2011.

Porter ME, Krammer MR (2006). Strategy and society: The link between competitive advantage and corporate social responsibility. Havard Business Review 84:78-92.

Reinhardt F (1999). Bringing the Environment down to Earth Havard Business Review July- August 1999: 149-157

Riddell R (1981).Eco development: Economics, Ecology,and Development: An Alternative to Growth Imperative Models. Gower London.

Sessions G, Devall B(1985). Deep Ecology: Living as if nature mattered, Peregine Smith Books Salt Lake City U.T

Speth JG (1989). A Luddite recants Amicus. 11(2):3-5.

Turral H, Burke I, Faures J(2011). Climate Change and Food Security, Water Report 36, Rome: FAO.

United Kingdom Met. Office (2010), Evidence, the State of the Climate, http://www.metoffice.gov.uk/media/pdf/m/6/evidence.pdf.

United Nations(2011), The Millennium Development Goals Report 2011, United Nations New York. http://www.un.org/millenniumgoals/11-MDG%20Report-EN.pdf.

Vitousek, PM, Ehrlich PR, Ehrlich AH, Matson PA (1986). Human Appropriation of the products of photosynthesis. BioScience 34: 386 - 373.

14

Community-environment relations and development of rural communities in Uyo, Nigeria

Ikurekong, E. E. and Atser Jacob

Department of Urban and Regional Planning, University of Uyo, PMB 1017 Uyo, Akwa Ibom State, Nigeria.

Primary production activity and environmental resources hold the key to rural development. The range of activities in primary production forms the basis with which rural communities relate with the environment and carry out rural development activities. This study examined the interactions in man-environment system and how such interactions could affect development processes. A multistage sampling framework was adopted in the selection of 400 respondents in eight rural communities in Uyo Local Government Area. Structured questionnaire was applied to collect data on 35 independent and 22 dependent variables. Factor analysis was applied to collapse the two sets of data. Both the 35 independent and 22 dependent variables yielded five new factors for each set of data. Step wise multiple regression analyses model was performed to regress all the five new x-components against each of the five y-components and the results showed strong and positive levels of relationships. The study recommended tripartite-P model to promote sustainable development in rural communities.

Key words: Environmental resources, exploitation, rural development.

INTRODUCTION

Human environment is integral to the overall process of development and includes the relationship and interdependences that exist between people and the natural resources. The environment in short is where we live and where development takes place. The two are inseparable yet their inter-relatedness causes enduring tensions (Evans, 1986). It is the resultant effects of this inter-relationship that indicates the level and trends of rural development in any community. According to Aniah (1995) man and his environment are intractably bound together. The land on which man lives, the air he breathes, the water he uses for domestic, industrial, transportation and other purposes, as well as the resources contained in these ecosystems all impinge on man and vice versa. Within each community, irrespective of the level of development, it is imperative that man determines the extent to which he can effectively exploit process and utilize the resources available to him. However, for such exploitation to be sustainable it has to be within the resources capacity for renewal. The underlying premise is that rural development effort should include the development of the rural population to enable them analyze their relationship with their environment and to raise awareness that natural resources should be exploited and utilized in a way that does not impair process of regeneration. Community-environment relations and rural development entail finding solutions to problems of poverty, malnutrition, health, roads, rural-urban migration, food and housing and planned exploitation of natural resources for better and more meaningful life.

In the developing countries of Africa, natural resource exploitation and utilization constitute the mainstay of rural economies. This ranges from activities such as farming, fishing and hunting to fuel-harvesting, lumbering, quarrying and rural craft. These ranges of activities form the basis with which community members relate with the environment and participate in rural development activities. Accordingly, Batten in 1957 asserted that "many communities have sought to perpetuate training in self-reliance among appropriate age groups all aimed at ensuring that the communities are appropriately equipped to change and control their environment" (Ekong, 2003). World Bank (1996) listed some of these factors that put rural environment at risk to include lack of capital, poverty and population growth. Environmental and socio-economic concerns that are consistently identified in rural areas include land degradation and deplorable state of infrastructure such as safe drinking water, paved roads, healthcare facilities and educational facilities.

The severity of these mutually re-enforcing constraints is the evidence of the increasing challenge of rural development when one considers the fact that up to 80% of people in the poorest countries still lives in rural areas and are dependant directly or indirectly on the land (Kohlmeyer, 2001). To the villager, for instance, wood fuel possesses the overwhelming advantage of being and it is commonly used as firewood. Being available source of income fuel, wood is harvested and carried over considerable distances into town for sale to urban dwellers (Sampson, 1992). Also, quarrying of sand gravel of building and road construction has contributed substantially to the degradation of the environment. NEST (1991) reported that one spectacular consequence of quarrying is the emergence of borrow pits along major highways and even within urban centres or their suburbs to mark to aid in the reshaping and modification of natural environment. On the other hand, while farming systems, technologies and cultural values and norms of rural communities have constructive and beneficial influence on the rural landscape they have also unleashed destructive measures that have seriously or permanently impaired the original productive power of the landscape. The obvious result of this negative impact is the increasing inability of the environment to provide the necessary sustenance to agricultural and rural development progammes.

In Uyo Local Government Area of Akwa Ibom State, rural communities occupy a major portion of the land mass. These communities depend directly or indirectly on the exploitation of land. Due to the changed status of Uyo to become a state capital and the gradual increase in population, the hitherto rural communities have increased in population, thereby exerting pressure on the rural land resources. The implication of this trend of development is that the quality of life of the rural dwellers is adversely affected. In most of the rural communities, literacy is still low, spatial inaccessibility is still high while basic social infrastructures such as portable water, electricity, healthcare and good housing are in deplorable state. As rural areas where productivity is predicated on the effecttiveness of man-environment nexus and in spite of the rural-urban linkages, the level of underdevelopment, poverty and social deprivation prevalent in the rural communities pose enormous challenges for rapid rural transformation.

Therefore, the increasing inability in community-environment interaction to provide the necessary impetus for rural development calls for investigation. It is hoped that the understanding of the complex linkages of human activities in the rural communities will lead to productive management of naturally occurring resources and in the process will also help in the reorganization of the rural development initiatives for enhanced rural income, rural commerce and environmental health. The extensive ravines and relief in some of the communities have a marked influence on farming (a predominant occupation of the people) and on rural development programmes. Farming in these areas has led to accelerated soil erosion and soil infertilities hence poor yield or output. On the other hand, infrastructural development has greatly been hampered.

Economic activities in these commu-nities are largely subsistent and localized in scope. Commercial farming activities are hampered by land fragmentation. This affects the socio-economic status of the people and their contribution to rural development hence negative impact on community-environment rela-tions and rural development. Another characteristic in the study area that impinges on community-environment relations and rural development is the beliefs attached to sacred forests and shrines. The general belief of the people is that the gods and the spirit of their ancestors resides in sacred forests and shrines, therefore attempts at clearing such forests and shrines for rural infra-structural development is always resisted. Restrictions placed on entry into these grooves also influence community-environment relations and development.

MATERIALS AND METHODS

The research design used is the survey approach, aims at assessing the effects of primary occupations on rural community development. This design is most appropriate because it permits an accurate assessment of the activities and characteristics of the population under investigation. Data required for the study were on socio-economic activities of the communities and development indicators. Data on socio-economic indicators include farming methods, size of farm, levels of exploitation of land resources, belief systems and cultural practices of the rural communities. Data on development indicators include education and health infrastructure, agricultural storage and marketing facilities, source of drinking water, type of houses and nature of rural roads and levels of income. The striking feature of man-development relationship in the context of rural development is that it places primary production activities at the centre of economic life of rural communities. To examine these relationships, two groups of variables representing socio-economic activities (independent variables) and development indicators (dependent variables) were identified and measured. The socio-economic activities and their specific effects on the environment have wide applicability for rural development hence the variables are interrelated and interdependent. Primary data were obtained from interaction with rural dwellers in the communities through structured questionnaire administration, semi-structured interview and observation.

This study employed a combination factor analysis and multiple regression analysis as statistical tools. The interest in multivariate analytical techniques arises from the fact that isomorphism, that is, the existence of one-on-one relationship is rare. An analysis of community-environment relations involves an understanding of causally related variables. The principal components analysis is a multivariate statistical technique which primarily facilitates the reduction of a large data matrix X1, X2, X3 ...Xn into a smaller one Y1, Y2, Y3 ...Yn without much loss of information. This it does by using an orthogonal transformation which converts a set of data of possible correlated variables into a set of uncorrelated data called factors. Multiple regressions technique was applied to measure the relationship between the dependent variable and a set of independent variables. It is a multivariate statistical tool which relates quantitatively variations in the dependent variable based on some independent variables. Uyo Local Government Area has 11 political wards. Three of these wards constitute the urban area while 8 wards constitute the rural areas. Due to the rural focus of the study, the 8 wards comprising 63 communities were used as population of the study. With a combined population of 129102 persons and a sample size of 400 respondents derived from Taro Yamane formula for finite population, an average of 6 household heads was sampled per community using proportional representation and systematic random sampling approach. The contribution of human activities to rural community development was investigated using the primary productive activities of the population (crop cultivation and extraction of sand, gravel, laterite and fuel wood). All the 35 socio-economic (independent) variables and 22 rural development (dependent) variables were subjected to factor analysis statistics so as to extract the major factor for X and Y as well as to achieve a parsimonious description of the sets of variables.

RESULTS

Extraction and naming of socio-economic factors

Of the 35 socio-economic (independent) variables (Table 1), the first 29 variables on socio-economic indicators were subjected to factor analysis and five major factors were extracted from the original 29 independent variables and thus, indicated the socio-economic activities and rural (Table 2) development initiatives of the people as shown in Table 3. Table 3 shows that the factor analysis procedure applied to the data set yielded a five-dimensional solution. The communalities, which are indicators of the importance of the variables selected for the study, were appropriate and relevant. The five factors altogether accounted for 93.9% of the total variance. The factors were named according to the high positive loadings of the variables that made up each of the factors.

Factor 1: Farm activities factor

This factor is so named as it has high positive loading on:
X25 = Cost of fuel wood, 0.95
X24= Quantity of fuel wood used, 0.92
X2 = Farm labour, 0.88
X7 = Farm additives, 0.88
X9 = Waiting period, 0.88
X6 = Fallow period, 0.74
X5 = Farm implements, 0.63

This independent X-variable accounted for 34.9% of the variation within the distribution of independent (X) variables and is so named because of its influence on productivity. The implication is that the farm activities factor could influence the volume of farm produce and thus, bring about increase in income and improved living conditions of rural dwellers.

Factor 2: Land cover factor

X18 = Sacred forest, 0.90
X26 = Income from hunting, 0.82
X19 = Sand/laterite extraction, 0.72
X1 = Farm size, 0.52

This independent variable accounts for 25% of the variation within independent (X) variables. The factor is so named because all the constituents facilitate conservation, and sustainability of resources. Where there is a cultural law that certain portion of land should not be cleared there is a regeneration of and conservation of forest resources including plants and protected habitat for endangered animal species. These areas could be reconfigured as tourism potential.

Factor 3: Complementary activities factor

X27 = Fishing income, 0.70

This factor accounts for 16.5% of the variation within independent (X) variables. Effective and efficient production calls for coordinated activities especially so in rural communities where there is no clearly defined direction of investment. The activities of the rural people are highly seasonal with the result that there is full utilization of labour during the peak period but there may be virtually unemployment during the slack season. Therefore, complementary activities could provide the population with possible areas of employment.

Factor 4: Labour factor

X3 = Age of farmers, 0.81

The factor is so named because of the high positive loading. This factor accounts for 10.1% of variation within

Table 1. Socio-economic activities (X- independent variables).

Variable		Parameter	Justification
X1	Farm size	Hectares (m^2)	The larger the farm size, the greater the volume of produce and vice-versa. Those having larger farm size cultivate more crops as against those with small farm size. This affects the level of income.
X2	Labour	Number	Availability and cost of labour determine how many rural farmers can afford to hire. Where labour is scares, the level of cultivation will be less hence reduces output.
X3	Age of farmer	Years	Farming activities in the rural area is stressful and energy intensive. The aged who are mostly engaged in farming in rural area lack the stamina hence productivity per unit of farmland is very low.
X4	Crops	Type and number	Certain types of crops being more income than others within a farming season. Thus the rates of earning overtime signify and influence the choice and type of crop and rural income.
X5	Farming implements	Type and number	Type and number on implement play major roles in farm output. In rural areas farming if rudimentary with the use of simple tools and implement. Farming practices where the hoe and matches are most use implement are not likely to yield much output compare to where mechanized implement are used.
X6	Land tenure period (fallow period)	Number of years	Fallow period affect rate of exploitation soil nutrient, hence farm output and rural income.
X7	Farm additives	Number	Influence soil fertility, output and income.
X8	Farm product sold	Percentage (%)	Determine rural income, savings and capacity building/utilization.
X9	Waiting period before harvest	Number of months	This represents slack period from farming and much influence the consumption pattern of the people thus affecting capital and any development projects.
X10	Price per bag of garri	Naira (₦)	Determine savings and capital accumulation.
X11	Price per bag of palm oil	Naira (₦)	Determine saving and capital accumulation.
X12	Price per bag of kernel	Naira (₦)	Determine saving and capital accumulation.
X13	Average cost of a plot of land	Naira (₦)	Affordability influences number of people in farming and income distribution. Increase contribution lead to development activities.
X14	Income from farming	Naira (₦)	Increase income may influence amount of money devoted or set aside for development projects.
X15	Loan received for farming	Naira (₦)	Facilitate volume of production, hence increased rural income.
X16	Farmer cooperative society	Number	Coordinates activities of rural farmers toward effective and efficient production. Also active on direction of investment.
X17	Average saving per farming	Naira (₦)	By savings a portion of income individual can afford to contribute substantially to the development of community.
X18	Sacred forest shrine etc.	Number	These could be reconfigured as tourism potential. Also conservation measures for sustainability of resources.
X19	Sand/laterite extraction sites	Number of trucks per month	These could reduce the land area for crop production besides the degradation effect on environment.
X20	Income from sand/literite per month	Naira (₦)	Quantity sold adds to the rural income and increased earning influences development.
X21	Amount of Osusu per month	Naira (₦)	This increases the amount of money available for investment by and individual overtime.
X22	Distance to farm	Km	The nearness to and from farm influences type of crops cultivated, care and yield
X23	Distance to market (accessibility)	Km	This influences what could be conveyed to the markets and the violence of sale hence rural income
X24	Fuel wood used per month.	Number of bindles	Determines level of exploitation and utilization of forest/vegetation resource and exposure of land to erosion.
X25	Price per bundles of fuel wood	Naira (₦)	As a source of rural income, the volume of sale contribute to rural economy and hence development
X26	Income from hunting	Naira (₦)	Determine what is set aside for development activities. Improves nutrient intakes and health of rural people

Table 1. Contd.

X27	Income from fishing	Naira (₦)	Determine what is set aside for development activities. . Improves nutrient intakes and health of rural people
X28	Income from cane craft (rural craft)	Naira (₦)	Determine what is set aside for development activities. Tourism potential hence rural income
X29	Income from sales of stakes for pumpkin	Naira (₦)	Determine what is set aside for development activities. Facilitates growth of related crops.
X30	Quantity of Sand extracted per week	Number of trucks in cubic meters	Determine level and rate of exploitation
X31	Quantity of gravel extracted per week	Number of trucks in cubic meter	Determine level and rate of exploitation
X32	Quantity of timber harvested per month	Number of logs	Determine level and rate exploitation
X33	Quantity of fire wood harvested per week	Number of bundle	Determine rate of exploitation
X34	Average quantity of farm produced per season	Number in the unit of crops	Determine rate of exploitation
X35	Quantity of 'ndisa' yam stakes harvested per season	Number of bundle	Determine rate of exploitation

Source: authors' field survey (2007).

Table 2. Rural development indicators (Y-dependent variables).

Variable		Parameter	Justification
Y1	School	Number	Development of human capital. Promotes literacy and education of the rural populace.
Y2	Drinking water	Distance in km	Enhance rural health hence improves economic activity and productivity. It affects quantity and quality available, which affect household hygiene and to a large extent affect environment health of the community.
Y3	Houses connected to electric		Electricity attracts rural enterprise, facilitates rural commerce and improve living condition. Create enabling environment for economic growth
Y4	Health facilities	Number	A healthy workforce is a productive one. Without good health no society or individual can develop. The richness of development depends on how healthy the populace is. Hence Millennium Development Goals number 4, 5 and 6 are directly link to the promotion of good health services in our society. Health care facilities is therefore essential for improve economic activities
Y5	Skill acquisition center	Number	Skill acquisition finds application Development Goals number 1with focus on poverty eradication, which focuses on development of skill to raise the component of people for income generation. The establishment of skill acquisition centres in rural areas is one means of empowering the people individually and collectively to contribute to rural economic development.
Y6	Community bank	Number	Create enabling environment for economics growth provides loan rural farmer. Efficient management of rural income. Also provides employment
Y7	Culture(sacred forest shrines etc)	Hectares (m^2) and Number and size	Biodiversity conservation could be reconfigured for tourism purpose

Table 2. Contd.

Y8	Agricultural storage facilities	Number	Efficient management and utilization of environmental resources. Enhance rural income. Provides employment
Y9	Co-operative development shop	Number	Co-ordinate rural commerce for efficient and effective exchange of goods and services. Improves employment
Y10	Length of paved roads	Km and number	Create enabling environment for economic growth facilitates movement of goods and persons hence improves rural investments and aesthetics.
Y11	Local industries	Number and type	The local industries offer employment especially for youth who otherwise will be idle and restive with adverse consequent to the community. As the saying goes" an idle man is a devil workshop" Local industries enhance economic growth. Ensures that some percentage of the bulk of the body of natural resources are retained or captured locally.
Y12	Adult literacy centres	Number	Provide education and training hence reducing literacy level. Improves rural living condition.
Y13	Motor park	Number	Facilitates movement of good, persons and services to and from rural areas. Enhance rural income and live ability. Farmer rely on these for easy movement production.
Y14	Markets	Number	Farmers rely on this for exchange of their products. Facilities rural commerce hence income.
Y15	Village hall	Number	Facilities effective administration and transmission of information. Rules and regulations within the community.
Y16	Children in school	Number	Indicates level and potentials human resources utilization for further development.
Y17	Income from farming	Naira (₦)	Indicate or measures development potentials of the community, as well as level of exploitation and utilization of resources.
Y18	Income from fuel wood	Naira (₦)	Indicate or measures development potentials of the community, as well as level of exploitation and utilization of resources.
Y19	Income from sand, laterite	Naira (₦)	Indicate or measures development potentials of the community, as well as level of exploitation and utilization of resources.
Y20	Income from hunting	Naira (₦)	Indicate or measures development potentials of the community, as well as level of exploitation and utilization of resources.
Y21	Income from fishing	Naira (₦)	Indicate or measures development potentials of the community, as well as level of exploitation and utilization of resources.
Y22	Income from rural craft	Naira (₦)	Indicate or measures development potentials of the community, as well as level of exploitation and utilization of resources.

Source: Authors' field survey (October, 2007).

the independent (X) variables. Rural development is handicapped by the quality and quantity of its human resources. With the pull exerted by the urban sector, rural areas are left with younger and older elements of the population who are less productive but more consumptive labour force. Fuller and effective utilization of available productive rural resources is therefore a function of the age of farmers.

Factor 5: Distance factor

X22 = Distance to farm, 0.56

This factor accounts for 7.3% of the variation within the independent (X) variables. This factor is so named because of its influence on primary production activities. The farm land activities are characterized by fragmented landholdings. Enormous time is wasted moving from one plot to another. The unprogressive nature of farm tenancy arrangement that is at work, the conditions of tenancy are fluid and informal and could involve plot allocation of several kilometers away from home. Therefore the distance to farm land does not only influence the number of people meaningfully engaged in farming but also the level of farmers' productivity.

Extraction and naming of rural development factors

Five major factors were extracted from the twenty two 22) original variables in the data set after being

Table 3. Rotated factor matrix for socio-economic activities in Uyo rural communities.

S/N	X-variable	F1	F2	F3	F4	F5	Communality
X1	Farm size	0.31755	0.53639	−0.70579	−0.15842	−0.29003	0.99591
X2	Labour	0.88429	0.16036	−0.12669	0.00800	0.35331	0.94863
X3	Age of farmers	−0.44526	−0.10423	−0.21707	0.81287	0.12724	0.93319
X4	Crops	0.70617	0.10213	0.56691	0.20923	−0.08073	0.88079
X5	Farming implements	0.63209	0.72180	0.21266	0.06726	−0.5435	0.97324
X6	Land fallow period	0.74333	0.00408	−0.35232	0.03794	−0.49928	0.92741
X7	Farm additives	0.88439	−0.16036	−0.12669	−0.00800	−0.35331	0.94863
X8	Farm product sold	−0.29484	0.46003	−0.66136	−0.44854	0.20536	0.97931
X9	Waiting period before harvest	0.88429	0.16036	−0.12669	0.00800	0.35311	0.94863
X10	Price per bag of garri	0.72180	0.35572	−0.52528	0.09238	0.24437	0.99169
X11	Price per ton of palm oil	−0.80995	0.31904	−0.45938	0.19013	0.29803	0.98874
X12	Price per bag of kernel	-0.11426	0.22861	-0.12075	0.05418	0.31106	0.90211
X13	Average cost of a plot of land	−0.60605	0.54806	0.34287	0.19013	0.29803	0.91020
X14	Income from farming	0.31755	0.53639	−0.70579	−0.15842	−0.29003	0.99591
X15	Loan received for farming	0.26625	0.41286	-0.14022	0.32171	-0.06441	0.84396
X16	Farmers cooperative society	−0.06448	0.72212	0.39988	0.42642	−0.02148	0.86781
X17	Average Savings per farming season	−0.43563	0.59174	0.29757	0.33431	−0.18072	0.77290
X18	Sacred forest shrine etc.	−0.11583	0.90470	0.11773	0.20242	−0.00237	0.88674
X19	Sand/laterite extraction sites	−0.04020	0.72394	0.46998	−0.16609	−0.17946	0.80638
X20	Income from sand/laterite per month	0.88429	0.16036	−0.12669	0.00800	0.35331	0.89533
X21	Amount of Osusu per month (saving)	−0.57400	0.73870	−0.24500	0.01255	0.24937	0.99775
X22	Distance to farm	0.43864	−0.06191	0.42490	−0.49444	0.56467	0.94010
X23	Distance to market (spatial access)	0.26299	−0.86524	0.19139	0.01619	−0.13458	0.87281
X24	Fuel wood used per month	0.92465	0.09509	0.08970	0.24101	0.21261	0.97353
X25	Price per bundle of fuel wood	0.95033	−0.09601	0.27968	−0.02683	0.08523	0.99855
X26	Income from hunting	0.47651	0.82390	−0.08573	−0.22757	0.12954	0.98180
X27	Income from fishing	0.27520	0.58130	0.69960	0.21882	−0.08434	0.95811
X28	Income from cane craft (rural craft)	0.78907	0.04023	−0.44354	0.36351	0.11376	0.96607
X29	Income from stakes for pumpkin	−0.32557	0.13147	0.46498	−0.78680	0.19742	0.99753
Eigen Value		9.42239	6.75068	4.45495	2.74039	1.97112	
Variance (%)		34.9	25.0	16.5	10.1	7.3	
Cumulative (%)		34.9	39.9	76.4	86.5	93.9	

Source: Authors' field data (2007).

(subjected to factor analysis. These were referred to as development factors and altogether accounted for 94.7% of the total variance in the data set of Y variables as Table 4 shows. The communalities are high and thereby indicating the importance of the variables considered in the study and were named according to the high positive loading of the variables.

Factor 1: Rural infrastructure factor

Y3– house connected to electricity, 0.92
Y11– Local industries, 0.90
Y4– Health facilities, 0.88
Y1– Schools, 0.77
Y10– Length o f paved roads, 0.79

This was so named because it loads highly and positively on all activities within the communities. This factor accounts for 44.8% of the total variation in the data set of Y-dependent variables it measures. An examination of this factor reveals lapses in the provision of rural infrastructures. The presence of adequate infrastructures enhances the production activities, promotes literacy and education in case of schools. It is also a measure of social satisfaction in the physical development of any community.

Factor 2: Rural income factor

Y21– Income from fishing, 0.74
Y9– Co-operative development shop, 0.68

Table 4. Rotated factor matrix for rural development indicators in Uyo rural communities.

S/N	Y-variable	F1	F2	F3	F4	F5	Communality
Y1	School	77088	-0.02276	-0.47269	0.23816	0.32761	0.98227
Y2	Drinking water	-0.37668	0.50684	0.66584	0.09368	0.37243	0.98960
Y3	House connected to electric	0.92394	-0.31659	0.10864	0.00102	0.04618	0.96783
Y4	Health facilities	0.88923	0.13129	-0.05704	0.32109	0.21019	0.95849
Y5	Skill acquisition centre	0.79814	-0.50953	0.28946	0.02881	0.13046	0.99828
Y6	Community bank	0.79814	-0.50953	0.28946	0.02881	0.13046	0.99828
Y7	Cultural centre (sacred forest, shrines)	0.73167	0.54044	-0.08939	-0.19931	-0.29808	0.96397
Y8	Agricultural storage facilities	0.34117	0.28401	-0.15721	0.29642	0.50162	0.91106
Y9	Co-operative development shop	0.36614	0.68142	-0.36414	0.39159	0.14475	0.90528
Y10	Length of paved road	0.79814	-0.80953	0.28946	0.02881	0.13046	0.90528
Y11	Local industries	0.90111	0.13453	-0.16535	-0.32165	-0.16698	0.98878
Y12	Adult literacy centres	0.79814	-0.50953	0.28946	0.02881	0.13046	0.99828
Y13	Motor park	0.14411	-0.42867	-0.24419	-0.27114	0.03886	0.80113
Y14	Markets	0.80661	31410	-0.10550	0.16953	0.08606	0.79655
Y15	Village hall	0.14393	0.21545	0.53037	0.61328	-0.32998	0.83343
Y16	Children in school	0.91603	-0.01998	-0.04117	-0.02869	-0.24846	0.90375
Y17	Income from farming	0.14389	0.27972	0.79775	-0.47634	-0.19003	0.99836
Y18	Income from fuel wood	0.71593	0.51084	0.15953	0.07999	-0.39105	0.95828
Y19	Income from sand, laterite	0.27302	0.16632	-0.10802	-0.83594	0.34616	0.93250
Y20	Income from hunting	0.46764	0.65838	0.35401	-0.29621	0.10917	0.97713
Y21	Income from fishing	0.41985	0.74274	-0.44112	-0.01537	0.18407	0.95663
Y22	Income from rural craft	-0.20150	0.45851	0.71585	0.24180	0.33398	0.93328
Eigen value		80.96382	30.89552	20.98492	10.95226	10.14234	
Variance (%)		440.8	190.5	140.9	90.8	50.7	
Cumulative (%)		440.8	640.3	790.2	890.0	940.7	

Source: Authors' field data (2007)

Y20– Income from hunting, 0.65
Y18– Income from fuel wood, 0.51

The factor accounts for 19.5% of the variables within the Y-independent variables. This measures level of income, the development potentials of the communities and the financial stability of the dwellers and ensures a level of livelihood that may be sustainable.

Factor 3: Rural employment factor

Y17– Income from farming, 0.79
Y2 – Income from rural craft, 0.71

This factor accounts for 14.9% of the variation within the Y-dependent variables. This is so named because it measures the level of stability and movement of human capital within rural communities. The entire population of labour force mostly the aged is involved in the factor. The introduction of modern farming techniques and adequate incentives may help to check rural-urban drift of the rural-farm labour force.

Factor 4: Rural administration factor

Y15– Village hall, 0.61

This factor accounts for 9.8% of the variance within the Y-independent variables and is so named because it demonstrates the capacity of rural communities to adapt to their cultural and traditional values with socio-economic development. The level and pattern of development of communities is thus a measure of the healthy relationship between families, communities and clans. Effective administrative set in rural areas could facilitate mobilization and sensitization of the people for collective developmental activities.

Factor 5– Agricultural storage facilities factor

Y8– Agricultural storage facilities, 0.50

This factor accounts for 5.7% of the variance within the Y-dependent variables. It loads highly and positively on the variable of development of rural communities.

Table 5. Impact of socio-economic activities on rural development.

Parameters	Y1	Y2	Y3	Y4	Y5
Multiple R	0.82403	0.86513	0.84130	0.93748	0.57854
R square	0.67903	0.74845	0.70779	0.87887	0.33470
Adjusted R^2	0.62554	0.11956	0.65909	0.57606	-1.32854
Standard Error	0.61193	0.93832	0.58388	0.65111	1.52596
F	12.69339	1.19012	14.53321	2.90235	0.20124
Sig. F	0.0119	0.5154	0.0088	0.2759	0.9352

Table 6. Level of resource exploitation and rural development.

Parameters	Y1	Y2	Y3	Y4	Y5
Multiple R	0.99976	0.75649	0.99976	0.71596	0.84003
R square	0.99952	0.57228	0.98433	0.51259	0.70565
Adjusted R^2	0.99663	0.50099	0.96891	0.43136	-1.06045
Standard Error	0.5806	0.70640	0.78236	0.75408	1.43543
F	345.941	8.02788	5.19384	6.31003	0.39955
Sig. F	0.0441	0.0298	0.3236	0.0458	0.8353

Agricultural storage facilities are vital to rural development. This is due to the perishability nature of rural produce. The provision of these facilities will reduce wastage during seasons of surplus harvest.

Impacts of socio-economic activities on rural development

This relationship was investigated using a multiple correlation/regression analysis. A multiple correlation/regression analysis of the five extracted X-independent variables on each of the five extracted Y-dependent variables defines the strength of association of the two sets of variables as well as the causal relationship between the variables. For the Y1 factor defined and named as Rural Infrastructure Factor, a multiple correlation coefficient of R = 0.82402 was obtained indicating that about 67.9% of this factor is impacted by the five extracted X- (components) factors. The square of the multiple correlations co-efficient R indicates the degree of predictability of the value of Y with all the X- components as inputs. This result of the multiple correlation/regression analysis is presented in Table 5.

In the same way, the multiple correlation/regression analysis for the second Y2 factor named as rural income factor showed strength of association of 0.86513. This means that 74.8% of the total value of Y is explained by the five extracted X independent factors. This indicates that the five X-component factors as inputs for factor 2 yields a very high significant. For the third Y3 factor (Rural Employment factor) the multiple correlation coefficient of 0.84130 indicates that about 70.8% of this factor is contributed by five extracted X-independent factors. The square of the correlation coefficient of 0.70779 indicates that about 70.7% can be predicted considering all the inputs of X. For the Rural Administration (Y4) factor, the multiple correlation coefficient of 0.93748 indicates that about 87.9% of this factor is contributed by five extracted X-independent factors. The square of the correlation coefficient of 0.87887 indicates that about 87.8% of Y can be predicted considering all the inputs of X. In the same way, the multiple correlation/regression analysis for the Y5 factor named as rural health reveals strength of association of 0.57854 as a multiple correlation coefficient R. This is about 33.5% contribution to the factor named as rural health factor. This is not very significant as F = 0.9352.

Level of resource exploitation and rural development

This relationship was investigated using the six X-variables (X30 – X35) to test the relationship between the rate of exploitation and rural development in the study area. Multiple regressions of the six X-independent variables were carried out on the five-y extracted development factors and the result is presented in Table 6. For Y1 factor defined and named as rural infrastructure, a high multiple correlation coefficient (R = 0.99976) was obtained. The square of the multiple correlations co-efficient is 0.99952 indicating that 99% of the total value of Y can be predicted with the six variables as inputs. This result is very significant as F = 0.0441. In the same way, the multiple correlation/regression analy-

Table 7. Number of rural infrastructure per community.

Infrastructure	Ifa Ikot Akpan	Ekpiri Nsukara	Nsukara Offot	Nung Asang	Ikot Ofong Ikono	Ndue Otong Ikono	Ndue Otong Oku	Iba Oku
School	3	2	1	3	1	1	2	2
Health	1	-	-	1	-	-	-	-
Skill Acquisition centre	-	-	-	1	-	-	-	-
Number of paved road	-	-	-	1	-	-	-	-
Village hall	1	1	1	1	1	1	-	1
Local industry	4	3	3	5	2	-	3	-
Adult literacy centre	-	-	-	1	-	-	-	-
Market	1	1	-	1	1	-	-	-
Total	10	7	4	13	5	2	5	3
Percentage (%)	20	14	12	26	10	4	10	6

Source: Authors' field Survey (October, 2007).

sis for all the factors yielded very high coefficients. The error term for Y5 was however positively high thereby indicating insignificant relationship with factors on the rate of resource exploitation.

DISCUSSION

The relationship between socio-economic activities and rural development in Uyo Local Government Area is expressed through the results of the factor analysis and multiple correlation/regression models.

Rural infrastructure

This can enhance efficient, effective and sustainable exploitation and utilization of rural resources. It also makes the society liveable. Data collected revealed the state of infrastructural development in the study area. Infrastructural development is generally poor in all the communities studied except for Nung Asang and Ifa Ikot Akpan that have some significant advantage (Table 7). This is likely due to abundance of natural resources. While infrastructure alone cannot induce economic growth, its absence is likely to constrain development especially in farm productivity and development of off-farm activities. This finding agrees with Aluko (1990) and Ekong (2003) who reported that infrastructures such as schools, health care facilities, transportation and roads will enhance rural dwellers production, distribution and consumption activities and ultimately the quality of their lives.

Rural income

The majority of rural dwellers derive their incomes from primary production, namely agriculture and extractive activities. Those with large farm sizes or plots tend to have income in a year. The reverse is true for those having smaller farm sizes. The agricultural system in rural communities of Uyo Local Government Area is dominated by (a) subsistence where market exchange is only incidental and (b) by farm size fragmentation due to land tenure system. Stating the obvious, most rural dwellers can rarely get rich as their incomes in many cases are not commensurate with the tasks involved in production and extractive processes. According to Olayide (1981), subsistence farming is oriented primarily to meeting the consumption requirements of the farmers and their households and as such any market exchange that takes place is peripheral and incidental. Similarly, Ekong (2003) and Anijah-Obi (2003) are of the view that subsistence farming is characterized by relatively isolated holdings, customs, rudimentary production techniques, tendency to self-sufficiency with regards to food production, and absence of any significant measure of economic calculations by producers.

Rural employment

Another development factor, which was significant in these communities, is rural employment. Data on employment distribution for farm activities in the study area was collected. Although every respondent is engaged in farming and other activities, the study reveals that farm labour is scarce and expensive (Table 8). The young and educated rural dwellers are not attracted to this form of employment given the relatively low status and prestige attached to farming occupation. Farming in these communities is dominated by aged who lacked the stamina to work the land effectively and thus, relied on hired labour to supplement family labour. Okoji (2000) in his study of Agriculture in South Eastern Nigeria agrees with this observation. Okoji maintained that the family constituted the major source of farm labour supply but

Table 8. Employment distribution in farming activity per community.

Community	Number of respondents (farmers)	Number employed in farming	Percentage of employed
Ifa Atai	76	228	25.3
Ifa Ikot Akpan	26	52	5.8
Ekpiri Nusara	55	57	6.3
Nsukara Offot	116	232	25.8
Nung Asang	36	108	12.0
Ikot Offong Ikono	19	57	6.3
Ndue Otong Oku	43	129	14.3
Iba Oku	18	36	4.0
Total	389	899	100

Source: Authors' field survey (October, 2007).

present supplies from this source have been drastically reduced and as a result farmers are increasingly dependent on hired labour, however, because of their lean financial resources, labour has become the most expensive and most restrictive of the factors of agricultural production in the region. Similarly, Todaro and Smith (2005) list three forces restricting the growth of output of rural subsistence farming to include labour scarcity during the busiest part of the growing season, planting and weeding times. They found out that the demand for workers during the early weeks of this season usually exceeds all available rural labour supplied.

Rural administration

Another development factor is rural administration. The rural communities' shared beliefs are nurtured and cherished by the people. This is expressed through restriction to sacred forests, streams and exploitation of palm fruits for some specified periods. There exist kinship relations, mutual trust and clean environment, built around local administrative rules and custom. An examination of this factor however reveals some lapses especially in the enforcement of restrictions covering the exploitation of certain naturally occurring resources. For instance, sand, gravels and laterite in some of these communities are extracted with little or no regards to its adverse effects on the environment. Streams, the main source of water supply in these communities, are polluted. Village roads are seriously damaged by heavy duty tippers which carry the sand and gravels. Also communities' forests are fast disappearing due to uncontrolled logging and firewood exploitation. The administration should strive to harmonize effective use of its natural resources with conservation and protection of its environment as cherished by its customs and tradition.

Rural health

The fifth factor, which is rural health, is crucial to human socio-economic activities. In fact, it constitutes a significant impact on primary production and other related activities. A healthy society is a productive society and the richness of development depends on how healthy the populaces are. Health indicator in Uyo rural communities is very poor as exemplified by the absence of orthodox health facilities. A close examination of the factor reveals a relative reflection and influence of belief and value system of rural dwellers. The rural communities rely heavily on indigenous and traditional health practices and their herbal knowledge than orthodox medicines. They prefer consulting the herbalists and prayer homes. This likely explains a low rating of health factor in socio-economic activities. However, a sub-group engaging in the extractive activities such as sand, gravel and logging seems to be more conscious of health need to maintain fitness for their strenuous work. This create in them appreciation for proper medical services. This subgroup tends to rely on the orthodox medicine; hence they visit the patent medicine shops where available for substitute to the herbal medicine. It should be noted that farming and extractive activities demand a great deal of energy and without good health there will be no stamina to carryout cultivation processes. Good health care facilities are therefore essential for any productive and sustainable development activities.

Conclusion and recommendations

The research focused on the numerous links between the community and environment that affects rural development. It highlighted major factors that significantly influence rural development in the study area using chosen statistical models. The implication of the results is that there are five major factors that could have significant influence on rural community development and thus, suggest the articulation of policies towards such issues. The farm activities factor has the potentials to influence rural development and therefore the existing customary land tenure practices, which are at variance with larger farm sizes, should be checked to permit acquisition of larger farm plots for agricultural activities. Other activities such as labour, farm activities, cropping, fuel wood and fallow period dominate this sector. It has a paramount influence and needs a focus of rural development policy. In the health sector, it is recommended that collaborative efforts involving the public and private sectors should be encouraged. Adequate provision should be made for safe drinking water supply in rural communities while access to healthcare facilities should be made free for the rural populace. Above all, Sanitary Inspectorate Department should be set up to monitor levels of sanitation in rural communities. Given that rural infrastructure generate a number of positive externalities and facilitate rural commerce and income, the study recommends public-private partnership in the provision and maintenance of basic rural infrastructure. The exploitation of common resources such as sand and gravel could improve rural economy and as such should be carried out under the supervision of village development committee in order not to undermine the sustainability of the environment in the process of resources exploitation as well as to channel the monetary gain into useful community oriented development projects that will enhance welfare of the community as a whole rather than individual gain.

REFERENCES

Aluko SA (1990). Industry in the Rural Setting. In Otite & Ikali (Eds) Readings in Nigerian Rural Society. Heinemann Educational Books (Nig.) Ltd; Ibadan.

Aniah EJ (1995). Environment and Development Perspectives. In Lawal N (Ed.) Education for Sustainable Development. Macmillan Nig. Pubs. Ltd; Ibadan.

Anijah-Obi FN (2003). Fundamentals of Environmental Education and Mgt. University of Calabar Press,Calabar.

Ekong EE (2003). Introduction to Rural Sociology. Dove Pubs; Uyo

Evans HZ (1986). Employers and the Environmental Challenges. ILO, Switzerland.

Kohlmeyer C (2001). Rural Development: A Reference Framework. Federal Ministry of Economic Coop. & Dev.

NEST (1991). Nigeria's Threatened Environment: A National Profile. NEST, Ibadan.

Okoji MA (2000). Agriculture. In Inyang I. B (Ed.) South-Eastern Nigeria: Its Environment. Abaan Press, Kaduna.

Olayide NO (1981). Elements of Rural Economics. University of Ibadan Press, Ibadan.

Sampson ES (1992). Energy Resources. In Gleav MB (Ed.) Tropical African Development: A geographical Perspective. Longman, London.

Todaro MP, Smith SC (2005). Economic Development. Pearson Educational Pubs; India.

World Bank (1996). Towards Environmentally Sustainable Dev. in Sub-Saharan Africa: A World Bank Agenda. Washington DC.

15

Exploring opportunities for climate change adaptation in semi arid areas of Tanzania: A case of Nzega District in Tabora region

Majule, A. E., Kauzeni, A. S and Mujwahuzi, M.

Institute of Resource Assessment (IRA), University of Dar es Salaam, Box 35097, Dar es Salaam, Tanzania.

Climate change and associated impacts are now widely acknowledged by most communities, institutions and organizations in the World to be affecting people's livelihoods. The University of Dar es Salaam in Tanzania through the Institute of Resource Assessment (IRA) implemented a three years research program to study the implications of climate change on natural and social systems in three agro ecolocilal zones in Tanzania. The present paper presents finding obtained from one of the zones studied namely semi arid central and western part. The study was conducted in two villages namely Upungwe and Mbogwe located in Nzega District, Tabora region. A sample size of 99 people; that is, 10% of the total number of households was used for household interviews. A total of 40 people strategically selected were involved in focus group discussion, 20 from each village. Findings show that different ethnic groups have moved into villages in response to climate change impacts in their areas of origin. Climate change impacts in particular increased pests associated with temper rise and resulted in loss on various crop yields ranging from 46 to 80%. Most crops affected are in the order cotton, rice, cassava, sweet potatoes, groundnuts and maize, respectively. Different adaptation options were also reported but the most important was found to be different forms of linkages such as rural-urban reported to be more viable by 39 to 68% of respondents while 48 to 62% reported such linkages to be viable on adaptation. To strengthen adaptation rural-urban, urban-rural and rural-rural linkages needs to be facilitated so that products and information can flow from all ends.

Key words: Adaptation, climate change, Mbogwe, Upungu, linkages.

INTRODUCTION

Global change, especially climate change, is one of the greatest challenges to human beings and natural systems in the 21th century (Hulme, 2001; IPCC, 2007). Although climate change is a global phenomenon, the distribution of impacts remains uneven, and will be severely felt by vulnerable communities with few safety nets and limited capacity to absorb climate variability and related shocks (Adger et al., 2005). In eastern and Southern African countries including Tanzania and Malawi, the National Adaptaion Programs of Action have identified agricultural sector to be one of the most vulnerable sectors to impacts associated with climate

Table 1. Sample size distribution used in this study.

Data collection tools	Village		Total
	Upungu	Mbogwe	
Household survey	48	51	99
Focus Group Discussion	20	20	40
Discussion with key informants	5	5	10
Transect walk	20	20	40
Total sample size			**189**

change (Lema and Majule, 2009; Kalinga-Chirwa et al., 2011).

Societal vulnerability to the risks associated with climate change may exacerbate ongoing social and economic challenges, particularly in those parts of societies dependent on rainfed agriculture that is more sensitive to changes in climate (Mongi et al., 2010). Risks are apparent in agriculture, fisheries and many other socio-economic components that constitute the livelihood of rural populations in developing countries (Watson et al., 1996). In most cases, documented vulnerability is in the form of severe and repeated droughts, unpredictable flood and irregular rainfall. For example droughts and floods result in crop damages and failure (Liwenga, 2003; Nelson and Stathers, 2009).

The traditional rain-fed subsistence agriculture is extremely vulnerable to changing patterns of climate variability such as through shifts in growing season conditions. Since most rural communities rely on their agricultural produce to generate income, poverty is directly coupled to agricultural production. It is however not well understood how households or communities cope or adapt to climate induced extreme variability. Similarly, sea water rise is a result of climate change and increased temperatures, which lead into the flooding of coastal lines often resulting in disasters.

Different studies have shown that all societies are fundamentally adaptive and there are many situations in the past when societies have adapted to changes in climate and to similar risks (Paavola and Adger, 2005; Gwambene, 2007). But some sectors are more sensitive than others and some groups in society are more vulnerable to the risks posed by climate change than others. Yet all societies need to enhance their adaptive capacity to face both present and future climate change outside their experienced coping range (Watson et al., 1996; McCarthy et al., 2001).

Experience suggests that the best way to address climate change impacts on the poor is by integrating adaptation responses into development planning. This is fundamental for achieving the Millennium Development Goals (World Bank, 2005). Effective adaptation strategies should build upon, and sustain, existing livelihoods and thus take into account existing knowledge and coping strategies of the poor. As pointed earlier, both social economic and biophysical characteristics as well as social infrastructure and linkages between two or more social economical settings determine societal vulnerability.

In agriculture sector in particular crop farming systems more recommendation has been made on using most drought resistant crops such as sorghum (Majule et al., 2009) without considering on other options such as the introduction of new crops.

In light of the issues raised, this study was conducted in a semi arid area of Tanzania in particular to examine implications of climate change on community livelihoods and the role of current existing adaption options in the agricultural sector as a response mechanism to climate change impacts.

MATERIALS AND METHODS

A multidisciplinary approach was used to generate relevant information from the different disciplines in the study area. Socio-economic data were collected through qualitative methods using participatory research methods including focus group discussions (FGD), discussion with key informants and household surveys in two selected villages.

The use of qualitative methods is among the evolving approaches in climate change studies, including perceptions, impacts, vulnerability, and adaptation (Gwambene, 2007; Majule et al., 2009). The various participatory rural appraisal (PRA) techniques used are the same used by Kangalawe et al. (2005). Before field work, a detailed literature review was undertaken on similar kind of work relevant to semi arid conditions. Table 1 presents the sampling size distribution employed in this study.

To understand the impacts of climate change and variability on community livelihoods, a sustainable livelihoods framework was used. The sustainable livelihoods framework (Frankenberger, 2000) presents the main factors that affect people's livelihoods, and typical relationships between these. Vulnerability of crops to climate change was considered within the framework in terms of trends, and seasonality affecting productivity.

This was explored particularly with respect to climate change and variability e.g. (drought, increasing incidence of crop pests and diseases) and associated impacts in terms of % loss of major crops over the last 20 years. In this case, average crop yield in bags per acre (each with 100 kg) over the last 20 and the last 5 years ago obtained through household surveys to a sample of 99 households

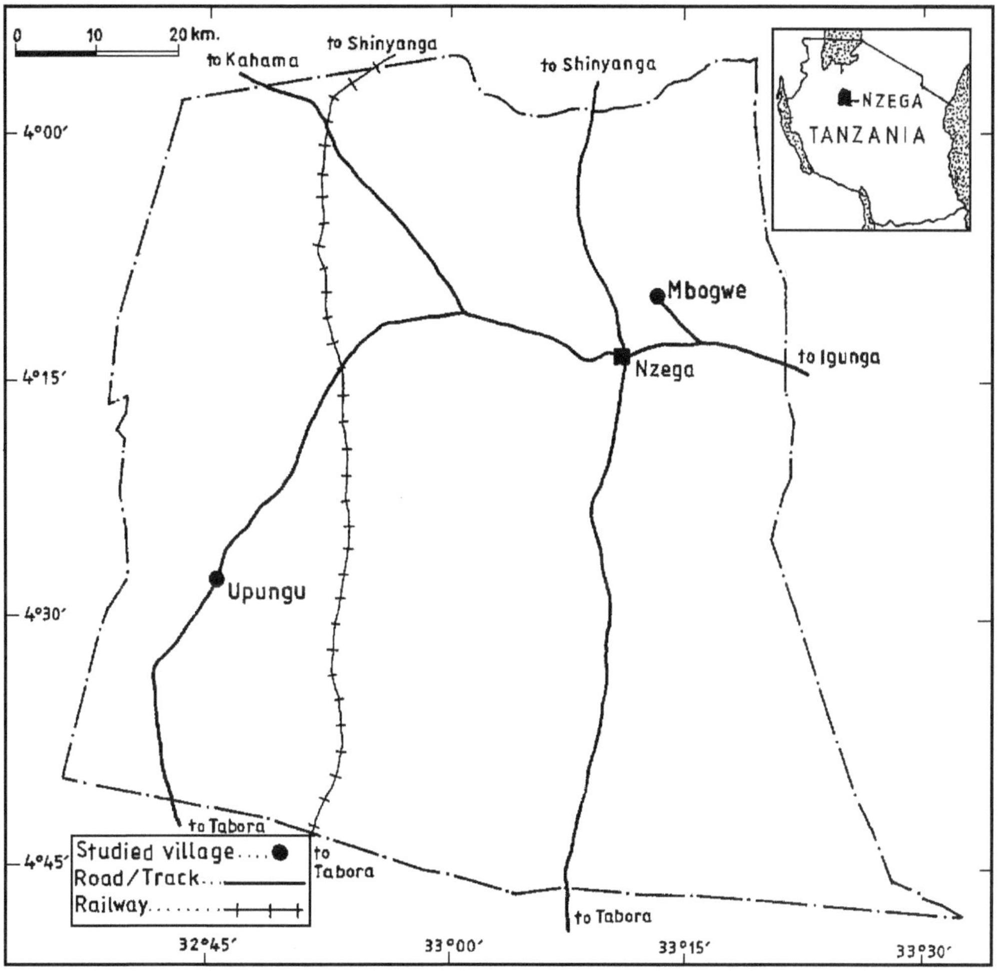

Figure 1. Location of study area in Nzega, Tanzania showing sample villages.

were analyzed and compared. In this case household members were requested to provide estimates of crop yields in relation pest incidences attributed by climate change. Household data was analyzed using appropriate statistical softwares commonly known as SPSS and results were presented in the form of Tables, Figures and descriptive statistics.

RESULTS AND DISCUSSION

Characteristics of the study area

Upungu village is located in Puge Ward, in Puge Division, Nzega District, Tabora Region located at 59 kilometers south of Nzega town on the main road to Tabora town. Upungu village is composed of five sub-villages namely Upungu East, Upungu West, Ntungulu, Ugawe and Kisumo. The village was established in 1974 under the villagilization (*Ujamaa*) program. Historically Nzega district is the six districts of Tabora region and it covers an area of 7,319 square kilometers, located between 32°30'-33°30' longitudes east of Greenwich and Latitude 3°45'-500' south of the equator. The district is located in semi arid areas of Tanzania where drought and poor soil conditions are limiting factors for agriculture development (De Pauw, 1984). A number of climate and environmental studies conducted in the area describe the area to be among the most vulnerable areas to climate change in Tanzania (Mongi et al., 2010; Kaijage, 2012; Muzo, 2012). Figure 1 presents the location of study area in Tanzania and specific villages used as a case study. the name "Upungu" is reported to have originated from the observed decrease in rainfall in the area (*Mvua kupungua*) which led to a food shortage (*Upungufu wa chakula*). This could BE a reflection of initial perception by communities on climate change and associated impacts.

On the other hand, Mbogwe village is located in

Table 2. Origin of community members in study villages.

Major reason	Upungu village		Mbogwe village	
	Frequency	%	Frequency	%
Born in the village	22	45.8	30	58.8
Born outside the village but within the district	12	25.0	16	31.4
Born outside the district but within the region	07	14.6	03	05.9
Born outside the region	07	14.6	02	03.9
Total	48	100.0	51	100.0

Table 3. Major reasons for in migration into study villages by %.

Major reason	Upungu village	Mbogwe village
Search for agricultural land	50.0	50.0
Search for business opportunities	33.3	96.2
Following relatives	71.0	90.0
Escaping unfavorable climatic conditions	57.3	50.0
Looking for employment	62.5	33.3

Mbogwe Ward, in Nyasa Division and is located 14 km east of Nzega town off the main road to the Dar es Salaam region. Mbogwe village is composed of ten sub-villages namely *Mbogwe kusini* (south), *Mbogwe kati* (middle), *Mwitinde*, *Mwakabilu, Kalangae*, *Budutu, Mwanubila, Mwangongo, Shapela* and *Bugegemela*. The village was established in 1929. The name "Mbogwe" is reported to have originated first from "*Mbogo*" (buffalo) who used to come and drink water in the Kilimi dam located within the village area. To date, the has dried off and buffalo are no longer coming due to environmental change and drying of the dam.

The socio economic profile of study villages

Ethnicity

A number of parameters constituting to this were studied and findings indicate that Upungu village is composed of three main ethnic groups namely Nyamwezi (66.7%), Sukuma (16.7%) and Tutsi (4.2) from Rwanda. Other minor tribes include Nyaturu, Chagga, Hehe and Kimbu and all together constitute 6.3% in total. There were 801 households in the village with a total population of 5,978 composed of 2,900 males and 3,078 females. On the other hand, 52 % in Mbogwe village were Nyamwezi while 44.2% and 3.8% were Tutsi from neighbouring country (Rwanda). Such a diversity of ethnic has enabled a range of livelihoods activities to exist in the study villages.

Family sizes and origins of community members

At community level, findings reveal that the number of family sizes varies in both villages as follows; i) those with between 1 and 3 members constitutes 31.3 and 25% in Upungu and Mbogwe villages respectively. Those with family sizes having four up to 6 members constituted 39.5 and 46.1% in Upungu and Mbogwe villages respectively. Those with more than six members were 29.2 and 28.9% in Upungu and Mbogwe villages respectively. Results suggest that the majority of the people in study areas had family sizes ranging from 4 to 6 members. On the other hand, family sizes in both villages follows the same pattern since families having more than 6 members are almost the same in both villages. The finding represents a typical African household size to support labor requirement based on subsistence farming as reported by Muzo (2012) and Kaijage (2012) in Nzega district.

With regards to origin of people in the study villages, some tribes have moved into the villages for different reasons (Table 2). Findings show that most of the community members (nearly 50%) were born within their villages and other migrated to the villages from different places due to several reasons. Discussion with key informants in Upungu revealed that those who migrated in the village came for the purpose of searching agricultural land and also for seeking employment to livestock keepers' especially grazing livestock.

However based on household survey, a number of reasons for migrating into the two villages emerged and are presented in Table 3. The reasons are generally common to both villages with few exceptions due to the location of a place and also existing opportunities. For example, 50% of respondents in both villages moved into the village for agricultural purposes while a significant large number of people moved into the Mbogwe village for business reasons. This is because Mbogwe village is

Table 4. Major livelihood activities in Mbogwe village.

Relevance	Major economic activities					
	Crop cultivation	Agro-pastoralism	Wage employment	Casual labour	Petty business	Crop trading
Very important	84.8	19.0	0.00	60.0	35.7	50.0
Important	15.2	52.4	6.30	20.00	28.6	50.0
Less Important	0.0	28.6	93.7	20.00	35.7	00.0
Total %	100.0	100.0	100.0	100.0	100.0	100.0

Table 5. Major livelihood activities in Upungu village.

Relevance	Major economic activities					
	Crop cultivation	Agro-pastoralism	Wage employment	Casual labour	Petty business	Crop trading
Very important	95.9	31.6	50.0	09.1	09.1	33.3
Important	04.1	63.2	50.0	54.5	36.4	66.7
Less Important	00.0	05.3	00.0	36.4	54.5	0.00
Total %	100.0	100.0	100.0	100.0	100.0	100.0

Table 6. Productivity of different crops over years in Upungu village.

Production pattern	Different crop types						
	Maize	Sorghum	Rice	Sweet potato	Groundnuts	Sunflower	Cassava
Increasing	13.1	25.0	08.3	07.7	03.1	20.0	00.0
Decreasing	65.2	50.0	75.0	53.8	75.0	20.0	61.1
Fluctuating	21.7	25.5	16.7	38.5	21.9	60.0	38.9
Total	100	100	100	100	100	100	100

very close to Nzega town which has large opportunity for making different small scale business.

Upungu village has large opportunity to invest in agriculture due to well developed vegetation cover including miombo woodlands and fertile soils which attracts agricultural activities to take place. The findings are not surprising because other studies have also indicated that in order to cope and adapt to climate change impacts, some community members have tended to move to suitable areas (Kangalawe et al., 2005).

Major economic activities and their relevance to adaption

The major socio-economic activities include agriculture and livestock keeping and small business activities. Major food crops produced are: maize, rice, groundnuts, bambaranuts, sweet potatoes, cassava, sorghum and vegetables (onions, tomatoes and spinach). Most of these crops and vegetable grown are local and were reported to be vulnerable to climate change impacts. Cash crops being produced are sunflower, tobacco, and cotton. Livestock kept include cattle, goats, sheep, pigs, chicken, ducks and donkeys as draught animals.

Non farm activities in the village include charcoal production, carpentry, bicycle repair, food vending, production of local brew and other small scale businesses such as running shops, selling of crops, provision of transport services (by use of motorcycles and bicycles), entertainment (by offering TV shows) and casual labor (Tables 4 and 5).

Vulnerability of common crops to climate change

Household interviews in Upungu village results showed that farmers have been experiencing fluctuating productivity of crops and mostly a declining trend in crop production (Table 6). Community's perceptions on productivity of different crops over the years are that this state of affairs has been caused by climate change and variability which lead to the breakup of pests and diseases attacking different crops.

Table 7. Perceptions on the productivity of different crops over years in Mbogwe village.

Production pattern	Different crop types						
	Maize	Sorghum	Rice	Sweetpotato	Groundnuts	Sunflower	Cassava
Increasing	10.4	40.0	00.0	04.8	00.0	25.0	00.0
Decreasing	79.2	38.0	80.8	66.7	75.0	50.0	83.3
Fluctuating	10.4	22.0	19.2	28.6	25.0	25.0	16.7
Total	100	100	100	100	100	100	100

Table 8. Impacts of crop pests on yields of different crops (kg/acre) Upungu village.

Type of crop	Average yield before pests*	Average yield with pests**	Loss incurred	% Loss
Maize	1131	435	696	62
Rice	2033	683	1350	66
Cassava	1525	725	800	52
Cotton	n/a	n/a	n/a	n/a
Groundnuts	1356	488	869	64
Sweet potato	1325	500	825	62

Source: Field data (2011).

Table 9. Impacts of crop pests on yields of different crops (kg/acre) Mbogwe village.

Type of crop	Average yield without pests	Average yield after pests	Loss incurred	% Loss
Maize	809	436	373	46
Rice	2336	647	1689	72
Cassava	480	155	325	68
Cotton	479	97	382	80
Groundnuts	763	313	450	59
Sweet potato	675	258	418	62

The findings for Mbogwe village are quite similar to those for Upungu village (Table 7). The analysis based on 100% of the house hold interviewed in this case showed that most vulnerable crops are maize and rice which require moisture availabitiy during their growth circle. Less vulnerable crops include sorghum and sunflower.

Impacts of pest on major crops

Climate change and variability have been reported to affect different cash and food crops in Tanzania (Lema and Majule, 2009). Studies conducted by Majule et al. (2009) in semi arid areas showed that major loss in crop production is associated to prolonged drought, attack by pests and diseases and also damage caused by flood. Household interviews in this study also revealed that recently increase pests and crop diseases has contributed significantly to a decrease in crop productivity in Nzega district. Table 8 presents rough estimates of crop loss in kg/acre and also the % loss due to crop pests for different crops. The impact of pest is assessed and discussed; it was easy to remember and mention a particular pest as compared to diseases.

In general (Table 9) but with an exception of maize yields in Mbogwe village, pest incidences result in a loss of more than half of the yields (i.e above 50%). Highest loss is on rice probably due to drought indicating increasing dry spells in both villages (see Mongi et al., 2010). Cotton crop is very sensitive to pest and thus a high loss in Mbogwe village.

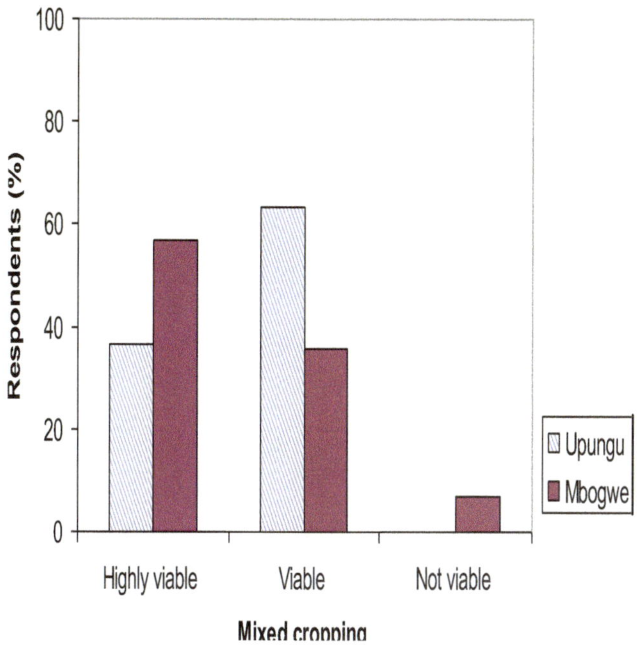

Figure 2. Viability of mixed cropping.

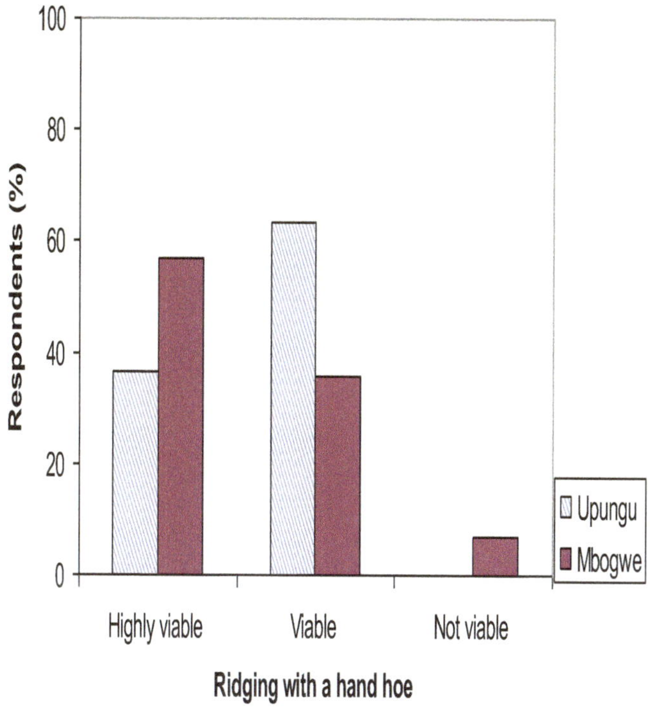

Figure 3. Viability of ridging.

Climate change adaptation and sustainability of strategies

Mixed cropping

Mixed cropping involves growing different types of crops on the same piece of land either on the same plot or different plots yet on the same piece of land. This is done in order to evade the risk loss of all crops to harsh climate change and variability impacts. The results from interviews conducted in Upungu and Mbogwe villages were that over 38 and 58% of respondents respectively said that mixed cropping was highly viable as an adaptation strategy in cases of climate change and variability impacts.

About 62 and 38% of respondents of the same villages respectively said that mixed cropping was a viable strategy for adaptation to climate change and variability impacts and 10% of respondents from Mbogwe village had the opinion that mixed cropping was not a viable strategy for adaptation to climate change impacts as indicated in Figure 2.

Ridging with a hand hoe

Ridging with a hand hoe is done by using cheap farming implements. Moreover ridges have a higher water retention capacity than flat cultivation which is ideal for areas with low rainfall records. During the interview, 38 and 59% of Upungu and Mbogwe villages respectively feel that ridging with a hand hoe is highly viable strategy for adaptation to climate change impacts as indicated in Figure 3 by over 61 and 38% of respondents in Mbogwe and Upungu villages respectively.

Small scale irrigation

Irrigation farming provides an opportunity for growing crops all-year round even when there is a drought or failure of the rains. About 50 and 59% of Upungu and Mbogwe village residents respectively pointed out that small scale irrigation is a highly viable strategy for adaptation to climate change and variability impacts, while about 40 and 18% of Upungu and Mbogwe residents respectively were of the opinion that small scale irrigation was a viable strategy for adaptation to climate change and variability impacts. However, 10 and 30% of Upungu and Mbogwe residents said that small scale irrigation was not a viable strategy for adaptation to climate change and variability impacts.

Crop rotation

Crop rotation which involves change of crops planted on same piece of land for in successive years can serve as

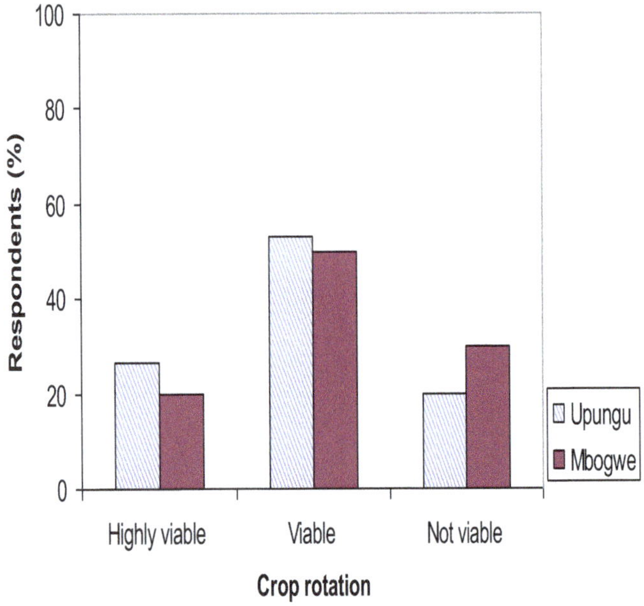

Figure 4. Viability of crop rotation

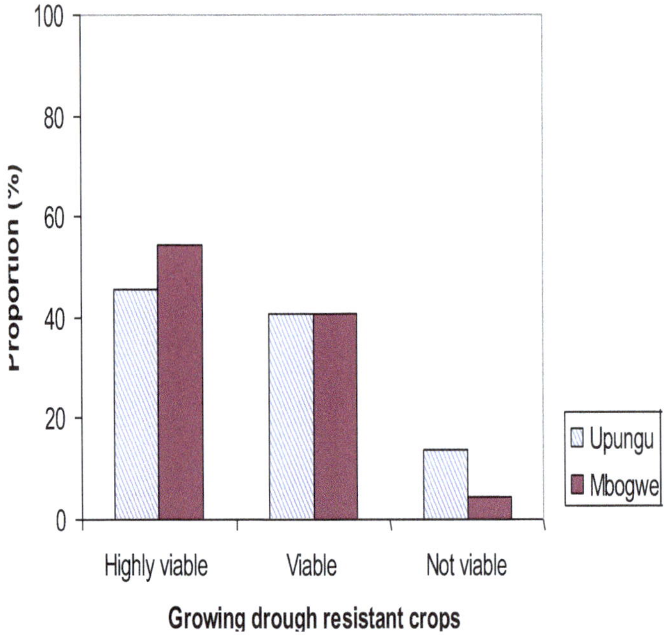

Figure 5. Viability drought resistant crops

a strategy for adaptation to climate change and variability impacts. About 23 and 20% of communities in Upungu and Mbogwe village respectively see crop rotation as a highly viable strategy for adaptation to climate change and variability impacts, while 55 and 50% see crop rotation as a viable strategy for adaptation to climate change and variability impacts as shown in Figure 4. At the same time, 19 and 30% of residents of the same village respectively do not regard crop rotation as a viable strategy to climate change and variability impacts. Crop rotation has also been reported to be a potential cropping practice in semi humid areas of southern highland Tanzania and other parts of east Africa (Majule and Mwalyosi, 2005; Maitima et al., 2009).

Adoption of drought resistant crops

This is also among potential adaptation option and farmers in those areas needs to grow drought resistant crops. The findings show that 43 and about 54% of Upungu and Mbogwe respondents respectively regard growing drought resistant crops as a highly viable strategy for adaptation to climate change and variability, while about 41% of respondents from each of Upungu and Mbogwe villages regard growing drought resistant crops as just a viable strategy for adaptation to climate change and variability (Figure 5). About 10 and 4% of Upungyu and Mbogwe village's residents respectively said growing drought resistant crops is not a viable strategy for adaptation to climate change and variability impacts.

Early planting and early maturing crops

The majority of respondents (more than 50%) were of the opinion that early planting is a higly viable strategy for enhancing adapatition of small farmers in study villages and this was further supported by approximately 40% of respondents who sees that practice to be viable (Figure 6). This suggests intervention with regard to early planting to be in place in collaboration with the meteorological deparment in Tanzania to inform communities on onset of rain. Figure 7 also clearly shows that planting early maturing crops or plants is a vaiable solution to farmers to escape a dry spell as reported by Mongi et al. (2010) in the region.

Dependence on wild food and non wood forest products

During years of serious food shortage or famine, people do not have a wide range of types of food to eat. Anything that is not poisonous but can sustain their lives will be consumed or eaten. During the interviews, about 21 and 10% of Upungu and Mbogwe residents respectively said that eating wild food is a viable strategy for adaptation to climate change and variability impacts (Figure 8). The same figure also shows that 62 and 43%

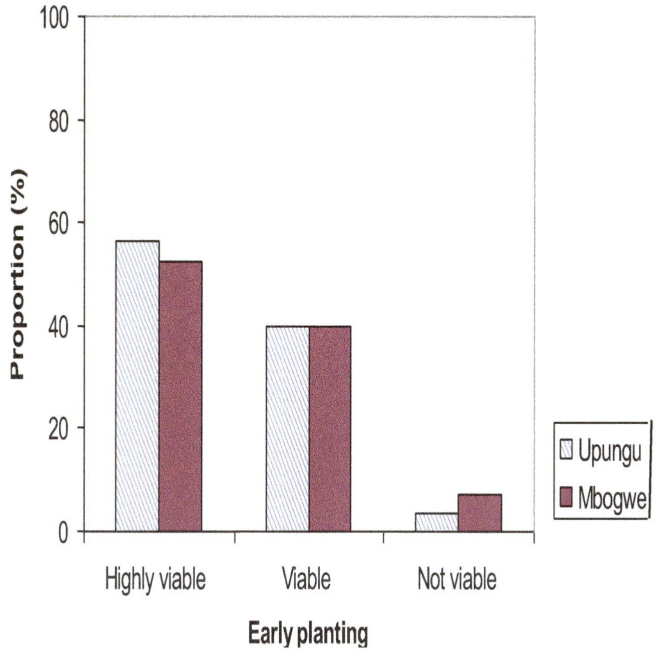

Figure 6. Viability of early planting.

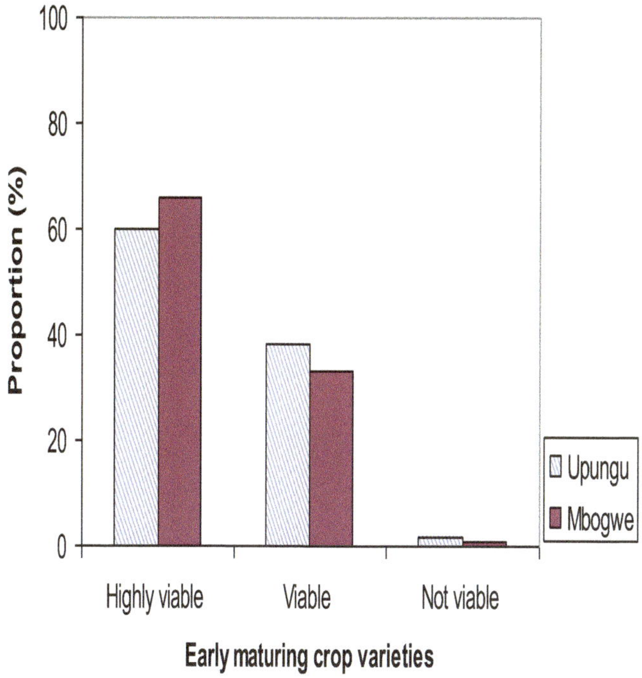

Figure 7. Early maturing crops

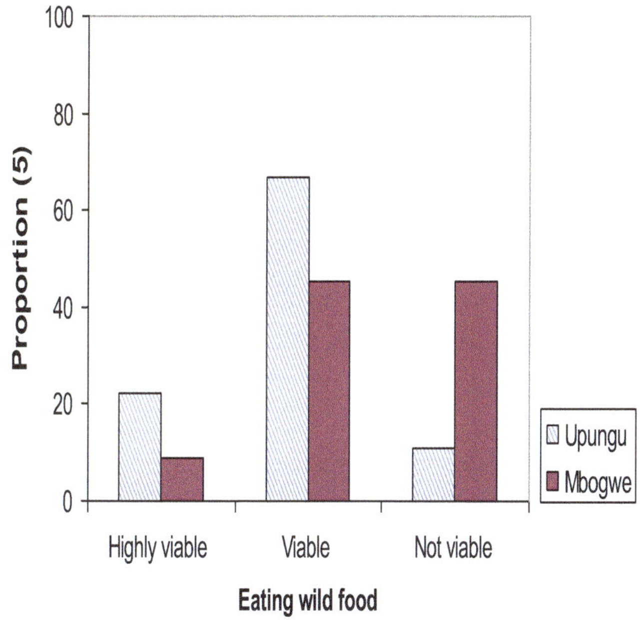

Figure 8. Viability of eating wild food.

of the respondents from Upungu and Mbogwe respectively were of the opinion that eating wild food is just a viable strategy for adaptation to climate change and variability impacts. More over 10 and 43% of Upungu and Mbogwe residents respectively said eating wild food is not a viable strategy for adaptation to climate change and variability.

Forests are major sources of timber and non-timber products. Non-timber products include edible fruits, tubers, mushrooms, etc which can be eaten during serious food shortage periods (Majule et al., 2009). Forests provide raw materials, maintain biodiversity, protect land water resources and play a role in climate change mitigation.

Use of forest products was investigated in both Upungu and Mbogwe villages with reference to viability as a strategy for adaptation to climate change and variability. Findings have shown that 8 and 80% of village residents in Upungu and Mbogwe respectively (Figure 9) said that use of forest products is a highly viable strategy for adaptation to climate change and variability impact, while about 50 and 20% of residents respectively of the same villages said that use of forest products is a viable strategy for adaptation to climate change and variability impacts.

Climate change adaptation linkages

Rural-Urban, Urban-Rural and Rural-Rural linkages observed in Upungu village involve an intricate interactions of social, economic, and political nature.

Rural-urban linkage

The movement of people from the rural areas to surround-

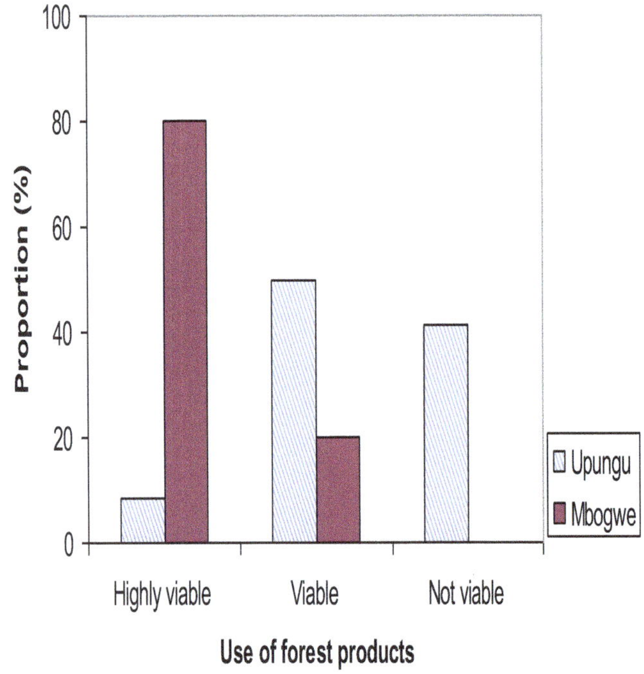

Figure 9. Use wild forest products.

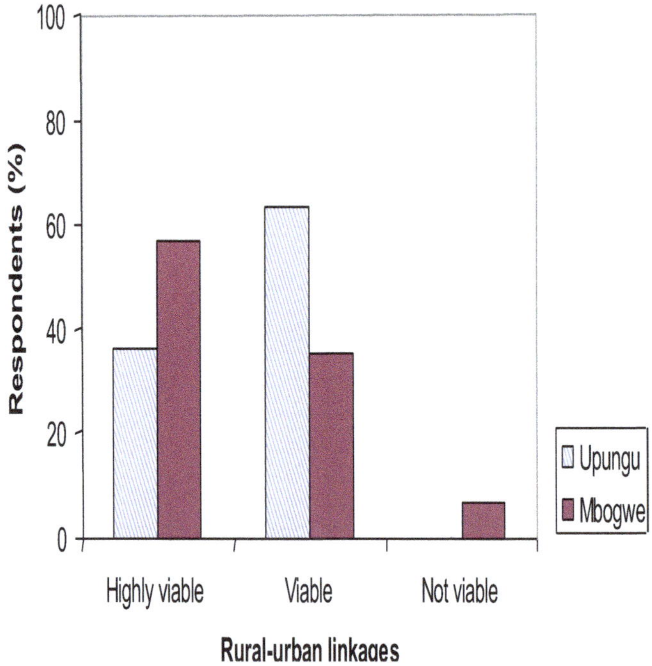

Figure 10. Viability of rural-urban linkage on adaptation.

ing town in search of employment and social services as well as social amenities can serve as a strategy for adaptation to climate change and variability impacts. Rural-urban linkages are regarded as a strategy for adaptation to climate change and variability.

About 39 and 68% of Upungu and Mbogwe village residents respectively concur that rural-urban linkages are highly viable as an adaptation strategy to climate change and variability impacts, while about 62 and 48% of residents respectively from the same village regard rural-urban linkages to be viable as adaptation strategies for climate change and variability impacts. From Mbogwe villages, less than 5% does not regard rural-urban linkages as adaptation strategies for climate change and variability impacts.

The rural-urban linkages involves: people moving from the rural areas to surrounding town in search of employment and social services as well as social amenities especially the youth; i) taking agricultural and livestock products (food and cash crops, chickens, goats, etc) and forest products (charcoal and firewood) to town for sale and ii) purchasing their requirements which are not available in their rural areas; conducting banking activities.

This study has established that such linkage provides an opportunity for strengthening involved parties or communities to adapt to impacts associated with climate change. Figure 10 presents responses from communities on the viability of such linkage on adaptation.

The movement of people from the rural areas to surrounding town in search of employment and social services as well as social amenities can serve as a strategy for adaptation to climate change and variability impacts. Rural-urban linkages are regarded as a strategy for adaptation to climate change and variability.

About 39 and 68% of Upungu and Mbogwe village residents respectively concur that rural urban linkages are highly viable as an adaptation strategy to Climate change and variability impacts, while about 62 and 48% of residents respectively from the same village regard rural-urban linkages as adaptation strategies for climate change and variability impacts. From Mbogwe villages, less than 5% does not regard rural-urban linkages as adaptation strategies for climate change and variability impacts.

Nearly 60% of respondents in Mbogwe village indicated that such linkage is highly viable and this was due to; i) selling of different products to nearby Nzega town; ii) transport to people to and from Nzega town as well as; iii) selling of different products from the village and also buying different products from urban area. On the other hand, in Upungu village such s linkage was also considered to be viable with more than 60% of respondents indicating so.

The urban-rural linkages

Based on FGD, the Urban-Rural linkages involve selling to rural communities industrial products which in the majority of cases are produced in towns such as agricultural inputs (inorganic fertilisers, pesticides, insecticides and implements), clothes, domestic utensils and drugs, etc); purchasing of products produced in the rural areas; offering social services (education, health services, extension services, etc) to rural communities; transmission of information (newspapers, radio and Televisions); financial remittance to relatives living in the rural areas and payments of salaries to workers in the rural areas.

The rural-rural linkages

Following discussions with key informants and FGD, the Rural - Rural linkages involve people moving from one place to another in search of pasture for their livestock; search of arable land for cultivation of crops; moving from densely settled area to sparsely populated areas; joining relatives living in other rural areas, marriages, etc.

There is great influx of particularly youth from rural areas to urban areas seeking for employments, casual labour and selling their rural produced products to urban markets. This movement is mainly seasonal. However, there are people moving into the community mainly for visiting relatives, establishing new economic activities, purchasing of agricultural and forest products. This movement is mainly seasonal and done mostly by business people (of different age groups). Percentage wise, nearly 5% of the population are those who are moving in and approximately 10% moving out of the village community.

Conclusions

Climate change and associated impacts is among factors controlling migration of people as they try to adapt. Impact of climate change are real and negatively affects crop yields in various magnitude rice and maize being the most vulnerable crops to drought and increased crop pests. Crops like sorghum and sunflower are less vulnerable in the area. A number of adaptation strategies does exists but are still weak.

Different linkages that exists tend to support adaptation in terms of information and products flows from urban-rural-urban or rural-rural. We recommend best adaptation practices such as growing more drought, pest and disease crop types and varieties to be grown in the area be promoted using innovation approaches in the area through existing District Agricultural Development Plans (DADPS) set up in the country.

Rural-urban, urban-rural and rural-rural linkages needs to be strengthen in the study area in order to ensure that information and products flow from and to both ends on time.

ACKNOWLEDGEMENTS

Authors thank the PITRO-UDSM program for the financial support to facilitate all research activities. Sincere appreciation is extended to the District council authority in Nzega for a very close collaboration with researcher during this study. Special thanks is also extended to the Mbogwe and Upungu villagers for availing their time during the data collection and consultative workshops. Finally, we express our gratitude to all those who provided support in one way or another.

REFERENCES

Adger WN, Arnell NW, Tompkins E L (2005). Successful adaptation to climate change across scales. Glob. Environ. Change 15:77-86.

De Pauw (1984). Soils, physiography and agro ecological zones of Tanzania. A Final Consultant Report submitted to the Ministry of Agriculture and Cooperative, United Republic of Tanzania.

Frankenberger TR (2000). A brief overview of sustainable livelihoods approaches, in: Proceeding from the Forum on Operationalizing Sustainable Livelihoods Approaches. Pontignano (Siena) 7-11 March 2000.http://www.fao.org/docrep/003/X9371e/x9371e22.htm#P0_0

Gwambene B (2007). Climate Change and Variability Adaptation Strategies and its Implication on Land Resources in Rungwe District, Tanzania. MSc. Dissertation, Dar es Salaam University Press

Hulme M (2001). Climatic Perspectives on Sahelian Desiccation: 1973-1998. Glob. Environ. Change 11:19-29.

IPCC (2007). Climate Change 2007: Impacts, Adaptation and Vulnerability. Summary for Policymakers. IPCC AR4.

Kaijage HR (2012) Impact of Climate Change on Groundwater Dynamics in Nzega District, Tanzania: Triangulation of Indigenous Knowledge and Empirical Method Scenario Analytical Approach. M.Sc Dissertation, University of Dar es Salaam, Dar es Salaam, Tanzania.

Kalinga-Chirwa R, Ngongondo C, Kalanda-Joshua M, Kazembe L, Pemba D, Kululanga E (2011). Linking rainfall and irrigation to clinically reported malaria cases in some villages in Chikhwawa District, Malawi. J. Phys. Chem. Earth 36(14–15):887–894.

Kangalawe RM, Majule AE, Shishira EK (2005). An Analysis of Land Use Dynamics and Land Degradation Process In the Great Rift Valley, Central Tanzania: A Case of Iramba District. Book Published by OSSREA. ISBN 1 904855 628

Lema MA, Majule AE (2009). Impacts of climate change, variability and adaptation strategies on agriculture in semi arid areas of Tanzania: The case of Manyoni District in Singida Region, Tanzania. Afr. J. Environ. Sci. Technol. 3 8):206-218.

Liwenga ET (2003). Food Insecurity and Coping Strategies in Semiarid Areas: The Case of Mvumi in Central Tanzania. Dissertation No. 11. Stockholm Studies in Human Geography, Department of Human Geography, Stockholm University.

Maitima JM, Mugatha SM, Reid RS, Gachimbi LN, Majule AE, Lyaruu H, Pomeroy D, Muthai S, Mugisha S (2009). The linkages between land use change, land degradation and biodiversity across East Africa. Afr. J. Environ Sci. Technol. 3(10):310-325.

Majule AE, Liwenga ET, Kangalawe RM, Yanda PZ (2009). Natural Resource Contribution to Community Livelihoods. Experience from Selected case Studies in Tanzania. Dar es Salaam University Press LTD. ISBN: 978-9976-60-486-6.

Majule AE, Mwalyosi RBB (2005). Enhancing Agricultural Productivity through Sustainable Irrigation. A case of Vinyungu Farming System in selected Zones of Southern Highland, Tanzania. A chapter in a Book Social and Environmental Impacts of Irrigation farming in Tanzania: Selected Cases: Edited by H.Sosovele, J. Boesen and F. Maganga. Dar es Salaam University Press. ISBN 9976 60 431 9.

McCarthy JJ, Canziani OF, Leary NA, Dokken DJ, White KS (2001).

Climate Change 2001: Impacts, Adaptaion and Vulnerabilty. Cambrige University, Uk.

Mongi H, Majule AE, Lyimo JG (2010). Vulnerability Assessment of Rainfed Agriculture to Climate Change and Variability in Semi-arid Tanzania. Tanzania. Afr. J. Environ. Sci. Technol. 4(6):371-381.

Muzo N (2012). Climate Change and its Impacts on Agricultural Land use Practices in Selected Viilages of Nzega District, Tanzania. MSc. Dissertation. MSc. Dissertation, Dar es Salaam University Press.

Nelson V, Stathers T (2009) Resilience, Power, Culture and Climate and New Research Directions: A case Study from Semi-Arid Tanzania. Gend. Dev. J. 17(4):81-94.

Paavola J, Adger TW (2005). Analysis of Fair Adaptation to Climate Change. Ecol. Econ. 56 (2006): 594-609.

Watson RT, Zinyowera MC, Moss RH (Eds.). (1996). *Climate Change 1995.* Impacts, Adaptations and Mitigation of Climate Change: Scientific-technical Analyses. Contribution of Working Group II to the Second Assessment Report of the Intergovernmental Panel on Climate Change. Cambridge University Press, Cambridge. pp. 879.

World Bank (2005). Global Monitoring Report available at http://siteresources.worldbank.org/GLOBALMONITORINGEXT/Resources/complete.pdf.

Vermiculture bio-technology: An effective tool for economic and environmental sustainability

Abdullahi Hussaini

Nigerian Meteorological Agency, Mallam Aminu Kano International Airport, Nigeria. E-mail: usaini2000@yahoo.com.

Vermicompost production and use is an 'environment friendly, protective and restorative' process as it diverts waste from ending up in landfills and also reduces emission of greenhouse gases (GHG) due to very small amount of energy used in its production process. Application of vermicompost in farm soil acts as soil conditioner and help to improve its physical, biological and chemical properties. Vermicompost production is also an 'economically productive' process as it 'reduces wastes' at source and consequently saves landfills space. Construction of engineered landfills incurs 20 to 25 million US dollars upfront before the first load of waste is dumped. Over the past five years, the cost of landfill disposal of waste increased from $29 to $65 per ton of waste in Australia. However, landfills have to be monitored for at least 30 years for emissions of GHG and toxic gases and leachate (Waste Juice) which also incur cost. During 2002 to 2003, waste management services within Australia costed $2458.2 million. Even in developing nations where there are no true landfills, dumping of wastes cost alot on government. This paper elaborates on the importance of this technology to environmental sustainability and economic empowerment.

Key words: Vermicompost, earthworm, nitrogen (N), phosphorus (P), potassium (K) and magnesium (Mg).

INTRODUCTION

Vermicomposting is a simple biotechnological process of composting, in which certain species of earthworms are used to enhance the process of waste conversion and produce a better end product. Vermicomposting differs from composting in several ways (Gandhi et al., 1997). It is a mesophilic process, utilizing microorganisms and earthworms that are active at 10 to 32°C (not ambient temperature but temperature within the pile of moist organic material). The process is faster than composting; because the material passes through the earthworm gut, a significant but not yet fully understood transformation takes place, whereby resulting earthworm castings (worm manure) rich in microbial activity and plant growth regulators, and fortified with pest repellence attributes. In short, earthworms, through a type of biological alchemy, are capable of transforming garbage into 'gold' (Vermi, 2001; Tara, 2003).

Earthworms consume various organic wastes and reduce the volume by 40 to 60%. Each earthworm weighs about 0.5 to 0.6 g, eats waste equivalent to its body weight and produces cast equivalent to about 50% of the waste it consumes in a day. These worm castings have been analyzed for chemical and biological properties. The moisture content of castings ranges between 32 to 66% and the pH is around 7.0. The worm castings contain higher percentage (nearly twofold) of both macro and micronutrients than the garden compost. From earlier studies it is also evident that vermicompost provides all nutrients in readily available form and also enhances uptake of nutrients by plants. Sreenivas et al. (2000) studied the integrated effect of application of fertilizer and vermicompost on soil available nitrogen (N) and uptake by ridge gourd (*Luffa acutangula*) at Rajendranagar, Andhra Pradesh, India. The available nitrogen in soil increased significantly with increase level of vermicompost and highest nitrogen uptake was obtained at 50% of the recommended fertilizer. Similarly, the uptake of nitrogen (N), phosphorus (P), potassium (K) and magnesium (Mg) by rice (*Oryza sativa)* plant was highest when fertilizer was applied in combination with vermicompost (Jadhav et al., 1997). The new economic theory of development today is 'Environmental-Economics' which advocates for judicious balance between 'economy and ecology' in all developmental programs including agricultural development and amalgamation of 'economic development' programs with 'ecological conservation' strategies to usher in the era of sustainable development. The new economic philosophy of development also stresses mankind to switch over from the 'fossil fuel (petroleum products) based economy' to 'renewable resource based and waste recycling based economy'. We have to understand that every natural resource, commodity, goods and services that we use from the environment has an environmental cost' (the hidden cost of environmental damage and repair while the raw material is procured from the earthly resources) other than its 'economic cost' (the cost of processing, manufacturing and trading) and only after adding the two costs, we arrive at the true cost of the product. There may be 'social cost' as well in the form of impaired human health and quality of life. We only pay for the cost of food grown in farms and its processing and transport. We never pay for the damage done to the environment due to production and use of chemical fertilizers and pesticides. Conventional economists are not bothered to deduct the cost of environmental damage (e.g. degradation of farmland and soil) and the cost of environmental repair and restoration (e.g. soil regeneration and management of degraded lands) from the GNP of nations. But the environmental-economists do (Tobey et al., 1996). The cost of production of vermicompost is simply insignificant as compared to chemical fertilizers.

Vermicompost is produced from a 'cheap raw material' (community wastes including farm wastes) which is in plenty all over the world and is growing in quantity with the growing human population while the chemical fertilizers are obtained from 'petroleum products' which are not only very 'costly raw materials' but also a 'vanishing resource' on earth. Vermicompost can be produced 'on farms' by all farmers, but the chemical fertilizers has to be produced in 'factories' at a high economic and environmental cost so vermicompost can be afforded by all farmers. The worms itself after producing the vermicompost can be sold to fishery, poultry, dairy and pharmaceutical industries for economic benefits.

VERMICULTURE

The production of degradable organic waste and its safe disposal is currently becoming a global problem. The rejuvenation of degraded soils by protecting topsoil and sustainability of productive soils is a major concern at the international level. Provision of a sustainable environment in the soil by amending with good quality organic soil enhances the water holding capacity and nutrient supplying capacity of soil with the increase of resistance in plants against pests and diseases. By reducing the time of humification process and by evolving the methods to minimize the loss of nutrients during decomposition, the fantasy becomes fact. Earthworms can serve as tools to facilitate these functions. They serve as "nature's plowman" and form nature's gift to produce good humus, which is the most precious material to fulfill the nutritional needs of crops. The utilization of vermicompost results in several benefits to farmers, industries, environment and overall national economy. The production of vermicompost and its uses are 'environmental friendly, protective and restorative' as it diverts wastes from ending up in landfills and also reduces emission of greenhouse gases (GHG) due to very small amount of energy used in its production process. Application of vermicompost to crop field soil's physical, biological and chemical properties.

Vermicompost production is also an 'economically productive' process as it 'reduces wastes' and consequently saves landfills space. Construction of engineered landfills incurs 20 to 25 million US dollars upfront before the first load of waste is dumped. Over the past five years the cost of landfill disposal of waste has increased from \$29 to \$65 per ton of waste in Australia. Landfills have to be monitored for at least 30 years for emissions of GHG and toxic gases and leachate (Waste Juice) which also incur cost. During 2002-2003, waste management services within Australia costed \$2458.2 million. Even in developing nations where there are no true landfills, dumping of wastes incurs high cost on government. Earthworms converts a product of 'negative' economic and environmental value that is, 'waste' into a product of 'highly positive' economic and environmental values that is,'highly nutritive organic fertilizer' (brown gold) which improve soil fertility and enhance farm productivity to produce 'safe food' (green gold) in farms. Vermiculture can maintain the global 'human sustainability cycle'-that is, producing food in farms back from food and farm wastes.

Vermicomposting is a self-promoted, self-regulated, self-improved and self-enhanced, low or no-energy requiring zero-waste technology, easy to construct, operate and maintain. It excels all other waste conversion technologies by the fact that it can utilize waste organics that otherwise cannot be utilized by others. It excels all other biological or mechanical technologies for production of 'bio-fertilizer' because it achieves 'greater utilization' than the rate of 'destruction' achieved by other technologies and the process becomes faster with time as the army of degrader worms and the decomposer microbes multiply in millions in short time (Sinha et al., 2008). Earthworms involve about 100-1000 times higher 'value addition' in any medium (composting wastes or soil) wherever it is present (Appellof et al., 2003). Some of the advantages of vermi-cultural technology are next presented.

Lower cost of food production

A matter of considerable economic and environmental significance is that the 'cost of food production'. The use of vermiculture could be significantly low by more than 60 to 70% as compared to chemical fertilizers and the food produced will be a 'safe chemical-free food' for the society. It is a 'win-win' situation for both producers (farmers) and the consumers. The cost of production of vermicompost is less as compared to chemical fertilizers. The former is produced from 'human waste'-a raw material which is plenty all over the world and the latter is obtained from 'petroleum products' which is a vanishing resource on earth. Vermicompost can be produced 'on-farm' at low-cost by simple devices, while the chemical fertilizers are high-tech and high-cost products manufactured in factories (Munroe, 2007). Vermicompost also helps the crops to attain maturity and reproduce faster, it shortens the 'harvesting time' (Sinha et al., 2008). It further cuts the cost of production and also adds to the economy of farmers as farmers can grow more crops every year in the same field.

Reduced usage of chemical pesticides and cut cost

Widespread use of chemical pesticides became an important requirement for the growth of high-yielding varieties of crops which was more susceptible to pests and diseases. Continued application of chemical pesticides induced 'biological resistance' in crop pests and diseases and consequently much higher doses are required to eradicate them. There has been considerable evidence in recent years regarding the ability of vermicompost to protect plants against various pests and diseases either by suppressing or repelling them or by 'inducing biological resistance' in plants to fight them or by killing them through pesticidal action. Pesticide spray was significantly reduced where earthworms and vermicompost were used in agriculture (Suhane, 2007). Studies also indicate that use of vermicompost help in disease control by almost 75% which significantly cut down the cost of food production.

Reduced usage of water for irrigation and cut cost

Vermicompost is able to retain more soil moisture thus reduces the demand of water for irrigation by nearly 30 to 40%, (Suhane, 2007).

Improvement of growth and higher yield

Studies indicate that smaller amount of vermicompost in fact promote better growth performance of crops. Subler et al. (1998) reported that in all growth trials the best growth responses were exhibited when the vermicompost constituted a relatively small proportion (10 to 20%) of the total volume of the container medium. Valani (2009), found that 200 g of vermicompost applied in pot soils performed better growth in wheat crop than those with 400 and 500 g of vermicompost. Singh (2009), found that in the farm plots where vermicompost was applied in the 2nd, 3rd and 4th successive years, the growth and yield of wheat crop increased gradually over the years at the same rate of application of vermicompost that is, at 20 Q/ha. In the 4th successive year the yield was 38.8 Q/ha which was very close to the yield (40.1 Q/ha) where vermicompost was applied at 25 Q/ha. Use of vermicompost in farm soil eventually leads to increase in the number of earthworm population in the farmland over a period of time as the baby worms grow out from their cocoons. It infers that slowly over the years, as the worms build up the soil's physical, chemical and biological properties, the use of vermicompost can slowly be reduced. The yield per hectare may also increase further as the soil's natural fertility is restored and strengthened. Webster (2005) found that single application of vermicompost increased yield of 'cherries' for three consecutive years after. Yield was much higher when the vermicompost was covered by 'mulch'. At the first harvest, trees with 5 and 20 mm vermicompost plus mulch yielded cherries of the value of AU $ 63.92 and AU $ 70.42 respectively. After three harvests, yield per tree were AU $ 110.73 and AU $ 142.21 respectively for the 5 and 20 mm vermicompost with mulch. The trees yielded cherries of AU $ 36.46 per tree with 20 mm vermicompost in the first harvest and after three harvest AU $ 40.48 per tree. Webster (2005) also studied the agronomic impact of compost in vineyards and found that the treated vines produced 23% more grapes due to 18% increase in bunch numbers. The additional yield in grapes was worth additional AU $ 3,400/ha.

As a commercial commodity

Vermiculture is a growing industry not only for managing waste and land very economically but also for promoting 'sustainable agriculture' by enhancing crop productivity both in quantity and quality at significantly low economic cost than the costly agrochemicals (Bogdanov et al., 1996). Earthworms not only converts 'waste' into 'wealth' it itself becomes a valuable asset as worm biomass. Large-scale production of vermicompost has the potential to replace chemical fertilizers, protein rich 'earthworms' can be a good business opportunity with awareness growing about the use of these products in agriculture and other allied industries (George, 2004). Municipal solid waste (MSW) is growing in huge quantities in every country with growing population and there will be no dearth of raw materials for production of vermicompost. Vermiculture has also enhanced the livelihood standard of poor's in India and have generated self-employment opportunities for the unemployed. In several Indian villages, Non Governmental Organization's (NGO) are freely distributing cement tanks and 1000 worms to encourage rural people to collect waste from villages and farmers, vermicompost them and sell both worms and vermicompost to the farmers (Hati et al., 2001). People are earning from Rupees 5 to 6 lakhs (Approx. AU $ 15-20 thousands) every year from sale of both worms and their vermicompost to the farmers. It is estimated that one ton of earthworm biomass on an average contain one million worms approximately. One million worms doubling after every two months can become 64 million worms at the end of a year. Approximately each adult worm (particularly *Eisenia fetida*) consume waste organics equivalent to its own body weight everyday, 64 million worms (weighing 64 tons) can consume 64 tons of waste everyday and can produce 30-32 tons of vermicompost per day at 40 to 50% conversion rate, (Visvanthan et al., 2009). In any vermiculture practice, earthworms biomass comes as a valuable by-product and they are good source of nutritive 'worm meal'. They are rich in proteins (65%) with 70-80% high quality essential amino acids 'lysine' and 'methionine' and are used as feed material in 'fishery', 'dairy' and 'poultry' industry. Also the by- product of vermicultural practices are used in the manufacture of pharmaceuticals and in the making of 'antibiotics' from the ceolomic fluid as it has anti-pathogenic properties.

VERMICULTURE AND CHEMICAL FERTILIZER

Chemical agriculture triggered by widespread use of agro-chemicals in the wake of 'green revolution' of the 1960s came as a 'mixed-blessing' rather a 'curse in disguise' for mankind. It dramatically increased the 'quantity' of the food produced but severely affected its 'nutritional quality' and also the 'soil fertility' over the years. The soil has become addict and increasingly greater amount of chemical fertilizers are needed every year to maintain the soil fertility and food productivity at the same levels. The early response to chemical fertilizers is 'levelling off' after a 3% annual increase within the period of 1950 to 1984. There is evidence that a plateau has been reached in global efforts to increase further the yield per hectare through agro-chemicals. Increased use of agro-chemicals has virtually resulted into 'biological droughts' (severe decline in beneficial soil microbes and earthworms which help to renew the natural fertility of soil) in soils in the regions of green revolution in world where heavy use of agro-chemicals were made. High use of agro-chemicals also demands high use of water for irrigation which causes severe stress on ground and surface water. Widespread use of chemical pesticides became a necessity for the growth of high-yielding varieties of crops which are highly 'susceptible to pests and diseases'. Continued application of chemical pesticides induced 'biological resistance' in crop pests and diseases. Studies indicate that there is significant amount of 'residual pesticides' contaminating our food stuff long after they are taken away from farms for human consumption. Vegetable samples were contaminated 100% with Hexachlorocyclohexane (HCH) and 50% with dichlorodiphenyltrichloroethane (DDT) (Rao, 1993). The Society for Research and Initiative for Sustainable Technologies and Institutions (SRISTI), Ahmedabad, India, analysed the residual pesticide in soil of croplands of Gujarat. They found that 41 out of 70 samples contained insecticidal residues of Phosphamidon, Methyl parathion, Malathion, Chlorpyriphos and three different pyrethroids. Rao (1993) also found residues of pest in chemically grown food, UNEP/GEMS (1992), the farmers today are caught in a 'vicious circle' of higher use of agrochemicals to boost crop productivity at the cost of declining soil fertility. This is also adversely affecting their economy as the cost of agrochemicals has been rising all over the world. Rao (1993) also reported residues of pesticides in meat, fish, eggs, butter, milk including in mother's milk and human fat. The contamination was 100% with HCH, 69% with DDT and 43% with aldrin. In human fat DDT residue ranged from 1.8 ppm in Lucknow to 22.4 ppm in Ahmedabad; HCH ranged from 1.6 ppm in Bombay to 7 ppm in Bangalore. Adverse effects of agro-chemicals on the agricultural ecosystem (soil, flora, fauna and water bodies in farms) and also on the health of farmers using them and the society consuming the chemically grown food have now started to become more evident all over the world. According to United Nation Environment Program (UNEP) and the World Health Organization (WHO) nearly 3 million people suffer from 'acute pesticide poisoning' and some 10 to 20 thousand people die every year from it in the developing countries, (UNEP, 1992). US scientists predict that up to 20,000 Americans may die of cancer, each year, due to this low levels of 'residual pesticides'.

ENVIRONMENTAL AND ECONOMIC BENEFITS OF VERMICULTURE

Sustainable agriculture is a process of learning new and innovative methods developed by both farmers and the farm scientist and also learning from the traditional knowledge and practices of the farmers and implementing what were good in them and also relevant in present times. Vermiculture was practiced by traditional and ancient farmers with enormous economic benefits for them and their farmlands. There is need to revive this 'traditional concept' through modern scientific knowledge-a 'Vermiculture Revolution'. Sir Charles Darwin called the earthworms as 'farmer's friends'. There is great wisdom in this statement of the great visionary scientist who advocated to use the earthworms, the 'nature's gift' in farm production. It is necessary to adopt and implement food and agriculture production system which must ensure:

(i) High productivity and stability of yield over the years.

(ii) Productivity with minimum use of water and even sustain dryness or heavy rainfall.
(iii) Preservation of crop diversity (biotopes).
(iv) Preservation of soil, water and air quality in the farm ecosystem.
(v) Preservation of benevolent organisms (predators) flora and fauna in the farm ecosystem.
(vi) Preservation of groundwater table.
(vii) Preservation of good health for all.
(viii) Reduction of water and energy use.

Sustained vermiculture practices and use of vermicompost in farm soil over the years would meet several of the above requirements for a truly sustainable agriculture. Vermicompost rich in microbial diversity and plant available nutrients; improve the moisture holding capacity of soil reduces water for irrigation. It also improves physical, biological and chemical properties of soil; soil porosity and softness. There are also ample opportunities in the reduction of uses of energy and GHG emissions in vermicompost production locally at farms by the farmers themselves, (Singh, 1993).

CONCLUSION

Planning global organic farming and sustainable agriculture can truly bring in 'economic prosperity' for the farmers, 'ecological security' for the farms and 'food security' for the people. This will require embarking on a 'Second Green Revolution' and this time through 'Vermiculture Revolution' by the earthworms. This practice will ensure economic viability and environmental sustainability.

REFERENCES

Appellof M (2003). Notable Bits; In WormEzine, Vol: 2(5). (Available at (http://www.wormwoman.com).

Bogdanov P (1996). Commercial Vermiculture: How to Build a Thriving Business in Redworms; VermiCo Press, Oregon. pp. 83. 180: 79-92.

George (2004). Feasibility of developing the organic and transitional farm market for processing municipal and farm organic wastes using large-scale vermicomposting; Pub. Of Good Earth Organic Resources Group, Halifax, Nova Scotia, Canada (Available on http://www.alternativeorganic.com).

Hati D (2001). 1000 Wriggling Worms and Rural Women. The Deccan Herald, 26th June, 2001, India.

Jadhav AD, Talashilkar SC, Pawar AG (1997). Influence of the conjunctive use of FYM, vermicompost and urea on growth and nutrient uptake in rice. J. Maharashtra Agric. Univ. 22(2):249-250.

Munroe G (2007). Manual of On-farm Vermicomposting and Vermiculture; Pub. of Organic Agriculture Centre of Canada, pp. 39.

Rao BN (1993). Pollution problems caused by pesticides. Symposium on Toxicity Evaluation in Biosystem; Academy of Environmental Biology (AEB), Indore, Nov. 7-9, 1993.

Singh K (2009). Microbial and Nutritional Analysis of Vermicompost, Aerobic and Anaerobic Compost. 40 CP Honours Project for Master in Environmental Engineering; Griffith University, Brisbane, Australia; (Supervisors: Dr. Rajiv K. Sinha & Dr. Sunil Heart).

Singh RD (1993). Harnessing the Earthworms for Sustainable Agriculture. Institute of National Organic Agriculture, Pune, India. pp. 1-16.

Sinha RK (2008). Organic Farming: An Economic Solution for Food Safety and Environmental Security; Green Farming-International J. Agric. Sci. 1(10-11):42-49.

Sreenivas C, Muralidhar S, Rao MS (2000). Vermicompost, a viable component of IPNSS in nitrogen nutrition of ridge gourd. Ann. Agric. Res. 21(1):108-113.

Subler S, Edwards C, Metzger J (1998). Comparing Vermicomposts and Composts. Biocycle 39:63-66.

Suhane RK (2007). Vermicompost (In Hindi); Pub. Of Rajendra Agriculture University, Pusa, Bihar; pp: 88 (www.kvksmp.org) (Email: info@kvksmp.org).

Tara Crescent (2003). Vermicomposting. Development Alternatives (DA) Sustainable Livelihoods. (http://www.dainet.org/livelihoods/default.htm)

Tobey JA, Henri S (1996). The Polluter Pay Principle in the Context of Agriculture and the Environment. The World Economy, Blackwell Publishers, 19 (1).

UNEP/GEMS (1992). The Contamination of Food. UNEP/GEMS Environment Library No. 5, Nairobi, Kenya.

Valani D (2009). Study of Aerobic, Anaerobic and Vermicomposting Systems for Food and Garden Wastes and the Agronomic Impacts of Composts on Corn and Wheat Crops; Report of 40 CP Honours Project for the Partial Fulfillment of Master of Environmental Engineering Degree, Griffith University, Australia (Supervisors: Dr. Rajiv K. Sinha and Dr. Sunil Herat).

Vermi C (2001). Vermicomposting technology for waste management and agriculture: an executive summary. (http://www.vermico.com/summary.htm) PO Box 2334, Grants Pass, OR 97528, USA: Vermi Co. 9. Appelhof, Mary, 1997. Worms Eat My Garbage; 2nd (Ed.). Flower Press, Kalamazoo, Michigan, U.S. (http://www.wormwoman.com).

Webster KA (2005). Vermicompost Increases Yield of Cherries for Three Years after a Single Application, EcoResearch, South Australia, (www.ecoresearch.com.au). 207. Weltzien, H.C., 1989. Some effects of composted organic materials on plant health. Agriculture Ecosystems.

Impairment in biochemical indices in sprayers exposed to commonly used organophosphorus and pyrethroids: A case study

Padmaja R. Jonnalagadda[1] , A. Y. E. Prasad[1], U. V. R. Vijaya Sarathi[2], G. Manohar[3] and K. Venkaiah[4]

[1]Food and Drug Toxicology Research Centre (FDTRC), National Institute of Nutrition, Indian Council of Medical Research, Department of Health Research, Hyderabad-500007, (A.P), India.
[2]National Centre for Mass Spectrometry, Indian Institute of Chemical Technology [IICT], Tarnaka, Hyderabad-607, India.
[3]Field Division, National Institute of Nutrition, Indian Council of Medical Research, Department of Health Research, Hyderabad-500007, (A.P), India.
[4]Department of Biostatistics, National Institute of Nutrition, Indian Council of Medical Research, Department of Health Research, Hyderabad-500007, (A.P), India.

This study was carried out to evaluate the effect of spraying organophosphate, carbamate and synthetic pyrethroid pesticides on biochemical parameters viz., red blood cell (RBC) acetylcholine esterase (AchE) and plasma butyrylcholinesterase (BChE) activities, oxidative stress parameters viz., plasma T-BARs, catalase activity and antioxidant levels such as reduced glutathione, and α-tocopherol and pesticide residues (analysed using the GC/LC-MS/MS) in sprayers cultivating cotton, chilly and paddy crops in Guntur district for a period of >10 years. The study population included sprayers (120) with a mean age of 37.8 ± 10.8 years and controls (60) with a mean age of 37.3 ± 12.2 years. A significant difference was observed for AchE, BChE activities, thiobarbituric acid reactive substances (TBARs), catalase activity, reduced glutathione and the concentration of pesticide residues in the sprayers cultivating different crops for a period of >10 years as compared to the controls. However, there is no such significant difference observed for α-tocopherol.

Key words: Pesticide residues, chronic exposure, biochemical indices, sprayers running head, biochemical indices, pesticide residues in sprayers

INTRODUCTION

Several groups of pesticides are being used by farmers worldwide for a long time now in order to protect crops from pests and to enhance food production (Issa et al., 2010). According to WHO (2009), approximately 25 million pesticide poisoning cases occur annually among agricultural workers in developing countries. Reports on the effects of pesticide exposure on the health of pesticide applicators have been studied earlier (Ismail et al., 2010). A review of these studies demonstrated a wide range of neuromuscular disorders, which include paresthesia, increased sensitivity to vibration, disturbed balance, tremors and acetylcholine esterase (AchE) inhibition and deterioration of the hematological, hepatic and renal functions (Ismail et al., 2010). Similarly, in our earlier studies, significantly lower level of AchE activity, oxidative stress and antioxidant levels in the cotton cultivating farmers engaged in spraying pesticides in cotton fields was observed. Such observations were not found in non- agricultural workers who were not occupationally exposed (controls) to the spraying operations during any time of their life (Jonnalagadda et al., 2010). Further, earlier studies indicated the alterations in biochemical parameters in pesticide exposed population (Bhatnagar et al., 2003; Khan et al., 2008). However, pesticide poisoning in agricultural farmers due to occupational exposure is poorly documented in developing countries (Smit et al., 2003).

Cultivation of commercial crops such as cotton and chilly is more prevalent in Guntur district of Andhra Pradesh, India and the use of pesticides for those crops is also high as compared to the other crops. The farmers are repeatedly exposed to different types of pesticides, either simultaneously or serially during their work in the field. There is paucity of data on the incidences of pesticides in blood samples of sprayers of Guntur District (AP) and their effect on biochemical parameters if any. Therefore, the present study was carried out with an objective to determine the pesticide residues in the serum of farmers cultivating cotton, chilly and paddy crops due to chronic exposure to multiple pesticides for a period of >10 years and their effect on biochemical parameters as compared to control subjects.

MATERIALS AND METHODS

Study population

A study was carried out in Guntur District (AP), India, targeting mainly the agricultural sprayers involved in pesticide spraying on three types of crops viz., cotton, chilly and paddy for short term (<5) and long term (>10) exposure (n=120). Control group (n = 60) matched for age and gender was also selected. It was ensured that they do not have any occupational history of spraying pesticides during their life time. The objective of the study was explained to the subjects and the informed consent was also obtained. Ethical clearance was obtained from Institutional Ethical Committee to include human subjects for drawing of blood samples. About 49% of the farmers were found to have used organophosphates viz., monocrotophos (dimethyl (E) -1-mehtyl-2-(methylcarbonyl) vinyl phosphate - $C_3H_{14}NO_5P$, dichlorovos, 2,2-dichlorovinyl dimethyl phosphate - $C_4H_7C_{l2}O_4P$, quinalphos (O,O-diethyl O-quinoxalin-2-yl phosphorothioate $C_{12}H_{15}N_2O_3PS$) etc. (supplied by M/S. Bayer India Pvt. Ltd., M/S. Dhanuka Pvt. Ltd., M/S. Rallies India Pvt. Ltd.,) and synthetic pyrethoroids (12%) such as (RS) - α cyano-3-phenoxybenzyl (1RS, 3RS; 1RS, 3SR)-3-(2,2-dichlorovinyl) -2,2-dimethyl cyclopropane carboxylate - $C_{22}H_{19}C_{12}NO_3$), Imidacloprid (18%) (6-chloro-3-pyridylmethyl)-N-nitroimidazolidin-2-ylideneamine- $C_9H_{10}ClN_5O_2$). About 11% have used each of spinosad $C_{41}H_{65}NO_{10}$ (spinosyn A) + $C_{42}H_{67}NO_{10}$ (spinosyn D) and a slightly hazardous group of carbamate (11%) like indoxacarb methyl (S)-N-[7-chloro-2,3,4a,5-tetrahydro-4a-(methoxycarbonyl)indeno[1,2-e][1,3,4]oxadiazin-2-ylcarbonyl]-4-(trifluoromethoxy-$C_{22}H_{17}ClF_3N_3O_7$).

Laboratory methods

Determination of erythrocyte AchE, α-tocopherol (vitamin E) and TBAR levels

About 10 ml (each) of the blood samples from all the three types of sprayers (n = 120) spraying pesticides for more than a period of 10 years was collected in EDTA (ethylenediamine tetra acetic acid) vials (used for reduced glutathione and AchE tests) and heparin vials for analyzing the RBC catalase activity, thiobarbituric acid reactive substances (TBARs) and α-tocopherol levels in plasma. All the samples were transported to the laboratory within 24 h in frozen condition using gel packs. Out of the 10 ml of the blood samples, 3 ml was taken in a separate tube for the purpose of obtaining serum. Serum was obtained by centrifuging 3 ml of the blood sample at 1500 to 2000 RPM and the same was used for the estimation of pesticide residues. The erythrocyte AchE levels and plasma BChE activity were measured immediately (Ellman et al., 1961). The samples were centrifuged at 2000 rpm for 15 min and the plasma and leukocyte layers were separated and used for other assays. The erythrocyte sediment in the tubes was washed thrice with 0.9% saline for carrying out the catalase and glutathione (GSH) assays. All the samples for the remaining tests were stored at -80°C until assayed. Catalase activity was determined spectrophotometrically and reduced glutathione was determined fluorimetrically by Op-Thalaldehyde (OPT) method (Aebi et al., 1974). Vitamin E was determined using high performance liquid chromatography (HPLC) method (Allard et al., 1998). Lipid peroxidation was quantified by plasma thiobarbituric acid reactive substances by spectrophotometric method (Bhat, 1991).

Extraction of pesticide residues from the human biological samples

One milliliter (each) of serum sample from all the farmers cultivating cotton, chilly and paddy crops was processed for the analysis of the pesticide residues using solid phase extraction (SPE) cartridges. Serum samples were loaded into the SPE cartridges after pre-conditioning with methanol and then the residues were eluted as per the standard procedures (Jonnalagadda et al., 2010). To determine the quality of the methodology, a recovery study was performed on 10 spiked replicates of blank human (not occupationally exposed) blood sera, which presented contamination levels below the detection limits. The spike studies were conducted for standard compounds of monocrotophos (36%) (dimethyl (E) -1-methyl-2-(methyl carbonyl) vinyl phosphate), $C_3H_{14}NO_5P$, dichlorovos (76%), quinalphos (25%) and synthetic pyrethoroids such as (RS) - α cyano-3-phenoxybenzyl (1RS,3RS;1RS, 3SR)-3-(2,2-dichlorovinyl) -2,2-dimethyl cyclopropane carboxylate $C_{22}H_{19}C_{12}NO_3$ etc., and a slightly hazardous carbamate like indoxacarb and a nicotinoid like spinosad. After extraction from the serum, one μL was injected into GC (for synthetic pyrethroid compounds) and LC-MS/MS (for organophosphorous compounds). It revealed that the mean recovery values ranged from 90.5 to 93.8%. To confirm the determination of synthetic pyrethroid compounds that corresponded to synthetic pyrethroid pesticide residue peaks, a gas chromatograph-mass spectrometer (GC-MS, Perkin Elmer and Turbomass with auto system XL-GC) was used. The peaks eluted were confirmed by comparing the results of mass spectra of substances from blood sera with those of standard compounds selecting specific ions obtained from an ion trap detector. The linearity curves were drawn for each compound. The minimum detection limit for the synthetic pyrethroid pesticide studied was 1.0 ng/mL. Simultaneously, linearity curves for organophosphorous, carbamate and nicotinoid pesticides were also drawn. The minimum detection limit for the compounds detected using LC.MS/MS ranged from 0.1 to 1.0 ng/mL.

Statistical analysis

Descriptive statistics such as mean and standard deviation of various biochemical parameters were calculated. The duration of exposure to spraying is divided broadly into two categories, that is,

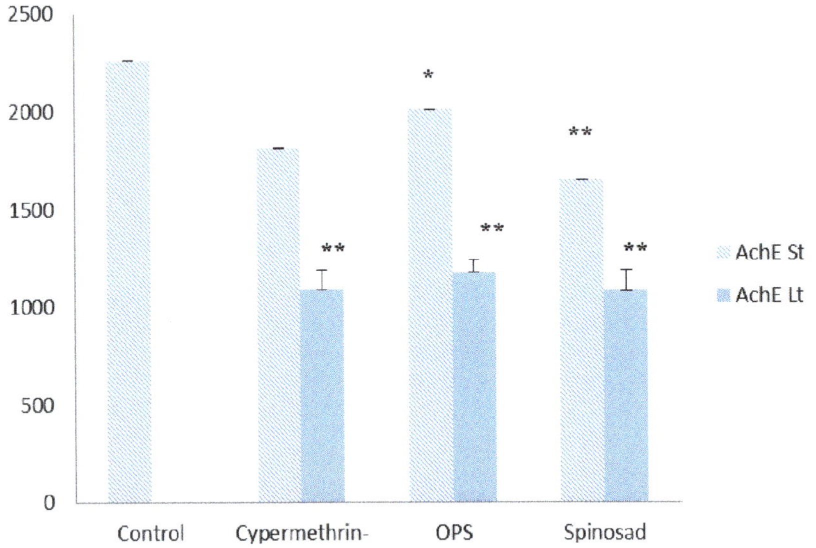

Figure 1. Comparison of mean AchE (U/L) among the sprayers exposed to cypermethrin, OPs and spinosad. * significant P<0.05, ** significant P<0.01.

short and long term of exposure depending on duration. Pesticide residues levels were compared between short, long term and controls using one way of ANOVA with post hoc analysis (least square difference). Since the variation is large between exposed and control subjects, non-parametric "μ" test was applied to assess the difference in pesticide residues between short and long term exposed groups. Significance was calculated at p = 0.05 and p = 0.01 level. The mean values of biochemical parameters were presented using bar charts.

Operating conditions of LC-MS/MS

Model: Agilent 6460 Triple Quad, Agilent Gradient Liquid chromatograph.
Mobile Phase A: 9:1 methanol: water containing 5 mM ammonium formate, Mobile Phase B: 9:1 water containing 5 mM ammonium formate: methanol. Mobile phase A, B were pumped in gradient mode at a flow rate of 0.6 ml/min, Column: Purosphere Star Column 4.6 x 150 mm, 5 μm,Column oven temperature 35°C, run time: 15 min. Reconstitution of the extracted sample has been done with the composition of mobile phase A and 5 μL was injected.

Operating conditions of GC

Model: Varian CP - 3800 equipped with electron capture detector (ECD). 5 phenyl, 95% methyl polysiloxane column was used with an an internal diameter of 0.50 μm, thickness of 0.53 μm and with 30 m length with split less system. GC-operating conditions: Column oven temp (°C) initially at 100 (hold for 2 min); 190 at 5°C per min (hold for 5min); 250 at 10°C per min (hold for 4 min); Injector port: 270°C, Detector temperature: 300°C using Iolar-1 nitrogen as a carrier gas with a flow rate of 1.5 ml/min with a total run time of 35 min.

RESULTS

The mean age of the exposed subjects was 37.8 ± 10.8, while that of the unexposed (control) subjects was 37.3 ± 12.2 years. Different biochemical parameters were estimated among the sprayers (exposed) cultivating different types of crops for short and long terms of exposure and the results were compared with controls who were not occupationally exposed to any type of the pesticides during any time of their lives. It was observed that the difference in the inhibition of AchE activity was highly significant (P<0.01) among the sprayers exposed to synthetic pyretheroid such as cypermethrin, organophosphorous pesticides (monocrotophos, dichlorvos, quinalphos) and significant to nicotinoid like spinosad for long term as compared to the control (Figure 1).

While the BchE activity inhibition was highly significant (P<0.01) among the sprayers exposed for long time to organophosphorous pesticide compounds like monocrotophos, dichlorovos, quinalphos when compared with the controls (Figure 2). It is also interesting to note that the reduction in the catalase activity was highly significant (P<0.01) among the sprayers exposed to cypermethrin and OPs for a long term exposure as compared to the controls (Figure 3). However, the reduced glutathione among the sprayers exposed for a long time to cypermethrin was significant while for OPs it was not significant as compared to the controls (Figure 4). Further, an increase in TBARs (P<0.01) for the sprayers exposed for long time to cypermethrin and organophosphorous compounds was observed when compared with the controls (Figure 5). It is interesting to note that there was no such significant difference observed for α-tocopherol.

The average levels of pesticide residues detected among the sprayers were high as compared to control subjects. However the levels of pesticide residues viz.,

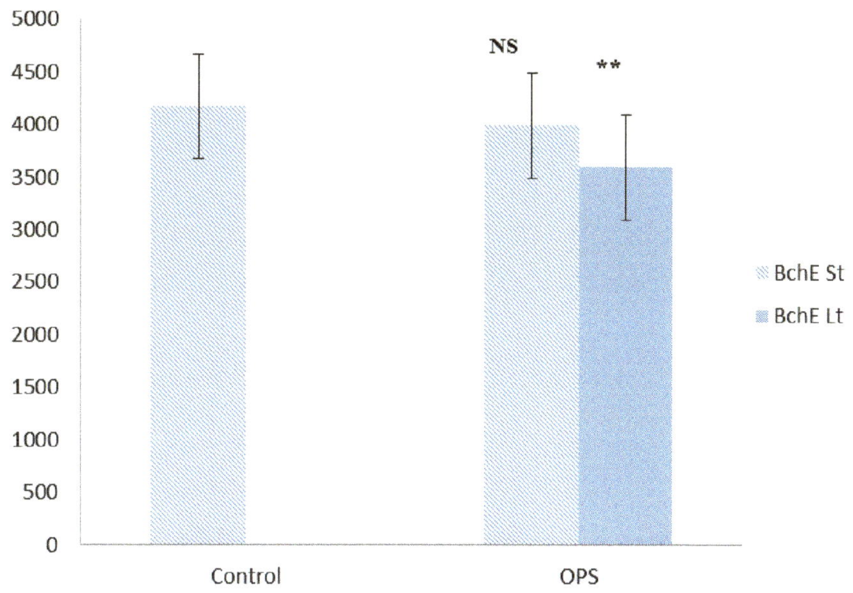

Figure 2. Comparison of mean BChE (U/L) among the sprayers exposed to OPs. NS. No sigificance. P>0.5, **significant P<0.01.

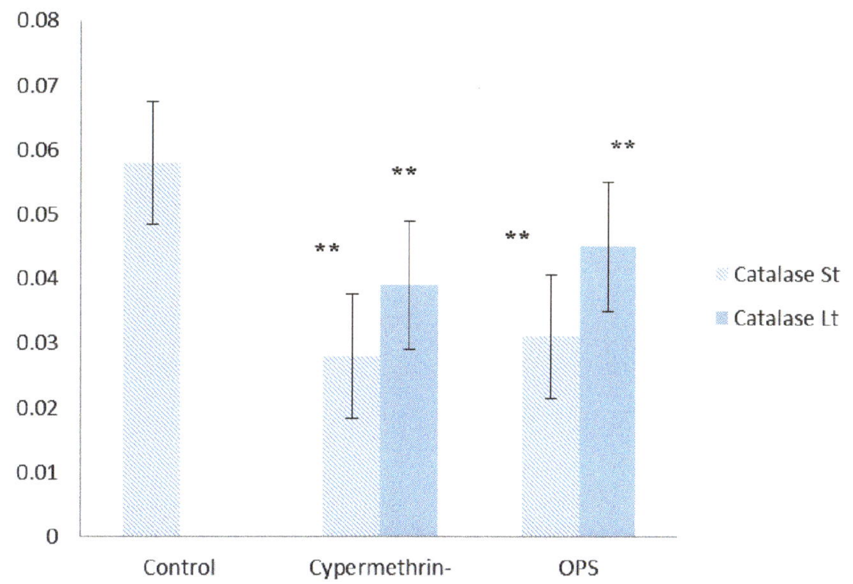

Figure 3. Effect of cypermethrin and OPs on mean catalase (mg protein/min) among the sprayers with control. ** significant P<0.01.

indoxacarb, spinosad and op group of compounds were high as compared to the rest of residues (Table 1).

DISCUSSION

Pesticides not only affect the ecosystem, but also affect the health of the sprayers, due to multiple exposures to different pesticides used by the sprayers against pests/weeds. The present study results indicate that chronic and multiple exposures to different complex mixture of pesticides suggest the accumulation of acetylcholine resulting in production of cholinergic effects and inhibition of red blood cell (RBC) AchE activity and significant inhibition of plasma cholinesterase in the sprayers could be due to the significant levels of pesticide residues in their blood/plasma as reported earlier (Khan et al., 2008) and are in conformity with earlier reports (Safi et al., 2005; Ismail et al., 2010; Issa et al., 2010).

Pesticides not only decreased the AchE and BchE activities, but also altered antioxidant and lipid peroxidation parameters in the sprayers spraying them for a

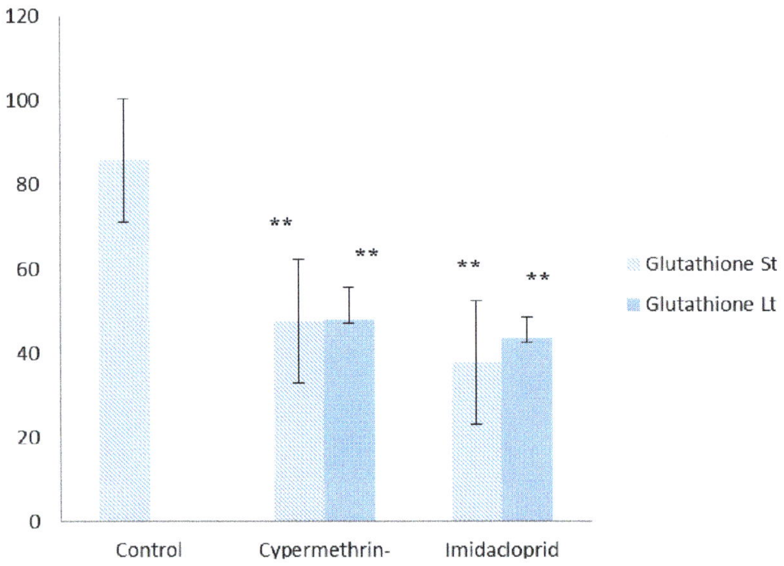

Figure 4. Comparison of mean glutathione (µg/mg protein) among the sprayers exposed to cypermethrin and imidacloprid with the control. ** significant P<0.01.

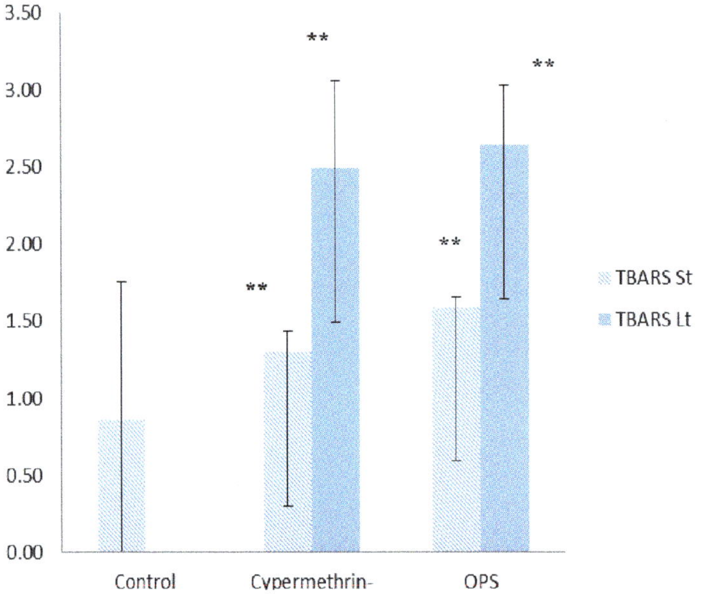

Figure 5. Comparison of cypermethrin and OPS on the mean plasma TBARs (nM/mL) among the sprayers and compared with control. **significant P<0.01.

Table 1. Serum concentrations (ng/ml) of different pesticide residues among the sprayers and control.

Pesticide	Control		Sprayer	
	no.	Mean ± SD	no.	Mean ± SD
Cypermethrin		0.0031±0.0017	14	15.93±12.89**
Monocrotophos, dichlorovos, quinalphos		0.0026±0.0016	59	20.35±16.08**
Imidacloprid	61	0.0027±0.0016	21	17.47±12.55**
Indoxacarb		0.003±0.0017	13	26.46±22.78**
Spinosad		0	13	23.87±15.96**

**Significant P<0.01 by using non parametric "µ" test.

prolonged period (Prakasam et al., 2001; Sulak et al., 2005; Gokalp et al., 2005; Ogutcu et al., 2006; Jonnalagadda et al., 2010). The increase in plasma TBARs among the sprayers was also accompanied by reduction in the catalase activity as a compensatory mechanism. Decrease in the antioxidant level such as reduced glutathione in the sprayers in the present study is also suggestive of increased utilization of antioxidants to scavenge the reactive oxygen species (Jonnalagadda et al., 2010). However, there was no significant difference in the levels of α-tocopherol in the sprayers engaged in spraying the pesticides for both short and long term exposure.

The multiple pesticide residues that were detected in the plasma of the sprayers, may be due to their extensive and repeated use for the crops for >10 years and were found in the order indoxacarb followed by spinosad, OP group of compounds, imidacloprid and cypermethrin (Klein and Karl, 1990). Pesticide exposure may produce biochemical changes even before the clinical health manifestations that may appear in the sprayers. However, low quantities of pesticide residues were detected in the control subjects indicating no significant alterations in the biochemical parameters (Jonnalagadda et al., 2010). Lack of knowledge and awareness on using the protective devices also would have increased the risk due to repeated exposure in the sprayers unlike in the developed countries. Whereas in India, a developing country extensive education and awareness are essential for sprayers, on the beneficial effects of using protective devices while spraying the pesticides so as to protect themselves from the adverse effects due to the exposure.

ACKNOWLEDGEMENTS

Financial assistance provided by Department of Science and Technology, Ministry of Science and Technology, New Delhi, India is gratefully acknowledged.

REFERENCES

Aebi ABH (1974). In: Bergmeyer, Editor Methods in Enzymatic Analysis. Academic Press, New York. pp. 643-684.

Allard JP, Aghdassi E, Chau J (1998). Oxidative stress and plasma antioxidant micronutrients in humans with HIV infection. Am. J. Clin. Nutr. 67:143-147.

Bhat KS (1991). Scavengers of peroxide and related oxidants in human brunescent cataracts. In:Gupta SK, Editors. Ocular Pharmacology-Recent Advance. Indian Ocular Pharmacol Soc:New Delhi. pp. 32-38.

Bhatnagar VK, Zaidi SSA, Kashyap R (2003). Pesticide residues in formulators and their relevance to certain biological indices. Toxicol. Int. 10:47-50.

Ellman GL, Courtney KD, Andres JV (1961). A new and rapid colorimetric determination of acetylcholinesterase activity. Biochem. Pharmocol. 7:88-95.

Gokalp O, Buyukvanli B, Cicek E (2005). The effects of diazinon on pancreatic damage and ameiolating role of vitamin E and vitamin C. Pestic. Biochem. Physiol. 81:123-128.

Ismail AA, Rohlman DS, Abdel RGM (2010). Clinical and Biochemical Parameters of Children and Adolescents Applying Pesticides. Int. J. Occup. Environ. Med. 3:132-143.

Issa Y, Sham'a FA, Nijem K, Bjertness E, Kristensen P (2010). Pesticide use and opportunities of exposure among farmers and their families: cross-sectional studies 1998-2006 from Hebron Governorate, occupied Palestinian territory. Environ. Health 63: 1-10.

Jonnalagadda PR, Prasad AYE, Ashok RK, Suresh C (2010). Biochemical alterations of certain health parameters in cotton growing farmers exposed to organophosphorous and pyrethroid insecticides. Afr. J Biotech. 49:8369-8377.

Khan DA, Bhatti MM, Khan FA, Naqvi ST (2008). Adverse effect of pesticides residues on biochemical markers in pakistani tobacco farmers. Int. J. Clin. Exp. Med. 1:274-82.

Klein O, Karl W (1990). Methylene - [14C] imidacloprid: metabolism part of the general metabolism study in the rat. California Department of Pesticide Regulation Registration Document No. 51950-021.

Mason HJ (2000). The recovery of plasma cholinesterase and erythrocyte acetylcholinesterase activity in workers after over-exposure to dichlorvos. Occup. Med. 50(5):343-347.

Ogutcu A, Uzunhisrcikli M, Kalender S, Durak D (2006). The effects of organophosphate insecticide diazinon on malondialdehyde levels and myocardial cells in rat heart tissue and protective role of vitamin E. Pestic. Biochem. Physiol. 86:93-98.

Prakasam A, Sethupathy S, Lalitha S (2001). Plasma and RBCs antioxidant status in occupational male pesticide sprayers. Clinica Chimica Acta. 310:107-112.

Safi JM, Abu MTA, Yassin MM (2005). Hematological biomarkers in farm workers exposed to organophosphorous pesticides in the Gaza Strip. Arch. Environ. Occup. Health 60(5):235-241.

Smit LAM, Berna N, Van-WJ, Dick H (2003). Neurological symptoms among Sri Lankan farmers occupationally exposed to Acetylcholinesterase inhibiting insecticides. Am. J. Med. 44: 254-264.

Suarez-Lopez JR, Jacobs DR Jr, John HH, Alexander BH (2013). Acetylcholinesterase Activity, Cohabitation with Floricultural Workers, and Blood Pressure in Ecuadorian Children. Environ. Health Perspect. 121:619-624.

Sulak O, Altuntas I, Karahan N, Yildrim B (2005). Nephrotoxicity in rats induced by organophosphate insecticide methidathion and ameliorating effects of vitamin E and C. Pestic. Biochem. Physiol. 83:21-28.

The WHO Recommended Classification of Pesticides by Hazard and Guidelines to Classification (2009) 6-81.

Lithofacies and paleo depositional environment of the rocks of Nkpuma-Akpatakpa, Izzi, Southeast Nigeria

Obasi A. I.[1] , Okoro A. U.[2] , Nweke O. M.[3] and Chukwu A.[4]

[1]Department of Geology, Ebonyi State University, Abakaliki, Ebonyi State, Nigeria.
[2]Department of Geological Sciences, Nnamdi Azikiwe University, Awka, Anambra State, State.

A detailed geological investigation of Nkpuma-akpatakpa, Izzi, Southeast Nigeria, done on a map scale of 1: 25,000, showed its lithofacies as brown shales, dark-grey shales, mudstones, siltstones and limestones. These lithofacies have been organized into two main lithostratigraphic units namely, brown shales (Unit A) and dark-grey shales (unit B). The brown colour of shales (unit A) suggests high content of ferric oxide due to intense weathering and low metamorphism caused by volcanism in the area. The dark colour of shale (unit B) suggests deposition in anoxic or reducing environment. Its dark-grey colour and the presence of bivalves suggest deposition in a low energy, marine environment. The abundance of illite which increases with depth indicates deeply buried shale not less than 3,500 m below surface. The presence of shelly limestone indicates shallow, warm, silt-free and clean marine environment. The intact shells of bivalves suggest deposition in a low energy protected shoreline where wave action is limited. Sedimentary structures like fissility and laminations, also suggest deposition in low energy marine setting. Pyroclastic rocks mapped in the area have been interpreted as volcanic tuffs and agglomerates interstratified with the shales.

Key words: Lithofacies, depositional environment, rocks, Nkpuma-akpatakpa, Izzi.

INTRODUCTION

Facies classification is one of the basic practices which define geology as a field of study. Initially, identification of rock facies was a common practice only to surface geological mapping. However, the story has changed in recent times as facies identification has become an integral part of the oil exploration/exploitation.

An understanding of depositional environment is quite important in appreciating the geology of an area. To put together the geologic history of a region, the depositional environments of its sedimentary rocks must be analyzed. By reconstructing depositional environments geologists are able to reconstruct the climates of the past, life forms of the past, and geography of the past, where the mountains, basins, large rivers, and bays of the ocean were. A depositional history of an area enhances the ability to locate energy deposits in the form of coal or oil, each of which originates in a certain type of depositional environment.

Presently, a search for hydrocarbon is incomplete without sequence stratigraphic analysis of core samples from appraisal wells and it involves detailed petrographic and geochemical studies. Several successful attempts of reconstruction of paleodepositional environment and facies have been made by Diessel (1986), Singh et al. (2010a, b, 2011, 2012a, b, c, 2013), on the basis of petro-chemical studies of coal and the associated rock formations.

Facies are a body of rock with specified characteristics (Reading, 1996). It is a distinctive rock unit that forms under certain conditions of sedimentation, reflecting

Figure 1. Map of Ebonyi State indicating the study area.

a particular process or environment. Gressly was the first to use the word in 1838 (Cross and Homewood, 1997). Establishment of sedimentary facies depends on the aspect of the sediment that is under study. If petrological characteristics are understood properly, lithofacies can be established; if paleontological characteristics are understood, biofacies are established. In this work, petrological characteristics of the rocks of Nkpuma-akpatakpa were used to establish lithostratigraphic units and their corresponding lithofacies. The objectives of this work include: i) To map and demarcate the lithologic units underlying the study area; ii) To make a detailed descry-ption of different rock types encountered; iii) To discuss the map, lithofacies units and interpret their depositional environment.

LOCATION, EXTENT AND ACCESSIBILITY

The study area is situated at about 5 km north of Abakaliki town, precisely in Izzi Local Government Area of Ebonyi State in the southeastern part of Nigeria (Figure 1). It is bounded on the north and south by latitudes 6°20^{1}N & 6°25^{1}N, and on the west and east by longitudes 8°10^{1}E and 8°15^{1}E (Figure 2). It covers an area of about 81 kilometre square (81 km^2). Accessibility to the area is enhanced by roads and footpaths which are common. Major roads leading to the area are the Abakaliki - Onuebonyi - Ogoja road, and the Onuebonyi - Sharon - Iboko road.

Physiography and geology

The study area has predominantly flat topography, with little undulation. The elevations are underlain by pyroclastic rocks (Figure 3). Two major rivers, Olilo and Ewe, drain the study area, along with their tributaries. While the Olilo River drains the northern part, the Ewe River drains the central - southern part of the area. The dominant drainage pattern is dendritic. The area is part of the rainforest vegetation of Nigeria with numerous species of trees, creepers and climbers. Most of the trees retain their green leaves throughout the year.

Geologically, the study area is part of the Abakaliki Formation of the Asu River Group (Reyment, 1965; Agumanu, 1989). It consists of fossiliferous shales, siltstones, limestones and minor fine to coarse-grained

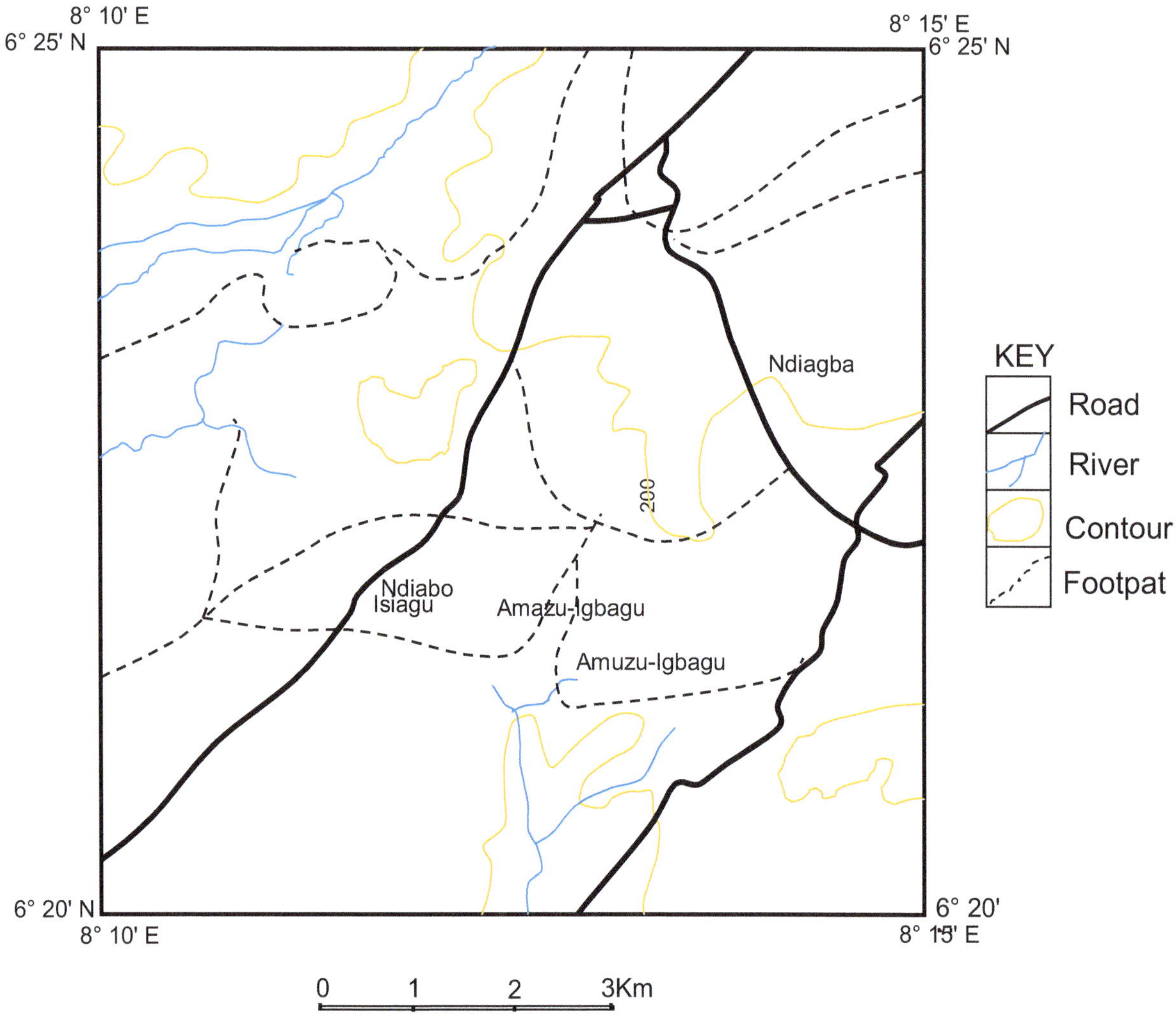

Figure 2. Topographic map of the study area.

Figure 3. The pyroclastic rocks underlying the Ezzaegu areas.

Table 1a. Lithostratigraphic units and their lithofacies.

Age	Group	Formation	Lithostratigraphic units	Lithofacies
Middle Albian	Asu River	Abakaliki	Dark-grey shales (unit B)	- Mudstone - Limestone - Shales
			Brown shales (unit A)	- Mudstone - Siltstone - Shales

Table 1b. Interpretation of depositional environment using features of the shales and fossil content (Modified from Obasi et al., 2011).

S/N	Features present	Depositional environment
1	Dark grey colour of shales	Anoxic, reducing environment such as in stagnant water columns.
2	Dark-grey Shales with pelecypod fossils	Low energy marine environment.
3	Shelly limestone	Shallow, well oxygenated, warm, silt free & clean marine environment.
4	Intact fossil shells	Low energy, protected shoreline where wave action is limited.
5.	Thick shelled *inoceramus species*	Shallow marine environment .

sandstones (Agumanu, 1989). They are associated with saline seepages, lead-zinc mineralization, basic intrusions and pyroclastics. These sediments were deformed during the Santonian tectonic phase producing numerous gentle folds. The anticlines of these folds are transected by northwesterly and northerly trending tensional faults and fractures (Olade, 1976). The folds, coupled with the identification of the igneous rocks such as andesites in the Abakaliki area led some workers to propose a compressional (subduction) rather than an extensional tectonic setting for the area (Farrington, 1952; Burke et al., 1971; Benkhelil, 1986). The thickness of the sediments varies from one point to another (Kogbe, 1976). However, no detailed geologic investigation has been done locally within the study area, hence, the need for the present study.

RESEARCH METHODOLOGY

The present work involves review of earlier literatures, field mapping and laboratory analysis. Compass and traversing method was adopted for the field mapping, which covered both rainy and dry seasons. In each location, samples were collected, described and numbered.

RESULTS AND DISCUSSION

Delineation of lithostratigraphic units

The rocks encountered in the study area are shales, siltstones, pyroclastics limestones, mudstones and pyroclastics. These lithofacies have been organized into two Lithostratigraphic units namely; "The brown shales" and 'the dark-grey shales" (Table 1a, b), while the pyroclastics are discussed separately as evidence of igneous activity in the area.

The brown shales (Units A)

This unit is older than the unit B and underlies. It occupies the central part of the study area (Figure 4). Outcrops are common along the river banks, channels and road cut in Ohagulode, Obeagu-Ikenyi and some parts of Amuzu-Igbeagu villages. Lithofacies were studied in the outcrops alone. Rock type is dominantly brown shales interbedded with mudstone and siltsone. The shale is generally grey to reddish brown in color, often with dark patches (carbonaceous stains) (Figure 5). It is indurated and severely jointed (Figure 6). Often, the joints break the shales into splintery blocks. They are composed of clay minerals and are micaceous along with the presence of some pyrites.

The induration of the shales can be attributed to baking effect due to volcanic activity in the area as evidenced by the presence of pyroclastic rocks interbedded with this shale unit (Figure 7).

The siltstone lithofacies of this unit is greyish-brown in colour, unconsolidated and fragile, often with mudstones and carbonaceous laminations. Its study on hand specimen indicates that it contains quartz, feldspars and muscovites.

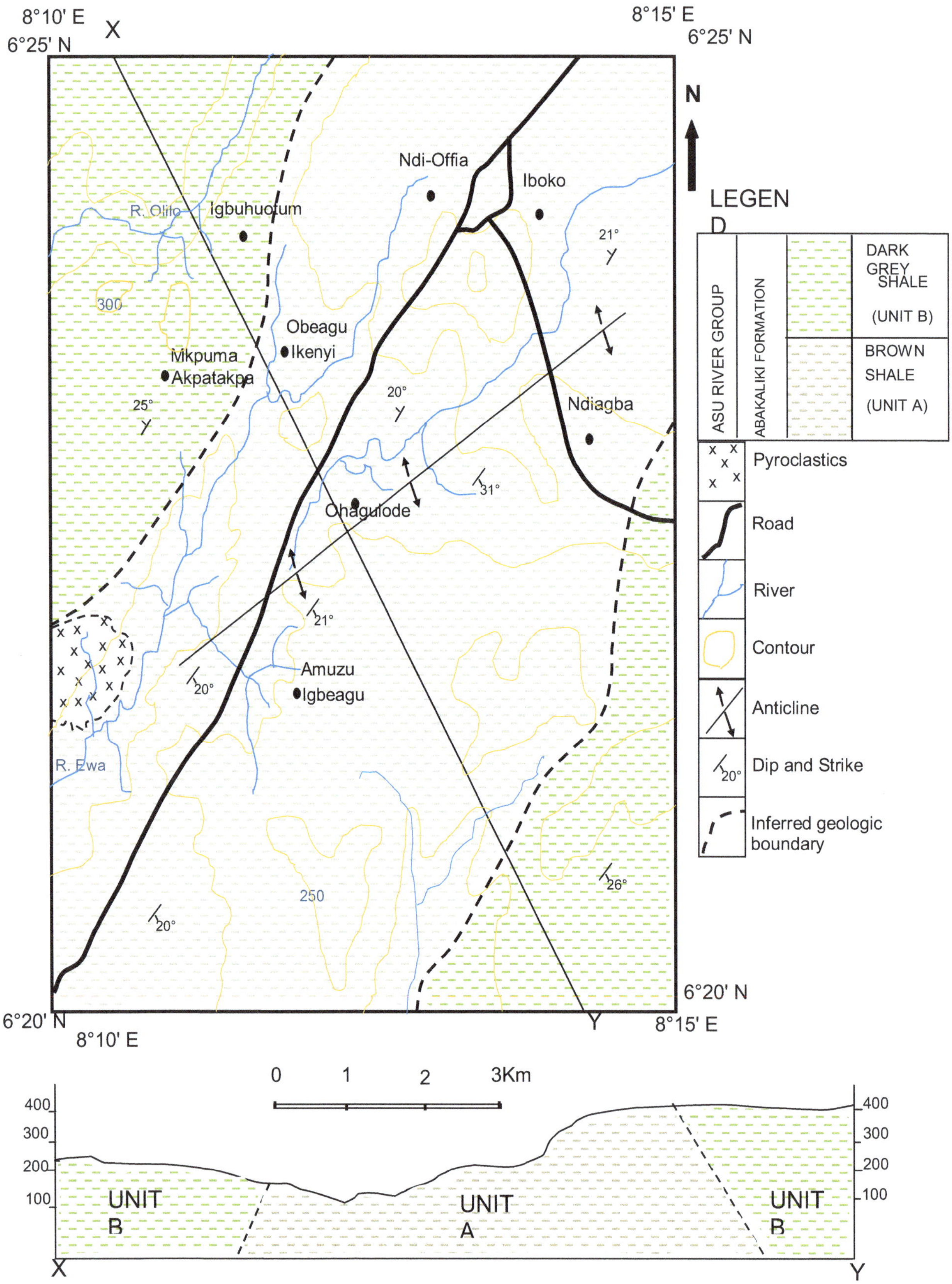

Figure 4. Geologic map of the study area.

Figure 5. Field occurrence of the brown shales at Obeagu - Ikenyi.

Figure 6. The joints in the brown shale unit.

The dark-grey shales (Unit B)

This unit consists predominantly of shales, which are dark-grey to black in colour and finely laminated, interbedded with limestones and mudstones. Outcrops are common in river channels at Ikenyi, Mkpuma-akpatakpa and west of Igbuhuotum. It occupies the northwest and northeastern parts of the study area (Figure 8). The dark-grey shales are hard, fissile and laminated. They are rich in clay mineral suites. Agumanu (1989) identified illite, kaolinite, chlorite and some smectites as the main minerals in the Abakaliki Shales. The illite and chlorite minerals increase with depth of burial. This could be due to diagenetic conversion of kaolinite to illite and chlorite at depth (Donuyer de Segonzac et al., 1967). Fissility could be brought about

Figure 7. The pyroclastic rocks interbedded with the shales at Ezzaegu. The yellow line shows sharp contact between the pyroclastic and shale.

Figure 8. Field occurrence of the dark-grey shales at Nkpuma-akpatakpa.

by parallel arrangement of micaceous minerals, organic matter content or diagenesis. However, fissility occurring in this shale is brought about by the mechanical arrangement of clay and micaceous minerals during diagenesis. Lamination as observed in the dark-grey shales is diagenetic in origin. They occurred as a result of the diagenetic cementation of organic matter contained in the rock.

The limestone lithofacies of this unit is dark-brown to yellow-brown in colour, and appears very light possibly due to weathering. Its outcrop was studied at Amuzu-Igbeagu where it interbedded with the dark-grey shales. It contains some bivalves and gastropods fossils. Identified bivalves are *Inoceramus* species and *Glycemeris* species. The fossil shells are intact, though few are broken.

The Ezzaegu pyroclastics

Igneous activity in the study area is represented by the occurrence of pyroclastic rocks in the southwestern part of the map (Figure 4). It crops out as elongate and domical body along the core of the anticline found in the area, forming part of the Abakaliki Anticlinorium. The structural relationship suggests they may be stratigraphically contemporaneous with the surrounding shales (Figure 7). Uzuakpunwa (1974) suggested that the pyroclastic rocks are older than the shales. However, field relationship in the study area does not support this. The shales have a depositional contact with the pyroclastics (Figure 7) which suggests that they are interstratified. Furthermore, they trend northeast -southwest alongside the surrounding shales. They are about 10 m thick, with its dominant lithologies as altered mafic agglomerates and tuffs. The agglomerates are composed of angular to sub-angular fragments (Olade, 1979). The tuffs are similar to the agglomerates in composition but are relatively finer grained. They are light to dark-grey in colour. They are calcitic as evidenced by their reaction to acid solution and also confirmed in thin section. The presence of pyroxenes and needles of plagioclases suggest a basaltic composition for the pyroclastics. It is a typical representation of the "Abakaliki Pyroclastics" (Okezie, 1965).

Burke et al. (1971) consider this pyroclastics as andesitic lavas and tuffs, while Olade (1975) considered the rock as altered basalts and tuffs with spilitic affinity. Olade (1978) believes that they are dominantly lapili tuffs and lapili stones of basaltic composition. Offoegbu and Amajor (1987) considered its primary mineralogy to comprise pyroxene which has been mostly replaced by chlorite and calcic-plagioclase that has been recrystallized and altered to albite, carbonate and epidote. Orajaka and Umenwaliri (1989) discussed the presence of zeolite minerals (Clinoptilolite and mordenite) and calcite in these rocks as evidence of diagenetic alteration.

Interpretation of depositional environment

The reddish brown colour of the brown shales indicate high content of ferric oxide (Fe^{3+}) which is due to high degree of weathering and low grade metamorphism associated with the volcanic activity in the area (Obiora and Charan, 2011). The dark colour of the dark-grey shales suggests deposition in an anoxic, reducing environment, such as in stagnant water columns (Zangerl and Richardson, 1963).

The dark-grey colour and the presence of bivalves suggest deposition in a low energy, marine environment. The abundance of illite which increases with depth indicates a deeply buried shale not less than 3,500 m below surface. The presence of shelly limestone indicate shallow, warm, well oxygenated, silt-free and clean marine environment. The intact shells suggest deposition in a low energy protected shoreline where wave action is limited. The presence of thick shelled *Inoceramus* species suggests deposition in a shallow marine environment (Obasi et al., 2011). The above interpretations are summarized in Table 1.

Correlation with a named formation

Owing to the worldwide distribution of the *Inoceramus* species in the Cretaceous period, the age of the lithologic unit in the map area has been considered as Cretaceous period. The lithologic and paleontologic characteristics of the lithostratigraphic units described above indicate similarity with the Abakaliki shales of the Asu River Group sediments (Reyment, 1965; Agumanu, 1989). They are therefore correlated to the Abakaliki shales of the Asu River Group.

Conclusion

The lithofacies and paleo-environment of the rocks of Nkpuma-Akpatakpa was determined based on their petrological characteristics and fossil content. The lithofacies identified are shales, which are dark grey and brown, siltstones, mudstones and limestones. The paleo-environment of the rocks has been interpreted as low energy shallow marine environment.

ACKNOWLEDGEMENTS

The authors remain grateful to Dr. P. N. Nnabo for his technical support. Favour Obia typesetted the manuscript. We thank our families for giving us a conducive atmosphere for academic work.

REFERENCES

Agumanu AE (1989). The Abakaliki and Ebonyi Formations: sub-divisions of the Albian Asu River Group in the the southern Benue trough, Nigeria. J. Afr. Earth Sci. 9(1):195-207.

Benkhelil J (1986). Structure and Geodynamic Evolution of the Intracontinental Benue Trough, Nigeria.

Burke K, Dessauvagie TFJ, Whiteman AJ (1971). Opening of the Gulf of Guinea and the geological history of the Benue depression and Niger Delta. Nat. Phys. Sci. 233:51-55.

Cross TA, Homewood PW (1997). Amanz Gressly's role in founding modern stratigraphy. Geol. Soc. Am. Bull.109(12):1617 - 1630.

Diessel CFK (1986). On the correlation between coal facies and depositional environments. - Proc. 20th Symp., Dep. Geol., Univ. Newcastle, N.S.W. pp. 19-22.

Farrington JLA (1952). preliminary description of the Nigeria lead-zinc field. Econ. Geol. 47:485-508.

Kogbe CA (1976). The Cretaceous and Paleogene sediments of Southern Nigeria In Kogbe, C. A (eds) Geology of Nigeria. Elizabeth Pub. Co. Lagos. pp. 275-281.

Obasi AI, Okoro AU, Ekpe II (2011). Determination of the Paleoenvironment of the lithologic units of Nkpuma-akpatakpa, Izzi, southeastern Nigeria, using their fossil content. Glob. Res. J. Sci. 1:9-11.

Obiora SC, Charan SN (2011). Geochemistry of regionally metamorphosed sedimentary rocks from the lower Benue rift: Implications for provenance and tectonic setting of the Benue rift sedimentary suite. S. Afr. J. Geol. 114:25-40.

Offoegbu CO, Amajor LC (1987). A geochemical comparison of the pyroclastic rocks from Abakaliki and Ezillo, Southern Benue Trough, Nigeria. J. Min. Geol. 23(1 and 2):45 -52.

Okezie CN (1965). The Abakaliki Pyroclastics. Rept. Geol. Surv. Nigeria, (Unpublished).

Olade MA (1975). Evolution of Nigeria's Benue Trough (aulacogen): a tectonic model. Geol. Mag. 112:575-83.

Olade MA (1978). Early Cretaceous Basalt volcanism and initial continental rifting in Benue Trough, Nigeria. Nature 273:458 -459.

Olade MN (1976). On the genesis of the Lead-Zinc deposit in Nigeria's Benue Rift (Aulacogen): A re-interpretation. J. Min. Geol. 13 (2):20-27.

Olade MN (1979). The Abakaliki Pyroclastics of the Southern Benue Trough, Nigeria: their petrology and tectonic significance. J. Min. Geol. 16 (I):17-25.

Orajaka IP, Umenwaliri S (1989). Diagenetic alteration of volcaniclastic rocks from Abakaliki area, southeastern Nigeria. J. Min. Geol. 25 (1&2):97 -102.

Reading HG (1996). Sedimentary Environments and Facies. Blackwell Scientific Publications.

Reyment RA (1965). Aspects of Geology of Nigeria. Ibadan University Press. p. 145.

Singh PK, Singh MP, Prachiti PK, Kalpana MS, Manikyamba C, Lakshminarayana G, Singh Alok K, Naik AS (2012c). Petrographic characteristics and carbon isotopic composition of Permian coal: Implications on depositional environment of Sattupalli coalfield, Godavari Valley, India. Int. J. Coal Geol. 90-91:34-42.

Singh PK, Singh GP, Singh MP (2011). Characterization of coal of seams II, III and IIIA from Ramagundam Coalfield, Godavari valley, Andhra Pradesh, India. Energy Sources Part A: Recovery, Utilization, and Environmental Effects, Taylor & Francis. 33(20):1863-1870.

Singh PK, Singh GP, Singh MP, Naik AS (2013). Petrology of coals from Rampur Seam-IV and Lajkura seam, Ib River coalfield, Mahanadi Valley, Orissa, India. Energy Sources Part A: Recovery, Utilization, and Environmental Effects. Taylor & Francis 35:1681-1690.

Singh PK, Singh MP, Singh AK (2010a). Petro-chemical characterization and evolution of Vastan Lignite, Gujarat, India. Int. J. Coal Geol. 82 (1-2):1-16.

Singh PK, Singh MP, Singh AK, Mukesh Arora (2010b). Petrographic characteristics of coal from the Lati Formation, Tarakan basin, East Kalimantan, Indonesia. Int. J. Coal Geol. 81:109-116.

Singh PK, Singh MP, Singh AK, Naik AS (2012b). Petrographic and geochemical characterization of coals from Tiru valley, Nagaland, NE India. Energy, Exploration Exploitation. 30(2):171-192.

Singh PK, Singh MP, Singh AK, Naik AS, Singh Vikas K, Singh Vijay K, Rajak PK (2012a). Petrological and geochemical investigations of Rajpardi lignite deposit, Gujarat, india. Energy, Exploration Exploitation. 30 (1):131-152.

Uzuakpunwa AB (1974). The Abakaliki pyroclastics, Eastern Nigeria: New age and Tectonic implications. Geol. Mag. III. pp. 65-70.

Zangerl R, Richardson ES (1963). The paleoecologic history of two Pennsylvanian shales, Fieldiana Memories 4, Field Museum of Natural History, Chicago. p. 352.

Health risks for population living in the neighborhood of a cement factory

Syed Sana Mehraj[1], G. A. Bhat[1], Henah Mehraj Balkhi[2] and Taseen Gul[2]

[1]Department of Environmental Science, University of Kashmir, J&K 190006, India.
[2]Department of Biotechnology, University of Kashmir, J&K 190006, India.

In order to assess the health risks associated with the manufacturing of Portland cement for the population living in the neighborhood of a cement industry in Khrew, Kashmir, India, particulate matter and trace gas samplings were done between March and December 2011 in the cement affected area. The main parameters considered for study included suspended particulate matter (SPM), respirable suspended particulate matter (RSPM), non-respirable suspended particulate matter (NRSPM), nitrogen oxides (NOx) and sulfur dioxide (SO_2). The population considered most suitable for study was the people residing in 2 - 3 km radius of the cement emission zone and for this, questionnaire based study was performed. The air temperature, air humidity, wind speed, wind direction and light intensity were studied at the sites for the sampling days. The results indicate that there was high level of air pollution with mean SO_2 concentration of 115.2 µg/m^3 at site I and 28.13 µg/m^3 at site II when compared. Similarly, NOx concentration at site I was 117.09 µg/m^3 when compared with control site II where it was found to be 19.46 µg/m^3 with high prevalence of diseases particularly, respiratory problems, 97% suffered from eye irritations and 95% suffered from dermatological problems among population living in the neighborhood of cement factory at site I. The assessment of oxidative and nitrosative stress among population was carried out by quantification of ROS and NO levels in serum of subjects. The results show that there was high level of air pollution in the area, adverse health impacts, over production of nitrogen species as well as ROS in subjects residing around cement pollution affected area.

Key words: Cement industry, trace gas samplings, human health, dermatological, respiratory problems, eye irritations.

INTRODUCTION

Different industrial activities degrade various environmental components like air, soil, water and vegetation (Dolgner et al., 1983; Sai et al., 1987; Mishra, 1991; Murugesan et al., 2004; Kumar et al., 2008). Cement industry is one of the 17 most polluting industries listed by the central pollution control board of India. It is the major source of particulate matter, SOx, NOx and CO_2 emissions. Cement dust contains heavy metals like chromium, nickel, cobalt, lead and mercury pollutants, hazardous to the environment with impact on human health, animal health, etc (Baby et al., 2008). Cement dust may cause carcinogenesis, decreased antioxidant capacity and causes acute respiratory symptoms and acute ventilatory effects (Aydin et al., 2010; Zeleke et al., 2010).

Kashmir valley, a world famous health resort once known for the clean and fresh environment, is also under the severe threat of atmospheric pollution. Major sources of air pollution in Kashmir valley are cement factories, brick kilns, stone crushers and automobile exhaust. In Kashmir valley there is a great potential to produce sufficient amount of cement for export and presently several factories are in operation in the areas like Khrew, Wuyan (administrative district Pulwama) and Khonmoh (administrative district Srinagar).

For the study, cement affected area of Khrew was taken into consideration. In 1979, a cement factory under the banner of Jammu and Kashmir Cements Ltd., with daily production of 600 tons which has risen to 1200 tons/year, was set up at Khrew. The factory releases an enormous amount of cement dust into the atmosphere apparently affecting the flora and fauna. Cement dust from the cement plant forms a thin deposition layer over the adjacent areas.

The adverse effects of cement pollution are well demonstrated in Khrew which has been experiencing such effects for the last three decades. The area is a live image of environmental degradation. A visit to the place is enough for one, of course without any pre-conceived notions and prejudices, to have a firsthand knowledge of the disastrous and continuous exposure to the cement pollution.

The increasing number of cement factories in Kashmir because of government promotion and local demand, is foreseen as blooming and major industry which is mainly operational near Khrew (Pulwama) and the main source for destroying world famous *Crocus sativus* (Saffron) and decreasing the health standard and thus pose threat to the flora (Jan and Bhat, 2008) and fauna (Tak and Bhat, 2009) of the region.

The present investigation has attempted to generate information on the apparent impacts of cement pollution on the health of local inhabitants with relation to air quality parameters using questionnaires, personal interviews to local habitants and workers, only non smokers were considered for study and air quality assessment was performed by using high volume air sampler enviroteck, for particulate matter cup vial method and filter paper method was used while for NO_2, Jacob and Hochheiser (1958) modified method was used and for SO_2 modified West and Gaeke (1956) method was used. The comparative study was conducted at cement polluted area of Khrew and non polluted area of Burzahama, Jammu and Kashmir, India.

MATERIALS AND METHODS

Study area

Site I

Site I was located at a distance of about 2 km from the cement factory (JK cements Ltd.) in southern direction, with dense human population. The area is covered with dwarf herbs and grasses with some scattered shrubs. Khrew is about 20 km from the main Srinagar city. Cement pollution in this area is apparent. The geographical co-ordinate of the site is 34°04′30′′ N 74°58′ 10′′ E. It lies at an altitude of 1649 m above mean sea level. This site was selected to study the impact of cement pollution on human population as the site was just two kilometers away from the factory (Figure 1).

Site II

Site II was taken as control site at Burzahama. This site is about 18 to 20 km away from the main Srinagar towards the northern side at geographical co-ordinates 34° 03′14.00′′N 075° 00′ 03.0′′E and lies at an altitude of 1632 m above mean sea level. The area remains under paddy cultivation with scattered horticultural orchards and it apparently do not receive any cement dust from the factory or any other kind of pollution so it was taken as control site. It is nearly 40 km away from cement effected area and is of archeological importance (Figure 1).

Air analysis

Ambient air quality was monitored from March to December during 2011 at Khrew near JK Cements Ltd. in Kashmir valley for priority parameters (total suspended particulate matter (TSPM), respirable suspended particulate matter ($PM_{10,}$), non-respirable suspended particulate matter (NRSPM) and nitrogen dioxide and sulphur dioxide). In selection of sampling points, priority was given to populated areas. Two sampling points were considered for one site.

Air quality parameters such as TSP, PM_{10}, NRSPM, SO_2 and NO_2 were monitored by using high volume respirable dust sampler (Enviro-tech Instrument APM-460NL). The sampling instrument was set up 3 m above ground and hourly values for all pollutants were measured at each site. The particulate matter (PM_{10}) collected on fiber-glass filter was determined by weighing the filter before and after exposure to ambient air. Non- Respirable particles (NRSPM) were determined by dust cup vial method. TSPM was determined from the sum of PM_{10} and particles larger than PM_{10}. The mass of PM larger than PM_{10} was determined from the initial and final weight of the dust cup vial. The collected samples (fiber glass filter) were properly stored and placed in vacuum desiccators and transported to the laboratory for analysis. The samples of nitrogen dioxide (NO_2) and sulphur dioxide (SO_2) were collected in glass impingers using sodium arsenate and sodium tetrachloro-mercurate absorption solutions, respectively. NO_2 in the samples was determined using Jacob and Hochheiser (1958) modified method, while SO_2 was determined using the modified West and Gaeke (1956) method. Samples were kept in a refrigerator until analysis to minimize volatilization.

Assessment of human health

In order to study the health hazards of cement factory on human health, the section of the population considered most suitable for study were the people residing in 2-3 km radius of the cement emission zone. All the randomly selected persons in 2-3 km vicinity of the emission zone were given a questionnaire, about 2000 subjects were taken into consideration for study, among them 1000 were male usually working as casual laborers or farmers and 1000 were females belonging to middle class families and majority of them answered frequently. Questionnaires were based on analysis of various standard organizations such as Department of Health And Ageing and Health Council (2002), WHO (1999), etc and according to the literature (Maureen et al., 1860; Lesliam et al., 2005; Winston et al., 2005; Sengupta, 2006; UNEP, 2008). In total, 10% of the population of the area was considered for study and administered questionnaires during every sampling month.

ROS and NO levels in serum of subjects were also quantified. NO is produced by various cells such as platelets, neutrophiles, macrophages, fibroblasts, in response to a number of pathological processes such as septic shock, inflammation, etc (Moncada et al.,1991; Nussler and Billiar,1993; Langrehr et al., 1992; Eizirik and

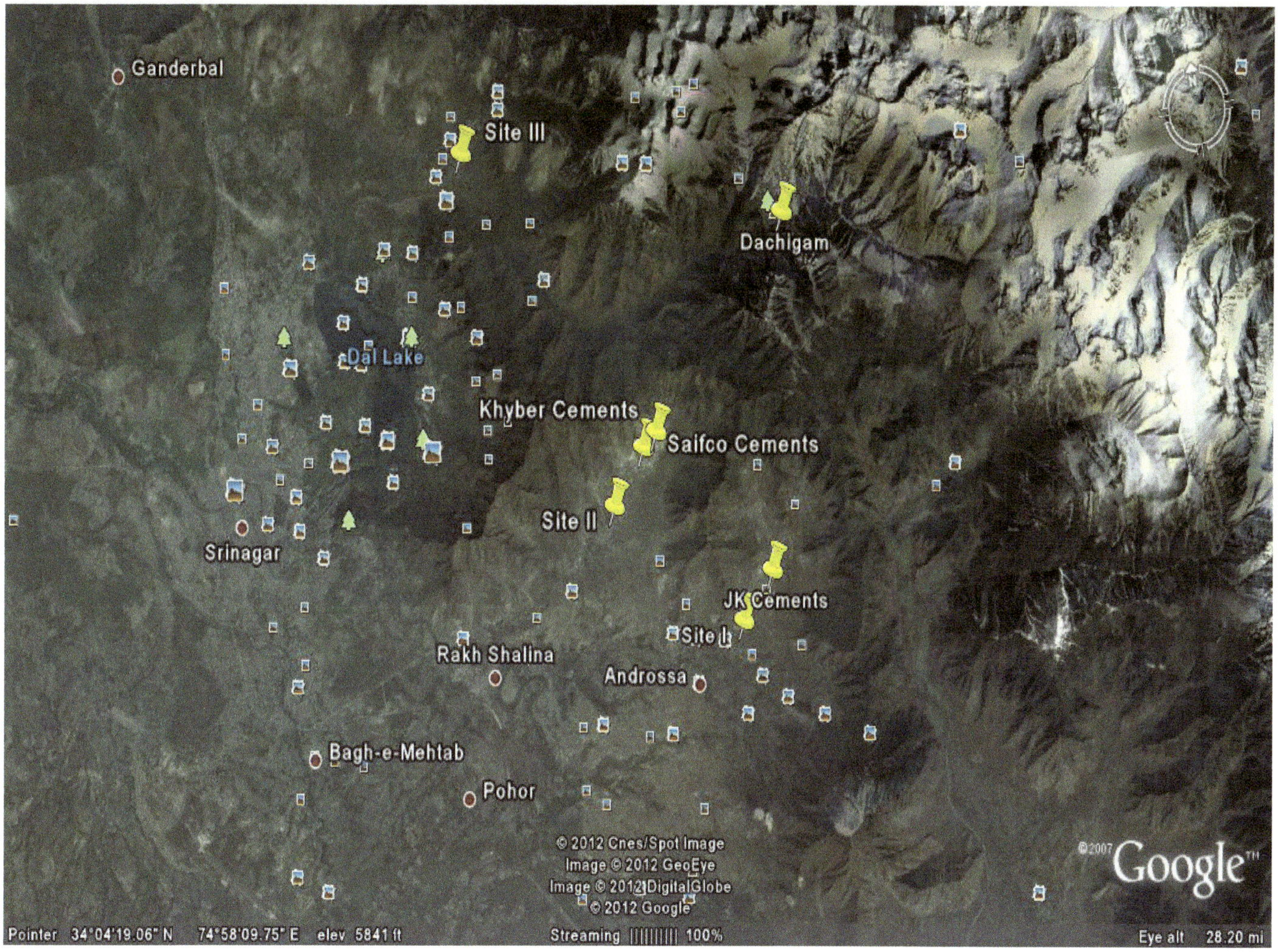

Figure 1. Study area map.

Leijerstam, 1994; Moncada, 1992; Nathan, 1992). NO is converted into nitrite (NO_2^{-}) and nitrate (NO_3^{-}). Excessive production of ROS has been reported to have an association with wide variety of clinical disorders associated with stress. For this purpose, twenty four blood samples of subjects residing permanently within 2-3 km radius of the cement emission zone were collected. While same number of control samples were taken from subjects living in area taken as control site. All these blood samples along with subject history were provided by health centers of respective area. Among affected individuals, 11 were male and 13 females belonging to three different age groups. While as among control samples 10 were males and 14 females, belonging to similar age groups as in affected group. Total NO production was calculated as sum of NO_2^- and NO_3^- as per Griess and Ber (1879) assay. The Griess Reagent System is based on the chemical reaction, which uses sulfanilamide and *N*-1-napthylethylenediamine dihydrochloride (NED) under acidic (phosphoric acid) conditions. A nitrite standard reference curve of sodium nitrite solutions with concentrations between 1-100 µM was prepared for each assay for accurate quantitation of NO_2^- levels in experimental samples. While ROS estimation of above serum samples was carried out by spectrophotometric assay of reactive oxygen species (Hayashi et al., 2007). Standard reference curve of hydrogen peroxide was with concentrations between 50-500 mg/l.

Meterological parameters

The wind speed in the area is usually found to be calm and direction of wind was found mostly in the south-east side. Air temperature, air humidity, wind speed, wind direction and light intensity were studied at the study site. For observing the air temperature, Celsius air thermometer was used while for measuring wind speed, Beaufort scale of winds was used, and the data for direction of winds and air humidity was collected from meteorological department for sampling days. Light intensity at various sites was measured by using Lux-meter.

Statistical analysis

Two statistical parameters were used in the study which are mean and standard deviation.

RESULTS

Questionnaire studies indicates that among residents, 21 types of ailments were (Table 5) recorded which included:

allergic reactions that interfere with breathing, chronic bronchitis, asthma, emphysema, lung cancer, pneumonia, tuberculosis, shortness of breath, cough, wheezing, seizures, chest pain, irregular heartbeat, swelling in your legs and feet (not caused by walking), high B.P, eye irritation, skin allergies, anxiety, fatigue, heart burn/indigestion not related to eating and neck and back disorders. 99% of the sampled people residing in Khrew complained of allergies that interfered with breathing. High blood pressure and maximum cases of chest pain were also found at Khrew.

Irregular heart beating in the residents of the cement affected area was 51%. 96% were found to be suffering from cough in the affected area. Chest pains were complained by 49% which was high among residents. Residents in cement affected areas were suffering from asthma (49%), 97% were suffering from eye irritation in the area. Swelling in legs and feet (not caused by walking) was 43% in Khrew area while 19% were suffering from tuberculosis in. The residents were also suffering from pneumonia. The prevalence of other various diseases was also found to be high.

Nitric oxide and ROS estimation

We measured NO levels and ROS among 21 effected individuals and 21 non effected individuals. High concentration of nitrite or nitrate was observed in the subjects living in vicinity of cement pollution affected area (P < 0.01) when compared with subjects from control site (Figures 2 and 3). Variations according to age or gender were statistically not significant, suggesting that the differences observed between groups in our study were unrelated to age or gender. The mean NO level among effected individuals, 51.349 ± 15 μM, mean ± S.D., was significantly higher than that among non effected individuals, 40.765 ± 11 μM, mean ±S.D; P< 0.0001. Similarly, mean ROS level among effected individuals, 198.2 ± 60.3 units (1 unit corresponds to 1mg/l H_2O_2, mean ±S.D. was significantly higher than that among non effected individuals, 98.9 ± 21 units, mean ± S.D. (P< 0.0001).

DISCUSSION

The fact that air pollution is hazardous to human health is well known. WHO estimates that worldwide, at least two million people every year die prematurely due to health effects caused by lack of clean air. Air is the basic necessity of human life but the quality of air is deteriorating continuously and it is being constantly polluted from different sources. One of the major sources of air pollution are automobiles and industries, as per estimates vehicular pollution is the primary cause of air pollution in urban areas (60%), followed by industries (20-30%) in India (Sivasamy and Srinivasan,1997). Number of industries in India is increasing with time so is increase in air pollution. Cement industry is one of the most important sources of pollution. The discharge of cement factories generally consist of particulate matter, sulphur dioxide and nitrogen oxides producing continuous visible clouds which ultimately settle on the surroundings as a result the whole ecosystem around the cement factory is subjected to extraordinary stress and neglect. Huge clouds of cement dust have been seen from generations in the area right from the establishment of these factories. This dust has been settling in the area around the cement factory to a considerable distance.

During the present investigation of various components of air samplings carried out during, March 2011 and December 2011 at the study site the concentrations of all the components, remained fairly higher than the prescribed standards for residential areas as prescribed by NAAQS (Table 4a). The average estimations of SPM pollution indicated 1208.78 μg/m^3 at site with maximum SPM in the month of December (1595 μg/m^3) at site.

RSPM

During the analysis, it was recorded that the average level of RSPM was 664.98 μg/m^3. When monthly values were compared the result indicates that the highest value of RSPM of 880 μg/m^3 was seen in the month of December as shown in Table 1.

NRSPM

The results indicated that the average NRSPM value was 543.85 μg/m^3 at site. Monthly maximum and minimum values of NRSPM were 833 μg/m^3 for the month of July and 309 μg/m^3 for the month of May as shown in Table 1.

Gaseous pollutants

Sulphur dioxide values also vary, the average value of 115.82 μg/m^3 was found at the site. Monthly maximum and minimum values of SO_2 for site were 129.34 μg/m^3 for the month of December and 101.1 μg/m^3 for the month of March (Table 2). Prevalence of relatively higher levels of SO_2 during winter months can be attributed to the burning of more coal (fuel) for the manufacturing of cement in the kilns of factory and also by the burning of more coal by the employees residing in nearby quarters in cold weather conditions. Lower levels of SO_2 values during summer can be attributed to the prevalence of high wind currents to which dispersion of the pollutants takes place, resulting in lower $S0_2$ levels. Similar observations were made by Reddy and Ruj (2003) while working on the ambient air quality status in an industrial area.

Nitrogen oxide (NO_x) is emitted from the combustion of

Table 1. Monthly, mean and S.D values of RSPM, NRSPM and SPM values (µg/m^3).

Month	RSPM (µg/m^3)		NRSPM (µg/m^3)		SPM (µg/m^3) (RSPM+ NRSPM)	
	Site I	Site II	Site 1	Site II	Site I	Site II
March	595.24	97	476.19	71	1071	168
April	562.5	73	416.67	78	979.17	151
May	619.05	91	309.52	77	928.57	168
June	619.05	89	476.19	93	1095.24	182
July	666.67	96	833.33	97	1500	193
August	777.78	78	555.56	72	1333.34	150
September	595.24	91	476.19	73	1071.43	164
October	583.33	99	625	91	1208.33	190
November	750	83	555.56	79	1305.56	162
December	880.95	81	714.29	75	1595.24	156
Mean	664.98	87.8	543.85	80.6	1208.78	168.4
SD	104.19	8.715	150.74	9.47	221.36	15.32

Table 2. Monthly, mean and S.D values of SO_2 µg/m^3 at different sites.

Month	Site I	Site II
March	101.1	21.44
April	103.12	23.6
May	117.88	26.9
June	121.4	28.2
July	118.2	22.64
August	114.16	31.12
September	111.11	30.64
October	128.25	33.44
November	113.64	31.23
December	129.34	32.12
Mean	115.82	28.13
SD	9.34	4.29

Table 3. Monthly, mean and S.D values of NO_2 µg/m^3 at different sites.

Month	Site I	Site II
March	97.56	26
April	101.61	22
May	103.16	13
June	121.95	21
July	99.11	24
August	131.64	17
September	119.45	13.65
October	123.43	19
November	131.62	21
December	141.44	18
MEAN	117.09	19.46
SD	15.71	4.19

fossil fuels and also from the oxidation of NO. The highest average value of 117.09 µg/m^3 was found at the site. The S.D value of NO_2 was 15.71. Monthly maximum and minimum values of NO_2 for site was 141.44 µg/m^3 for the month of December and 97.56 µg/m^3 for the month of March (Table 3). Therefore, highest value for NO_2 was found in winter, similar observations were made by Anglauf et al. (1986).

The month wise variations in the estimations of SO_2, NO_2, SPM, RSPM, NRSPM are in general attributable to variations in wind speed and wind direction as this remained 2 km and calm, respectively, during most of sampling months so pollutants do not disperse too much and rest of the months also appeared to be well related to the variations in meteorological data of wind direction and wind speed.

Human health

The aerodynamic diameter of cement particles makes it a potential health hazard as these are respirable in size and reaches internal organs particularly lungs leading to occupational lung diseases. This size distribution would make the trachea-bronchial respiratory zone, the primary target of cement deposition. The main route of entry of cement dust particles in the body is the respiratory tract and/or the gastrointestinal tract by inhalation or swallowing, respectively (Green, 1970). Both routes, especially the respiratory tract are exposed to numerous potentially harmful substances in the cement mill environment. Besides cement dust, various gaseous pollutants are also contributed by cement factories which cause pollution and ultimately affect human health.

Table 4a. National ambient air quality standards (NAAQS) CPCB.

Sensitive area	Time weighted average	Sensitive area (μg/m^3)	Industrial area (μg/m^3)	Residential, rural and other areas (μg/m^3)	Method of measurement
Suspended particulate matter (SPM)	Annual*	70	360	140	High volume sampler
	24 h**	100	500	200	
Respirable particulate matter (RSPM)	Annual*	50	120	60	High volume sampler
	24 h**	75	15	100	
Sulphur dioxide (SO_2)	Annual*	15	80	60	West and Gaeke Method.
	24 h**	30	120	80	
Oxides of nitrogen (NO_x)	Annual*	15	80	60	Jacob and Hochheiser Modified method.
	24 h**	30	120	80	

*Annual arithmetic mean of minimum 104 measurements in a year taken twice a week 24 hourly at uniform interval. **24 hourly/8 hourly values should be met 98% of the time in a year. However, 2% of the time, it may exceed but not on two consecutive days.

Table 4b. Meteorological parameters including temperature, humidity, wind speed, wind direction and light intensity at the study site.

Month	Temperature (°C)		Humidity (%)		Wind speed		Wind direction		Light intensity (lux, range: 2000)	
	Site I	Site II	Site I	Site II	Site I	Site II	Site I	Site II	Site I	Site II
March	15.4	19	76	54	2 m/s	Calm	NW	Calm	326	399
April	7.6	15.5	79	79	2 m/s	Calm	NE	Calm	389	421
May	25.6	24.4	46	73	2 m/s	2 m/s	S	SW	471	931
June	30.8	31.5	46	40	2 m/s	Calm	S	Calm	547	1025
July	28.3	26.8	58	79	Calm	Calm	Calm	Calm	651	957
August	32.2	32.6	65	61	Calm	4 m/s	Calm	SSE	518	901
September	31	32.2	69	58	1 m/s	Calm	SE	Calm	501	943
October	28.4	23.6	57	86	2 m/s	2 m/s	SE	SE	376	614
November	17.6	12.6	75	82	2 m/s	2 m/s	SE	SE	398	621
December	13.3	8.2	76	87	Calm	Calm	SE	Calm	314	452
Mean	23.03	22.64	64.7	69.9	2 m/s	Calm	SE	Calm	449.1	72.64

SE, South east; SSE, South-southeast; SW, South-west.

Among the residents, total of 21 diseases were found and possibly among 21, 20 may be associated with the cement dust exposure, these includes: Allergic reactions that interfere with breathing, chronic bronchitis, asthma, emphysema, lung cancer, pneumonia, tuberculosis, shortness of breath, cough, wheezing, chest pain, Irregular heartbeat, swelling in legs and feet (not caused by walking), high B.P, eye irritation, skin allergies, anxiety, fatigue, heart burn/indigestion not related to eating, neck and back disorders. Most of the residents complained about allergic reaction that interfere with the process of breathing followed by cough, wheezing, seizures, chest pain, bronchitis and even 1% cases of lung cancer were also found in the cement affected area of Khrew. The clear difference in the prevalence of diseases can be estimated from the results (Table 5) which clearly shows the impact of cement pollution. Similar observations were reported (Dockery and Pope, 1993; Al-Neaimi et al., 2001; Schwartz, 2002) in cement affected areas. According to residents, they had most times visited ENT specialist and dermatologist as well as general physician.

Some diseases which are mostly related to cement pollution such as asthma, bronchitis and emphysema patients were much more in number at Khrew which is because of the abundance of huge dust clouds in the area which is the outcome of cement factory. Cement affected area had high levels of pollution and thus low health status. The high levels of NRSPM, RSPM, SO_2, NO_x and cement dust levels in the ambient air at the cement affected areas were responsible for deterioration of health status in the study area. Therefore, there is a direct impact of ambient air quality on health status (Hart, 1970; Yhdego, 1992; Sivicommar et al., 2001; Pope and Dockery, 2006; Sabah, 2006; Adak et al., 2007).

NO and ROS quantification results showed over production of nitrogen species as well as ROS in subjects residing around cement pollution affected area. Under

Table 5. Relative occurrence of diseases with clinical features in the residents of cement polluted and control areas.

Disease	Khrew (%)	Burzahama (%)
Allergic reactions that interfere with breathing	96	3
Chronic bronchitis	57	0
Asthma	49	1
Emphysema	9	0
Lung cancer	1	0
Pneumonia	21	1
Tuberculosis	19	1
Shortness of breath	96	10
Cough	96	15
Wheezing	96	21
Seizures	7	1
Chest pain	49	11
Irregular heart beat	51	13
Swelling in legs and feet (not caused by walking)	43	0
High B.P	85	14
Eye irritation	97	12
Skin allergies	95	11
Anxiety	89	12
Fatigue	91	17
Heart burn/ indigestion not related to eating	58	11
Neck and back disorders	59	10

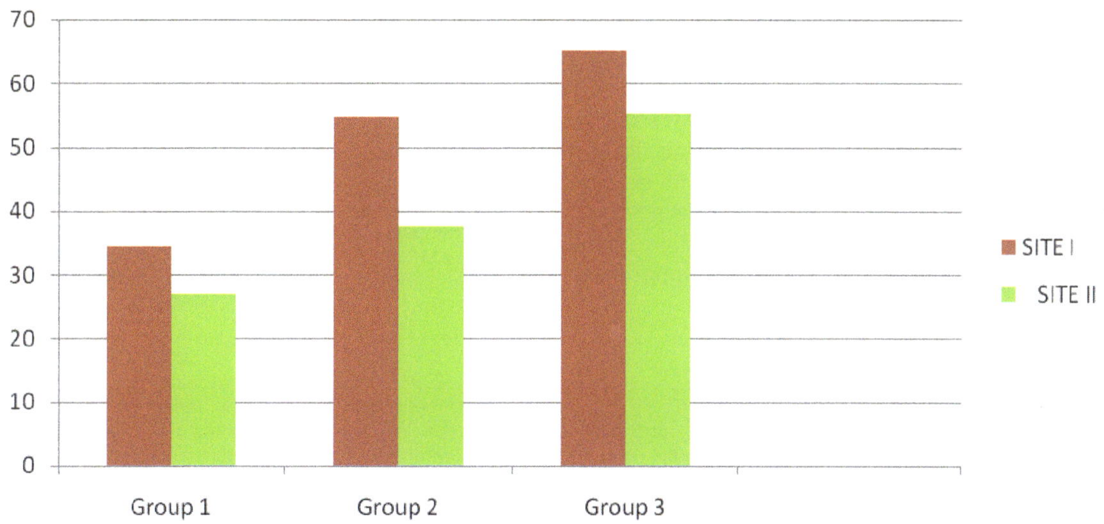

Figure 2. Serum NO levels (µm) among subjects from Site I and II. Group 1- up to age 25 years, Group 2- 25-45 years, Group 3- above 45 years.

normal conditions, NO and ROS activity is very low but it is stimulated during stress or in sustained inflammatory scenario. This is consistent with results from questionnaires and clinical data of subjects from the area, which show a high frequency of allergic and irritational complications associated with vital organs such as lungs, eyes and skin.

Conclusion

From the study, it is concluded that there was high level of gaseous and particulate pollutant load at cement polluted environment when compared with the control site and the standards prescribed by the Central Pollution Control Board of India. The present study clearly

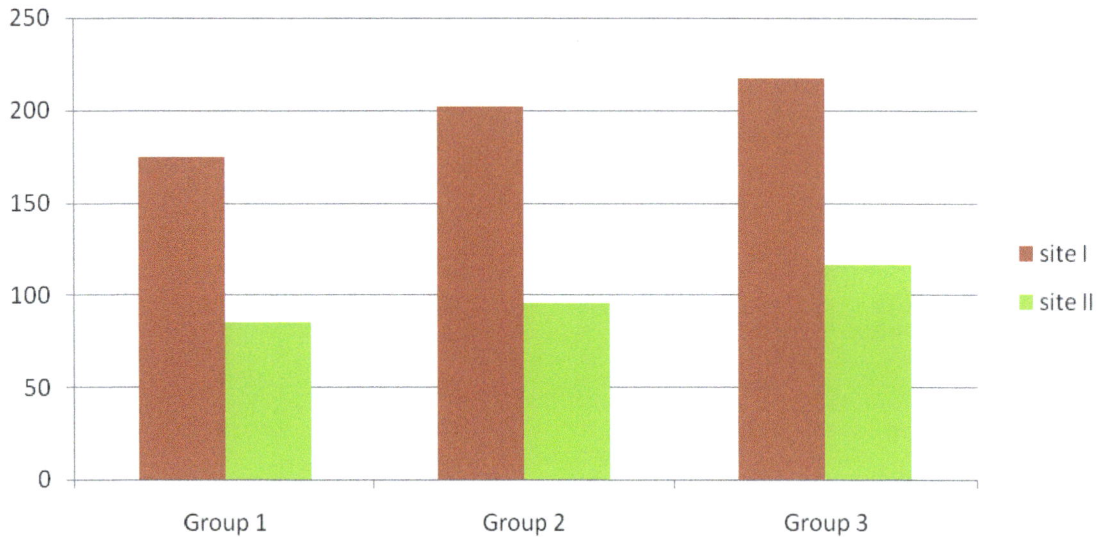

Figure 3. Serum ROS levels (units) among subjects from Sites I and II. Group 1- upt o age 25 years, Group 2- 25-45 years, Group 3- above 45 years.

revealed that cement dust is not only the major cause of environmental pollution in the study area but also a threat to health of local habitants and insightful analysis can show that the cement industries has a significant adverse impacts.

ACKNOWLEDGMENTS

The authors thank all medical, paramedical staff for their support, residents of the area who participated in this study and shared their perceptions, as well as those who helped to recruit participants and who offered help through various stages of the study. We specially thank Dr Ehtishamul Haq, Sr Assistant professor, Department of Biotechnology, University of Kashmir.

REFERENCES

Adak MD, Adak S Purohit KM (2007). Ambient air quality and health hazards near min-cement plants. Pollut. Res. 26(3):361- 364.

Al-Neaimi YI, Gomes J, Lloyd OL (2001). "Respiratory illnesses and ventilatory function among workers at a cement factory in a rapidly developing country." Occup. Med. 51(6):367-373.

Anglauf KG, Boltenheim KA, Brice. Wiebe HA (1986). A comparison of summer and winter measurements of atmospheric nitrogen and sulphur compounds. Water, Air and Soil Pollut.(30):163-160.

Aydin S, Aydin S, Croteau G, Sahin I, Citil C (2010). "Ghrelin, nitrite and Paraoxonase/Arylesterase concentrations in cement plant workers." J. Med. Biochem. 29(2):78-83.

Baby S, Singh NA, Shrivastava P, Nath SR, Kumar SS, Singh D, Vivek K (2008). Impact of dust emission on plant vegetation of vicinity of cement plant. Environ. Eng. Manage. J. 7(1):31-35.

Department of health and ageing and enHealth council (2002). Environmental health risk assessment- guidelines for assessing human health risks from environmental hazards.

Dockery DW, Pope CA (1993). Acute respiratory effects of particulate air pollution. Anim. Rev. Public Health. 1(5):107-132.

Dolgner R, Brockhaus A, Ewers U, Weigand H, Majewski F Soddemann H (1983). Repeated surveillance of exposure to thallium in a population living in the vicinity of a cement plant emitting dust containing thallium. Int. Arch. Occup. Environ. Health (52):79-94.

Eizirik DL, F Leijerstam (1994). The inducible form of nitric oxide synthase in insulin-producing cells. Diabetes Metab, 20:116-122.

Green GM (1970). The J. Burns Amberson lecture. In defense of lung. Am. Rev. Rep. Dis. (102):691-703.

Griess JP, ZAHH Bemerkungen (1879). Über einige Azoverbindungen. Ber. Deutch Chem. Ges.12:426-428.Environmental questions (ACE), department of health and ageing and enHealth council.

Hart RH (1970). The concept of APS: Air pollution syndromes. J. South Carolina Med. Assoc. 66:71-73.

Hayashi IY, Morishita K, Imai M, Nakamura K, Nakachi T, Hayashi (2007). High-throughput spectrophotometric assay of reactive oxygen species in serum. Mutation Res./Gen. Toxicol. Environ. Mutagen. 631:55-61.

Jacobs MB, S Hochheiser (1958). Continuous sampling and ultra micro determination of nitrogen oxide in air analyst. Chem 30:426-428.

Jan A, Bhat GA (2008). Impact of cement dust pollution on Agriculture crops with special reference to Saffron cultivation in Pampore area.

Kumar SS, Singh NA, Kumar V, Sunisha B, Preeti S, Deepali S, Nath, SR (2008). Impact of dust emission on plant vegetation in the vicinity of cement plant. Environ. Eng. Manage. J. 7(1):31-35.

Langrehr JM, Murase N, Markus PM, Cai X, Neuhaus, Schraut PW Hoffman RA (1992). Nitric oxide production in host-versus-graft and graft-versus host reaction in the rat. J. Clin. Invest. 90:679-683

Lesliam QMS, Awelka GW (2005). Assessment of cement dust concentrations and noise levels in a cement plant in Nicaragua.

Maureen LC, Nathalie BS, Anna A, Sharma PK. (1860). The health effects of air pollution in Delhi, India.

Mishra GP (1991). Impact of industrial pollution from a cement factory on water quality parameters at Kymore. Environ. Ecol. 9(4):876-880.

Moncada S (1992). The L-arginine:nitric oxide pathway. Acta Physiol. Scand. 145:201-227.

Moncada S, Palmer RM, Higgs EA (1991). Nitric Oxide: Physiology, pathophysiology and pharmacology. Pharmacol. Rev. 43:109–42.

Murugesan M, Sivakumar A, Jayanthi N, Manonmani K (2004). "Effect of cement dust pollution o physiological and biochemical activities of certain plants. Pollut. Res. 23(2):375-378.

Nathan C (1992). Nitric oxide as a secretory product of mammalian cells. FASEB J. 6:3051-3064.

Nussler AK, TR Billiar (1993). Inflammation, immunoregulation and

inducible nitric oxide synthase. J. Leukocyte Biol. 54:171-178.
Pope CA, Dockery WD (2006) Health effects of fine particulate air pollution. Air and Waste Manage Assoc. (56):709-742.
Reddy GS, Ruj B (2003). Ambient air quality status in Ranignaj Asansol area, India. J. Eviron. Protect. 10(2):105-112.
Sabah A. Abdul-Wahab (2006). Impact of fugitive dust emissions from cement plants on nearby communities. J. Ecol. Modeling 195:338-348.
Sai VS, Mishra MP, Mishra GP (1987). Effect of cement dust pollution on trees and agricultural crops. Asian Environ. 9(1):11-14.
Schwartz J (2002). Particulate air pollution and chronic respiratory diseases. Environ. Res. (62):7-13.
Sengupta B. (2006) Design of air quality monitoring systems to enable health impact studies, Central Pollution Control Board Ministry of Environment and Forests New Delhi.
Sivasamy N, Srinvasan V (1997). Environment pollution and its control by tree employment news New Delhi. (52):1-2.
Sivicommar R, Jayabalou R, Subrahmanyam Y, Jothikumar N, Swarnalatha S (2001). Air pollution in stone crushing industry and associated health effect. Indian J. Environ. Health (4):169-73.
Tak S, Bhat GA (2009). Assessment of epigeal invertebrate community in cement polluted and non - polluted areas. J. Res. Dev. 9:45-52.
UNEP (2008) Environmental pollution and impacts on public health by united nations environment programme.
West PW, Gaeke GC (1956). Fixation of sulphur dioxide as sulfitomercurate (11) and subsequent colorimetric determination. Anal. Chem. 28:1916-1819.
WHO (1999). Monitoring ambient air quality for health impact assessment WHO Regional Publications, European Series, No. 85 World Health Organization Regional Office for Europe Copenhagen 1999.
Winston HH, Joan D (2005) A guide to health risk assessment California Environmental protection agency office of environmental health hazard assessment.
Yhdego M (1992). Epidemiology of industrial environmental health in Tanzania. Environ. Int. 18(4):381-387.
Zeleke Z, Moen B, Bratveit M (2010). Cement dust exposure and acute lung function: A cross shift study. BMC Pulm. Med. 10(1):19.

Assessing the contribution of Community-Based Natural Resources Management Programme to environmental sustainability in Ondo State, Nigeria

Adisa, Banji O.

[1]Department of Agriculture Extension and Rural Development, Obafemi Awolowo University, Ile-Ife, Osun State, Nigeria.

This study assessed Community-Based Natural Resources Management Programme (CBNRMP) for environmental sustainability in Ondo State, Nigeria. Data were gathered through a structured interview schedule from 120 rural dwellers participating in CBNRMP. Data collected were described with descriptive statistical tools such as frequency counts, percentage, mean and standard deviation while correlation was used for the inferential to test the hypothesis set. The study showed that the mean age of rural dwellers participating in CBNRMP in the study area was 56.2±16.8 years, many (56.7%) of the respondents were married and spent an average of 12.3±9.5 years in formal schooling. The majority (83.3%) of the respondents took farming as a main occupation; information about CBNRMP was sourced through extension workers (66.7%) and television/radio (60.0%). In addition, majority (81.7%) of the respondents asserted that they are well aware of the programme before its commencement and it has improved their socio economic status. There was a positive significant relationship ($r=0.578$; $P \leq 0.01$) between improvement in socio economic status of the participating communities and their perception about the programme. Since improvement was noticed in the socio economic status of participating communities, CBNRMP is recommended for other states outside the Niger Delta region.

Key words: Assessment, community-based, natural resources, socio-environmental sustainability, management.

INTRODUCTION

A resource is a source or supply from which benefit is produced, also resources are assets that are transformed to produce benefit and in the process may be consumed or made unavailable. Social benefits of resource utilization may include increased wealth, meeting needs or wants, proper functioning of a system, and/or enhanced well-being (Miller and Spoolman, 2011). Resources have three main characteristics: utility, limited availability and potential for depletion or consumption. There are many categories of resources and 'natural resources' are part of them.

Natural resources occur within environments that are relatively undisturbed by humans. A natural resource is often characterized by amounts of biodiversity and geodiversity in various ecosystems. A natural resource may exist as a separate entity such as fresh water, and air, as well as a living organism, or it may exist in an alternate form which must be processed to obtain the resource such as metal ores, oil and most forms of energy. There is much debate worldwide over natural resource allocations; this is partly due to increasing scarcity (depletion of resources) but also because the exportation of natural resources is the basis for many economies (particularly for developed nations such as Australia) (Ricklefs, 2005; United Nations, 2002).

Some natural resources such as water in its cycle, sunlight and air can be found everywhere, and are known as ubiquitous resources. However, most resources only occur in small sporadic areas, and are referred to as localized resources. There are very few resources that are considered inexhaustible (will not run out in the foreseeable future) these are solar radiation, geothermal energy, water in its cycle and air (though access to clean air may not be available at all the time and in all circumstance especially in industrialized nations). The vast majority of resources are exhaustible, which means they have a finite quantity, and can be depleted if managed improperly (Salvati and Marco, 2008; Schilling and Chiang, 2011).

Natural resource management is a discipline in the management of natural resources such as land, water, soil, plants and animals, with a particular focus on how management affects the quality of life for both present and future generations. Natural resource management involves identifying the resources by researchers, determined who has the right to use the resources and who does not for defining the boundaries of the resource (Salvati and Marco, 2008). The resources are managed by the users according to rules governing when and how the resource is used, and local condition (Van Dyke, 2008; United Nations Development Programme, 2005).

The successful management of natural resources should engage the inhabitants (men and women) of the specific community because of the nature of the shared resources; individuals who are affected by the rules need to participate in setting or changing them. This implies that members of a specific location are the custodians of the natural resources; they utilize them for their own benefits and would be interested in their sustainability.

Globally, community-based natural resources management (CBNRM) is an approach under which communities become responsible for managing natural resources (forests, land, water, biodiversity) within a designated area. It implies that the community is expected to assist in the planning and management of the resources within their locality and can be monitored by outside technical specialists. They are utilized and protect natural resources within established guidelines, following mutually agreed plan. The active participation of stakeholders in natural resource decision making and use increases economic and environmental benefits. According to Brown et al. (2002) critical investment areas include the introduction of viable management systems, securing legal control over resources and resource utilization, improving environmental governance, and information management. It has also been established that population growth and economic development are increasing pressure on land, water, forest, and biodiversity resources. Government attempts to conserve natural resources through top-down regulatory systems have often failed. Limited government capacity to enforce laws and regulations compounds management problems, particularly when regulations are inappropriate to the social, cultural and ecological context. The role of local people in managing their natural resources is often suggested by environmentalist as the most appropriate solution to sustainable natural resources management.

CBNRM as a sustainable approach to natural resources management gives communities full or partial control over decisions regarding natural resources, such as water, forests, pastures, communal lands, protected areas, and fisheries. In his own contribution, Brown (1999) opinion that the extent of CBNRM control can range from community consultations to joint management or to full responsibility for decision making and benefit collection, using tools such as joint management plans, community management plans, stakeholder consultations and workshops, and communal land tenure rights. By this, community-based institutions are key to any CBNRM project, and selecting and building the capacity of local institutions is critical. This ensures stakeholder participation, increases sustainability and provides a forum for conflict resolution.

Community-Based Natural Resources Management Programme (CBNRMP) in Ondo State: Explorative review

The Community-Based Natural Resource Management Programme (CBNRMP) in Ondo State, Nigeria, carries out projects based on the needs and aspiration of the people in the selected communities using a bottom-top approach. The approach allows the community to identify their pressing needs that the programme can proffer solution to them. The programme was promoted by the International Fund for Agricultural Development (IFAD) and Federal Government, but funded by the IFAD, Federal Government, Niger Delta Development Commission (NDDC), participating States and Local Government Areas for a period of eight (8) years. The programme started in the year 2005 in River State but took off in Ondo State in the year 2006. There are Community Base Animator Teams (CBAT) which consist of six (6) people: threemale (including one youth) and 3 female (including one youth). The CBAT are to coordinate and monitor the project(s) in their community in order to ensure performance.

The selection of Local Government Areas (LGAs) which is the closet arm of government to the people to benefits from the programme in the study area were based on their poverty level. By this, the communities that will participate are the extreme poor communities in terms of basic social and economic facilities. Nine (9) out of eighteen (18) LGAs in Ondo State were purposively selected for the study, and covers 27 communities. The objectives of the Community-Based Natural Resources Management Programme as outlined by IFAD, (2006) are as follows:

1. Provision of improved standard of living and quality of life.
2. Provision of development for women and youth in the economy.
3. Provide market information dissemination.
4. Provide empowerment and development for farmer and fishermen.
5. Provide community management of natural resources (woodlands, rivers, watershed, fishing area).
6. Provision of improved planting materials.
7. Provision of community access improvement (small culvers, village access road, land sites).

The importance of rural communities cannot be over emphasized in the developing nation's economy because rural areas predominantly provide food for a growing population and raw materials for agricultural based industries. Also rural areas serve as a place of refuge during political crises and most urban dwellers visit rural areas to relax. In general, the rural areas provide primary activities that form the foundation of any developing nation's economy. Unfortunately, in some developing nations like Nigeria, rural areas have suffered long time neglect in such a way that has created wide gap between the rural and urban areas most especially with regards to social and economic opportunities, physical development and available infrastructural facilities.

Poor social well-being especially the provision of basic social and infrastructural facilities and deplorable working condition calls for development that involves the transformation of rural areas into a socially, economically, politically, educationally and materially desirable condition, with a purpose of improving the quality of life of the rural population (Jibowo, 2000; Ekong, 2010).

Participating rural dwellers have recently taken a central role in project identification and implementation. This role change was put in place between the year 2005 and 2013 in selected LGAs of Ondo State. After over a decade of the programme, it was considered necessary to ascertain the contribution of the programme to overall development of the study area.

The main objective of this study was to assess the CBNRMP in Ondo State. The specific objectives of this study are to i. Describe the socio-economic characteristic of CBNRMP's participating rural dwellers in the study area; ii. Examine the degree of awareness of rural dwellers about the programme; iii. Examine rural dwellers perception of CBNRMP.

Hypothesis for the study

The following null hypothesis was set for the study: There is no significant relationship between rural dwellers perceptions of the programme and improvement in the socio economic status since the inception of the programme.

METHODOLOGY

The study was conducted in Ondo State, one of the NDDC members in Nigeria. Multi stage sampling technique was used to select respondents for the study. At the first stage, four (4) out of nine (9) participating LGAs were purposively selected. The selected LGAs were Idanre, Ondo- East, Okitipupa and Ile-Oluji/Okeigbo. At the second stage, fifty percent of rural communities participating in the programme were proportionally selected making a total of twelve communities. The selected communities were Oloruntedo, Kajola-ojurin, Ebijaw, Ikota, Ibutitan, Elemo, Araromi Fasawe, Ayede-oja, Kajola-usama, Owena Egbeda, Abalaka and Oniyewu. At the final stage, one hundred and twenty rural dwellers (120) were proportionately selected and interviewed for the study. Duly validated and pretested structured interview schedule was used to collect the data. Test re-test method was used. The instrument was given to five CBAT in a Local Government Area which was not included in the final sampling frame. The instrument was pre-tested at an interval of two weeks in January 2012. The data from the pre-test was subjected to Spearman Rank Correlation Coefficient and a value of 0.865 was obtained which was higher than the empirical and acceptable coefficient of 0.84 (Ogunfiditimi, 1986) and are regarded as good enough to measure the validity of the instrument. Necessary correction and adjustments that was observed/ suggested during the pre-test were incorporated to the final instrument. Data were summarized with percentages, means and standard deviation, while Chi-square and Pearson Product Moment Correlation (PPCM) were used to analyse the data using SPSS 16 statistical package.

Measurement of variables

Dependent variable

The dependent variable for this study was conceptualized as level of improvement in socio economic status of the participating rural communities. It was measured by listing and scoring the extent of improvement in health, education, economy and agriculture observed in the communities as a result of the introduction of CBNRMP against a 4-rating scale of large extent (4), some extent (3), little extent (2), no extent (1) as used by Adisa and Adeloye (2012). The perception of the people was measured using declarative statement which was rated against a four-point rating scale: SA = strongly agreed, A = agreed, DA = disagreed, SD = strongly disagreed. These indicated their decisive position as regards their perception of the project.

RESULTS AND DISCUSSION

Results in Table 1 reveal that majority (62.5%) of the

Table 1. Distribution of respondents according to their socio-economic characteristics. n = 120.

Variable	Frequency	Percentage	Mean	Standard deviation
Age (years)				
< 30	9	7.5		
30-50	36	30.0	56.2	16.8
> 50	75	62.5		
Sex				
Male	58	48.3		
Female	62	51.7		
Years of formal education				
>12	53	44.2		
7-12	33	27.5	12.3	9.5
1-6	21	17.5		
No formal education	13	10.8		
***Major occupation**				
Farming	100	83.3		
Trading	52	43.3		
Public service	15	12.5		
***Source of information about CBNRMP**				
Extension workers	80	66.7		
Television and radio	72	60.0		
Neighbors	55	45.8		
Marital status				
Married	68	56.7		
Widowed/widower	24	20.0		
Single	28	23.4		

Source: Field survey, 2012; *Multiple choices.

respondents were over 50 years of age, while 7.5% were less than 30 years of age; the mean age of the respondents was 56.2 with a standard deviation of 16.8. These results suggests that the respondents comprise few young people, which might be as a result of high rate of rural-urban migration of able bodies in search of white-collar jobs in big cities. Furthermore, it was revealed that there was a marginal (48.3:51.7) difference between the number of male (48.3%) and female (51.7%) respondents. Table 1 also shows that the mean of years respondents have spent in formal education is 12.3 years with a standard deviation of 9.5; this result reveals that the majority of respondents could read and write which could affect their participation in CBNRMP positively. The vast majority (83.3%) of respondents were farmers by occupation while few engaged in trading and public service; in addition, many (66.7%) of the respondents had information about the programme from extension workers. The table also revealed that many (56.7%) of the respondents was married. This finding is in consonance with earlier reports of Adisa and Jibowo (2006) who reported that a high percentage of married people in the rural communities of Osun State are involved in community based development projects.

The result in Table 2 revealed that the respondents had high degree of awareness about the programme, this might be connected with the channels (extension workers and mass media) used in disseminating information concerning the programme in the study area.

The result in Table 3 reveals that perception means

Table 2. Distribution of respondents according to their degree of awareness about CBNRMP. n = 120.

Degree of awareness	Scores	Frequency	Percentage
High degree	> 35.6	98	81.7
Low degree	< 35.6	22	19.3

Mean score = 35.6; Standard deviation= 7.2; Source: Field survey, 2012.

Table 3. Distribution of the respondents by perception about the programme. . n = 120.

	SA	A	DA	SD
	F (%)	F (%)	F (%)	F (%)
Information dissemination about the project is inadequate.	-	4 (3.3)	49 (0.8)	67 (55.8)
The project implementation is good the way it has always been carried out.	30 (25)	86 (71.7)	4 (3.3)	-
The project agency has carried the people in the community along properly.	54 (45)	61 (50.8)	5 (4.2)	-
The project needs improvement in some areas.	26 (21.7)	57 (47.5)	34 (28.3)	-
There is proper utilisation and monitoring of the project by the rural people.	58 (48.3)	58 (48.3)	4 (3.3)	3 (2.5)
The project is relevant to community needs and aspiration of the people.	41(34.2)	71 (59.2)	8 (6.7)	-
The consultation with members of the community was not enough on the choice of project	-	5 (4.2)	44 (36.7)	-
The project has impact positively on the well being of the community.	69 (57.5)	50 (41.7)	1 (0.8)	-
The staff agencies are not easily accessible and capable of ensuring project success.	6 (5)	15 (12.5)	70 (58.3)	71 (59.2)
The project is a waste of resource by the government	-	95 (79.2)	25 (20.8)	29 (24.2)

SA = Strongly Agreed, A = agreed, DA = disagreed, SD = strongly disagreed; Mean = 72.7 Standard deviation = 1.8. Source: Field survey 2012

score was 72.7 with standard deviation of 1.8. This analysis shows that many (71.7%) of the respondents have a favourable perception towards CBNRMP. It could be deduced that their favourable perception of the programme would lead to their active participation in various projects embarked upon by CBNRMP. This result is similar to that of Adisa et al. (2003) which reported similar favourable perception to community-based development among rural dwellers in Osun State.

The result in Table 4 revealed that the improvements grand mean score was 3.35. Agriculture has the highest grand mean of 3.5, this was followed by economy and education, 3.4 and 3.3, respectively.

The grand mean for health was 3.2. Table 4 also shows that the communities participating in CBNRMP had recorded an increase in level of improvement in socio economic status as a result of their participation in the CBNRMP. Results further revealed that the highest level of improvement (3.5) was recorded in agriculture, followed by economy (3.4), education (3.3), and health (3.2).

This finding might be connected to the fact that majority of the study area were farming communities that need improvement in their farming activities.

Hypothesis testing

Result in Table 5 show that there is a positive and significant relationship ($r = 0.578$; $P \leq 0.01$) between the level of socio economic status of the participating communities and their perception towards CBNRMP. The contribution of respondents' perception towards CBNRMP to their improvement in socio economic status was 33.4% ($r^2=0.3341$). This suggests that the more favourable the respondents' perception is towards the CBNRMP, the higher their socio economic status.

Conclusion and recommendation

Based on the findings of this study, it was concluded that the CBNRMP has contributed to agricultural, economical, educational transformation of the benefiting communities in Ondo State.

It also established that active participation of the benefiting communities contributed to the success recorded since improvement was noticed in the socio economic status of the study area, CBNRMP is then recommended for other states outside Niger Delta region.

Table 4. Distribution of extent of improvement on the socio economic status of the participating communities as a result of programme. n = 120.

	Large extent F (%)	Some extent F (%)	Little extent F (%)	No extent F (%)	Grand mean
Health					
Provision of portable water	33(27.5)	46(38.3)	32(26.7)	9(7.5)	
Access to quality health facilities	39(32.5)	33(27.5)	30(25)	18(15)	3.2
Reduction in mortality rate	26(21.7)	33(27.5)	35(29.2)	26(21.7)	
Education					
Access to quality education	29(24.2)	19(15.8)	36(30)	42(35)	
Reduction in illiteracy level	22(18.3)	27(22.5)	37(30.8)	34(28.3)	3.3
Empowerment opportunities for the youth	22(18.3)	21(17.5)	32(26.7)	45(37.5)	
Economy					
Provision of processing industries	59(49.2)	46(38.3)	15(12.5)	0	
Access to good transportation network	68(56.7)	41(34.2)	10(8.3)	1(0.8)	3.4
Provision of marketing facilities	66(55)	41(34.2)	12(10)	1(0.8)	
Agriculture					
Provision of mproved seeds and seedlings for planting	71(59.2)	36(30)	13(10.8)	0	3.5
Provision of agricultural implement like tractor	66(55)	47(39.2)	7(5.8)	0	
Access to agric. loan	66(55)	33(27.5)	21(17.5)	0	

Source: Field survey, 2012.

Table 5. Correlation analysis showing the relationship between level of improvement in the socio economic status of the participating communities and their perception towards CBNRMP. n = 120.

Variables	Correlation coefficient (r)	Coefficient of determination (r^2)
Perception	0.578**	0.3341

Source: Field survey, 2012; **Significant at the 0.01 level.

REFERENCES

Adisa BO, Adeloye KA (2012). Analysis of farmer field school as an extension approach to cocoa production in Osun State, Nigeria. World J. Agric. Sci. 8(4):421-428

Adisa BO, Jibowo AA (2006). Effect of community variables on participation of community based organisation in development projects in Osun State, Nigeria. Niger. J. 6(1&2):83-94

Adisa BO, Oladoja MA, Adeokun OA (2003). Participation of Community Based Organisations in Environment Protection Projects in Osun State, Nigeria. Niger. J. Rural Sociol. Assoc. 2(3):294

Brown D (1999). Principles and Practice of Forest Co-Management: Evidence from West-Central Africa. EU Tropical Forestry Paper No. 2. London: Overseas Development Institute (ODI).

Brown D, Malla Y, Schreckenberg K, Springate-Baginski O (2002). "Community forestry, from supervising 'subjects' to supporting 'citizens': recent developments in community forestry in Asia and Africa." Natural Resource Perspectives 75. Overseas Development Institute (ODI), London.

Ekong EE (2010). An Introduction to rural sociology. Dove Educational Publisher Uyo, Nigeria. pp. 20-37, 132-133.

International Fund for Agricultural Development (IFAD)/ Federal Government Community Based Natural Resources Management Programme (2006) Activity Report.

Jibowo AA (2000). Essentials of rural sociology. Gbem Sodipo Press Limited, Abeokuta. pp. 223-225.

Miller GT, Spoolman S (2011). Living in the Environment: Principles, Connections, and Solutions, 17th ed. Brooks-Cole, Belmont, CA. ISBN 0-538-73534-1.

Ogunfiditimi TO (1986). Community Survey Methods, Statistical Techniques and Computer Analysis. Ibadan. Marygrant (Educational) Publishers, Nigeria. pp. 7-43.

Ricklefs RE (2005). The Economy of Nature, 6th edition. WH Freeman, New York, NY.ISBN 0-7167-8697-4.

Salvati L, Marco Z (2008) Natural resource depletion and economic performance of local districts: suggestions from a whithin-country analysis. J. Sustain. Dev. World Ecol. 15(6):518-523

Schilling M, Chiang L (2011). The effect of natural resources on sustainable development policy: The approach of non- sustainable externalities. Energy Policy 39: 990-998.

United Nations (2002). Conservation and management of resources for development. Earth Summit Agenda 21. The United Nations programme for action from Rio.

United Nations Development Programme (2005). The wealth of the poor: Managing ecosystem to fight poverty. The World Bank and world resources institute.

Van Dyke Fred (2008). Conservation Biology: Foundations, Concepts, Applications, 2nd ed. Springer Verlag. pp. 478. ISBN 978-1-4020-6890-4 (hc).

21

Monitoring the land-use and vegetation cover changes in the Kainji Lake Basin, Nigeria (1975-2006)

Ikusemoran, Mayomi[1] and Olokor, J. O.[2]

[1]Department of Geography, University of Maiduguri, Borno State, Nigeria.
[2]Hydrology Unit, National Institute for Freshwater Fisheries Research, New Bussa, Niger State, Nigeria.

This paper demonstrates the use of remotely sensed data and geographic information system (GIS) techniques for monitoring the land-use and vegetation cover changes in the Kainji Lake Basin between 1975 and 2006- a period of thirty one (31) years. Since 1968 when Kainji Lake was constructed, various activities such as agriculture, deforestation, irrigation, fishing and construction of roads and bridges have taken place. It is all these activities coupled with other natural factors that cause environmental degradation and the damage of the ecosystem of the lake basin. Landsat TM image of 1975, Landsat MSS image of 1986 and Landsat ETM+ images of 2001 and 2006 were acquired, classified and analysed with the use of Ilwis 3.3 and Idrisi 32 to study the landuse and vegetation changes in the Lake Basin between 1975 and 2006. Area calculations of Idrisi 32 were used to derive the trends, rates and magnitudes of changes, while map overlay was employed for assessing the nature and location where the changes have taken place. Map overlay technique was also used to create a matrix to discuss the location of the changes within the study periods. The study reveals that the Lake Reservoir and settlements around the lake were gradually increasing; intensive agriculture was capturing the basin at alarming rate, while the lake surface itself is recently been occupied by weeds. The woodland vegetation in which the first and the most popular National Park in the country where it is located is also discovered to be changing to other land-use types. Various conservation and control measures were suggested for the sustainable utilization of the Lake Basin.

Key words: Remote sensing, geographic information system (GIS), image classification and land-use and land-cover, change detection.

INTRODUCTION

The Kainji Lake basin is one of the most important inland basins in Nigeria, it is the home of the first and the largest hydro-electricity station in the country, the woodland vegetation of the basin is also the home of the first National Parks in Nigeria. Agboarumi (1997) described the basin as blessed, mighty, great and the central hub upon which the entire social economic activities of the nation depends. It is the realization of this fact that the government of Nigeria in conjunction with the Global Environmental Facility (GEF) financed the sustainable management of the protected areas in the country with the main objective of strengthening the capacity of the relevant institutions at all levels in maintaining the biodiversity of this areas, Kainji Lake basin being among these protected areas (Papka, 2004).

Rivers, lakes, streams and water bodies have complex ecosystem ranging from mountain torrent to quiet, still lowland water which may be deep or shallow, large or small (Haslam 1978). The most important physical variable which affect the land-use and vegetation aquatic ecosystem include water movement and the quantity of flow, the soil or substance of the bed of the water course, the width and the depth of the channel, the general position of the channel in the river or lake system, the drainage orders, the slope (gradient of the channel and human intervention). Ikusemoran and Adesina (2009), reported that generally whenever a dam is constructed along a river channel, riparian communities around such locations are often affected directly and indirectly. For instance, the places that are inhabited by man are usually flooded, since the surface area of the river channel increases because of the dam construction. Also, the backward effect of the lake water creates some disturbances to the human population around the river channels. Moreover, human activities are also subjected to changes.

According to Ikusemoran and Adesina (2009), the impoundment of Kainji Lake on River Niger has converted the river to a lake ecosystem and has also changed the land cover around the formed lake. The forest formation has changed overtime, which could be due to the change in the orientation of the riparian communities through temporal displacement that took place after the creation of the lake reservoir. Any nation with sustainable utilization of her resources in mind must have adequate information on many complex interrelated aspects of its activities in order to make decisions. Land-use is only one of such aspects. The knowledge about land-use and land-cover has become increasingly important as every nation plans to overcome the problems of haphazard, uncontrolled development, deteriorating environmental quality, loss of important wetlands, and loss of fish and wildlife habitat. Land-use data analysis are needed in the analysis of environmental processes and problems that must be understood if living conditions and standards are to be improved or maintained at current levels.

One of the prime prerequisites for better use of land is information on existing land-use patterns and changes in land-use through time. The knowledge on land-use such as agricultural, recreational as well as information on their changing proportions is needed by legislators, state and local government officers in determining better land-use policies, identification of future development on pressure points and areas, and implementation of effective plans for regional development. In this dynamic situation, accurate and meaningful current data on land-use are essential for inventory of water resources, flood control, water supply planning and wastewater treatment. Many federal agencies use current and comprehensive inventories of existing and changing uses of land to improve the management of public lands. Federal agencies also need land-use data to access the environmental impact resulting from the development of energy resources, to manage wildlife resources and minimize man-ecosystem conflicts, to make national summaries of land-use patterns and changes for national policy formulation and to prepare environmental impact statements and assess future impacts on environmental quality.

Remote sensing and/or GIS have been applied to land-use and land-cover changes of Lake Reservoir within and outside Nigeria. Okhimanhe (1993) used the combination of Spot HRV imagery of 1986 and aerial photographs of 1974 to study the environmental impact assessment of Burumburum/Tiga dam in Kano State, Nigeria.

The study reveals that the construction of the dam contributed to the depletion of the vegetation that could have minimized desertification. Adeniyi and Omojola (1999) also used aerial photographs, Landsat MSS, Spot XS/panchromatic image transparency and topographical maps to study landuse and landcover changes in Sokoto and Guronyo dams, Nigeria between 1962 and 1986. Their studies revealed that settlement covered most part of the area after the construction of the dam.

Therefore, because of the central role of provision of electricity as well as conservation of flora and fauna, by the Kainji Lake, proper monitoring of the Lake Basin becomes important for sustainable utilization and development.

Objectives of the study

1. To create land-use/land-cover maps of the Kainji Lake basin from 1975-2006 using remotely sensed data and GIS techniques and using the maps to assess the trends, magnitudes, nature and locations of the land-use and vegetation cover changes of the lake basin within the study period.
2. To assess the actual land areas which have been lost or gained by the principal features in the lake basin, that is, the lake reservoir, intensive agriculture and the woodlands vegetation within the study period.
3. To evaluate the environmental and social economic implications of the land-use and land-cover changes in the lake basin.

The study area

According to National Electric Power Authority (NEPA) Diary (1995), the genesis of Kainji Lake power station dates back to 1951 when the demand for electricity was rising faster than supply due to the growth of industries and rapid urbanization in Nigeria. In order to meet the increasing demand for electricity, and consequent upon the realization that bulk supply of electricity could be cheaper through the utilization of hydro power technology, the former Electricity Corporation of Nigeria (ECN) began the exploitation of water resources of River Niger upstream of Jebba. Nedeco and Balfour (1961) reported that the reason for the choice of Kainji as the best site for the Lake were many among which are; rock foundation which was tested and was found to be capable of

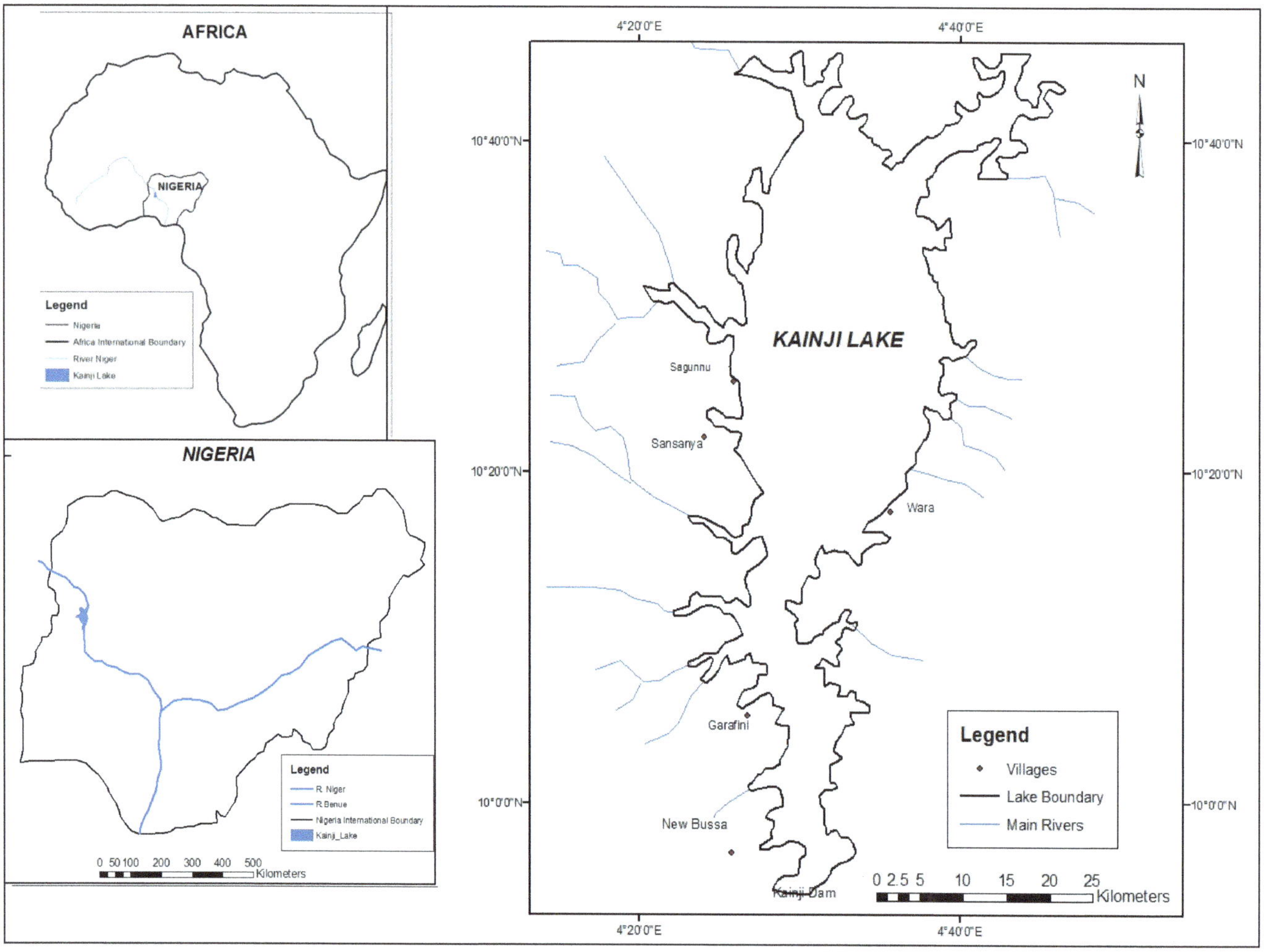

Figure 1. The study area.

holding the enormous height of the dam, it is also the point where the river valley is not too wide, the physical features of upstream of the dam valley also allows for a large reservoir. Agboarumi (1997) observed that the Kainji dam is today a pride of nature, providing cheap and abundant means of electricity for the continuously growing population and industries, sources of revenue, fishing, irrigation, cattle crossing, tourism, employment, international recognition, man-power training and many more.

The lake is located between latitudes 9° 50″ to 10° 42″N and longitudes 4° 20″ to 4° 42″E (Figure 1). Olokoh (1995) reported that construction of the Kainji dam which began in March 1964, was completed in December, 1968 and was officially commissioned on 15th February, 1969. The impounding of the lake started exactly on the 2nd August, 1968 and the water level rose to 140.2 m on the 19th of October the same year - a period of seventy eight days. The lake that was formed sunk most parts of Kainji Island on which the dam was constructed. NEPA Diary (1995) stated that the lake covers an area of 1250 km^2 with a maximum depth of 54.9 m. The lake extends to about 136.8 km upstream of Jebba beyond Yelwa in the North. The lake gets its water from two sources: the river Niger with its headwater in Guinea, and local rivers around the lake basin which flow directly into the lake or into River Niger before entering into the lake.

According to Nedeco and Balfour (1961), the soil depth of the area increases with slope and a gentle undulating topography for the area with red to brown well drained soils differing in texture from sandy loam to clay loam. Most of the basin area have dry season of five months starting from early November to middle of March. Rainfall increases from the month of April, reaching the peak in August and starts declining in September. Rainfall decreases with decrease in latitudes within the basin and also increases with increasing altitude (Olokoh 1995).

METHODOLOGY

Sources of data

The first four data sets (Table 1) which cover a total period of 31 years (1975-2006) were the main images that were acquired

Table 1. Sources of data

Data type	Source	Date
Landsat TM	http:glcffap.umd.edu:8080/esdi/index.jsp	08/12/1975
Landsat MSS	http:glcffap.umd.edu:8080/esdi/index.jsp	21/10/1986
Landsat ETM+	http:glcffap.umd.edu:8080/esdi/index.jsp	22/10/2001
Landsat ETM+	http:glcffap.umd.edu:8080/esdi/index.jsp	05/11/2006
Landuse/Landcover/ Vegetation Maps	Forestry Monitoring and Evaluation Unit (FOMECU) Abuja	Nov.1978 & Nov. 1995

for the monitoring of the Lake Basin. The first two data sets have large gap of more than ten years each, while the last two have close gaps of five years each. This was designed to reflect the impact of dam in the past years of low population and the recent years when the population of the country has been increasing tremendously. All the four images were acquired between late October and early December which is the transition period between the dry and wet season, the climatic conditions during this period is the same all over the lake basin (Olokor, 1995), hence the vegetation cover and other land-use types appear the same on the image regardless of the year when they were obtained. The fifth data sets (Table 1) were acquired from Forest Monitoring, Evaluation and Coordinating Unit (FOMECU) which was established in 1987 as an organ of the Department of Forestry in the Federal Ministry of Agriculture and Natural Resources with its primary mandate of coordination and monitoring of the implementation of Forestry II programs which contains aforestation, reforestation and desertification control projects (Federal Republic of Nigeria, 1999). The maps were used as complimentary and guide during the classification and ground truthing of the imageries.

Image classification

Three bands of infrared, red and green were selected from each of the four imageries to perform colour composition operations using Ilwis 3.3 Academic software. The Imageries were then "submap" individually through the use of the coordinates of the four corners of the lake basin. Colour separation was then performed in order to generate "map lists" while, the sample set module of the Ilwis was also used to create the "domains" which were used for the image classifications. The Maximum Likelihood Classification techniques of the Ilwis 3.3 were used to classify each of the four images. Accuracy assessment was conducted as error matrix commonly referred to as confusion matrix. The following overall accuracy results were obtained for the four images: 90.88% (1975), 87.44% (1986), 82.04% (2001) and 89.63% (2006). Each of the images was filtered through the filtering module of the Ilwis 3.3. The results of the filtered images are shown in Figure 2a-d. The processes of ground-truthing was carried out with the use of Germin 76 Global Positioning System which was used to take the coordinates of settlements, intensive agricultural area, extensive agricultural area around the dam site through Kainji town (see location in Figure 1), as well as some areas within the settlement of Kainji town (Figure 1), the recorded coordinates was linked with the classified image and while other landuse types were found to be correct, the differences between intensive and extensive agriculture on the image was initially difficult until several readings were made to determine the actual positions of each class. The acquired tailor-made map of 1995 processed from satellite images by FOMECU Nigeria, were also used as guide during the ground-truthing. In fact, it was the map in conjunction with the sampled coordinates of freshwater marsh and freshwater swamps taken at the off shore of the Lake at Gaski that was used to ground-truth the two features, that is, freshwater marsh and freshwater swamp. The lake itself was conspicuous on all the images.

Change detection techniques

Three main change detection methods which have been previously applied by several researchers (Adeniyi and Omojola, 1999; Ikusemoran, 2009a) were employed in this paper, they are:

Change detection by area calculation

There are three steps in calculating change detection by area calculation

1. The first step is the calculation of the magnitude of change, which is derived by subtracting observed change of each period of years from the previous period of years.
2. The second step was the calculation of the trends, that is, the percentage change of each of the land-use, by subtracting the percentage of the previous land-use from the recent land-use divided by the previous land-use and multiplied by 100 (B-A/Ax100).
3. The last is the calculation of the annual rate of change by dividing the percentage change by 100 and multiplied by the number of the study years, that is, thirty-one (31) years (1975-2006).

Change detection by nature

The nature of change was derived by map overlay. Each of the four classified images in Figure 2a-d was exported using the TIF. (Geo Tiff) TIFF of Ilwis 3.3 to Idrisi 32 for map overlay and analysis. The output images were overlain to generate the matrix tables (Tables 4a-c) using the addition module of the Idrisi. The area of the features in each image were then calculated through the area module of the Idrsi to generate the magnitude, trends and the percentage change of each of the features in each image.

In generating the matrix table (Table 4a-c) the first image was assigned values from 1 to the maximum class, that is, 10 (as in 2001 and 2006) and nine and eight in 1986 and 1975, respectively. While the second image was assigned 20, 30, 40 and so on until the maximum class of that image (for example 110 in 2001/2006 and 100 in 1986) was reached. (Table 4a-c) The essence of using these figures was that, since the addition of the overlay module was used, when two of these numbers in rows and columns were added, no two additions would produce the same result.

In the matrix tables (Table 4) all the numbers in red colour are the pixel values of each class when the maps were overlain. Those in blue colour were the pixel numbers of each class before they were overlain. The black numbers represents the areas of each of the overlain classes. The purple colour figures are the totals of each

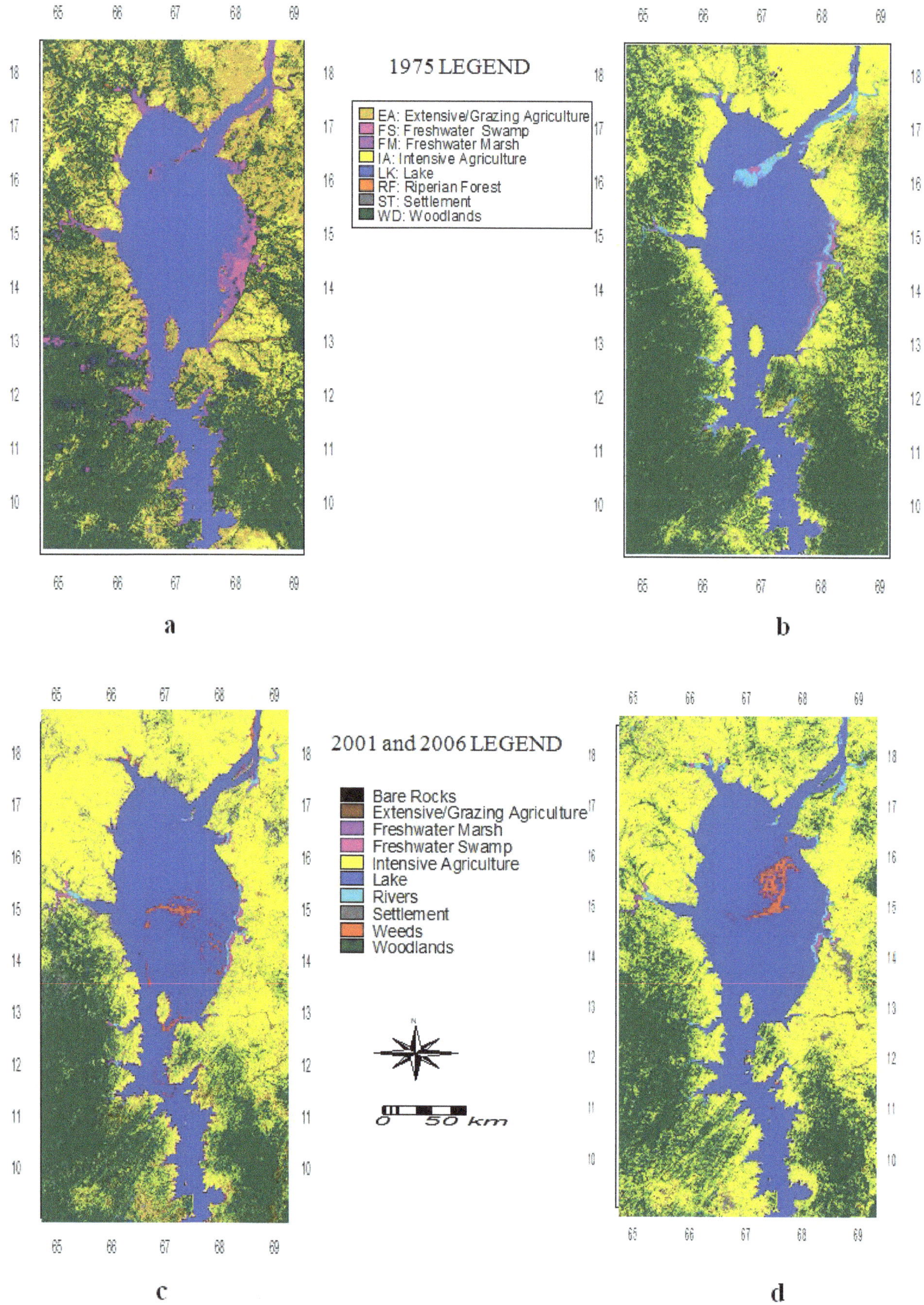

Figure 2. Classified Kainji Lake Basin. **a,**1975; b, 1986; **c,** 2001; **d,** 2006.

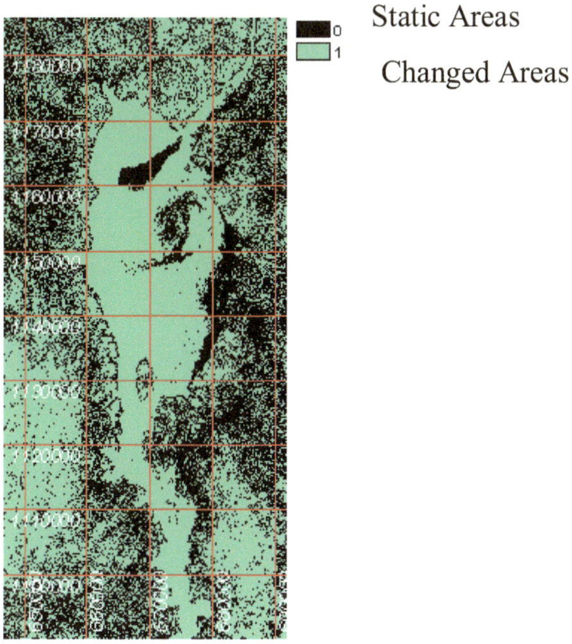

Figure 3. Changed and static areas (1975-2006).

of the classes. Finally, the green colour represents the total area of the study area in square kilometres.

There are three important classes of information that could be generated from the change detection by nature through the matrix table; the first information is that it shows the areas that have not recorded any change at all throughout the study period. This is represented by all the figures (bold) in the diagonals, that is, all the classes with the values in the diagonal have not recorded any change.

The second information that could be extracted were the areas or classes that have changed from their initial land-use or land-cover to another land-use or land-cover type, that is, the areas that have lost to other land-use/land-cover type. These areas were represented by all the classes with values along the columns in the matrix tables. The third was the areas that have gained from other land-use/land-cover type, that is, the areas with all the values along the rows of the matrix tables.

Change detection by location

The reclass module of the idrisi software was used to create the map that shows the areas that have experienced changes (changed areas) and the areas that have not (static areas) within the study periods. All the classes with the values in the diagonals, (the static areas) were assigned "1" while the values of the other classes, that is, those at the rows and columns were assigned "0" (changed areas). The output map shows the static (1) and the changed areas (0) (Figure 3). The area of the output map was also automatically calculated in Idrisi environment.

Mapping and calculations of the extent of changes of each class

The logic expression of the image calculator and the area module of the Idrisi were used to map and calculate the extent to which each of the features in the basin has changed respectively.

However, only three features of major concern were selected for this purpose: the lake reservoir, the woodland vegetation and the intensive agriculture areas. The result was to produce the following:

1. Areas that were not Lake Reservoir in 1775 but have changed to lake reservoir in 2006 so as to know whether the lake is expanding or not as has been the fear of Nigerians.
2. Areas that were of woodland vegetation type in 1975 but have changed to other land-use types in 2006. This was done to determine the land areas of the woodlands (home of National Park) that have been lost within the study periods to other landuse types.
3. Areas that were not intensive agriculture in 1975 but have changed to intensive agriculture in 2006 so as to assess the rate of land captured by intensive agriculture within the study periods.

RESULTS

The result of the trends of changes of the landuse and landcover of Kainji Lake is presented in Table 2, while the magnitudes, percentage changes and annual rate of changes are presented in Table 3a-c. The spatial locations of the changes in the lake area between 1975 and 2006 are presented in Table 4a-c, while Table 5 shows the extent of changes in the Lake area within the study period. Moreover, Figure 2a-d represents each of the four classified images of 1975, 1986, 2001 and 2006, respectively, while Figure 3 shows the parts of the lake that have not been subjected to changes (static) between 1975 and 2006. Figure 4a represents the total areas captured by the reserviour between 1975 and 2006, Figure 4b shows the area that was captured by intensive agriculture within the study period, while Figure 4c shows the areas that have been lost by woodland vegetation between the same period.

DISCUSSION

Trends, magnitudes, percentage changes and annual rate of changes

Table 2 shows the trends in terms of the area coverage and the percentage of each class of the basin area from 1975 to 2006. It was revealed that the Lake Reservoir that covered 910.6224 km^2 in 1975 has increased to 1094.2302 km^2 (addition of 1053.3750 km^2 lake reservoir and 40.8602 km^2 weed coverage) in 2006. The woodland vegetation of the Basin area which is the home of the first National Park in the country has also reduced in size from 41.41% in 1975 to 24.38% in 2006. Moreover, intensive agriculture that covered only 5.59% in 1975 has increased to 40.95% in 2006. There was the presence of riparian forest in 1975 which had disappeared in the lake basin before 1986. Human settlement area that covered only about 12 km^2 in 1975 has risen to 175 km^2 in 2006. Finally, between 1975 and 1986, there were no weeds on the lake, but the weeds has not only appeared on the lake but was rapidly capturing the lake surface which has been a great concern to the country as noted by Ayeni et

Table 2. The trends of the landuse/landcover changes of the lake basin (1975-2006).

Landuse/Landcover	1975		1986		2001		2006	
	Area (km^2)	Area (%)	Area (km^2)	Area (%)	Area (km^2)	Area (%)	Area (km^2)	Area (%)
Lake Reservior	910.6224	21.07	955.8086	22.12	1010.4105	23.38	1053.3760	24.38
Woodlands	1789.2332	41.41	1925.3175	44.55	1366.9067	31.63	1146.5369	26.53
Intensive Agriculture	241.4720	5.59	1031.4233	23.87	1549.6968	35.86	1769.5118	40.95
Extensive/ Grazing	1131.4632	26.18	200.4712	4.64	119.1034	2.76	59.8514	1.39
Settlement	12.3040	0.28	69.9070	1.62	140.7425	3.26	175.5531	4.06
Freshwater Swamp	187.6231	4.34	40.9016	0.95	9.0850	0.21	7.4500	0.17
Freshwater Marsh	41.8048	0.97	12.0611	0.28	8.3117	0.19	6.1934	0.14
Riparian Forest	6.6992	0.15	-	-	-	-	-	-
Rivers	-	-	84.4552	1.95	51.5014	1.19	61.4718	1.42
Bare Surface	-	-	0.8732	0.02	0.3219	0.01	0.4142	0.01
Weeds	-	-	-	-	62.1358	1.51	40.8602	0.95

Table 3a. The magnitudes, percentage and annual rate of changes of the lake basin (1975-1986).

Landuse/Landcover	1975	1986	Magnitude of change	Percentage change	Annual rate of change	Remark
LakeReservoir	910.6224	955.8086	45.1862	2.03	+4.97	Increase
Woodlands Vegetation	1789.2332	1925.3175	136.0343	6.11	+7.61	Increase
Intensive Agriculture	241.4720	1031.4233	789.9513	35.45	+327.14.	Increase
Extensive/Grazing	1131.4632	200.4712	930.992	41.78	-82.28	Decrease
Settlement	12.3042	69.9070	57.603	2.59	+468.16	Increase
Freshwater Marsh	187.6231	40.9016	146.7215	6.58	-78.20	Decrease
Freshwater Swamp	41.8084	12.0611	29.7437	1.33	-71.15	Decrease
Riparian Forest	6.6962	-	6.6962	0.30	-100	Decrease
Rivers/Freshwater	-	84.4552	84.4552	3.79	+8445.52	Increase
Bare Surface	-	0.8732	0.8732	0.04	+87.32	Increase
Total	4321.2189	4321.2187	2228.3066	100	9009.09	

Table 3b. The magnitudes, percentage and annual rate of changes of the lake basin (1986-2001).

Landuse/Landcover	1986	2001	Magnitude of change	Percentage change	Annual rate of change	Remarks
LakeReservoir	955.8086	1010.4105	54.6019	3.85	+5.71	Increase
Woodlands Vegetation	1925.3175	1366.9067	558.4108	39.39	-29.00	Decrease
Intensive Agriculture	1031.4233	1549.6968	518.2735	36.55	+50.25	Increase
Extensive/Grazing	200.4712	119.1034	81.3678	5.74	-40.59	Decrease
Settlement	69.9070	140.7425	70.8355	5.00	+101.33	Increase
Freshwater Marsh	40.9016	9.0850	31.8166	2.24	-77.79	Decrease
Freshwater Swamp	12.0611	8.3117	3.7494	0.26	-31.09	Decrease
Rivers/Freshwater	84.4552	51.5015	32.9538	2.32	-39.02	Decrease
Bare Surface	0.8732	0.3249	0.5513	0.04	-63.14	Decrease
Weeds	-	65.1359	65.1358	4.59	+651.36	Increase
Total	**4321.2187**	**4321.2189**	**1419.6964**	**99.98**	**528.02**	

al. (1994), Obot et al. (1995) Adesina et al. (1998) and Mbagwu et al. (2002).

In Tables 3a-c, the total land area that was subjected to changes (magnitude) in the lake basin was revealed to

Table 3c. The magnitudes, percentage and annual rate of changes of the lake basin (2001-2006).

Landuse/Landcover	2001	2006	Magnitude of change	Percentage change	Annual rate of change	Remark
LakeReservoir	1010.4105	1053.3760	42.9655	6.98	+4.25	Increase
Woodlands Vegetation	1366.9067	1146.5369	220.3698	35.81	-16.12	Decrease
Intensive Agriculture	1509.6969	1769.5118	219.8149	35.72	+14.18	Increase
Extensive/Grazing	119.1034	59.8514	59.252	9.63	-49.75	Decrease
Settlement	140.7425	175.5531	34.8106	5.66	+24.74	Increase
Freshwater Marsh	9.0850	7.4500	1.635	0.27	-18.00	Decrease
Freshwater Swamp	8.3117	6.1934	2.1183	0.34	-25.49	Decrease
Rivers/Freshwater	51.5015	61.4718	9.9703	1.62	+19.36	Increase
Bare Surface	0.3219	0.4142	0.0928	0.02	+28.67	Increase
Weeds	65.1359	40.8602	24.2757	3.95	-37.27	Decrease
Total	4321.2189	4321.2188	615.3044	100	-55.45	

Table 4a. The matrix table created from the 1975 and 1986 overlay.

	1986									
Landuse/ Landcover 1975 ↓	Lake 20	Woodlands 30	Intensive Agric. 40	Extensive/ Grazing 50	Settlement 60	Freshwater Marsh 70	Freshwater Swamp 80	Rivers 90	Bare Surface 100	Total
Lake 1	21 **860.0672**	31 1.6351	41 12.9424	51 0.5239	61 8.6951	71 6.0439	81 1.5352	91 19.1756	101 0.0041	910.6225
Woodlands 2	22 8.2687	32 **1338.4333**	42 296.5595	52 95.1436	62 8.6927	72 8.5479	82 3.5479	92 21.1956	102 0.4240	1789.2332
Intensive Agric 3	23 0.1316	33 83.2523	43 **136.7763**	53 16.0322	63 4.7008	73 0.1218	83 0.0097	93 0.3095	103 0.1381	241.472
Extensive/ Grazing 4	24 19.7831	34 457.7723	44 539.6454	54 **78.1327**	64 24.9840	74 4.2408	84 0.1852	94 6.4411	104 0.2786	1131.4632
Settlement 5	25 0.3891	35 1.8113	45 8.5847	55 0.3696	65 **1.0803**	75 0.0268	85 -	95 0.0162	105 0.0260	12.304
Freshwater Marsh 6	26 67.0398	36 18.8986	46 31.9482	56 3.5179	66 12.7686	76 20.1308	86 2.1308	96 30..5991	106 0.0016	187.6232

Table 4a. Contd.

Freshwater Swamp 7	27 0.1291	37 18.8344	47 3.7298	57 6.2600	67 0.6141	77 1.6245	87 **4.0335**	97 6.5743	107 -	41.8047
Riparian Vegetation 8	28 -	38 4.6802	48 1.2371	58 0.4914	68 0.0967	78 0.0203	88 0.0260	98 **0.1438**	108 0.0008	6.6963
Total	955.8036	1925.3175	1031.4235	200.4113	69.9074	40.9016	12.0611	84.4552	1.2732	4321.22

Table 4c. The matrix table created from the 2001 and 2006 overlay.

	2006										
Landuse/ Landcover 2001 ↓	**Lake 20**	**Woodlands 30**	**Intensive Agriculture 40**	**Extensive/ Grazing 50**	**Settlement 60**	**Freshwater Marsh 70**	**Freshwater Swamp 80**	**Rivers 90**	**Bare Surface 100**	**Weeds 110**	Total
Lake 1	21 **964.4217**	31 0.1129	41 1.3394	51 0.0227	61 5.2975	71 0.0008	81 0.0812	91 6.8546	101 0.0008	111 32.2788	1010.4104
Woodlands 2	22 1.2444	32 **867.8331**	42 453.8906	52 16.1589	62 18.9546	72 1.9039	82 1.2874	92 5.4015	102 0.0252	112 0.2071	1365.9147
Intensive Agric 3	23 4.4503	33 235.0917	43 **1146.0406**	53 32.2625	63 115.9486	73 2.2954	83 0.8878	93 11.6915	103 0.2729	113 0.7554	1550.0967
Extensive/ Grazing 4	24 0.2087	34 21.8463	44 80.8440	54 **10.0093**	64 5.2309	74 0.1340	84 0.1470	94 0.5515	104 0.0536	114 0.0780	119.1033
Settlement 5	25 7.6693	35 16.7632	45 80.0991	55 1.2070	65 **24.1368**	75 0.8090	85 0.3736	95 9.1102	105 0.0487	115 0.5255	141.2872
Freshwater Marsh 6	26 0.3078	36 2.5781	46 1.0194	56 0.0439	66 0.2372	76 **1.1469**	86 1.3118	96 2.4254	106 -	116 0.0146	9.0852
Freshwater Swamp 7	27 1.2371	37 0.4776	47 0.1949	57 0.0114	67 0.0796	77 0.6100	87 **1.5287**	97 4.1668	107 -	117 0.0057	8.3119
Rivers 8	28 24.6737	38 1.5985	48 3.1328	58 0.0707	68 1.9267	78 0.5483	88 0.5491	98 18.4202	108 0.0008	118 0.5808	50.5022

Table 4c. Contd.

Bare Surface 9	29	39	49	59	69	79	89	99	109	119	0.3211
	0.0016	0.0065	0.1746	0.0089	0.1178	-	-	0.0041	**0.0106**	0.0008	
Weeds	30	40	50	60	70	80	90	100	110	120	65.1358
10	49.1614	0.2291	2.7763	0.0560	3.6234	0.0016	0.0268	2.8461	0.0016	**6.4135**	
Total	1053.376	1146.537	1769.5117	59.8513	175.5531	7.4499	6.1934	61.4719	0.4142	40.8602	4321.22

Table 5. The extent of changes of the lake reservoir, woodlands and intensive agriculture (1975-2006).

	Land-use/ Land-cover class	Area (km^2)	Area (%)
1	LakeReservoir in 2006 but notLake Reservoir in 1975	205.2586	4.75
2	Intensive Agriculture in 2006 but not Intensive Agriculture in 1975	1587.7759	36.75
3	Woodlands in 1975 but not woodland in 2006	943.2975	21.83

be 2228.3066 km^2 between 1975 and 1986 which decreased to 1419.6964 km^2 before 2001 and reduced drastically to 615.3044 km^2 in 2006. This means that some parts of the lake have been subjected to changes. This was also revealed in Figure 3.

Extensive agriculture/grazing land and intensive agriculture had the highest percentage changes of 41.78 and 35.45%, respectively, between 1975 and 1986, though, while extensive agriculture changes positively, intensive agriculture recorded negative changes.

From 1986 through 2001 to 2006, extensive agriculture/grazing land reduced drastically which is a result of the policy of the Federal Government against grazing in the basin area in order to protect the National Park.

This policy has always resulted in communal classes between the communities, the park rangers and the nomads (Mayowa and Omojola, 2005; Ikusemoran, 2009b; Henry et al., 2013; Wahab and Adewumi, 2013).

The nature of the changes of the lake basin

The matrix tables (Table 4a-c) show the nature of the land-use and land-cover changes in the basin area. Between 1975 and 1986 (Table 4a), the reservoir had static area of 860.0672 km^2 but gained heavily from freshwater marsh (67. 0398 km^2), extensive agriculture (19.7831 km^2) and woodlands (8.2687 km^2). Within this period, parts of the lake were also lost to other classes especially 19.1756 and 12.9424 km^2 to rivers/-freshwater and intensive agriculture, respectively. Woodland vegetation, though gained 457.7723 km^2 from extensive agriculture and a sizeable areas from other classes but also lost 296.5595 km^2 to intensive agriculture, 95.1436 km^2 to extensive agriculture and little losses to other classes.

However, the gain was more than the losses which might be due to the then few population in the basin area coupled with the strict protection of the Kainji Lake National Park that was just established and shouldered the responsibility of the protection of the basin.

Riparian forest that was found especially along the main rivers had disappeared before 1986 as it was mostly captured by woodlands. In Table 4b, the Lake basin, though, was still capturing much of the freshwater marsh (29.8412 km^2) but the rivers/freshwater class was the worst victim as the lake gained a total land area of 48.2208 km^2 from the class.

Weeds that suddenly appeared on the lake after 1986 as reported by Daddy et al. (1995) and water hyacinth *(Eichhorinia crassipes)* was first reported in 1989 on Kainji Lake and since then the plant continued to increase and spread to every part of the lake. Hence the dominance of the weeds on the lake as 42.7381 km^2 of the surface of the lake was captured by the weeds between 1986 and 2001. Intensive agriculture gained much from woodlands, and extensive agriculture during this period (644.2672 and 132.0717km^2 respectively) but also lost some parts to other land-use

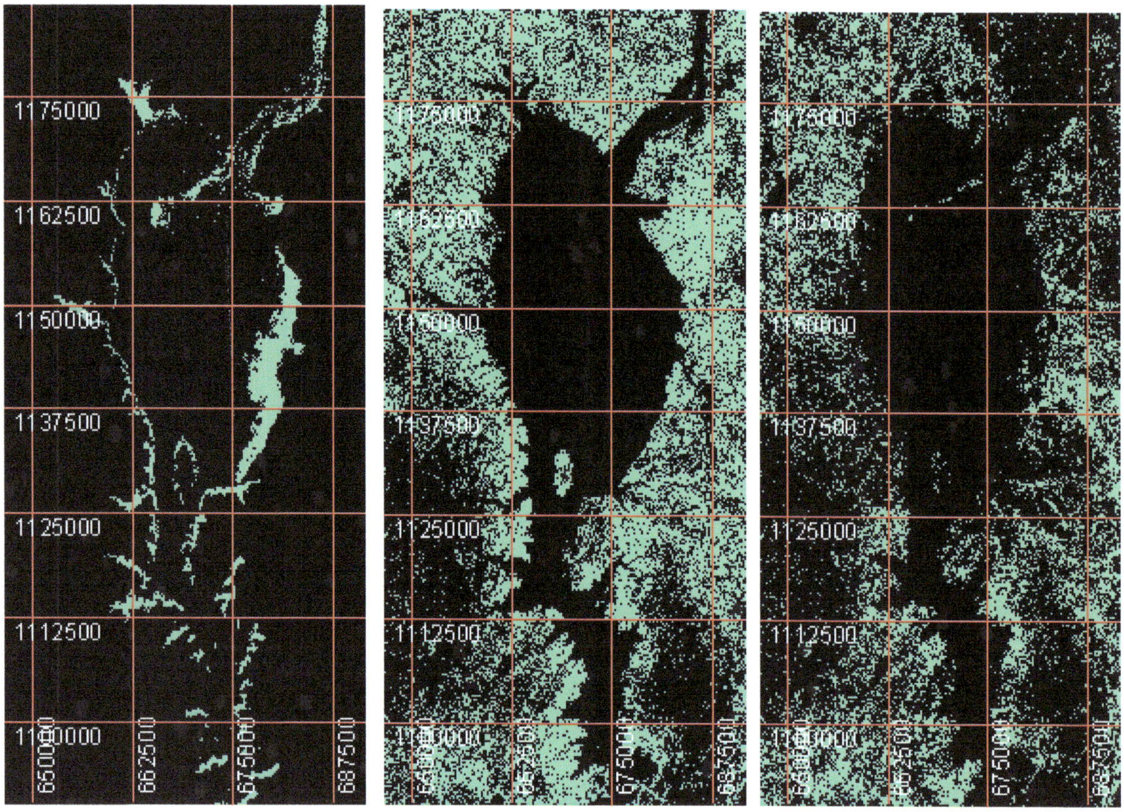

a b c

Figure 4. a. Area captured by the reservoir between 1975 and 2006; **b.** Area captured by intensive agriculture between 1975 and 2006; **c.** Area loss by woodlands between 1975 and 2006.

types as 155.4621, 78.0515 and 31.1660km^2 to woodlands, settlement and extensive agriculture, respectively.

Table 4c shows the nature of the changes of the lake basin between 2001 and 2006. Due to the control measures especially by the National Institute of Freshwater Fisheries Research (formerly Kainji Lake Research Institute). Daddy et al. (1995) reported that the institute evolved mechanical, biological and manual control measures in controlling the water hyacinth on the lake. They reported that 45% of the estimated coverage by the weeds has been effectively removed. It was these efforts that made the lake to regain 49.1614 km^2 from the weed coverage area within this period

The spatial location of the changes in the lake basin

Spatial location procedures show the actual points or areas where changes have really taken place within the study period. The spatial locations of changes between 1975 and 2006 were derived by the overlay of 1975 and 2006 images and creating a matrix table from them in the same way it is shown in Figure 4a-c. The result of the reclass processes using the figures in the diagonal as 1 and other figures in the rows and columns as 0 gives an output map as shown in Figure 3.

The area module of the Idrisi was used to calculate the area of the output map. Out of the total basin area of 4321.22 km^2, it was revealed that 1638.9273 km^2 representing 37.93% has been subjected to changes, while 2682.2915 km^2 (62.07%) of the lake area has not been subjected to any change (Figure 3). The large static area is due to the large area coverage of the lake and the protected woodland area by the National Parks of Nigeria.

The extent of changes of each land-use/land-cover

It is important to know the extent of changes of each land-use type so as to assess the area coverage of the areas in any desired unit and percentages for proper land-use planning. The image calculator of the Idrisi software was used to determine the extent of the changes on the Lake Reservoir, woodland vegetation and intensive agriculture. These three classes are of great concern to the country as the lake has been a subject of discussion as having the possibility of dam collapse, some environmentalists (Ikusemoran, 2009a; Mbagwu et al., 2002) have also been emphasizing the loss of the woodland vegetation in the basin area which may affect the ecosystem of the lake basin, no wonder Wahab and Adewumi (2013) noted the presence of 14 sawmills within 40 km buffer around the park boundary between Wuromakoto and Gidan Aboki. Finally, human activities especially intensive agriculture was singled out to be the main activity responsible for the degradation of the lake basin (Ikusemoran, 2009a). The result of the image calculator on the different images as used is presented in Figure 4a to c. The values of the area of each class were also used to create Table 5.

A total area of 205.2558 km^2 representing 4.75% (Table 5) of the lake basin has been captured by the reservoir between 1975 and 2006. This means that the lake is actually expanding as has earlier been reported by Abiodun (2009) and Ikusemoran (2009a). The woodland vegetation has also lost a total area of 943.2975 km^2 (21.83%) between 1975 and 2006. Intensive agriculture is the degrading factor of the basin as 1587.7759 km^2 (36.75%) of the basin area have been lost to intensive agriculture between 1975 and 2006 (Table 5). This conforms to the earlier reports of Henry et al. (2013) that increased human and cattle population is continuously putting more pressure on the Kainji Lake Area and has ultimately cause fragmentation and degradation of wildlife habitats.

Implications of the land-use changes on the lake basin

1. Communal classes: Communal classes among the farmers, communities and the nomadic have recently become a problem especially in the savanna belt of Nigeria because of classes on land for their desired interests. For instance, extensive/grazing land has reduced drastically from 1131.4632 km^2 in 1975 to 200.4712 km^2 in 2006 and yet more animals are being reared in this belt since it is the main occupation of the people in the area, hence there is much pressure on the land and government efforts in controlling the parks have been faced with stiff opposition by the nomads who have no other occupation and no other home than the savanna where pastures are available for their animals. Henry et al. (2013) have reported on this syndrome that the effects of all these pressure on land culminate in increased man and animal conflicts lead to revenge killings and poaching.
2. Dam collapse: The area coverage of the lake reservoir has been increasing. The lake that covered 910.6224 km^2 in 1975 has increased to 1075.5464 km^2 (weed area coverage inclusive) a difference of 164.924 km^2. Many individuals, organizations and NGOs have been reporting that if the lake is actually increasing in size, then dam failure is inevitable unless solid maintenance is put in place, but how effective the maintenance will be in a country where maintenance culture is not valued remains a great question that calls for immediate answer.
3. Loss of Agricultural Land: With the increasing settlement area (12.3040 to 175.5531 km^2 between 1975 and 2006), and with the high rate of increase of intensive agriculture, (241.4720 km^2 in 1975 to 1769.5118 km^2 in 2006) (Table 2), if unchecked, there would soon be no land to use for agriculture. At present from Figure 4b, the only places that are yet to be utilized for agriculture are the lake reservoir and some parts of the preserved woodland areas.
4. Reduction or loss of park land: The reduction of the size of the woodland vegetation area as reflected in Table 3a-c coupled with the increasing capturing of the woodland area especially by intensive agriculture (Table 4a-c). According to Amusa et al. (2010), reduction of the park land could be attributed to the fact that the primary occupation of the Kainji Lake area is agriculture, which in effects reduced the park land area. This means that sooner or later the woodland vegetation which is the home of the first national park in Nigeria will become a history. Moreover, due to poaching and other human interference, many animals, plants and other micro organisms in the basin may become extinct.
5. Land exposure/desiccation: Continuous usage of land for agriculture may lead to agricultural land been converted to open, non-cultivated type such as open grassland or sandy bare surfaces as noted in Table 3a-c where bare surface that was not present in the basin in 1975 has been appearing though at a slow rate since 1986. Adeniyi and Omojola (1997) asserted that land exposure and desiccation are noted for increasing the local rainfall run-off and reducing infiltration. This factor affects the total water balance of a drainage basin, reduce soil matter by exposure to agents of erosion and consequently hamper land cultivation. Ogunwusi (2012) also reported that over 90% of the natural vegetation in Nigeria had been cleared and over 350,000 ha of forest and natural vegetation are lost annually.
6. Flooding: The expansion of the lake water has resulted into serious flooding especially at the western bank (where much land has been captured by the lake (Figure 4a) and the downstream which is the receiving end (Abiodun, 2009). The sudden appearance of weeds on the lake has also been attributed to another possible cause of flood of the lake (Olokor et al., 1998; Mbagwu, 2002).

In the work of Okhimanhe (1993) highest susceptibility of the flood in the Kainji Lake occurs within the first 3 km from the reservoirs including the river channels and surface waterbody lying between 69 and 162 m above sea level, while the lowest is at 15 km away from the reservoir and greater than 193 m above sea level. The serious flood at the downstream of this lake in this year (2012) has caused many displaced settlements, destructions of farmlands and properties (Adedeji et al., 2012; Nkeki et al., 2013).

7. Reduction in fish catch: The degradation of the lake area coupled with other factors such as excessive water (floods) and weed encroachment no doubt have collectively impart negatively on the annual fish catch in the lake area as Omojowo et al. (2010) reported that total fish catch reduced from 38,346 mt in 1996 to 13, 361 mt in 2001.

Conclusion

In this paper, the use of remotely sensed data and GIS techniques for environmental monitoring has been demonstrated. The Kainji Lake basin which is the home of the Kainji dam (the first and the largest hydro-electricity dam in Nigeria which provides electricity to the entire country and other neighbouring countries such as Niger Republic), the home of the first National Parks in Nigeria, and the base of various research institutions such as National Institute of Freshwater Fisheries Research, College of Wildlife, Fisheries College and many more is so important to the country that all hands must be on deck to see to its sustainable utilization. The lake water has been revealed to be expanding- causing fear of dam collapse, the woodland vegetation is rapidly decreasing- putting the existence of the National Park in suspense, water hyacinth has recently captured some parts of the lake, and despite numerous efforts by the government and research institute, the weeds as at 2006, still covered more than 40 km^2 of the lake surface, intensive agriculture that was very minimal in 1975 was discovered to have captured more than 40% of the lake basin. The lake basin therefore, needs nothing but proper monitoring if only the survival of the lake is in the interest of the country. Human interventions especially agricultural practices was discovered to be the major cause of the changes in the basin, this calls for further studies in the preservation of the basin from human interference. Dam collapse, communal classes, flooding, land desiccation and reduction in park and agricultural lands have also all been discussed as the possible implications of the changes on the lake basin. Remote sensing and GIS techniques can be used to monitor land-use and land-cover changes to guarantee quality environment which would provide for future needs of the nation.

Recommendations

The Kainji Lake basin is so much important to Nigeria that its sustainable utilization should not be compromised. The basin which is the central hob of Nigeria's economy should be well protected from environmental degradation. The lake water is rapidly expanding which may cause serious problems to the inhabitants especially at the downstream, hence, the government should do all they can to resettle the people at the sides and the downstream of the lake. Furthermore, government should discourage new settlement in the basin area so as to reduce pressure on the land area. To complement the efforts of the National Parks in the protection of the basin, Kainji Lake Basin Management Authority should be established to be solely responsible for the protection of the Lake Basin. The protection of the woodland vegetation of the basin should be of primary concern to the country as the depletion of the vegetation may lead to the end of the first and the most popular national park in the country. There should be proper land-use and land-cover planning of the basin area to reduce land degradation or loss of wetlands, provision of modern irrigation facilities for efficient utilization of the lake water instead of bastardizing the basin land surface, provision of improved seedlings that would yield better output from minimal land area, extensive training on lake basin, wetlands and general environmental management and finally ensuring public awareness and action to guarantee quality environment and making the lake basin a recreation and eco-tourism site. Remote Sensing and GIS techniques, though, still very new in most parts of Nigeria, is second to none in environmental monitoring, hence, the techniques is recommended for use in environmental monitoring and management of our already fragile environment.

REFERENCES

Abiodun AA (2009). Waters of Lake Kainji - Hydrological Predictions and Perfomance. Hydrol. Sci. J. 18:(3) 321-327.

Adedeji OH, Odufuwa BO, Adebayo OH (2012). Building Capabilities for Flood Disaster and Hazard: Preparedness and Reduction in Nigeria. J. Sustain. Dev. Afr. 14(1):45-48.

Adeniyi PO, Omojola AS (1999). Landuse and Landcover Changes in Sokoto-Rima Basin of North-West Nigeria, Based on Archival Remote Sensing and GIS Techiques. An African Association of Remote Sensing of the Environment (AARSE) on Geo-information Technology Applications for Resource and Environmental Management of Africa.

Agboarumi B (1997). Borgu: Past, Present and Future. Matanmi Press, Ilorin, Nigeria.

Amusa TO, Aridanzi P, Haruna M (2010). Ethnobotany and Conservation of Plant Resources in the Kainji Lake National Park. Ethnobotany Res. Appl.. 8:181-194.

Ayeni JSO, Daddy F, Mdalhi M, Obot EA (1994). Control and Utilization of Aquatic Weeds: Water Hyacinth and Niger Grass on Lake Kainji. Annual Report of the Federal Ministry of Agriculture, Water Resources and Rural Development.National Institute for Freshwater Fisheries Research. pp. 20-24.

Daddy F, Pitan F, Aina E, Abubakar A (1995). Establishment of NeochetinaEichomiaeand N. BruchiPopulatios for Water Hyacinth Control on Lake Chad. Annual Report of the Federal Ministry of Agriculture, Water Resources and Rural Development: National Institute for Freshwater Fisheries Research.Forestry Monitoring and Evaluation Unit (FOMECU), Federal Ministry of Environment, Abuja, Nigeria

Federal Republic of Nigeria (1999). Combating Desertification and Mitigating the Effects of Drought in Nigeria: National Report on the Implementation of the UN Conventions to Combat Desertification in those Countries Experiencing Serious Drought and or Desertification

Particularly in Africa: Third Session of the Conference of the Parties, Racita, Brazil. Nov. 1999

Haslam SM (1978). River Plants.The macrophyte Vegetation of Water Courses.Cambrdge University Press, London. http:glcffap.umd.edu:8080/esdi/index.jsp. Retrieved on Oct.10th, 2010

Henry MI, Augustine UO, Damilola O (2013). Analysis of Poaching Activities in the Kainji Lake National Park, Nigeria. Environ. Nat. Resour. Res. 3(1): 51-61

Ikusemoran M (2009a).Landuse and Landcover Change Detection of Kainji Lake Basin (1978-1995), Nigeria.: Remote Sensing and GIS Approach. Bayero J. Pure Appl. Sci. 2(1): 83-91.

Ikusemoran M (2009b). Logistic Regression and GIS for Predicting Woodland Vegetation Changes of the Kainji Lake Basin, Nig. J. Environ. Sci. University of Jos, Nigeria. 8(1):54-65

Ikusemoran M, Adesina O (2009). Landuse and Landcover Change Detection of Jebba Lake Basin, Nigeria: Remote Sensing and GIS Approach. University of Benin, Nigeria. Knowl. Rev. Multidiscip. J. 18(1): 122-132

Mayowa JF, Omojola AS (2005). Climate Change, Human Security and Communal Classes in Nigeria. A Paper Presented at an International Workshop on Human Security and Climate Change. Asker, USA 21-23 June 2005.

Mbagwu IG, Ladu BMB, Amadi AO (2002). Contributions of the Seasonal and Long Term Flood Patterns of River Niger to the Development and Control of Water Hyacinth in Kainji Lake Nigeria.Annual Report of the Federal Ministry of Agriculture, Water Resources and Rural Development. National Institute for Freshwater Fisheries Research.

Nedeco C, Balfour B (1961). River Studies and Reformation on Improvement of River Niger and Benue. North Holland Publishing Company, Amsterdam.

NEPA Diary (1995). National Electric Power Authority, Federal Ministry of Power, Nigeria

Nkeki FN, Henah PJ, Ojeh VN (2013). Geo-Spatial Techniques for the Assessment and Analysis of Flood Risk along the Niger - Benue in Nigeria. J. Geogr. Inf. Sys. 5:123-135.

Obot EA, Ayeni JSO, Daddy F, Olokor JO (1995). Water Hyacinth in Kainji Lake: Its Effect on the Use of the Lake Resources.Annual Report of the Federal Ministry of Agriculture, Water Resources and Rural Development. National Institute for Freshwater Fisheries Research. pp. 16-21.

Ogunwusi AA (2012). The Forest Products Industry in Nigeria. African Research Review. Afr. J. Online 6(4):191-205.

Okhimanhe AO (1993). Assessment of Environmental Impact of Dam Construction in Nigeria: A Case Study of Tiga Dam, Kano State. An Unpublished M.Tech Disertation in Remote Sensing, Department of Geography, Federal University of Technology, Minna, Nigeria.

Olokor JO (1995). The Climate of Kainji Lake Area.Annual Report of the Federal Ministry of Agriculture, Water Resources and Rural Development. National Institute for Freshwater Fisheries Research.

Olokor JO, Daddy F, Adesina O (1998). Mapping the Spatial Spread of Water Hyacinth on KainjiLake. Annual Report of the Federal Ministry of Agriculture, Water Resources and Rural Development.National Institute for Freshwater Fisheries Research. pp. 15-17.

Omojowo FS, Olowosegun T, Omojowo TMa (2010). Fish Consumption Pattern in Kainji Lake Areas. World Rural Obs 2(1):75-79.

Papka PM (2004). Local Empowerment and Environmental Management Project (LEEMP) Information Handbook.Federal Ministry of Environment, Nigeria.

Wahab MKA, Adewumi AA (2013). Assessment of Community Participation in Protected Area: A Case Study of Kainji Lake National Park, Nigeria. Int. J. Econ. Finance Manage. 2(1):60-64.

Heavy metal content of selected African leafy vegetables planted in urban and peri-urban Nairobi, Kenya

Mutune, A. N., Makobe, M. A. and Abukutsa-Onyango, M. O. O.

Jomo Kenyatta University of Agriculture and Technology, P.O. Box 62000 - 02000. Nairobi, Kenya.

African indigenous vegetables planted along Nairobi Rivers are suspected to absorb metals from industrial and domestic effluent. Ten (10) of the commonest vegetables in Kenyan markets grown along these rivers and the soils in their rhizosphere from 25 sites were analyzed for Pb, Cu, Zn, Cd and Cr. Soils were air-dried then leached using 0.05 M EDTA, filtered and analysed in AAS. Vegetables were oven-dried, ground, ashed then analysed in AAS. In both soil and vegetables, the metal concentration was generally in the order Zn > Cu > Pb > Cd >Cr. In soil the concentrations were Pb 0.57 - 20 mg/kg, Cu 3.59 - 75.37 mg/kg, Zn 14.62 - 198.3 mg/kg, Cr 0.03 - 1.4 mg/kg and Cd 0 - 2.6 mg/kg. In vegetables the values were Pb 0 - 2.4 mg/kg, Cu 0.52 - 21.34 mg/kg, Zn 20.13 - 89.85 mg/kg, Cd 0 - 3.02 mg/kg and Cr 0 - 1.24 pp. There were significant differences within vegetables at each site (P<0.05). In most sites, there was a positive correlation of soil metal content with that in vegetables. The metal concentrations in soil were within permissible levels allowable by WHO/FAO except for a few instances in Cd. In vegetables all metals except Cu were in a few sites higher that the recommended limits. Government clean-up activities and monitoring of waste disposal is recommended for potential agricultural land.

Key words: African Indigenous vegetables, heavy metals, pollution, urban and peri - urban agriculture, Nairobi.

INTRODUCTION

Nairobi, like other cities in the developing countries is expanding tremendously without proper planning of utilities, safety and regard to environment, making residents dump waste into rivers and on land (Yebpella et al., 2011). Wastes may contain heavy metals among other contaminants (Olowoyo et al., 2011; Tiwari et al., 2011; Sinha et al., 2010). Toxic metal sources include ordinary activities of industrialization, civilization, agriculture and natural sources (Olowoyo et al., 2011; Kisamo, 2003). Heavy metal contaminants in soil can be absorbed by plants that are consumed by humans (Mutune et al., 2013; Ghosh and Singh, 2005; Salt et al., 1995). Heavy metals such as cadmium, chromium, lead, copper and zinc have been obtained in vegetables planted or consumed within urban areas sometimes in more than allowable limits (Yebpella et al., 2011; Salariya et al., 2002). Some plants such as *Brassica* vegetables have been shown to hyperaccumulate metals in their edible parts (Kumar et al., 2007). Copper and zinc are essential micronutrients obtained from vegetables (Sinha et al., 2010; Iyaka, 2007) but at high levels they cause oxidative stress through redox reactions (Sinha et al., 2010; Ghosh and Singh, 2005). Lead also causes oxidative stress and in young children it causes mental retardation (ATSDR,

Table 1. Sites from where vegetables and soils were collected for metal analysis.

Code	Site description	Code	Site description
S1 (IA)	Mukurwa Kwa Njenga along river	S14 (RR)	Githurai Bridge
S2 (IA)	Kware	S15 (SA)	Nairobi West Prison Farm
S3 (IA)	Donholm bridge	S16 (SA)	Langata Estate
S4 (SA)	Kayole farm	S17 (OF)	Kawangware
S5 (RR)	Komarock Junction road	S18 (OF)	Kikuyu town
S6 (OF)	Saika	S19 (OF)	Karura forest
S7 (SA)	Njiru bridge	S20 (OF)	Kiambu town
S8 (SA)	Ruai	S21 (SA)	Ruiru river near town
S9 (SA)	Ruai near treatment plant	S22 (SA)	Ruiru river near town
S10 (RR)	Kariobangi bridge	S23 (RR)	Juja bridge
S11 (OF)	Korogocho	S24 (OF)	Juja farm
S12 (RR)	Mwiki River bridge	S25 (OF)	KM
S13 (OF)	Mwihoko		

IA - Industrial area; SA - sewer area; RR - road reserve; OF - ordinary farm.

2007). Chronic exposure to cadmium causes glomerula damage and also alters Zn, Cu and Se metabolism (ATSDR, 2012) while some oxidative forms of chromium (Cr VI) are carcinogenic.

In addition to toxicity, heavy metal deficiencies also occur therefore knowledge of their amounts in vegetables for dietary supply is imperative (Iyaka, 2007). The high population of the Nairobi City and high cost of living has forced urbanites to farm within the city for consumption and income generation (Shackleton et al., 2009). Water and soil in cities are rich in nutrients such as N, P, K, S and Mg from waste which makes vegetables large and appealing. Making urban agriculture environmentally and health wise sustainable is a major constraint. There is need for research to address the constraint and help to inform relevant policy interventions. The present study aims at investigating the concen-trations of selected heavy metals in ten common vegetables and the soil where they are planted in.

METHODOLOGY

The chemicals used during this study were of analytical grade and water used for rinsing glassware was distilled.

Collection of vegetables

The vegetables collected were African nightshade (*Solanum villosum* V1), Swiss chard (*Beta vulgaris* V2), African kale (*Brassica carinata* V3), spiderplant (*Cleome gynandra* V4), Amaranths (*Amaranthus sp* V5), pumpkin leaves (*Curcubita moschata* V6), cowpea leaves (*Vigna unguiculata* V7), jute mallow (*Corchorus olitorius* V8), slenderleaf (*Crotalaria* sp. V9) and kale (*Brassica oleracea* var. Acephala V10). Collection was done in 25 sites which were ranked into four categories: industrial area (IA), road reserve (RR), sewage area (SA) and ordinary farm (OF). The industrial area was along Nairobi River and its tributaries, and all vegetables planted there were watered from these rivers which are adjacent to high manufacturing and other industrial activities. The road reserve sites were areas near roads, at least within 100 m from the roads. Sewage area sites were evidently watered with river water that contained lots of sewage effluent. Ordinary farms were far from any source of contamination, where activities are low. The sites are represented in Table 1. The vegetables were packed in polythene bags and packed in cooler boxes. In the laboratory, samples were thoroughly washed and rinsed with distilled water before drying at 70°C for three days, then powdered using a miller.

Soil collection

Top soil (0-15 cm depth) samples were obtained from each of the points where vegetables were collected. Sampling was done in a zigzag (Okalebo et al; 2002) of 5 cores across the field, avoiding areas of obvious interference like litter, manure and rocks. The sites ranged from 50 to more than 100 m^2. The five samples from each site, each about 0.5 kg, were composted and air dried at 25°C for about 1 month, sieved to <2 mm, then 3 replicates were obtained from each composite for analysis.

Digestion of samples

Soil mineral extraction was done using 0.05 M EDTA disodium salt according to Gupta (1999). Five grams of soil were vortexed with 25 ml EDTA for 60 min at 120 rpm. The supernatant was filtered through Whatman No. 42 paper, and topped up to 50 ml using distilled water and stored in a cold room at -20°C. Soil was extracted using EDTA because only extractable metal is bioavailable and mobile in reference to plants (Sinha et al., 2010). EDTA has a strong chelating ability for different heavy metals (Tandy et al., 2004).

Five grams of vegetable powder was weighed into a silica crucible. It was charred to remove carbon in a block digester at 200°C

Table 2. Heavy metal content (ppm) of soils from the sites where vegetables were collected.

Site	Pb	Zn	Cu	Cr	Cd
S1 (IA)	1.70^{e*}	42.96^{e}	6.43^{d}	0.03^{c}	0.222^{d}
S2 (IA)	20.94^{a}	196.00^{a}	70.96^{a}	1.41^{a}	2.644^{a}
S3 (IA)	16.33^{b}	198.30^{a}	75.37^{a}	0.23^{b}	1.731^{b}
S4 (SA)	9.43^{c}	190.05^{a}	45.61^{b}	0.12^{c}	0.2279^{d}
S5 (RR)	0.66^{e}	29.42^{e}	9.22^{d}	0.05^{c}	0.5571^{d}
S6 (OF)	0.57^{e}	43.45^{e}	8.29^{d}	0.09^{c}	0.0333^{d}
S7 (SA)	0.94^{e}	40.34^{e}	10.47^{d}	0.14^{bc}	0.00^{d}
S8 (SA)	0.67^{e}	13.01^{h}	5.77^{e}	0.18^{b}	0.2629^{d}
S9 (SA)	5.40^{d}	105.08^{c}	18.17^{c}	0.17^{b}	0.7146^{c}
S10 (RR)	2.40^{e}	110.14^{c}	21.97^{c}	0.09^{bc}	0.1279^{d}
S11 (OF)	1.88^{e}	127.87^{b}	12.61^{d}	0.07^{c}	0.0108^{d}
S12 (RR)	0.78^{e}	16.63^{h}	11.18^{d}	0.06^{c}	0.0758^{d}
S13 (OF)	2.42^{e}	88.77^{d}	18.08^{cd}	0.06^{c}	0.07^{d}
S14 (RR)	2.83^{e}	47.65^{f}	3.59^{e}	0.11^{bc}	0.08^{d}
S15 (SA)	1.28^{e}	49.81^{f}	6.54^{d}	0.18^{b}	0.1671^{d}
S16 (SA)	2.18^{e}	125.79^{b}	18.04^{c}	0.07^{bc}	0.0084^{d}
S17 (OF)	1.22^{e}	14.62^{h}	10.58^{d}	0.10^{bc}	0.0129^{d}
S18 (OF)	0.75^{e}	55.68^{g}	15.70^{c}	0.07^{bc}	0.00^{**d}
S19 (OF)	1.75^{e}	42.99^{g}	6.38^{d}	0.25^{b}	0.00^{d}
S20 (OF)	2.12^{e}	117.72^{b}	17.45^{c}	0.08^{bc}	0.00^{d}
S21 (SA)	2.20^{e}	109.13^{c}	22.00^{c}	0.16^{bc}	0.105^{d}
S22 (SA)	1.91^{e}	133.57^{b}	15.62^{cd}	0.20^{b}	0.041^{d}
S23 (RR)	0.86^{e}	20.20^{h}	12.56^{d}	0.13^{bc}	0.0458^{d}
S24 (OF)	0.83^{e}	17.42^{h}	11.98^{d}	0.21^{b}	0.0675^{d}
S25 (OF)	1.93^{e}	111.62^{c}	13.78^{d}	0.15^{bc}	0.1154^{d}
S.E.	0.043	3.28	1.18	0.03	0.09

*Alphabetical letters indicate significance of variation within each metal among the collection sites; **0 value means below detection limit.

until all smoke had gone. The sample was then dry-ashed at 550°C in a muffle furnace for 14 h (method adapted from Okalebo et al., 2002). To the ash, 15 ml of 6 N HCl was added and the mixture heated for about 10 min at 250°C on a hot plate in order to digest the minerals in the ash. The resulting digest was filtered using Whatmans No. 42 filter paper. The filtrate was topped up to 50 ml using distilled water and stored in plastic bottles in a cold room till analysis.

Heavy metal analysis

Atomic Absorption Flame Emission Spectrophotometer (Shimadzu AAS-6200) was used to quantify heavy metal content for soil and vegetable samples. Calibration was done by running standards at concentrations of 0, 1, 1.5, 2 and 2.5 mg/kg. Hydrochloric acid (6 N) was used as a blank to zero the AAS.

Data analysis

Data on metal content obtained from the AAS was analyzed for variance using ANOVA and multiple mean comparisons were done using Tukeys at 5% level. Pearson's Product Moment was used to correlate the amount of metal in soil to that in the plant.

RESULTS AND DISCUSSION

Heavy metals in soils

The soil metal concentrations for each element and site are listed in Table 2. Soil metal concentration was significantly different in the different sites. The trend for the mean percentage for metals was Zn > Cu > Pb > Cd > Cr. EDTA-extractable Pb ranged from 0.57 to 20.94±0.43 mg/kg. The highest levels were in industrial area (sites 2, 3 and 4). This may be explained by occasional fuel spills and other waste flowing from the industrial area. Factories in this area include paint, vehicle batteries that may contain lead oxide, glass that may contain lead, and fuel some of which still contains lead. Sewage area where the effluent from the main sewer line of the city is let into the river contained up to 5.4 mg/kg. Soil Pb levels above 20 mg/kg are toxic to plants (Audu and Lawal, 2006), the highest obtained in this study.

Cu values were 3.59 to 75.37±1.18 mg/kg, with industrial area sites registering the highest levels. Copper has extensive use in industries, domestic items and farms such as coin money, electrical wiring, water and fabric treatment and fungicides. Taber (2009) suggests a 120 mg/kg higher limit in soil. It is found in soil either from natural or anthropogenic sources through water or dumping. Soil Cu content was much less as compared to earlier studies by Kimani (2007) (0 - 198 mg/kg) working on a dumpsite area in Kenya, but higher than Yebpella et al. (2011) (2.41 - 7.39 mg/kg) working on sites along a Nigerian river. EDTA-extractable Zn content ranged between 13.01 and 198.03±3.28 mg/kg. Industrial area sites had very high Zn contents as compared to other sites. Zinc industrially is used for galvanization, in dry cell batteries, cosmetics and medicines (ATSDR, 2005). Although some soils in ordinary farm sites had deficient Zn content, none had higher content than the recommended 30 - 300 mg/kg (Taber, 2009). Values in the study were lower than that of Kimani (2007) who analyzed soil around a dumpsite (175 - 1150 mg/kg).

Extractable Cd values were 0 to 2.64±0.09 mg/kg. Similarly, the highest values were at the industrial area region. Cadmium has wide usage in batteries, PVC plastics, cigarettes and paint pigments. Chromium concentration ranged from 0.03 to 1.4±0.03 mg/kg with industrial areas as the highest; none was above the recommended 100 mg/kg. Both Cd and Cr were however below detection limit in many sites. From the results, the highest values for all metals were at industrial area, followed by sewage area implying that metal content was

Table 3. Pb content of vegetables collected from Nairobi urban and peri-urban sites.

Site	Vegetable									
	V1	V2	V3	V4	V5	V6	V7	V8	V9	V10
S1 (IA)	0.97^{f**}	1.20^{hi}		0.86^{e}	0.61^{cdefg}		0.65^{abc}	0.67^{ab}	0.55^{a}	0.82^{bcde}
S2 (IA)	*	1.88^{j}	1.17c	2.10^{g}	2.41^{k}	2.56^{k}				0.66^{abcde}
S3 (IA)		0.70^{defg}	0.77b	2.44^{g}	1.74^{j}	1.48^{j}	1.63^{de}			0.47^{abc}
S4 (SA)	0.14^{a}	0.79^{efgh}	0.35a		1.56^{j}	0.56^{bcdefg}	0.58^{abc}	0.62^{ab}		0.61^{abcde}
S5 (RR)	0.28^{ab}	0.28^{abcd}			0.50^{bcdefg}	0.63^{cdefg}	0.78^{abc}	0.73^{ab}	0.54^{a}	0.39^{abc}
S6 (OF)	0.46^{abcd}	0.17^{ab}		0.85^{de}	0.41^{bcde}	0.60^{cdefg}	0.61^{abc}	0.95^{b}	0.58^{a}	0.94^{cdef}
S7 (SA)	0.40^{abc}	0.50^{abcdef}		0.86^{de}	0.65^{cdefg}		0.57^{abc}			0.29^{ab}
S8 (SA)	0.30^{ab}	0.92^{fghi}		0.87^{e}	0.98^{i}	1.80^{j}	0.93^{bc}	0.25^{a}		0.68^{abcde}
S9 (SA)	0.55^{bcde}	1.20^{hi}	0.61^{ab}	1.32^{f}	1.69^{j}	0.85^{ghi}	0.95^{bc}	1.97^{c}	0.94^{b}	0.37^{abc}
S10 (RR)	0.43^{abcd}	0.25^{abcd}		0.30^{ab}	0.09^{a}	1.71^{j}	2.02^{e}	1.49^{c}	0.58^{a}	1.46^{fg}
S11 (OF)	0.86^{ef}	1.04^{ghi}	1.22^{c}	1.27^{f}	1.51^{j}	1.12^{i}	1.02^{bcd}			0.55^{abcde}
S12 (RR)		0.43^{abcde}		0.24^{a}	0.27^{ab}	0.59^{bcdefg}	0.72^{abc}	0.52^{ab}		1.56^{g}
S13 (OF)	0.77^{def}	0.66^{defg}	0.42^{a}		0.41^{bcde}	2.53^{k}	0.25^{a}			
S14 (RR)	0.93^{f}				0.63^{cdefg}	0.76^{cdefg}	0.99^{bc}			0.74^{abcde}
S15 (SA)	0.49^{abcd}	0.17^{ab}		0.48^{abc}	0.41^{bcde}	0.42^{abcde}	0.64^{abc}			0.41^{abc}
S16 (SA)		1.36^{i}		0.60^{bcde}	0.96^{hi}					0.44^{abc}
S17 (OF)	0.40^{abc}	0.09^{a}			0.80^{ghi}	0.16^{a}				0.51^{abcd}
S18 (OF)	0.30^{ab}	0.41^{abcde}			0.66^{defgh}					1.05^{defg}
S19 (OF)	0.86^{ef}	1.19^{hi}	1.22^{c}	1.27^{f}	1.51^{j}	1.12^{i}	1.09^{bcd}		0.48^{a}	0.55^{abcde}
S20 (OF)	0.16^{a}	$0.55b^{cdef}$	0.45^{a}	0.29^{ab}	0.37^{abcd}	1.19^{i}	0.53^{ab}			0.18^{a}
S21 (SA)	0.24^{ab}		0.44^{a}		0.68^{efghi}					0.42^{abc}
S22 (SA)	0.72^{cdef}	0.20^{abc}			0.68^{efghi}	0.14^{a}	1.18^{cd}	0.75^{ab}		0.85^{bcde}
S23 (RR)	1.03^{f}	0.63^{cdefg}	0.39^{a}	0.65^{cde}	0.75^{fghi}	0.83^{ghi}	1.18^{cd}	0.82^{b}	0.50^{a}	1.09^{efg}
S24 (OF)	0.44^{abcd}	0.70^{defg}		0.52^{abcd}	0.34^{abc}		0.46^{ab}			1.09^{efg}
S25 (OF)	0.33^{ab}	0.43^{abcde}			0.49^{bcdef}	1.18^{hi}	0.52^{ab}	0.75^{ab}		1.56^{g}
S.E.	0.043	0.103	0.186	0.076	0.069	0.067	0.144	0.113	0.071	0.128

*Blanks indicate that the vegetable was missing in a site; **alphabetical letters indicate significance of variation within each vegetable among sites; ***0 value means below detection limit.

significantly lowered by dilution as water flows downstream and progressively reduced dumping of industrial waste.

Heavy metals in vegetables

Metal content in vegetables had a similar trend with that in soil (Zn > Cu > Cd >Pb > Cr). There was a significant difference in the amount of Pb within vegetables at all sites. The highest Pb content (Table 3) of 2.4 mg/kg was in spiderplant and Pumpkin from sites 3 and 2 (industrial area). The lowest was 0.09 mg/kg in spinach from an ordinary farm and pumpkin from sewage area.

Copper concentration was also significantly different in vegetables from different sites (Table 4). The highest was 21.34 mg/kg in pumpkin from industrial area and the lowest was 0.5 mg/kg in cowpea from a road reserve. According to Taber (2009) and Mills and Jones (1991), normal Cu content in plants should range from 2 - 20 mg/kg, and CAC (2001) sets it at 40 mg/kg therefore the concentrations in this study were within acceptable range. Many vegetables contained lower than 2 mg/kg which is the recommended lower limit of Cu in plant tissue because some soils were deficient; or when they had sufficient Cu, a pH imbalance or an excess of other nutrients such as phosphorous which were not determined in this study could have limited absorption by vegetables (Mills and Jones, 1991).

Zinc concentration in vegetables also varied significantly (Table 5). The highest was 89.85 mg/kg in African Kale from site 2 (industrial area) and the lowest was 20.13 mg/kg from Amaranth and cowpea from ordinary farms. Normal Zn content in vegetables ranges should range within 25 - 300 mg/kg (Taber, 2009). These results were slightly higher than those of Jung (2008) who found

Table 4. Cu content of vegetables collected from Nairobi urban and peri-urban sites.

Site	Vegetable									
	V1	V2	V3	V4	V5	V6	V7	V8	V9	V10
S1 (IA)	1.86^{a}	2.72^{bcd}		4.82^{d}	11.35^{i}		7.11^{hi}	4.86^{b}	7.35^{c}	8.10^{j}
S2 (IA)		11.60^{h}	11.03^{e}	18.00^{h}	21.34^{l}	19.93^{i}	12.28^{k}			8.76^{i}
S3 (IA)		12.24^{h}	13.29^{f}	14.42^{g}	18.89^{k}	20.26^{i}				11.74^{k}
S4 (SA)	1.91^{a}	1.75^{ab}	1.17^{a}		2.22^{bc}	4.74^{d}	0.93^{abc}	6.47^{c}		3.28^{e}
S5 (RR)	7.74^{f}	9.40^{g}			6.79^{h}	8.16^{f}	9.21^{j}	9.81^{d}	11.34^{d}	5.11^{fg}
S6 (OF)	7.87^{f}	9.56^{g}		3.37^{c}	2.92^{cd}	2.36^{bc}	2.50^{efg}	1.91^{a}	7.31^{c}	1.97^{bc}
S7 (SA)	3.43^{bc}	7.79^{f}	4.40^{c}	6.41^{e}	1.60^{ab}					2.31^{cd}
S8 (SA)	14.88^{h}	9.11^{fg}		10.89^{f}	15.54^{j}	16.84^{h}	14.07^{l}	18.01^{f}		14.77^{l}
S9 (SA)	6.78^{e}	9.37^{g}	3.84^{c}	10.48^{f}	1.47^{ab}	10.48^{g}	1.40^{abcde}	3.92^{b}	2.5^{a}	1.22^{ab}
S10 (RR)	3.93^{c}	3.44^{cde}		2.82^{bc}	3.55^{def}	3.91^{d}	1.96^{cde}	8.70^{d}	3.24^{b}	8.66^{i}
S11 (OF)	1.72^{a}	2.07^{abc}	1.26^{a}	1.88^{ab}	2.38^{bc}	2.45^{bc}	1.88^{bcde}			5.57^{g}
S12 (RR)		1.23^{a}		10.12^{f}	1.34^{a}	0.75^{a}	0.52^{a}	3.65^{b}		2.83^{cde}
S13 (OF)	9.13^{g}	1.72^{ab}	2.16^{b}		3.92^{ef}	7.94^{f}	3.58^{g}			
S14 (RR)	5.88^{d}				3.99^{ef}	3.45^{cd}	0.63^{a}			1.96^{bc}
S15 (SA)	7.33^{ef}	2.83^{bcde}		1.32^{a}	3.09^{cde}	1.82^{ab}	0.77^{ab}			6.59^{h}
S16 (SA)		11.27^{h}		1.22^{a}	1.26^{a}					2.47^{cde}
S17 (OF)	7.21^{ef}	3.69^{de}			3.91^{ef}	6.47^{e}				2.07^{bc}
S18 (OF)	2.97^{b}	4.29^{e}	1.26^{a}		5.73^{g}					4.64^{f}
S19 (OF)	1.72^{a}	2.07^{abc}		1.88^{ab}	2.38^{bc}	2.45^{bc}	1.88^{bcde}		3.37^{b}	5.57^{g}
S20 (OF)	1.66^{a}	8.05^{fg}	6.12d	1.08^{c}	2.84^{cd}	1.67^{ab}	1.35^{abcd}			1.35^{ab}
S21 (SA)	7.80^{f}				7.54^{h}		3.40^{fg}			3.18^{de}
S22 (SA)	5.73^{d}	4.26^{de}			5.73^{g}	1.86^{ab}	6.42^{h}	15.79^{e}		0.77^{a}
S23 (RR)	9.13^{g}	1.07^{a}	3.84c	11.27^{f}	1.24^{a}	7.49^{ef}	11.27^{k}	3.79^{b}	2.37^{a}	9.13^{i}
S24 (OF)	3.80^{bc}	3.91^{de}		1.72^{ab}	4.40^{f}		7.93^{i}			9.13^{i}
S25 (OF)	1.72^{a}	2.07^{abc}			2.38^{bc}	2.45^{bc}	2.30^{def}	15.79^{e}		5.57^{g}
S.E.	0.072	0.35	0.065	0.249	0.2109	0.326	0.2556	0.331	0.173	0.2

*Blanks indicate that the vegetable was missing in a site; **alphabetical letters indicate significance of variation within each vegetable among sites; ***0 value means below detection limit.

Table 5. Zn content of vegetables collected from Nairobi urban and peri-urban sites.

Site	Vegetable									
	V1	V2	V3	V4	V5	V6	V7	V8	V9	V10
S1 (IA)	31.23^{f}	44.43^{i}		54.46^{e}	33.76^{e}		38.12^{fgh}	43.01^{a}	38.45^{d}	54.48^{j}
S2 (IA)		59.33^{k}	89.85^{g}	81.15^{h}	80.55^{l}	61.38^{i}	55.76^{l}			55.6^{j}
S3 (IA)		43.86^{hi}	82.45^{f}	53.56^{e}	44.34^{g}	64.41^{k}				56.49^{j}
S4 (SA)	26.32^{cde}	60.62^{l}	54.43^{c}		22.44^{b}	47.76^{gh}	31.45^{cd}	32.27^{c}		33.14^{d}
S5 (RR)	29.24^{e}	47.32^{j}			43.88^{f}	44.5^{fg}	50.24^{k}	47.7^{f}	30.64^{b}	33.73^{d}
S6 (OF)	35.24^{gh}	43.76^{hi}		52.11^{d}	53.44^{i}	42.43^{ef}	45.67^{ijk}	46.54^{f}	40.52^{e}	31.11^{c}
S7 (SA)	23.15^{b}	34.02^{bcd}		26.75^{a}	26.82^{cd}		45.79^{ijk}			31.32^{c}
S8 (SA)	25.05^{bc}	34.56^{cd}		87.42^{i}	37.47^{f}	45.73^{g}	40.99^{ghi}	30.27^{c}		55.79^{j}
S9 (SA)	21.3^{a}	26.81^{a}	52.22^{c}	56.77^{f}	62.08^{k}	39.44^{cde}	35.88^{defg}	24.55^{a}	32.22^{c}	41.83^{g}
S10 (RR)	31.98^{f}	54.35^{l}	50.68^{bc}	53.88^{e}	48.73^{ij}	42.69^{ef}	44.87^{ij}			38.57^{f}
S11 (OF)	38.73^{i}	39.19^{e}	70.44^{e}	56.25^{f}	45.11^{h}	76.7^{k}	37.11^{efgh}			44.81^{h}
S12 (RR)		37.18^{de}		58.45^{f}	25.18^{c}	22.42^{a}	36.95^{efgh}	23.63^{c}		52.76^{i}
S13 (OF)	33.08^{fg}	45.68^{ij}	47.46^{b}		32.44^{de}	32.63^{b}	26.44^{cd}			33.45^{d}
S14 (RR)	35.05gh				52.55i	27.13c	32.65cde			35.77e

Table 5. Contd.

S15 (SA)	35.14^{gh}	41.16^{ef}		48.46^{c}	41.33^{g}	43.56^{ef}	22.05^{b}			41.22^{g}
S16 (SA)		43.61^{ghi}		74.32^{g}	63.56^{k}					44.22^{h}
S17 (OF)	47.51^{k}	32.83^{bc}			45.56^{h}	42.46^{ef}				32.36^{cd}
S18 (OF)	35.11^{gh}	41.27^{efg}			21.56^{a}					37.39^{f}
S19 (OF)	31.98^{f}	54.35^{l}	79.88^{f}	49.63^{c}	51.39^{i}	50.24^{h}	28.41^{d}		30.28^{b}	50.38^{i}
S20 (OF)	27.24^{cde}	36.34^{d}		44.75^{b}	37.69^{f}	$41.16d^{ef}$	35.5^{e}	43.12^{e}	32.71^{c}	33.46^{d}
S21 (SA)	37.15^{hi}		56.27^{cd}		22.47^{b}					58.28^{k}
S22 (SA)	26.92^{cde}	59.08^{S}			28.54^{d}	35.48^{c}	20.45^{a}	36.56^{d}		21.91^{a}
S23 (RR)	33.73^{fg}	36.19^{d}	60.43^{d}	57.72^{f}	25.18^{c}	24.15^{b}	36.76^{efgh}	32.65^{c}	28.31^{a}	35.51^{e}
S24 (OF)	27.23^{cde}	43.03^{fgh}		45.68^{b}	20.13^{a}		41.46^{hi}	35.06^{d}		35.51e
S25 (OF)	42.50^{j}	31.98^{b}	61.45^{d}		50.48^{i}	50.33^{h}	38.96^{fgh}			50.38^{i}
S.E.	0.654	0.547	0.606	0.401	0.515	1.003	1.163	0.306	0.173	0.3462

*Blanks indicate that the vegetable was missing in a site; **alphabetical letters indicate significance of variation within each vegetable among sites; ***0 value means below detection limit.

that vegetables had a Zn range of 18 - 52 mg/kg, and was highest in mining sites. Mean Zn values were very high as compared to Iyaka (2007) who got 7 mg/kg Zn in spinach in a similar study, and Yebpella et al. (2011), who got the highest value of 26.7 mg/kg while working on a site irrigated with contaminated river water. Zinc is least toxic among the heavy metals in this study; its deficiency could be more detrimental than toxicity (Kumar et al., 2007). CAC (2001) sets the limit of its intake at 60 mg/kg while Nair et al. (1997) puts it at 150 mg/kg, therefore the content in this study may be acceptable.

Cadmium concentration in vegetables was significantly different in vegetables from different sites (Table 6). The highest was 3.02 mg/kg noted in African Kale followed by 2.5 mg/kg in spider plant both from Industrial area. In all vegetables, the Cd concentration was consistently highest in Industrial area and sewage area.

Although in most vegetables, Cd was not above the dietary recommended limit in vegetables (0.2 mg/kg) (Yebpella et al., 2011), continued low-intake of oral Cd may cause negative gastrointestinal, musculoskeletal, renal and neurological effects (Kumar et al., 2007). Chromium concentration was the lowest of all the metals in the vegetables (Table 7). The highest noted was 1.24 mg/kg in spider plant followed by 1.19 mg/kg in spinach both from Industrial area.

Correlation of soil and vegetable heavy metal concentrations

Simple correlation was carried out to establish whether the heavy metal concentration in vegetables was a function of the amount in soil where they were grown. At $P < 0.05$ of most metals there was a significant positive linear relationship between the heavy metal in soil and that in vegetable leaves (Table 8). There was a stronger linear relationship in Cu, Cd and Cr for most vegetables and soils. Jung (2008) through a multiple regression using several factors determined that soil metal content was the principal determinant of plant tissue metal content.

In cases where soil metal was less than vegetable content other metal sources such as dumped items could be implicated. In a few instances such as Zn in African Nightshade and Jute mallow, and Pb in kale the relationship was negative implying that the plant accumulated the metals in leaves or the amount of the metals in the field at some point were higher than at the time of sampling. Factors that were not tested in this study that would account for the negative correlation include soil pH, cation exchange capacity, organic matter content, types and varieties of vegetables, and vegetable age (Jung, 2008).

Conclusion

From this study, the most elevated soil metal concentrations were found nearest to industrial area with average of 20.94, 198.3, 75.37, 2.64 and 2.1 mg/kg for Pb, Zn, Cu, Cd and Cr, respectively. Most vegetables planted in these sites of the study area are within safe limits although some exceed.

This notwithstanding, longterm metal exposure by regular consumption of such locally grown vegetables poses potentially health problems to animal and humans. Further cleaning of the rivers needs to be done, and curtailing of all effluents and dumping of solid waste should be done to minimize contaminant leakage into water and soil that may be used for agriculture.

Table 6. Cd content of vegetables collected from Nairobi urban and peri-urban sites.

Site	Vegetable									
	V1	**V2**	**V3**	**V4**	**V5**	**V6**	**V7**	**V8**	**V9**	**V10**
S1 (IA)	0.15^{bcd}	0.24^{h}		0.27^{cd}	0.16^{e}		0.11^{bc}	0.14^{d}	0.05^{a}	0.05^{ab}
S2 (IA)		1.95^{j}	2.04^{c}	2.54^{f}	1.77^{i}	1.81^{g}	1.53^{g}			0.36^{fg}
S3 (IA)		2.07^{k}	3.02^{d}	2.01^{e}	1.52^{h}	1.66^{f}				1.58^{h}
S4 (SA)	0.07^{abc}	0.09^{cde}	0.11^{a}		0.04^{abc}	0.06^{ab}	0.13^{cd}	0.12^{cd}		0.34^{fg}
S5 (RR)	0.21^{de}	0.32^{i}			0.32^{f}	0.42^{e}	0.21^{ef}	0.15^{d}	0.10^{b}	0.21^{de}
S6 (OF)	0.02^{a}	0.03^{ab}		0.04^{ab}	0.04^{abcd}	0.06^{ab}	0.03^{a}	0.02^{a}	0.12^{b}	0.04^{ab}
S7 (SA)	0.03^{a}	0^{a}		0^{a}	0^{a}		0.00^{a}			0.00^{a}
S8 (SA)	0.23^{de}	0.11^{cdef}		0.02^{ab}	0.02^{ab}	0.11^{bc}	0.05^{ab}	0.09^{bc}		0.00^{a}
S9 (SA)	0.22^{de}	0.33^{i}	0.39^{b}	0.43^{d}	0.42^{g}	0.24^{d}	0.19^{de}	0.32^{f}	0.31^{c}	0.40^{g}
S10 (RR)	0.03^{a}	0.16^{fg}		0.20^{bc}	0.14^{de}	0.06^{ab}	0.03^{a}	0.05^{ab}	0.07^{a}	0.05^{ab}
S11 (OF)	0.04^{a}	0.07^{bcd}	0.06^{a}	0.10^{abc}	0.15^{e}	0.03^{ab}	0.00^{a}			0.08^{ab}
S12 (RR)		0.15^{fg}		0.11^{abc}	0.16^{e}	0.19^{cd}	0.25^{f}	0.21^{e}		0.25^{ef}
S13 (OF)	0.06^{abc}	0.14^{efg}	0.06^{a}		0.08^{abcde}	0.10^{abc}	0.21^{ef}			
S14 (RR)	0.06^{ab}				0.03^{ab}	0.13^{bcd}	0.14^{cd}			0.07^{ab}
S15 (SA)	0.07^{abc}	0.09^{cde}		0.15^{abc}	0.27^{f}	0.17^{cd}	0.14^{cd}			0.20^{cde}
S16 (SA)		0.26^{h}		0.17^{abc}	0.17^{e}					0.13^{bcd}
S17 (OF)	0.30^{e}	0.03^{ab}			0.04^{abc}	0.00^{a}				0.00^{a}
S18 (OF)	0.03^{a}	0.06^{bc}			0.03^{ab}					0.00^{a}
S19 (OF)	0.01^{a}	0^{a}	0.00^{a}	0^{a}	0^{a}	0.00^{a}	0.00^{a}		0.15^{b}	0.03^{ab}
S20 (OF)	0.06^{abc}	0.00^{a}	0.00^{a}	0^{a}	0.00^{a}	0.00^{a}	0.00^{a}			0.00^{a}
S21 (SA)	0.10^{abc}		0.07^{a}		0.11^{bcde}				0.10^{b}	0.08^{ab}
S22 (SA)	0.03^{a}	0.08^{cd}			0.08^{abcde}	0.09^{abc}	0.05^{ab}	0.09bc		0.03^{ab}
S23 (RR)	0.03^{a}	0.08^{cd}	0.05^{a}	0.07^{ab}	0.10^{bcde}	0.06^{ab}	0.01^{a}	0.07b	0.04^{a}	0.05^{ab}
S24 (OF)	0.03^{a}	0.13^{defg}		0.09^{abc}	0.13^{cde}		0.06^{ab}	0.07b		0.04^{ab}
S25 (OF)	0.16^{cd}	0.16^{g}			0.13^{cde}	0.09^{abc}	0.10^{bc}			0.09^{abc}
S.E.	0.032	0.01232	0.0646	0.04104	0.0224	0.024	0.0142	0.0094	0.017	0.0261

*Blanks indicate that the vegetable was missing in a site;**alphabetical letters indicate significance of variation within each vegetable among sites; ***0 value means below detection limit.

Table 7. Cr content of vegetables collected from Nairobi urban and peri-urban sites.

Site	Vegetable									
	V1	**V2**	**V3**	**V4**	**V5**	**V6**	**V7**	**V8**	**V9**	**V10**
S1 (IA)	0^{a}	0.02^{a}		0.02^{a}	0.02^{a}		0^{a}	0.02^{a}	0.02^{a}	0.03^{ab}
S2 (IA)		1.19^{e}	1.08^{b}	1.24^{c}	0.96^{i}	0.26^{g}	0.34^{g}			0.07^{bcd}
S3 (IA)		0.12^{abc}	0.12^{a}	0.20^{a}	0.16^{cdef}	0.12^{cde}				0.00^{a}
S4 (SA)	0.11^{bc}	0.18^{bcd}	0.14^{a}		0.29^{g}	0.13^{def}	0.08^{bcde}	0.11^{cd}		0.07^{bcd}
S5 (RR)	0^{a}	0.05^{ab}			0.08^{abc}	0.08^{bcd}	0^{a}	0.12^{cd}	0.07^{b}	0.06^{bc}
S6 (OF)	0.1^{abc}	0.07^{ab}		0.07^{a}	0.10^{abcd}	0.11^{cde}	0.04^{abc}	0.06^{ab}	0.13^{c}	0.06^{bcd}
S7 (SA)	0.04^{ab}	0.24^{cd}		0.03^{a}	0.05^{ab}		0.05^{bcd}			0.07^{bcd}
S8 (SA)	0.023^{ab}	0.05^{ab}		0.23^{a}	0.18^{def}	0.07^{abc}	0^{a}	0.03^{a}		0.08^{cde}
S9 (SA)	0.06^{ab}	0.14^{abc}	0.50^{ab}	0.13^{a}	0.17^{cdef}	0.13^{cdef}	0.05^{bcd}	0.14^{d}	0.04^{ab}	0.04^{abc}
S10 (RR)	0.02^{ab}	0.16^{abcd}		0.03^{a}	0.13^{bcde}	0.05^{ab}	0.03^{ab}	0.06^{ab}	0.12^{c}	0.05^{bc}
S11 (OF)	0.06^{ab}	0.04^{ab}	0.00^{a}	0.06^{a}	0.10^{abcd}	0.09^{bcde}	0.04^{abc}			0.04^{abc}
S12 (RR)		0.04^{ab}		0.02^{a}	0.12^{bcd}	0.07^{abcd}	0.09^{de}	0.03^{a}		0.05^{bc}
S13 (OF)	0.03^{ab}		0.05a		0.08^{abc}	0.07^{abcd}	0^{a}			
S14 (RR)	0.09abc	0.04ab			0.17cdef	0.14ef	0.03ab			0.06bc

Table 7. Contd.

S15 (SA)	0.17^{c}	0.14abcd		0.18^{a}	0.21^{efg}	0.18^{f}	0.16^{f}			0.10^{def}
S16 (SA)		0.04^{ab}		0.09^{a}	0.11^{abcd}					0.03^{ab}
S17 (OF)	0.03^{ab}	0.13^{abc}			0.14^{bcdef}	0.05^{ab}				0.35^{g}
S18 (OF)	0.05^{ab}	0.07^{ab}			0.09^{abcd}					0.00^{a}
S19 (OF)	0.09^{abc}	0.19^{bcd}	0.10^{a}	0.17^{a}	0.23^{fg}	1.09^{h}	0.06^{bcde}		0.08^{b}	0.11^{def}
S20 (OF)	0.08^{abc}	0.17^{abcd}	0.10^{a}	0.75^{b}	0.08^{abc}	0.08^{bcd}	0.03^{ab}			0.03^{ab}
S21 (SA)	0.04^{ab}		0.09a		0.23f^{g}					0.07^{bcd}
S22 (SA)	0.09^{abc}	0.29^{d}		0.10^{a}	0.47^{h}	0.01^{a}	0.10^{e}	0.11^{cd}		0.13^{f}
S23 (RR)	0.53^{d}	0.13^{abc}	0.16^{a}	1.22^{c}	0.10^{abcd}	0.05^{ab}	0.07^{bcde}	0.09^{bc}	0.07^{b}	0.05^{bc}
S24 (OF)	0.09^{abc}	0.09^{ab}			0.09^{abcd}		0.08^{bcde}	0.13^{d}		0.08^{cde}
S25 (OF)	0.08^{abc}	0.10^{abc}			0.11^{abcd}	0.12^{cde}	0.09^{cde}			0.12^{ef}
S.E.	0.023	0.0123	0.065	0.05	0.0218	0.014	0.0114	0.0094	0.0093	0.001

*Blanks indicate that the vegetable was missing in a site; **alphabetical letters indicate significance of variation within each vegetable among sites; ***0 value means below detection limit.

Table 8. Correlation between heavy metals in soil and vegetables.

Heavy metal	V1	V2	V3	V4	V5	V6	V7	V8	V9	V10
Pb										
R	-0.0192	0.551	0.4326	0.8249	0.7064	0.6122	0.4571	0.0304	0.4024	-0.2055
P value	0.9341	0.0059	0.2118	<0.001	<0.001	0.0053	0.0427	0.9293	0.5018	0.3469
Cd										
R	0.5777	0.9518	0.887	0.9858	0.9752	0.9692	0.9639	0.7537	0.8874	0.949
P value	0.0061	<0.001	<0.001	<0.001	<0.001	<0.001	<0.001	0.0074	0.0446	<0.001
Cr										
R	0.2679	0.9683	0.9199	0.6866	0.9173	0.2115	0.8964	0.6736	0.5583	0.1188
P value	0.2404	<0.001	<0.001	0.0033	<0.001	0.3848	<0.001	0.0231	0.3281	0.5892
Cu										
R	0.6121	0.6358	0.9324	0.7538	0.8518	0.9053	0.6356	0.5962	0.8119	0.5817
P value	0.0032	0.0011	<0.001	<0.001	<0.001	<0.001	0.0026	0.0529	0.0951	0.0036
Zn										
R	-0.4961	0.0385	0.6797	0.49	0.3507	0.174	0.4965	-0.0357	0.1459	0.5341
P value	0.0222	0.8615	0.0306	0.054	0.0856	0.4762	0.026	0.9171	0.8149	0.0087

P is at 5% and R is the correlation coefficient.

ACKNOWLEDGEMENT

This research was financed by the National Council of Science and Technology, Ministry of Education, Kenya.

REFERENCES

Agency for Agency for Toxic Substances and Disease Registry (ATSDR) (2012). Toxicological profile for Cadmium. U.S. Department of Health and Human Services. Public Health Service.

Agency for Agency for Toxic Substances and Disease Registry (ATSDR) (2007). Toxicological profile for Lead. U.S. Department of Health and Human Services. Public Health Service.

Agency for Toxic Substances and Disease Registry (ATSDR) (2005). Toxicological profile for Zinc. U.S. Department of Health and Human Services. Public Health Service.

Audu AA, Lawal AO (2006). Variation in Metal Contents of Plants in Vegetable Garden Sites in Kano Metropolis. J. Appl. Sci. Environ. Manag.10 (2):105 - 109

CAC (Codex Alimentarius Commission, FAO/WHO) (2001). Food additives and contaminants. Joint FAO/WHO Food Standards Programme, ALINORM 01/12A:1-289.

Ghosh M, Singh SP (2005). A review of phytoremediation of heavy metals and utilization of its by-products. Appl. Ecol. Environ. Res. 3(1):1-18.

Gupta A (1999). Soil, Plant, Water and Fertilizer analysis. Agro Botanica, New Delhi, India.

Iyaka YA (2007). Concentration of Cu and Zn in some fruits and vegetables commonly available in North-Central zone of Nigeria. Electronic J. Environ. Agric. Food Chem. 6(6):2150 - 2154.

Jung MC (2008). Heavy metal concentration in soils and factors affecting metal uptake by plants in the vicinity of a Korean Cu-W mine. Sensors 8:2413-2423

Kimani NG (2007). Environmental Pollution and Impact to Public Health; Implication of the Dandora Municipal Dumping Site in Nairobi, Kenya. a pilot study report In cooperation with The United Nations Environment Programme (UNEP)

Kisamo DS (2003). Environmental Hazards Associated with Heavy Metals in Lake Victoria Basin (East Africa) Tanzania. In: Africa Newsletter on Occupational Health and Safety. 2003; 13:67-69

Kumar NJI, Soni H, Kumar RN (2007), Characterization of heavy meals in vegetables using inductive coupled plasma analyser (ICPA). J. Appl. Sci. Environ. Manag. 11(3) 75-79.

Mills HA, Jones JB (1991). Plant analysis handbook II: a practical sampling, preparation, analysis and interpretation guide. MicroMacro publishing,Georgia, USA.

Mutune AN, Makobe MA, Abukutsa-Onyango MO (2013). Impact analysis of lead, copper and zinc content in selected African Indigenous and exotic vegetables from Nairobi Markets. Afr. J. Environ. Sci. Technol. 7(6) 540 - 545.

Nair M, Balachandran KK Sankarnarayan VN Joseph T (1997). Heavy metals in fishes from coastal waters of Cochin, South West Coast of India. Indian J. Mar. Sci. 26:98-100.

Okalebo RJ, Gathua, KW Woomer PL (2002). (2nd ed). Laboratory methods of soil and plant analysis. A working manual. TSBF-CIAT and SACRED Africa, Nairobi, Kenya.

Olowoyo JO, Okedeyi OO, Mkolo NM, Lion GN Mdakane STR (2011). Uptake and translocation of heavy metals by medicinal plants around a waste dump site in Pretoria, South Africa. South Afr. J. Bot. 78:116-121.

Salariya AM, Rehman ZU Ashraf M. (2002). Effect of polluted water on accumulation of heavy metals in commonly consumed vegetables. J. Chem. Soc. Pak. 25(3).

Salt DE, Blaylock M, Kumar NP, Dushenkov V, Ensley BT, Chet I, Raskin I (1995). Phytoremediation: a novel strategy for the removal of toxic metals from the environment using plants. Biotechnology 13:468-474.

Shackleton CM, Pasquini MW Drescher AW (eds). 2009. African indigenous vegetables in urban agriculture. Earthscan, London.

Sinha S, Sinam G, Mishra K Mallick S (2010). Metal accumulation, growth, antioxidants and oil yield of Brassica juncea L. exposed to different metals. Exoticol. Environ. Saf. 73(6):1352 - 1361.

Taber HG (2009). Plant Analysis Sampling Procedures and Micronutrient Characteristics with Emphasis on Vegetable Crops. Commercial Vegetable, Horticulture. Iowa State University.

Tandy S, Bossart K, Mueller R, Ritschel J, Hauser L, Schulin R, Nowack B (2004). Extraction of heavy metals from soils using biodegradable chelating agents. Environ. Sci. Technol. 38:937-944.

Tiwari KK, Singh NK, Patel MP, Tiwari MR, Rai UN (2011). Metal contamination of soil and translocation in vegetables growing under industrial waste water irrigated agricultural field of Vadodara, Gujarat, India. Ecotoxicol. Environ. Saf. 74 (2011):1670-1677

Yebpella GG, Magomya AM, Udiba UU, Gandu I, Amana SM, Ugboaja VC Usman NL (2011). Assessment of Cd, Cu, Mn and Zn Levels in Soil, Water and vegetable Grown in Irrigated Farm along River Kubani, Zaria, Nigeria. J. Appl. Environ. Biol. Sci. 1(5):84-89.

Evaluation of the performance of improved sweet potato (*Ipomoea batatas* L. LAM) varieties in Bayelsa State, Nigeria

C. Wariboko and I. A. Ogidi

Department of Crop Production Technology, Niger Delta University, Bayelsa State, Nigeria.

This study was conducted using randomized complete block design with three replications each in two locations (Amassoma Wilberforce Island and Yenagoa, Bayelsa State) to evaluate the performance of improved sweet potato varieties (Ex-Igbariam, TIS 8164, 199004-2 and TIS 87/0087 including Kukunduku local) from March to June 2010. There were significant differences among varieties at both locations and across locations but locations and location x variety interaction were non-significant for sweet potato root yields. Ex-Igbariam and TIS 87/0087 had higher fresh root yields of 7.39 and 4.17 t ha^{-1}, respectively, than others across locations. Regarding trailing characteristic (soil surface cover), too, varieties were significantly different at both locations and across locations but location and location x variety interaction were non-significant with Ex-Igbariam and TIS 87/0087 having best soil surface cover, and consequently, best weed suppressants. There was incidence of diseases but that of insects was low. For fresh root phenotypic characteristics, Ex-Igbariam and 199004-2 had yellow flesh, indicative of the presence of vitamin A precursor. Since Ex-Igbariam, TIS 87/0087 and a few others showed real promise in yield and carotene content, carrying out a multi-locational trial would, hopefully, enable selection of high - yielding varieties for commercial production to improve farmers' yields and income in the different agro-ecological zones of Bayelsa State.

Key words: *Ipomoea batatas* L. Lam, improved variety, yield, Nigeria.

INTRODUCTION

Sweet potato (*Ipomoea batatas L. Lam*), the only member of the genus Ipomoea whose roots are edible, is one of the world's most important food crops because of its high yield, nutritive value and capacity to tolerate marginal environmental conditions (Hahn, 1977; Date and Eronico, 1987). It contains about 70, 1.5, 0.5, 25 and 1 water, fat, carbohydrate, fibre and ash, respectively (Martins and Leonard, 1955). Of the carbohydrates, 3-4.5% is sugar and the remainder mostly starch while 0.2 and 0.5% of the ash are calcium and phosphorus, respectively. Also, it contains carotene and fair quantities of ascorbic acid and B vitamins.

Regarding utilization, sweet potato cultivation is for human consumption (Hahn, 1977; Horton, 1988; Nwokocha and Okwowulu 1996; Tewe et al., 2000; Odebode, 2004: Odebode et al., 2008), livestock feed (Tewe and Ologhobo, 1983; Scott and Wall, 1992; Woolfe, 1992; Asuquo et al 1992; Asuquo and Anuebunwa, 1993; Okoroji et al., 1996; Onwubuemeli et al., 1996; Okposen et al., 1996; Ojeniyi and Tewe, 2003), industrial processing to make alcohol and starch (Martins and Leonard, 1955; Prain et al., 1997) including confectioneries (Obieri et al., 2004; Igbokwe et al., 2005; Uzoka et al., 2005; Odebode, 2004; Odebode et al., 2008). The green leaves could be consumed by humans and animals (Okposen et al., 1996) including fish (Madugba et al., 2005) while the whole plant is a natural weed suppressant (Udealor et al., 1993; Nnawuchi et al., 2001; Melifonwu and Ikeorgu, 2003; Ajagu et al., 2005). Also, potato root intake is medicinal in combating blindness and diabetes in humans (Low et al., 1996; Carey et al., 1999).

Its large starchy sweet-tasting roots are an important root vegetable (Purseglove, 1992; Woolfe, 1992) and, sometimes, the young leaves and shoots are eaten fresh. Although the leaves and shoots are edible, the starchy tuberous roots are by far the most important product. In some tropical areas, they are a staple food crop.

In West African countries (Guinea, Sierra Leone and Liberia) as well as in Northeastern Uganda and East Africa, the consumption of young leaves and vine tips of sweet potato as a vegetable is common. In 1990, FAO reported sweet potato leaves and shoots as good source of vitamins A, C and B2 (riboflavin) (FAO, 1990). However, sweet potato varieties with dark orange flesh have more beta carotene than those with light coloured flesh and their increased cultivation is being encouraged in Africa where vitamin A deficiency is a serious health problem (Hagenimana et al., 1999b).

Right now, sweet potatoes are useful in Africa to combat a widespread vitamin A deficiency that results in blindness and even death for 250,000 - 500,000 African children yearly. About two - thirds of the children developing xerophthatmia, the blindness - inducing disease resulting from lack of vitamin A, die within a year of losing their sight (Low et al., 1996; Carey et al., 1999; Bourke and Vlassak, 2004). Though for long it has been a staple food of the continent's diet, the African sweet potato contains white flesh which has no beta carotene (a precursor to vitamin A) unlike its sweeter, yellow-orange fleshed relative.

Although researchers long believed that African states would reject the coloured variety, an International Potato Centre (CIP) project conducted in Eastern and Southern Africa over the past ten years has identified a yellow-orange variety palatable to Africans (Woolfe, 1992). The researchers complemented their findings with the development of a nutritional education programme that has been successful in motivating African mothers to accept new varieties of sweet potato. Now researchers intend to focus their efforts on developing a more productive crop and expanding the reach of their education programme (Wardle et al., 1991).

Globally, sweet potato is a very important food crop and ranked fourth in terms of consumption as the world's most important. FAO (2004) reported that approximately 129,536,275 million tons were produced in more than 100 countries and Asia as the world's largest sweet potato - producing region with 114 million tons annually as the leading supplier of sweet potatoes. However, nearly half of the sweet potatoes in Asia are for animal feed with the remainder primarily for human consumption either as fresh or processed products. In contrast, African farmers produce only about 12 million tons of sweet potatoes annually but most of the crop is for human consumption. Also, they indicated the world potato hectarage in 2004 as 9074459 ha comprising 4874180, 945000, 602000 and 500,000 ha from China, Nigeria, Uganda and Tanzania, respectively. However, the major obstacles to sweet potato growth in the tropics are insects and diseases such as the sweet potato weevils and viral diseases.

In China, recent research by CIP personnel has shown that sweet potato yield could increase by as much as 30 - 40% without additional fertilizer, pesticide or genetic improvement (Prain et al., 1997). In a five year project in the province of Anhui and Shandong, using a procedure that eliminates viral diseases from planting materials, scientists had virus-free cuttings that developed into healthy plants. Extending this to all of China's sweet potato-growing regions, benefits exceeding USD$1.5 billion could be realized and would considerably reduce the country's reliance on cereal imports for livestock feed. In Kwara State, sweet potato enjoys the high cultural status of yams (*Dioscorea* sp.) in Southern Nigeria celebrating with feasts and cultural dances. Although Nigeria produces only 0.2% (about 0.26 million tons per year on 20 - 25 thousand ha) of the world total (Horton, 1988), there are indications of the crop turning into a life saver and foreign exchange earner for Nigeria. It has a 4-month growth cycle and its cultivation could be two to three times with supplementary irrigation.

On-shelf technology from National Root Crops Research Institute, Umudike, Nwokocha and Okwowulu (1996) reported mean total yields of 8.41, 15.18 and 11.16 t ha^{-1} in 1993 including 9.15, 14.32 and 19.85 t ha^{-1} in 1995 for improved varieties TIS 8441, 87/0087 and 8164, respectively. Later, Ojeniyi and Tewe (2003) indicated that the International Institute of Tropical Agriculture (IITA), Ibadan and the National Root Crops Research Institute (NRCRI), Umudike had reported high agronomic yield potential of sweet potato as a food security crop in Nigeria but the high potential was yet to be converted into increased output under the present cropping system.

However, one of the reasons identified for the failure to achieve increased sweet potato production in Nigeria was the bad agronomic system of cultivation. Also, the devastating pests notably *Cylas* sp. (sweet potato weevil), viruses and rots need control. After harvest, up to 80-100% of the roots could be lost within one to two weeks and so effective storage methods are necessary. The dark/brown discoloration of the sweet potato flour on or after reconstitution needs elimination.

Cylas sp., the most economically important field pest which devastates the root tubers of sweet potato has been a major constraint in production. *Acraea acerata* is a pest reported for the first time at Umudike in 1985 and population dynamics of *A. acerata* showed it as mainly a dry season pest occurring mainly from mid-November to mid-December (Anioke et al., 1987). However, the incidence of pests is, generally, low due to prolonged rainy season which is inimical to the development of insect pests (Anioke et al., 1990). Also, they reported TIS 81/647 as having the highest percentage of *Cylas* sp. damage on its tuber but TIS 2498 and TIS 87/0087 were found to be tolerant to the weevil *A. acerata* Hew damage on foliage. In addition, they reported the incidence of young instars of *Zonocerus variegatus* which caused serious damage to leaves. In 1991, NRCRI reported TIS 87/0087 as having good growth habit and tolerance to virus and *Cylas* sp. when harvested at four months. Based on the fact that sweet potato creeps, it is a natural weed suppressant, and given the right plant density and appropriate varieties, can develop a thick cover fast (Nwokocha, 1992). However, it has been observed that a mixture of Alachlor (Lasso) and Cloramben mixed at the rates of 1.92 and 2.88 kg ai/ha, respectively, has served in NRCRI, Umudike as an effective pre-emergence herbicide (Nwokocha, 1992).

Since sweet potato is important in Bayelsa State, it features in the farming system with growing interest. Usually, farmers cultivate local varieties which are not only low-yielding and less nutritive but, also, susceptible to diseases including insects and, some late-maturing, making evaluation of improved varieties for cultivation necessary.

The objective of this study, therefore, was to evaluate the performance of improved sweet potato varieties that would enable selection of high-yielding ones for commercial production to improve farmers' yields and income in Bayelsa State, Nigeria.

MATERIALS AND METHODS

A field experiment was conducted from March to June 2010 at two locations (Amassoma and Yenagoa, Bayelsa State) both in the Niger Delta humid tropics with bimodal rainfall pattern. The experiment was conducted on a two-year fallow land previously cultivated with maize (*Zea mays*). Four improved varieties (Ex-Igbariam, TIS 8164, 199004-2 and TIS 87/0087) from National Root Crops Research Institute (NRCRI) Umudike and a locally grown variety Kukunduku (control from Amassoma) in a randomized complete block design with three replications. After taking pre-cropping soil samples for analysis and land preparation without liming since smallholder farmers usually do not apply lime, planting of sweet potato was with four node vines plots at 100 x 30 cm spacing giving a population density of 33,333 plants per hectare. Weeding was at 4 and 8 weeks after planting (W.A.P) and basal application of 100 kg/ha NPK 20:10:10 fertilizer after the first weeding. A total of 5 treatments were used and these were replicated thrice. Plot size was 5 x 4 m. The treatments are: T_1- Ex-Igbariam; T_2- TIS 8164; T_3- 199004-2; T_4- TIS 87/0087; T_5- Kukunduku local (control).

All these potato varieties are, however, phenotypically viny and trailing varieties. Sweet potato fresh root yield was determined in tonnes per hectare including soil surface cover (trailing characteristic) on a scale of 1-5 with 1 as very good, 2 good, 3 average, 4 bad and 5 very bad.

Insect and disease incidence was on a scale of 0-5 with 0 as absent and 5 high as well as fresh root phenotypic characteristics (skin and flesh colour).

Statistical analysis obtained

The data obtained were subjected to analysis of variance (ANOVA) according to procedures of Statistical Analysis System (SAS 1999) and split plot arrangement (combined analysis) by Gomez and Gomez (1984). Differences between means were determined using least significant difference (LSD) statistic ($p \leq 0.05$).

RESULTS

Chemical and physical properties of soil at the experimental sites

The soil was clay, acidic (low pH) as well as low in organic matter and total nitrogen including essential nutrients (Table 1).

Sweet potato fresh root yields

The results obtained show that the treatments (varieties) were significantly different at Amassoma and Yenagoa locations. In Amassoma location, TIS 8164, 199004-2, TIS 87/0087 and Kukunduku local were not significantly different but were significantly different from Ex-Igbariam which had the highest fresh root yield (12.81 t ha^{-1}) followed by TIS 87/0087, TIS 8164, 199004-2 and Kukunduku local which had 7.48, 7.01, 5.99 and 5.67 t ha^{-1}, respectively (Table 2). Although the trend was the same at Yenagoa, mean fresh root yields were higher at Amassoma.

In a combined analysis using split plot arrangement with locations as main plot and treatment sub plot, the results obtained (Table 3) show means for fresh root yields of the sweet potato varieties across locations as 7.39, 4.19, 3.93, 3.28 and 3.09 t ha^{-1} for Ex-Igbariam, TIS 87/0087, TIS 8164, 199004-2 and Kukunduku local, respectively, still, with the same trend.

Soil surface cover (trailing characteristics)

In Table 4, there are means for the soil surface cover (trailing characteristic) and the results obtained at Amassoma and Yenagoa locations showed that treatments (varieties) were significantly different. In Amassoma, TIS 87/0087 had the highest (best soil surface cover) followed by Kukunduku local, TIS 8164, Ex-Igbariam and 199004-2 with 4.3, 3.8, 3.3 and 2.7, respectively, whereas at Yenagoa, TIS 87/0087 and Kukunduku local were highest and non-significantly different but significantly different from TIS 8164, Ex-Igbariam and 199004-2 which had 3.8, 3.3 and 2.6, respectively. In a combined analysis using split plot

Table 1. Soil characteristics of Amassoma, Wilberforce Island and Yenagoa locations.

Parameter		Amassoma, W. Island	Yenagoa
Organic matter (%)		1.81	1.9
pH*		4.8	4.32
Mechanical	Sand	23	25
Analysis (%)	Silt	31	34
	Clay	46	41
Soil texture		Clay soil	Clay soil
Total N (%)		0.07	0.09
Available P (mg/kg)		78.1	86.6
CEC (cmo1/kg)		2.0	2.02
	Ca	1.16	1.41
Exchangeable	Mg	10.0	12.0
Bases (cmo1/kg)	K	47.0	48.0
Base saturation		42	51

*pH in 1:10 distiled water.

Table 2. Mean fresh root yields (t ha^{-1}) of sweet potato varieties at Amassoma, Wilberforce Island and Yenagoa.

Treatment	Amassoma, W. Island	Yenagoa
Ex-Igbariam	12.81	11.81
TIS 8164	7.01	6.08
199004 - Z	5.99	4.96
TIS 87/0087	7.48	6.44
Kukunduku local	5.67	4.64
LSD .05	2.31	2.44

Table 3. Mean fresh root yields (t ha^{-1}) of sweet potato varieties across locations.

Treatment	Fresh root yields
Ex-Igbariam	7.39
TIS 8164	3.93
199004 - Z	3.28
TIS 87/0087	4.19
Kukunduku local	3.09
LSD .05	1.61

Table 4. Mean soil surface cover* of sweet potato varieties at Amassoma, Wilberforce Island and Yenagoa locations.

Treatments	Amassoma	Yenagoa
Ex-Igbariam	3.5	3.1
TIS 8164	3.8	3.8
199004 - Z	2.7	2.6
TIS 87/0087	4.9	4.3
Kukunduku local	4.3	4.3
LSD 0.05	0.42	0.42

*Indicates soil surface cover (1-5) 1 as very good, 2 as good, 3 average, 4 bad and 5 very bad.

Table 5. Mean soil surface cover* of sweet potato varieties across locations (Amassoma, Wilberforce Island and Yenagoa) in Bayelsa State.

Treatments	Soil surface cover
Ex-Igbariam	1.92
TIS 8164	2.28
199004 - Z	1.58
TIS 87/0087	2.58
Kukunduku local	2.58
LSD .05	0.27

arrangement, locations and location x varieties interaction were non-significant while there were significant differences among varieties for this characteristic across locations (Table 5).

Insect and disease incidence

The results obtained (Table 6) showed no incidence of diseases but that of insects was low.

Fresh root phenotypic characteristics (skin and flesh colours)

Among the sweet potato varieties involved (Table 7), visual observation of fresh root cuttings showed that Ex-Igbariam and 199004-2 had yellow flesh indicative of the presence of vitamin A precursor (carotene).

Table 6. Incidence of diseases and insects* on sweet potato varieties at Amassoma, Wilberforce Island and Yenagoa locations.

Treatment	Amassoma		Yenagoa	
	Disease	Insect	Disease	Insect
Ex-Igbariam	0	1	0	1
TIS 8164	0	1	0	1
199004 - Z	0	1	0	1
TIS 87/0087	0	1	0	1
Kukunduku local	0	1	0	1

*Incidence of diseases and insects with 0 as absent and 5 high.

Table 7. Fresh root phenotypic characteristics of sweet potato varieties at Amassoma, Wilberforce Island and Yenagoa locations.

Treatment	Amassoma, W. Island		Yenagoa	
	Skin colour	Flesh colour	Skin colour	Flesh colour
Ex-Igbariam	Orange	Yellow	Orange	Yellow
TIS 8164	Light pink	White	Light pink	White
199004 - Z	Pink	Yellow	Pink	Yellow
TIS 87/0087	Light orange	Cream	Light orange	Cream
Kukunduku local	Pink	White	Pink	White

DISCUSSION

Considering that the tolerable pH level for sweet potato growth and performance is 6.7 - 8, the pH level of 4.8 appeared low. This was expected because most soils in the state are acid and, traditionally, farmers do not apply lime. The basis for using these varieties is because they are early maturing and high yielding.

Except for slightly lower fresh root yields, the trend was the same in Yenagoa. In a combined analysis using split plot arrangement, locations and location x variety interaction were non-significant but varieties were significantly different with the same trend when compared with the results of Nwokocha and Okwowulu (1996) who reported mean total yields of 8.41, 15.18 and 11.16 t ha^{-1} in 1993 including 9.15, 14.32 and 19.85 t ha^{-1} in 1995 for improved varieties TIS 8441, 87/0087 and 8164, respectively, the results obtained with Ex-Igbariam having fresh root yield (12.81 and 11.81 t ha^{-1} at Amassoma and Yenagoa, respectively) would be regarded acceptable under our experimental conditions of low pH. Therefore, selection of an improved variety for increased yield and commercial production appears feasible.

The observations obtained for soil surface cover (trailing characteristic) showed, therefore, that TIS 87/0087 and Kukunduku local provided better soil surface cover and suppressed weeds best indicating that selection of a good weed suppressant is feasible.

Since there was no incidence of diseases and that of insect was low because only a few grasshoppers and dragon flies were observed, the results were similar to those of Anioke et al. (1990) who reported, generally, low incidence of pests because of rainy season which is inimical to the development of insects.

Due to the fact that visual observation of fresh root cuttings showed that Ex-Igbariam and 199004-2 had yellow flesh, the results were similar to the reports of other scientists (Wolfe, 1992; Hegenimana et al., 1999) who indicated that sweet potato varieties with dark orange flesh have more beta carotene than those with light coloured flesh and their increased cultivation is being encouraged in Africa where vitamin A deficiency is a serious health problem.

Conclusion

Generally, the study shows that the improved varieties are higher in root yields with Ex-Igbariam, TIS 87/0087 and few others showing real promise in yield and carotene content. It is, therefore, recommended that carrying out a multi-locational trial would, hopefully, enable selection of high-yielding varieties for commercial production to improve farmers' yields and income in the different agro-ecological zones of Bayelsa State.

ACKNOWLEDGEMENT

The authors are grateful to Mr. E. W. Kumoinein of Department of Crop Production Technology, Niger Delta University for his contribution.

REFERENCES

Ajagu CO, Onyishi GC, Ano AO, Nwodo O (2005). On-farm evaluation of sweet potato inter crop as weed control in ware yam production in Enugu State. Proc. Of the 20th Ann. Zonal Res. Ext. Farmer Input Linkage System (REFILS) workshop, S.E. Agro-ecological Zone of Nigeria. pp. 129 - 134.

Anioke SC, Agbo FMO, Abazie AC, Azoke SO (1990). Genetic resource conservation, evaluation and utilization. pp. 5-8.

Anioke SC, Ogbonna JC, Ikeorgu G (1987). Field evaluation of ten sweet potato cultivars for resistance to Cylas sp. And Acraea accerata Hew. NRCRI 1987 Annual Technical Report, Umudike, Nigeria. pp. 25-27.

Asuquo BO, Anuebumwa FO, Nyiam-Bisong FMA (1992). Determination of optimum combination level of pelleted mash and forage(potato leaves) in rabbit feeding (weaning to market). Proc. of the 7th Ann. Farming Systems Res. and Ext. Workshop in Southeastern Nigeria. pp. 47-52.

Asuquo BO, Anuibunwa FO (1993). Evaluation of nutritional potentials of four common forages for rabbit feeding. Proc. Of the 8th Ann. Farming Systems Research and Ext. Workshop in Southern Nigeria. pp. 49-53.

Bourke RM, Vlassak V (2004). Estimates of food crop production in Papua New Guinea, Anu Canberra, 2004.

Carey EE, Hagenimana V, K'asambo L, Oyunga MA, Benesi I, Smit N, Ocittip'Obwoya C, Turymureeba G, Ndolo PJ, Low J (1999). Using orange-fleshed sweet potato varieties to combat Vitamin A deficiency and enhance market opportunities for smallholder farmers in Sub-Saharan Africa. Proceedings of the Scientific Workshop of the Southern African Root Crops Research Network (SARRNET), held at Lusaka, Zambia. p. 476.

Date ES, Eronico PS (1987). Storage performance of some newly developed sweet potato hybrids. Radix 9(1):3-5.

FAO (1990). Production Statistics for 1990. FAO, Rome, Italy.

FAO (2004). Production Year Book for 1988. FAO, Rome, Italy.

Hagenimana VMA, Oyunga J, Low SM, Njoroge S, Guchuki, Kabira J (1999b). The effects of women farmers' adoption of orange flesh sweet potatoes: Raising vitamin A intake in Kenya. ICRW/OMNI Res. Programme, Res. Report Series 3. International Centre for Res. on women, Washington DC. p.24.

Hahn SK (1977). Sweet potato ecophysiology of tropical crops, Academic Press Inc; New York. pp. 248, 327.

Horton D (1988). Underground crops. Winrock International Institute for Agric. Dev., Morrilton, U.S.A. p. 22.

Igbokwe EO, Onyemaobi BA, Nwauzor EC (2005). On-farm evaluation of the acceptability of sweet potato/wheat composite flour blends in confectionery products (cake, chin-chin and doughnuts) Proc. Of the 20th Ann. Zonal Res. and Ext. Farmer Input Linkage System (REFILS) workshop, S.E. Agroecological zone of Nigeria. pp. 82 - 87.

Low J, Gichuki PK, Hagenimana V, Oyunga MA, Kabira J (1996). Combating Vitamin A deficiency through the use of beta-carotene rich Sweet potato varieties: Phase I of an action research project in South Nyanza, Kenya. CIP-Nairobi.

Madugba PA, Onyemaobi BT, Nwauzor EC (2005). On farm evaluation of partial substitution of fish feed with sweet potato leaves on the performance of Tilapia in Abia State. Proc. of the 20th Ann. Zonal Res. and Ext. Farmer Input Linkage System (REFILS) workshop, S.E. Agroecological zone of Nigeria. pp. 79 - 81.

Martins JH, Leonard WH (1955). Principles of field crop production. The Macmillan Co., New York. U.S.A. pp. 1118-1119.

Melifonwu AA, Ikeorgu JE (2003). Use of four cover crops for management of *Imperata Cylindrica* fallow followed by cassava/maize of inter crop in forest savanna transition and wet savanna zones of Nigeria. Proc. 8th Technical Symposium of the International Society for Tropical Root Crops - Africa Branch held at International Institute of Tropical Agric. (IITA), Ibadan, Nigeria 12 - 16 Nov, 2001. pp. 287 - 292.

Nnawuchi PC, Ubbaonu CN, Ezinwoke SO, Eke Okoro ON (2001). On-farm evaluation of the acceptability of high ratio (40 and 50%) sweet potato/wheat flour composite ability in bakery products (buns, chin-chin and doughnuts) in Imo State. Proc. Of the 16th Ann. Zonal Res. Ext. Farmer Input Linkage System (REFILS) workshop, South East/South South zones of Nigeria. pp. 113 - 119.

NRCRI (1991).Sweet potato Programme Annual Report 1991. NCRI, Umudike, Nigeria. pp. 1-2.

Nwokocha HN (1992). Agronomy of sweet potato Root Crops Research and Technology Transfer Training Course. Provisional manual NRCRI, Umudike. pp. 17-84.

Nwokocha HN, Okwuowulu PA (1996). On-shelf technologies from Natural Root Crops Res. Inst., Umudike. Proc. of the 10th Ann. Farming Systems Research and Ext. Workshop in Southeasthern Nigeria. pp. 40-44.

Obieri AU, Onyichi GC, Ano AO, Nwodo O (2004). On-farm evaluation of the acceptability of sweet potato/wheat composite flour blends in confectionery products. Proc. Of the 19th Ann. Zonal Res. Ext. Farmer Input Linkage System (REFILS) workshop, S.E. Agro-ecological zone of Nigeria. pp. 76 - 83.

Odebode SD (2004). Acceptability of sweet potato "sparri" and its potentials for enhancing food security and economic empowerment of rural woman in Southeastern Nigeria. Nig. J. Rural Sociol. 1(2):104-112.

Odebode SO, Egeonu N, Akoroda MO (2008). Promotion of sweet potato for industry in Nigeria. Bulg. J. Agric. Sci. 14(3): 300-308.

Ojeniyi ET, Tewe OO (2003). Processing and utilization of sweet potato for food and livestock in Nigeria. Proc. of the 8th Triennial Symposium of the of the International Soc. For Tropical Root crops-African Branch (1STRC-AB) pp.100-103.

Okoroji CN, Anyanwu CA, Nwokoro U, Nwagu JU, Echebiri TO (1996). Comparative evaluation of the profitability of raising weaner rabbits using mash-supplemented forages and the farmers' practice in Imo state. Proc. Of the Ann. Farming systems Res. and Ext. Workshop in Southeastern Nigeria. pp. 125 - 127.

Okposen OEB, Meyen IW, Ufur OJ (1996). Comparative Evaluation of the nutritional potentials of feeding rabbits with sweet potato leaves and centrosema (each supplemented with 50g mash). Proc. of the 10th Ann. Farming System Workshop in Southeastern Nigeria. pp. 119 - 121.

Onwubuemeli CN, Agbakoba AM, Ikeorgu JEO (1996). Evaluation of the performance of weaner rabbits fed with sweet potato vines and a mixture of other forages of the 10th Ann. Farming Systems Res. and Ext. Workshop in Southeastern Nigeria. pp. 122-124.

Prain G, Wheatley L, Nguyen DD (1997). The potential of root crop processing for rural development in Vietnam. In: International Potato Center (CIP) Programme Report 1995-1996, Lima, Peru. International Potato Center.

Purseglove JW (1992). Tropical Crops. Diocotyledon. Vol. Longman Nigeria. p.710.

Scott GJ, Wall PE (1992). Sweet Potato in African Food System. In Product Der. For Root and Tuber Crops. Vol. m.

Tewe OO, Abu OA, Ojeniyi EF, Nwokocha NH (2000). Status of Sweet potato production, utilization and marketing in Nigeria. Proceedings of the 7th Triennial Symposium of the International Society for Tuber and Root Crops-Africa Branch (ISTRC-AB), held at Cotonou, Benin. pp. 65-74.

Tewe OD, Ologhobo AO (1983). Performance, carcass characteristics and nutritive utilization of broilers fed sweet potato/soybean-based ration. The 8th Ann. Conf. of Nigerian Institute of Food Science and Technology, University of Ife, Ife, Nigeria.

Udealor A, Edokobi BI, Amalu U, Chinedu SAND, Umeh D (1993). Weed control in yam/maize/cassava alternate row intercropping system using low-growing crops in Enugu and Anambra State. Proc. Of the 8th Ann. Farming Systems Res. and Ext. Workshop in Southeastern Nigeria. pp. 22 - 27.

Uzoka IG, Nwadukwe PO, Udealor A (2005). On-farm evaluation of the acceptability of sweet potato/wheat composite flour blends in confectionery products (cake, chin-chin and doughnuts) Proc. Of the 20th Ann. Zonal Res. Ext. Farmer Input Linkage System (REFILS) workshop, S.E. Agroecological zone of Nigeria. pp. 109 - 113.

Wardle P (1991). The vegetation of New Zealand Cambridge, England: Cambridge University Press.

Woolfe J (1992). Sweet potato: An untapped food resource. Cambridge University Press, Cambridge, UK.

Evaluation of allelopathic effect of *Eclipta alba* (L.) Hassk on biochemical activity of *Amaranthus spinosus* L., *Cassia tora* L. and *Cassia sophera* L.

Aasifa Gulzar and M. Badruzzaman Siddiqui

Department of Botany, Aligarh Muslim University, Aligarh, India.

A study was conducted to assess the magnitude of suppressing ability of *Eclipta alba* (L.) Hassk. on some selected weeds of Aligarh namely, *Amaranthus spinosus* L., *Cassia tora* L. and *Cassia sophera* L. The research finding was conducted to evaluate the effect of aqueous leaf leachate and organic fractions of donor plant on biochemical activities (carbohydrate content, chlorophyll content and protein content). The result shows that aqueous leachate and organic fractions reduced the level of biochemical activities. Carbohydrate content was increased in treated plants as compared to control while chlorophyll content and protein content were reduced as compared to the control. Aqueous leachate showed maximum toxicity on weeds than organic fractions. The reduction may be due to the allelochemicals present in *E. alba* leaves.

Key words: Allelochemicals, aqueous leachate, organic fractions, chlorophyll content, carbohydrate content, *Eclipta alba*.

INTRODUCTION

Weed infestation is one of the major causes of yield reduction in crops. The incidence of allelopathic effect of weeds on growth of crops has become increasingly widespread. When the two plant species grow together, they interact with each other by either inhibiting or stimulating their growth or yield through direct or indirect allelopathic interaction (Rice, 1984; Nandal et al., 1994; Abendroth and Elmore, 2007; Kumari et al., 2006). Several reports have documented the deleterious effect of decaying weed residues on the growth and yield of subsequent crops in the field (Guenzi and McCalla, 1966; Shaukat et al., 1985, 2003; Burhan and Shaukat, 1999). The effects of decaying weed residues depend upon the release of allelochemicals from them into the soil. These chemicals may be washed directly from the residues, or may result from microbial activity during decomposition (Putnum and Duke, 1978; Lynch and Cannell, 1980; Kumar et al., 2006). The effect of allelopathic chemicals tends to be highly species-specific (Stowe, 1979; Melkania, 1983). Normally, the effect is harmful, but beneficial effect is also possible (Newman, 1978).

Eclipta alba is a small branched annual herbaceous plant, occasionally rooting at nodes, cylindrical or flat, rough due to the presence of white hairs, nodes are distant, greenish but occasionally brownish. The allelopathic potential of *E. alba* has been revealed by many researchers (Yonli et al., 2010; Nisar and Hussain, 1992; Pawinde et al., 2008).

The present study was to evaluate the aqueous leaf leachate and organic fractions effect on biochemical activity of some weed plants (*Amaranthus spinosus*, *Cassia tora* and *Cassia sophera*).

MATERIALS AND METHODS

Collection of material

Leaves of *E. alba* were collected locally from AMU campus, Aligarh. Healthy and freshly collected leaves were surface cleaned, dried and powdered. Seeds of the weed species, *A. spinosus*, *C. tora* and *C. sophera* were collected from the road sides of the Aligarh Muslim University, Aligarh, U.P., India.

Preparation of leachable allelochemical

Based on the methods devised by Kumari et al. (1985), healthy and freshly collected leaves of *E. alba* were cut roughly into pieces after clearing their surface and their dry weight per unit fresh weight were determined by desiccating the tissue in the oven. The weighed amount of fresh leaf pieces of the plant was soaked in requisite amount of pure water for a period of 20 h at room temperature. It was filtered completely through double layer of muslin cloth followed by Whatman No. 1 filter paper and the requisite concentration was made with distilled water. One half of this filtrate referred to as the aqueous leachates was used, while the other part was chilled and subjected to acid hydrolysis using pre-chilled, 3 N HCl. The precipitates so formed were recovered through centrifugation (2000 rpm). These were washed 5-6 times with pure water. Every time the recovery was made through mild centrifugation. For experimental purpose, requisite amount of the precipitate was dissolved in a few drop of ethyl alcohol and the final volume was made with pure water. A drop of Tween 20 was added to it, to serve as surfactant. This is referred to as aglycone or aglyconic or organic component of aqueous leachates.

Extraction of organic fraction

The healthy leaves of the donor plant were freshly collected, surface cleaned, then dried under shade and powdered. The powder was immersed in petroleum ether (60-80°C) for 20 h. The liquid was separated residue (marc), through mild centrifugation (500 rpm for 2 min). From the liquid portion the solvent (petroleum ether) was recovered on a hot water bath. Requisite amount of the residue so obtained was weighed and a few drops of xylene, a part from a drop of Tween-20 (to act as surfactant) were added to it. Final volume was made with pure water. This was termed petroleum ether fraction. The marc (residue from petroleum ether suspension) was suspended in methanol for 20 h and filtered, from one half of the filtrate, methanol was covered on a hot water bath. The residue, so obtained was dissolved in a drop of methanol and the final volume was made with pure water. It has been called methanol fraction. From another half of methanol filtrate, the solvent was removed and the residue was partitioned between chloroform and water (1:1 v/v). The two layers so formed were separated in a separating funnel. The chloroform was recovered over a hot water bath. To the requisite amount of residue, a few drops of methanol were added and the final volume was made with pure water. This has been termed as the chloroform fraction. The water from the aqueous layer after separating chloroform fraction was dried under low pressure on a rotary flash evaporator. The solution made with water was termed the water fraction.

Treatment for mature plants

For the estimation of various macro-molecular content (carbohydrate contents, protein content and chlorophyll content), nine plants of each test plant (*A. spinosus*, *C. tora* and *C. sophera*) were sprayed with 100 ml of the treatment solution per plant daily for five for five days. On the sixth day, the estimation of the carbohydrate content, protein content and chlorophyll content was done from the freshly plucked leaves.

Determination of carbohydrate content

The methodology employed by Loweus (1952) was followed for this purpose.

Estimation of chlorophyll content

The total chlorophyll content from leaves of treated or control plants were extracted in Di-methyl sulphoxide (DMSO) following the method of Hiscox and Israelstan (1979). Finely cut uniform discs (100 mg fresh weight) were made from fully expanded leaves of test plants. Dry weight equivalents of each of the treated samples were determined by keeping 100 mg fresh weight discs in an oven. The weighted material (100 mg fresh weight leaf disc) was suspended in 10 ml of di-methyl sulphoxide (DMSO) incubated at 65°C for one hour (the period of incubation was found sufficient for the complete extraction of chlorophyll). The DMSO was recovered by thorough decantation. The final volume was corrected to 10 ml with fresh DMSO.

The extinction of chlorophyll thus recovered in DMSO was measured at dual wave-length of 645 and 663 nm on spectrophotometer against DMSO as blank. The extinction values were read and the amount of chlorophyll was calculated according to the equation given by Arnon (1949), with modification by Hiscox and Israelstan (1979).

Estimation of total soluble proteins

The method as given by Lowry et al. (1951) was adopted for this purpose.

Statistical analysis

Each experiment was performed in completely random design block and results were mean of three replicates. All results were statistically analyzed through LSD.

RESULTS AND DISCUSSION

It is very clear from the result of this experiment that *E. alba* leaves exert a very negative influence on the acid soluble and water soluble carbohydrates of the weeds *A. spinosus*, *C. tora* and *C. sophera*. It is very well depicted that an increased amount of carbohydrates content exerts its influence mainly through its aqueous leachates. Effect of different treatments on carbohydrate content is in the following order AL>CF>PF>WF>MF (Table 1).

An increased amount of carbohydrates points out to the fact that the plant is under stress and it is gathering up its energy reserves to meet any condition of adversity. The results are in line with radish where appreciable increase in the increased concentration of soluble sugars in response to leaf extracts of heliotrope (*Heliotropium foertherianum*) was reported (Abdulghader et al., 2008). Similar increase in soluble sugars of maize in response to leaf extracts of *Acacia* and *Eucalyptus* has been reported

Table 1. Effect of aqueous leachates and organic extract fractions derived from the leaves of *E. alba* on total carbohydrate content of the leaves of *Amaranthus spinosus* L., *Cassia tora* L. and *Cassia sophera* L.

Treatment solution	*Amaranthus sinosus*		*C. tora*		*C. sophera*	
	Acid soluble carbohydrates (mg/gdry wt.)	**Water soluble carbohydrate (mg/g dry)**	**Acid soluble carbohydrate (mg/g dry wt.)**	**Water soluble carbohydrate (mg/g dry wt.)**	**Acid soluble carbohydrate (mg/g dry wt.)**	**Water soluble Carbohydrate (mg/g dry wt.)**
Control	49.45±0.65	52.12±0.31	58.47±0.36	56.79±0.49	52.83±0.57	64.83±0.33
Aqueous leachates AL (1% g/ml fresh weight)	76.45±0.12	79.63±0.37	65.34±0.15	69.11±0.15	81.04±0.67	80.37±0.52
Petroleum ether fraction PF (0.09w/v)	67.34±0.15	70.11±0.15	65.34±0.15	69.11±0.15	75.31±2.57	77.84±2.65
Methanolic fraction MF (0.09% w/v)	52.63±0.92	64.52±0.39	51.63±0.92	64.52±0.39	69.02±1.53	81.05±1.14
Chloroform fraction CF (0.09% w/v)	74.12±0.29	82.02±.0.61	72.15±0.13	70.19±0.92	79.32±0.52	70.21±0.43
Water fracrtion WF (0.09% w/v)	64.71±0.54	59.45±0.37	47.36±0.20	61.29±0.38	52.21±0.81	69.67±2.32
LSD at 5%	6.28	6.59	6.55	7.39	5.96	7.10
LSD at 1%	8.94	9.38	9.32	10.51	8.48	10.10

± Represents standard deviation.

Table 2. Effect of aqueous leachates and organic extract fractions derived from the leaves of *E. alba* on total chlorophyll content of the leaves of *Amaranthus spinosus* L., *Cassia tora* L. and *Cassia sophera* L.

Treatment solution	Total chlorophyll content (μg/mg fresh wt.)		
	A. spinosus	*C. tora*	*C. sophera*
Control	2.85±0.13	3.46±0.09	3.20±0.11
Aqueous leachates AL (1%g/ml fresh weight)	1.83±0.16	3.20±0.66	2.11±0.10
Petroleum ether fraction PF (0.09w/v)	1.63±0.42	2.74±0.04	2.01±.0.37
Methanolic fraction MF (0.09%w/v)	1.64±0.19	2.34±0.65	2.26±1.75
Chloroform fraction CF (0.09%w/v)	1.39±0.73	1.69±0.03	2.61±0.33
LSD at 5%	0.270	0.321	0.339
LSD at 1%	0.384	0.456	0.482

± Represents standard deviation.

(Sahar et al., 2005).

The chlorophyll content was also reduced as compared to the control (Table 2). Aqueous leachate and organic fraction shows different level of inhibition on different weeds. In *A. spinosus,* maximum inhibition was seen in aqueous leachate treatment. The reduction in the chlorophyll content in this experiment may be due to the fact that allelochemicals either inhibit the synthesis of chlorophyll or perhaps they breakdown the chlorophyll molecule by acting on the pyrrolic ring and the phytol chain (Blum et al., 1985; Colton and Einhelling, 1980; Yang et al., 2002, 2006). Hence the allelo-chemicals act by inhibiting the process of photosynthesis which ulti-mately can lead to the death of plant. Allelopathic effect of *Croton bonplandianum* cause reduction in chlo-rophyll content as reported by Sarkar and Chakraborty (2010) on *Triticum uestivam* and *Brassica campestris*; Liu et al. (2009) on *Lycopersicon esculentium* and Abu-Romman et al. (2010) in *Euphorbia hierosolymitana*. Recently, there has been increase in research on the role of the demand for photo-assimilates in regulating photo-synthesis through changes in carbohydrate partitioning and accumulation under stress condition (Paul and Foyer, 2001; Paul and Driscoll, 1997; Nielsen et al., 1998; Osmond et al., 1987; Levitt, 1982).

The decrease in protein content is shown in Table 3.

Table 3. Effect of aqueous leachates and organic extract fractions derived from the leaves of *E. alba* on total protein content of the leaves of *Amaranthus spinosus* L., *Cassia tora* L. and *Cassia sophera* L.

Treatment solution	Total protein content (mg/g dry wt.)		
	A. spinosus	*C. tora*	*C. sophera*
Control	58.17±0.17	58.78±2.21	55.94±0.99
Aqueous leachates AL (1% g/ml fresh weight)	32.44±0-73	45.52±1.14	44.69±0.17
Petroleum ether fraction PF (0.09% w/v)	34.17±0.76	42.62±0.65	47.21±0.65
Methanolic fraction MF (0.09% w/v)	46.24±0.23	33.81±0.82	35.27±0.19
Chloroform fraction CF (0.09% w/v)	47.16±0.31	48.4±0.84	36.27±0.12
Water fracrtion WF (0.09% w/v)	32.17±0.12	39.02±0.32	33.93±0.32
LSD at 5%	3.99	4.38	4.24
LSD at 1%	5.68	6.22	6.03

± Represents standard deviation.

Our findings are supported by the results of Duhan et al. (1995) who demonstrated significant decrease in the level of soluble proteins in legume crops in response to *Acaccia nilotica extracts*. Baziramakenga et al. (1997) demonstrated that phenolic acids reduced the incorporation of certain amino acids into proteins and thus reduce the rate of protein synthesis. Maize has been reported to contain 3 phenolic acids (Iman et al., 2006), which might have resulted in decreasing the protein content of soybean leaves. The phenolic acids have been shown to be toxic to activities of many enzymes (Hopkins, 1999).

REFERENCES

Abdulghader K, Nojavan M, Naghshbandi N (2008). Chemical stress induced by heliotrope (*Heliotropium europaeum* L.) allelochemicals and increased activity of antioxidant enzymes. Pak J. Biol. Sci. 11(6):915-919.

Abu-Romman S, Shatnawi M, Shibli R (2010). Allelopathic effects of spurg (*Euphorbia hierosolymitana*) on wheat (*Triticum durum*). Am-Euras. J. Agric. Environ. Sci. 7(3):298-302.

Arnon DI (1949). Copper enzymes in isolated chloroplasts phenol oxidase in *Beta vulgaris*. Plant Physiol. 24:1-15.

Baziramakenga R, Leroux GD, Simard RR, Nadeau P (1997). Allelopathic effects of phenolic acids on nucleic acid and protein levels in soybean seedlings. Can. J. Bot. 75:445-450.

Blum U, Dalton BR, Shann JR (1985). Effect of various mixtures of ferulic acid and some of its microbial metabolic products on cucumber leaf expansion and dry matter in nutrient culture. J. Chem. Ecol. 11(5):619-642.

Burhan N, Shaukat SS (1999). Allelopathic potential of *Argemone mexicana* L., a tropical weed. Pak. J. Biol. Sci. 2:1268-1273

Colton CE, Enhalling FA (1980). Allelopathic mechanism of velvet leaf (*Abutilon theophrasti* Medic.) on soyabean. Am. J. Bot. 67:1407-1413.

Duhan A, Khetarpaul N, Bishnoi S (1995). Variability in nutrient composition of newly evolved pigeon pea cultivars. Legume Res. 18(2):93-99

Abendroth L, Elmore R (2007). Allelopathy: A cause for yield penalties in corn following corn. Integr. Crop Manag. 12(1)16-17.

Guenzi WD, McCalla TM (1966). Phenolic acids in oats, wheat, sorghum, and corn residues and their phytotoxicity. Agron. J. 58:303-304.

Hiscox JD, Israelstan JF (1979). A method for extraction of chlorophyll from leaf without maceration. Can. J. Bot. 57:1332-1334.

Hopkins WG (1999). Introduction to Plant Physiology. 2nd Edn. Johon Wiley and Sons, New York. pp. 267-281.

Iman A, Wahab Z, Rastan SOS, Ridzwan MAH (2006). Allelopathic effect of sween and vegetable soybean extracts at two growth stages on germination and seedling growth of corn and soybean varieties. J. Agron. 5:2006.

Kumar M, Lakiang JJ, Gopichand B (2006). Phytotoxic effects of agroforestry tree crops on germination and radicle growth of some food crops of Mizoram. Lyonia 11:83-89.

Kumari A, Kohli RK, Saxena DB (1985). Allelopathic effect of *Parthiniumhysterophorus* L. leachates and extracts on *Brassica campestris* L. Ann. Biol. 1(2):189-196.

Levitt J (1982). Stress Terminology. In: N.C. Turner and P.J. Kramer (eds). Adaptations of plants to water and high Temperature Stress.Wiley at two growth stages on germination and seedling growth of corn and soybean varieties. Wiley-Interference New York. pp. 437-439.

Liu JX, Bin HH, Xin W (2009). Allelopathic effect of aqueous extract from cucumber (*Cucumissativus* L.) aboveground part on tomato (*Lycopersiconesculentum* Mill). Chin. J. Eco-Agric. 17:312-317.

Loweus FA (1952). Improvement in Anthrone method for determination of carbohydrates. Anal. Chem. 24(1):219.

Lowry OH, Rosebrough NJ, Farr AL, Rendall RJ (1951). Protein estimation with Folin Phenol reagent. J. Biol. Chem. 193:265-275.

Lynch JMC, Cannell RQ (1980). Plant residues. In:Crop seed and soil environment. Agric. Dev. Advisory Serv. (U.K.) 321:26-37.

Melkania NP (1983). Influence of certain selected tree species on ground flora. Ph.D. Thesis, Kumaun Unviersity, Nainital. p. 442.

Nandal DPS, Birla SS, Narwal SS, Koushik JC (1994). Allelopathic interactions in agroforestry systems. In:Allelopathy in Agriculture and Forestry, Jodhapur. pp. 93-130.

Newman EI (1978). Allelopathy:Adaptation or accident? In:Harborne JB (ed) Biochemical aspects of plant and animal coevolution, Ann. Proc. of the Phytochemical society of Europe 15, Academic Press, London. pp. 327-34.

Nielsen TH, Krapp A, Rooper-Shwarzand U, Stitt M (1998). The sugar-mediated regulation of gene encoding the small subunit of Rubisco and the regulatory subunit of ADP glucose pyrophosphorylase is modified by phosphate and nitrogen. Plant Cell Environ. 21:443-454.

Nisar S, Husain SI (1992). Allelopathic effect on of some plants on larval emergance of *Meloidogyne incognita* (Kafoid and white) Chit wood. Indian J. Appl. Pure Biol. 129-130

Osmond CB, Austin MP, Berry JA, Billings WD, Boyer JS, Dacey JWH, Nobel DP, Smith SD, Winner WE (1987). Stress Physiology and the distribution of plants. Biosci. 37:38-48.

Paul MJ, Driscoll SP (1997). Sugar repression of photosynthesis: The role of carbohydrates in signaling nitrogen deficiency through source source: sink imbalance. Plant Cell Environ. 20:110-116.

Paul MJ, Foyer CH (2001). Sink regulation of photosynthesis. J. Exp. 52:1383-1400.

Pawinde EZ, Sereme P, Leth V, Sankara P (2008). Effect of aqueous extract of *Acacia gourmaensis* A. Chev and *Eclipta alba* (L.) Hassk. on seed length, seedling vigour and grain yield of Sorghum and Pearl Millet. Asian J. Plant Pathol. 2(2008) 40-47.

Putnum AR, Duke WB (1978). Allelopathy in agroecosystems. Ann. Rev. Phytopathol. 16:431-451.

Rice EL (1984). Allelopathy 2nd edition. Orlando, Florida: Academic Press Inc., p.422.

Sahar A, El-Khawas, Shehata MM (2005). The allelopathic potential of *Acacia nilotica* and *Eucalyptus rostata* on monocot (*Zea mays* L.) and dicot (*Phaseolus vulgaris*) plants. Biotechnology 4:23-34.

Sarkar E, Chakraborty P (2010). Allelopathic effect of *Crotonbonplandianum*Baill.on mature growth phases of Wheat and Mustard. IUP J. Life Sci. 4:25-32.

Shaukat SS, Ghazala P, Khan D, Ahmed M (1985). Phytotoxic effects of *Citrullus colocynthis* L., Schard on certain crop plants. Pak. J. Bot. 17:235-246.

Shaukat SS, Tajuddin Z, Siddiqui IA (2003). Allelopathic potential *of Launaea procumbens* (Roxb.) Rammaya and Rajgopal: A tropical weed. Pak. J. Biol. Sci. 6(3):225-230.

Stowe LG (1979). Allelopathy and its influence on the distribution of plants in an Illinois old field. J. Ecol. 67:1065-1085

Yang CM, Lee CN, Chou CH (2002). Effects of three allelopathicphenolics on chlorophyll accumulation of rice (*Oryza sativa*) seedlings: I. Inhibition of supply orientation. Bot. Bull. Acad. Sin. 43:299-304.

Yang GQ, Wan FH, Liu WX, Zhang XW (2006). Physiological effects of allelochemicals from leachates of *Ageratina adenophora* (Spreng.) on rice seedlings. Allelopathy J. 18(2):237-245.

Yonli D, Traore H, Sereme P, Sankara P (2010). Use of local plant aqueous extract on potential bioherbicides against *Striga hermonthica* (Del.) Benth., In Burkina Faso. Asian J. Crop Sci. 2:147-154.

A new model of dynamic of plant biodiversity in changing farmlands: Implications for the management of plant biodiversity along differential environmental gradient in the Yellow River of Henan Province in the spring

Bing-Hua Liao

[1]The Key Laboratory of Ecological Restoration in Hilly Areas, Forestry Department of Henan Province, Ping-ding-shan University, Ping-ding-shan, Henan Province 467000, China.
[2]Department of Environment and Geography, Ping-ding-shan, Henan Province 467000, China.
[3]Institute of Ecological Science and Technology, College of Life Sciences, Henan University, Kaifeng 475001, China.

The intermediate disturbance hypothesis (IDH) is often used as a model or investigating the linkages between disturbance intensity and biodiversity. However, the relationship between plant diversity and disturbance gradient makes it difficult to compare data from different researches and draw general models. It is importance to assess the relationship between disturbance gradients and herbaceous species diversity in the farmland center/farmland ridge/farmland boundary on varying spatial scales along the Yellow River watershed ecosystem of Henan Province in the arctic. Using community ecology techniques and quantitative measurements of disturbance, results show a linear relationship between weighted values of disturbance intensity and herbs species diversity (Shannon index), which were significantly, negatively correlated (*P<0.01*) in the differential ecosystem types (farmland center/farmlandridge/farmland boundary) along differential disturbance gradient. Thus, understanding a linear relationship between herbs diversity dynamics and disturbance gradient in the differential type's landscapes is important for further research of local ecosystem functions and the goal of agricultural sustainable development in the context of biodiversity conservation. These results indicated that a synthesis quantitative index of disturbance (e.g. weighting values of disturbance) is the most essential environmental factors affecting herbs species diversity in the ecosystem to building a model. Therefore, this model plays a vital part in conserving global biodiversity and maintaining global ecosystem function.

Key words: Model, herbs species diversity, disturbance gradient, significantly, the Yellow River.

INTRODUCTION

Ecological model of the relationship between species diversity and environment is important along environment gradient (Boyd, 2012; Liao and Wang, 2010; Keeling et al., 2000). Ecosystems are composed of large numbers

Table 1a. The relationship between IDH and biodiversity from the different perspectives.

The relationship between IDH and biodiversity from the different perspectives	Author
Water level fluctuations, fire and grazing are essential for maintaining plant diversity.	Keddy (2005)
Increased diversity at intermediate disturbance was due primarily to increased evenness.	Aronson and Precht, (1995)
Diversity and soil properties were best at intermediate disturbance levels.	Zhang et al. (2010)
The IDH pattern was obtained for low frequency dependence and low immigration.	Cordonnier et al. (2006)
Generate a period-bubbling bifurcation structure and population dynamics that are most variable at intermediate disturbance frequencies.	Reluga (2004)

Table 1b. The physical geographic conditions of the Yellow River basin nature reserve.

Precipitation (mm)	Annual mean temperature(°C)	Latitude (°):	Longitude (°):	Elevation (m)†
800 - 1100	12.4 - 14.8	34.58 - 36.07	116.06 - 110.41	34 - 412

†Above sea level.

of plant species along environmental gradient, making species-centered studies of systemic processes and functions extremely difficult, if not out rightly impossible to carry out (Liao et al., 2011a, b). The intermediate disturbance hypothesis (IDH) suggests that species diversity will be maximal at intermediate levels of disturbance (Connell, 1978). Moreover, many experiments have assessed the relationship between plant species diversity and disturbance intensity from IDH perspective along different disturbance gradients in theoretical ecology (Leis et al., 2005). For example, Leis et al. (2005) found that disturbance up to intermediate levels can be used to maintain biodiversity by enriching the plant species pool. Further, Biswas and Mallik (2010) suggest that plant species richness and diversity, functional richness and diversity reached peaks at moderate disturbance intensity in riparian and upland plant communities.

Therefore, the relationship between plant species diversity and intermediate disturbance intensity has many real world applications in the field of ecological conservation and restoration (Table 1a).

Here, using plant community ecology techniques, this study investigated three key questions regarding the relationship between herbaceous plants species diversity and intermediate disturbance levels in the ecosystem types (e.g. farmland center, farmland ridge/farmland boundary, etc.). First, this study asked if the relationship between herbaceous species diversity dynamics and disturbance gradient from IDH perspective can be compressed into a general model that is universally applicable to ecosystem studies. Therefore, this study shows how to link herbaceous species diversity and disturbance along disturbance gradient and can be applied to real world scenarios in order to prevent biodiversity loss and help in the management of degrading or degraded ecosystems.

MATERIALS AND METHODS

The dynamics of plant diversity in Henan Province section of the Yellow River basin are the results of the historical natural and anthropogenic activities (Table 1b, Figure 1).

Quantitative assessments mainly depended on disturbance of human-driven and season driven changes in herbaceousdiversity and were based on a careful choice of landscape dynamics (Table 2).

Using community ecology techniques, we examined the influences of different disturbances on dynamics of herbaceous species diversity and disturbance interactions along the different environmental gradients in the Yellow River in the Henan Province from May 2010 to January 2011. Thirteen plots were established per 10 km length. A total of 30 plots were set. Each study plot, consisted of one 20 × 20 m tree layer plot, five (the center and four corners of the study plot) 2 × 2 m shrub layer quadrates and five 1 × 1 m herbaceous layer quadrates. Thus, these plots include 150 herbaceous layer quadrates in the farmland center, 150 herbaceous layer quadrates in the farmland boundary, and 150 herbaceous layer quadrates in the ridge of farmland from May 2010 to January 2011 (Table 3).

Traditional indices

These are the most widely used indices: Shannon index (Margalef, 1958); Shannon index of diversity is generally calculated as the

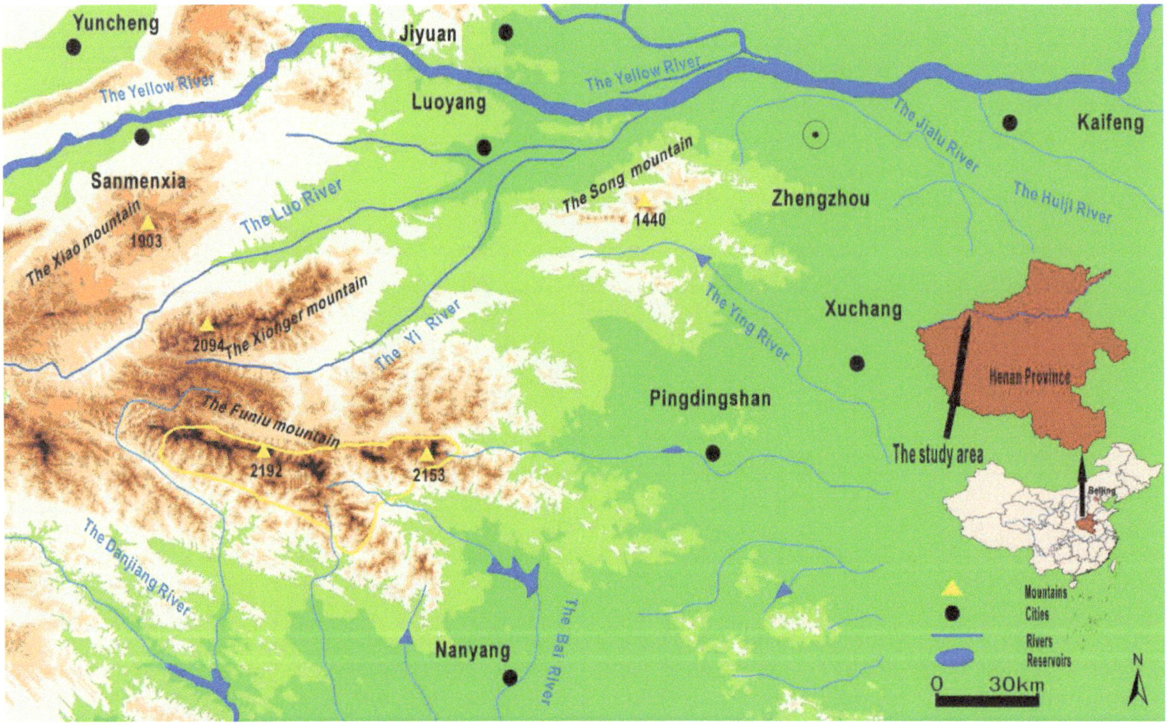

Figure 1. A digital cadastre map in the Yellow River of Henan Province.

Table 2. The selection of the weighting values of disturbance.

Quantitative measurements of disturbance to select and identify different disturbance	**Differential agro-ecosystem types**
Values of weighting of disturbance=Usage amount of herbicide	
+100/Return interval of using herbicide	Farmland center
+ 100/Return interval of tillage	Farmland ridge
+100/ The width of different types	Farmland boundary

Table 3. Investigation index along the elevation gradient variable.

Investigation	**Layer**	**Community**	**Species**	**Height**	**Crow**	**Diameter**
Community Investigation	Tree/shrub /herbaceous	Coverage /community's age structure	Species/ individual number	Layer's Height	Crow height/ width	Basal diameter

complement of H: $H' = -\Sigma\, p_i \log(p_i)$. Where $_i$ is the number of species and p_i is the proportion of the sample belonging to the i th species.

RESULTS

Disturbance is an important environmental factor affecting weed species diversity in anthropogenic activities in the farmland center/the farmland ridge areas/the farmland boundary areas. The relationship between values of weighing disturbance and the weed species diversity investigated vary significantly in the herbaceous layer along the environmental gradient on varying disturbance scales in the different disturbance intensity farmland (Figures 2 to 4).

These results which shows that herb diversity (e.g. Shannon index) is expected to increase when weighting values of disturbance is reduced along disturbance gradients in the farmland areas of the Yellow River in the

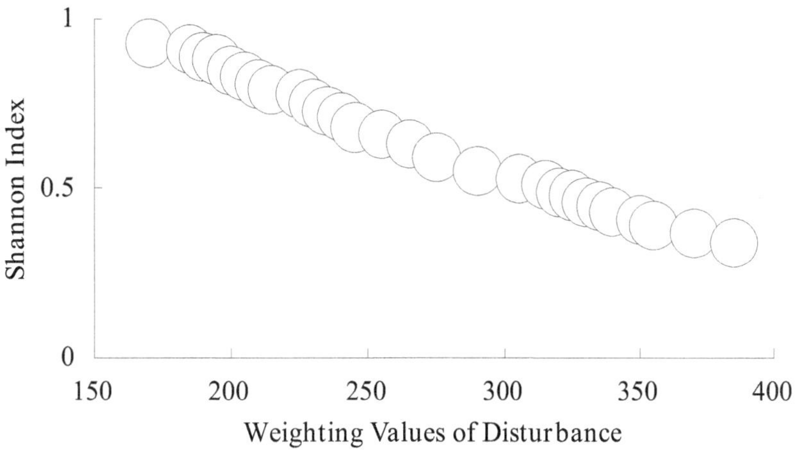

Figure 2. The relationship between Shannon index and disturbance in the farmland center areas.

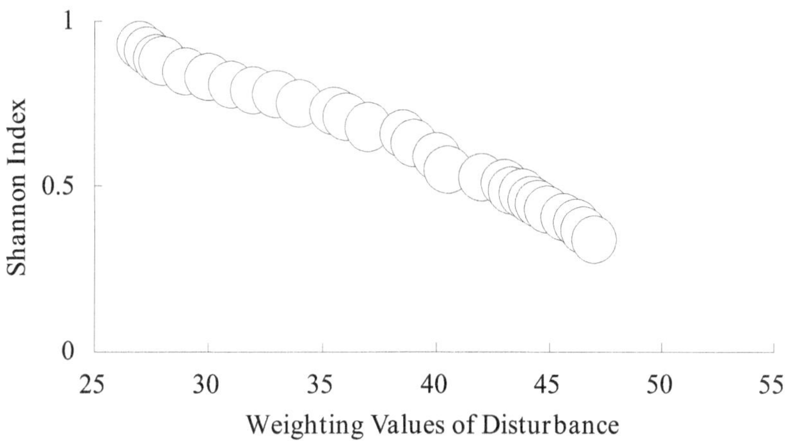

Figure 3. The relationship between Shannon index and disturbance in the farmland boundary areas.

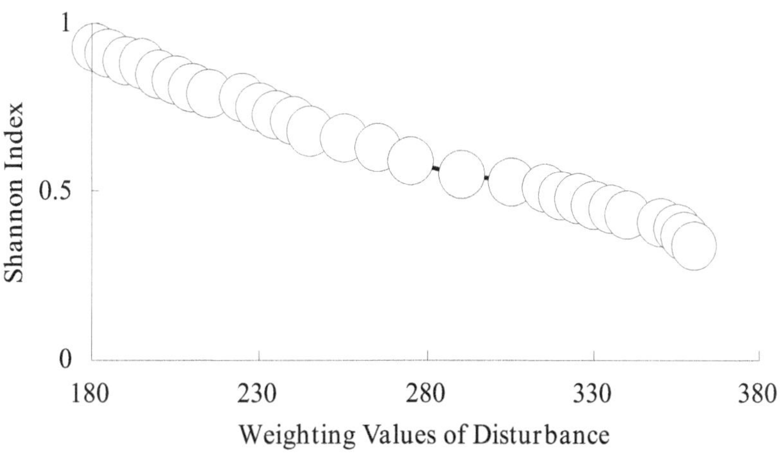

Figure 4. The relationship between Shannon index and disturbance in the farmland ridge.

Table 4. The negative correlation between Shannon index and weighting values of disturbance.

Differential ecosystem types	Shannon index
Farmland center	-0.974**
Farmland ridge	-0.983**
Farmland boundary	-0.989**

Henan Province (Figures 2 to 4). Secondly, herbaceous diversity (e.g. Shannon index) is expected to decrease when weighting values of disturbance is increased along disturbance gradients in the farmland. Thus, our result implies that weak disturbance should lead to higher herbs diversity in the farmland.

These results suggest that herb species diversity (e.g. Shannon indicator) is expected to increase when weighting values of disturbance is reduced along disturbance gradients in the farmland boundary areas and the ridge of farmland of the Yellow River in the Henan Province (Figures 3 and 4). In addition, herb species diversity is expected to decrease when weighting values of disturbance is increased along disturbance gradients in the farmland boundary areas and the ridge of farmland (Figures 3 and 4). Therefore, this result implies that weak disturbance should lead to higher herbs species diversity in the farmland boundary areas. However, is there relationship between the species biodiversity and disturbance gradient in the Yellow River of Henan Province? To show this, the correlation between species biodiversity and the weighting values of disturbance was then analyzed (Table 4).

This results suggest a linear relationship between weighting values of disturbance and herb species diversity (Shannon index), which were significantly, negatively correlated ($P<0.01$). The results indicated that disturbance, largely determined by values of weighting of disturbance, are the most important environmental factors affecting herb species diversity (Table 4).

A general model of the links between herbs diversity and disturbance gradient from IDH perspective

The IDH is used as a framework for investigating the linkages between disturbance and species diversity. Unfortunately, the various IDH makes it difficult to compare data from different studies in order to draw general conclusions (Connell, 1978). It is therefore important to develop the model which is universally applicable across a range of agro-ecosystems so that the relationship between herbaceous species diversity and disturbance gradient can be more accurately quantified with quantitative measurements of disturbance intensity. Here, we proposed the use of model framework that incorporates model (Figures 2 to 5) operating on varying spatial-temporal and disturbance scales for in-depth studies of the relationship between herbs species diversity and disturbance gradient (e.g. anthropogenic disturbance and natural disturbance) (Figures 5).

The model can help identify the relationship between herbaceous species diversity and disturbance intensity most relevant to tolerating environmental fluctuations or recovering from IDH perspective (Figure 5). The model is useful for ecosystem sustainable development in the context of biodiversity conservation. For example, Bartels and Chen (2010) explained that whether resource quantity or resource heterogeneity is the determinant of under story plant diversity in individual studies, was dependent on stand succession stage(s), presence or absence of intermediate disturbance, and forest biome within which the studies were conducted. Carvalheiro et al. (2011) suggest that presence of weeds allowed pollinators to persist within sunflower fields, maximizing the benefits of the remaining patches of natural habitat to productivity of this large-scale crop.

Applying general model to species conservation and ecosystem management

The model can use more theoretical ecology (e.g. diversity dynamics, species composition, food web structure, diversifying plant-microbial system, ecosystem stability/functioning, organic agriculture, biodiversity law, watershed ecosystem management, assessment of ecosystems and biodiversity, scale dependence) (Stutter et al., 2009; Shennan, 2008; Menalled et al., 2001, Bartels and Chen, 2010; Crowder et al., 2010; Bai et al., 2004; Davis et al., 2007; Bernez et al., 2002; Marris, 2010; Crawley and Harral, 2001). For instance, Menalled et al. (2001) suggested that agricultural management systems can have both immediate and long-term effects on weed species density, abundance and diversity. Shennan (2008) proposed that important elements for understanding biotic interactions include consideration of the effects of diversity, species composition and food web structure on ecosystem processes; the impacts of timing, frequency and intensity of disturbance; and the importance of multitrophic interactions in agro-ecosystem. Stutter et al. (2009) proposed a mechanism whereby the establishment of vegetated buffer strips between cropland and watercourses on previous agricultural land causes a diversifying plant-microbial system. Bartels and Chen (2010) explained that whether resource quantity or resource heterogeneity is the determinant of under-story plant diversity in individual studies was dependent on stand succession stage(s), presence or absence of intermediate disturbance, and forest biome within which the studies were conducted.

Therefore, it will nonetheless be a substantial challenge

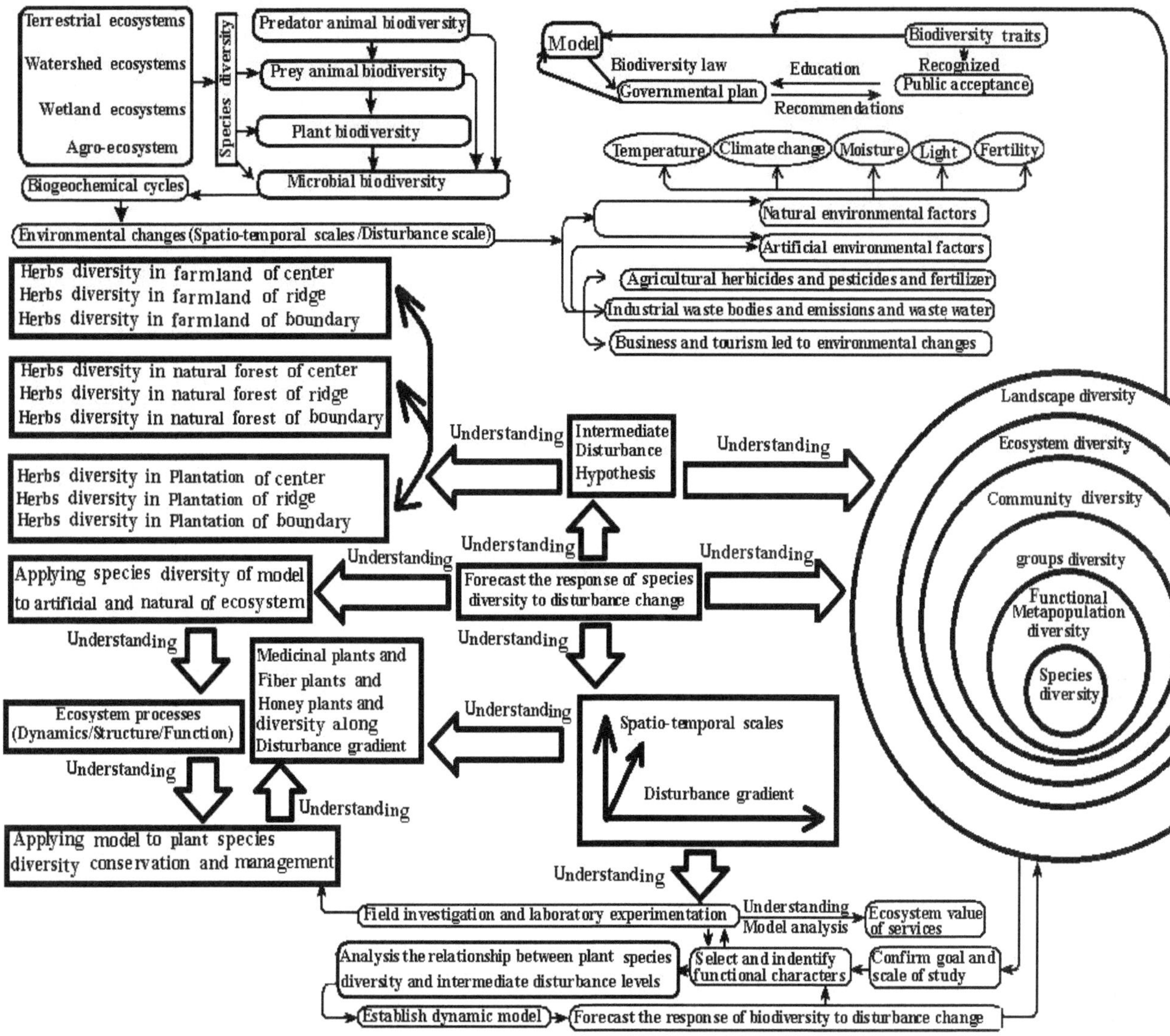

Figure 5. The proposed framework of relationship between herbs species diversity and disturbance gradient that incorporates herbs species diversity dynamic from the intermediate disturbance level, and operating on varying spatial-temporal and disturbance scales. The model was based on some studies (Liao and Wang, 2010; Liao et al., 2011a, b).

to apply this model to specific real world policy problems (Gilbert, 2010; James and Vorhies, 2010; Marris, 2010).

Conclusion

Progress in three key areas will substantially increase efforts to gain a rigorous understanding on how the links between herbaceous species diversity (Shannon index) and disturbance intensity, and their interactions, influence the response of ecosystem properties to changing biodiversity. First, better understanding of the relationship between herbaceous species diversity and disturbance intensity is correlated, particularly with respect to the predominant forces of human activities. Secondly, the model based on both field investigation of community ecology techniques of 450 herbs layer quadrates and scientific analysis, can use more and more theoretical ecology. Finally, the model can be useful in understanding the plant species diversity and disturbance intensity from IDH perspective. Understanding the model is essential for further research of local ecosystem functions and the goal of agricultural sustainable development in the context of herbs biodiversity conservation in natural or artificial disturbance ecosystem. Therefore, the model may help to ameliorate this situation of losses of plant species diversity (e.g. medicinal plants, honey plants, fiber plants, etc.).

Future work

The watershed ecosystems are results of the historical, natural and anthropogenic activities along differential environmental gradient in the Yellow River of Henan Province. However, there is an important question in the future: what are the change patterns for watershed landscape, habitat and biodiversity in the region.

Conflict of Interests

The author(s) have not declared any conflict of interests.

ACKNOWLEDGEMENTS

This work was supported by the Forestry Department of Henan Province Science Foundation of China (No.201004044), the National Natural Science Foundation of China (No.41071118), Henan Province Educational Foundation of China (No.12B180022), Ping-Ding-Shan City Societal Science Foundation of China (No.454) and the ideas of some researchers of "1st Biotechnology World Congress" is appreciated.

REFERENCES

Aronson RB, Precht WF (1995). Landscape patterns of reef coral diversity: a test of the intermediate disturbance hypothesis. J. Exp. Mar. Biol. Ecol. 192(1):1-14.

Bai YF, Han XG, Wu JG, Chuo ZZ, Li LH (2004). Ecosystem stability and compensatory effects in the Inner Mongolia grassland. Nature 431:181-184.

Bartels SF, Chen HY (2010). Is understory plant species diversity driven by resource quantity or resource heterogeneity? Ecology 91(7):1931-1938.

Bernez I, Haury J, Ferreira MT (2002). Downstream effects of a hydroelectric reservoir on aquatic plant assemblages. Sci. World J. 2:740-750.

Biswas SR, Mallik AU (2010). Disturbance effects on species diversity and functional diversity in riparian and upland plant communities. Ecology 91(1):28-35.

Boyd IL (2012). Ecology The Art of Ecological Modeling. Science 337(6092):306-307.

Carvalheiro LG, Veldtman R, Shenkute AG, Tesfay GB, Pirk CW, Donaldson JS, Nicolson SW (2011). Natural and within-farmland biodiversity enhances crop productivity. Ecol. Lett. 14(3):251-259.

Connell JH (1978). Diversity in tropical rain forests and coral reefs. Science 199:(4335):1302-1310.

Cordonnier T, Courbaud B, Franc A (2006). The effect of colonization and competition processes on the relation between disturbance and diversity in plant communities. J. Theor. Biodivers. 243(1):1-12.

Crawley MJ, Harral JE (2001). Scale Dependence in Plant Biodiversity. Science 291:864-868.

Crowder DW, Northfield TD, Strand MR, Snyder WE (2010). Organic agriculture promotes evenness and natural pest control. Nature 466:109-112.

Davis SR, Brown AG, Dinnin MH (2007). Floodplain connectivity, disturbance and change: a palaeoentomologi -cal investigation of floodplain ecology from south-west England. J. Anim. Ecol. 76(2):276-288.

Gilbert N (2010). Biodiveristy law counld stymie reserch. Nature 463: 598.

James AN, Vorhies F (2010). Green development credits to forest global biodiversity. Nature 465:869.

Keddy P (2005). Putting the plants back into plant ecology: six pragmatic models for understanding and conserving plant diversity. Ann. Bot. 96(2):177-189.

Keeling MJ, Wilson HB, Pacala SW. (2000). Reinterpreting Space, Time Lags, and Functional Responses in Ecological Models. Science 290(5497):1758-1761.

Leis SA, Engle DM, Leslie DM Jr, Fehmi JS (2005). Effects of short- and long-term disturbance resulting from military maneuvers on vegetation and soils in a mixed prairie area. Environ. Manage. 36(6):849-861.

Liao BH, Ding SY, Liang GF, Guo YL Tian L, Shu S, Zhang Y, Hu HX (2011a). Dynamics of plant functional groups composition along environmental gradients in the typical area of Yi-Luo River watershed. Afr. J. Biotechnol. 10(65):14485-14492.

Liao BH, Ding SY, Liang GF, Hu N, Gu YF, Lu XL, Liang GF, Liu J, Fan YL, Zhai YJ, Ding SP, Ding S (2011b). Dynamics of plant functional groups composition along environmental gradients in the typical area of Yi-Luo River watershed. Afr. J. Biotechnol.10(82): 18939-18947.

Liao BH, Wang XH (2010). Plant functional group classifications and a generalized hierarchical framework of plant functional traits. Afr. J. Biotechnol. 9(54):9208-9213

Margalef RD (1958). Information Theory in Ecology. General Systems pp. 36-71.

Marris E (2010). UN body will assess ecosystems and biodiversity. Nature 465:859.

Menalled FD, Gross KL, Mark H (2001). Weed aboveground and seedbank community responses to agricultural management systems. Ecol. Appl. 11:1586-1601.

Reluga TC (2004). Analysis of periodic growth-disturbance models. Theor. Popul. Biol. 66(2):151-161.

Shennan C (2008). Biotic interactions, ecological knowledge and agriculture. Philos. Trans. R. Soc. Lond. B Biol. Sci. 363(1492):717-739.

Stutter MI, Langan SJ, Lumsdon DG (2009). Vegetated buffer strips can lead to increased release of phosphorus to waters: a biogeochemical assessment of the mechanisms. Environ. Sci. Technol. 43(6):1858-1863.

Zhang C, Xie G, Fan S, Zhen L (2010). Variation in vegetation structure and soil properties, and the relation between understory plants and environmental variables under different Phyllostachys pubescens forests in southeastern China. Environ. Manage. 45(4):779-792.

Comparison of plant nutrient levels between compost from *Sky loo* and *Fossa alterna* toilets

I. B. M. Kosamu[1] , D. Mlelemba[1], H. W. T. Mapoma[1], P. Masache[1], C. C. Kaonga[1] and K. Harawa[2]

[1]Department of Physics and Biochemical Sciences, The Polytechnic, University of Malawi, Private Bag 303, Blantyre 3, Malawi.
[2]Water for People, P.O. Box 1207, Blantyre, Malawi.

Recent scholars have highlighted the benefit of harvesting compost from eco-san toilets for application as plant nutrients. However, levels of nutrients in eco-san compost may vary depending on the type of toilet and also the type of top soil in a particular geographical region. This study compared levels of nitrogen, phosphorous, potassium, calcium and sodium between compost from *Sky loo* and *Fossa alterna* toilets. Samples were collected from Zimora and Ng'ombe villages in Chikhwawa district, southern Malawi. Statistical analysis using SPSS showed significantly higher levels of nitrogen, calcium and moisture in compost from *Fossa alterna* toilets than in *Sky loo* ($p < 0.05$). However, there were no significant differences in the levels of potassium in the two types of eco-san compost studied. This study has revealed that compost from *Fossa alterna* is generally a richer source of plant nutrients than that from *Sky loo* thereby presenting poor, local communities in sub-Saharan Africa with a cheaper alternative to commercial fertilizer.

Key words: Eco-san toilets, compost, Malawi, plant nutrients, soil quality.

INTRODUCTION

With a Human Development Index rank of 171 out of 187 countries, Malawians living in rural areas face numerous challenges that include lack of access to safe drinking water, rising poverty, rapid population growth and poor sanitation (United Nations Development Programme, 2012). The growing numbers of poor households means that the majority of smallholder farmers cannot afford commercial fertilizers due to high prices. Such poor communities suffer from periodic famine due to persistent drought, and depreciating soil quality among many other external factorsg (Sugden, 2009). Eco-san compost has the potential of improving soil quality thereby increasing food security for poor households. Eco-san technology provides a safe way of recycling human excreta through composting. However, meaningful and comprehensive knowledge on variations of plant nutrient composition between compost from different eco-toilets is still scare.

There are three types of eco-san latrines, namely: Arborloo, Fossa alterna and Sky loo. Arborloo, sometimes known as "fertility pit", does not provide mechanism for harvesting manure; instead, a tree seedling (preferably a fruit tree) is planted on the spot when the pit becomes full. Sky loo, which is also known as "dry toilet", has a permanent super structure and it is a urine diversion system. The urine and stool are collected separately. Fossa alterna is a kind of toilet in which urine and faeces are mixed in one vault (Snel and Smet, 2006).

Although, eco-sans toilets provide an environmentally friendly waste management technology for disposing of faeces and urine and their recovery through composting for possible use as manure, it faces some resistance due to socio-cultural issues. Adoption of eco-san technology has made little progress in Malawi despite a decade of advocacy and promotion by non-governmental organizations such as Water for People (Malawi). Such negative attitude stems from a general reluctance of people to use of their excreta as a valuable resource. Traditionally, issues of human excreta are considered a taboo in most parts of the world and are not usually discussed in public. For this reason, sharing of information and willingness to learn practical and environmental benefits of recycling nutrients in excreta back to soil seem remote.

Across the world, eco-san is increasingly being seen as an alternative for providing safe sanitation, reducing health risks associated with poor sanitation, protecting water resources, soil fertility and optimizing resources management at the same time (Water for People, 2009). Eco-san technology can improve social and economic conditions especially for impoverished communities. Similar technologies or 'dry box toilets' have been used successfully for many years in a number of countries, such as Vietnam, China, Mexico, El Salvador, Guatemala, Ethiopia, Zimbabwe and Sweden (Esrey and Anderson, 2001). In Asia, containers of untreated human excrement are set outside residences during the night and taken to crop fields. The contents of these containers are called 'night soil' because the boxes are collected at night for application (Nawab et al., 2006).

Soil is the primary natural resource base for agriculture, but most of the soils in Malawi have low levels of nitrogen. This means that they cannot support crop production without supplementation of key nutrients such as nitrogen, phosphorous and potassium from external sources. Nitrogen, potassium, phosphorous, calcium, magnesium and sulphur are collectively known as macronutrients. These macronutrients are vital for the rapid growth of plants, increased seed and fruit production and improving the quality of leaf and forage crops. Ferrallitic soils are widely prevalent in Malawi. These soils have levels of both nitrogen (0.05 - 0.12 %) and organic matter (0.4 - 1.6 %) in low amounts; indicating poor soil fertility. Available phosphorous is between 0 - 22 ppm while potassium ranges from 0.11 - 0.36 mE/100g). Enhancement of soil productivity is very important to the sustainability of agriculture and to meeting basic food needs of the growing population.

Commercial inorganic fertilizers in Malawi costed about US$40 for a 50 kg bag in 2012. This is a huge expense in a country where almost seventy percent of the population live on less than 1US$ per day (United Nations Development Programme, 2012). Conversely, human excreta are a freely available resource in all societies. They can be utilized as soil amendments to boost crop production. Eco-san technologies are important in econo-mic emancipation of poor masses because compost from eco-san toilets is an alternative to commercial fertilizers and have potential to reduce reliance on inorganic fertilizers. Compost has the ability to improve the soil structure and water holding capacity and is comparatively better off than commercial fertilizers (Centre for Community Organisation and Development, 2011).

The concept of eco-san technology was first introduced in Malawi in 2001 by Water Aid in Salima district and the Church of Central Africa Presbyterian (CCAP) at Embangweni, in Mzimba district, before being taken up by Community Water Sanitation and Health (COMWASH) in Phalombe and Thyolo districts (Morgan, 2010). Despite attaining a decade since its inception, it is still not widely used. Recent reports indicate that 50% of the communities in Ng'ombe and Zimora villages in Chikhwawa district have adopted eco-san toilets as a result of promotion activities by Water for people in Malawi (Water for People, 2012). However, this progress is minimal and there remains a lot of work to be done before a real paradigm shift in sanitary provision can occur whereby people do not simply equate sanitation to toilets but rather to total and ecologically sound sanitation with an agronomic and economic perspective. This study was therefore conducted to compare levels of essential plant nutrient between compost from *Sky loo* and *Fossa alterna* toilets in Chikhwawa district, southern Malawi.

MATERIALS AND METHODS

This study was conducted between May and August 2012. Three *Sky loos* and three *Fossa alternae* which had been in use for at least one year were identified in the area as sampling points. Samples were collected from well decomposed and harvested compost (approximately 500 g of compost was collected in triplicates into sealed plastic zipper bags) and were transported to Polytechnic Chemistry Laboratory in Blantyre City, Malawi. Samples were air-dried by thin spreading on clean sheets of paper in the laboratory for five days. Dry samples were sieved over 2 mm mesh and then stored in air tight polyethylene zipper bags for further analysis. The standard analytical methods used were from American Public Health Association series of Standard Methods of Examination of Water and Effluent (APHA, 1998) and AOAC Official Methods of Analysis (AOAC International, 1998).

Data analysis

The data collected were subjected to student's independent t-test (two tailed) using SPSS to validate the hypothesis that the two

Table 1. Comparison of plant nutrient levels in *Fossa alterna* and *Sky loo* compost.

Compost type	mc (%)	N (%)	P (%)	K (%)	Na (%)	Ca (%)	S
Sky loo	5.03 ± 0.21	0.06 ± 0.00	0.05 ± 0.01	0.04 ± 0.00	0.01 ± 0.00	0.16 ± 0.02	BDL*
Fossa alterna	11.39 ± 0.44	1.88 ± 0.08	0.06 ± 0.00	0.05 ± 0.00	0.01 ± 0.00	0.10 ± 0.03	BDL*

*BDL = below detection limit.

Table 2. Nutrient levels in natural topsoil, compost from *Fossa alterna* and *Sky loo for studies done in Zimbabwe.* Source: Morgan, 2002.

Source	N (mg/kg)	P (mg/kg)	K (mEq/100g)
Natural dry land topsoil	38	44	0.94
Fossa alterna compost	275	292	4.51
Sky loo compost	232	297	3.06

Table 3. Chemical properties of natural soils in Chikhwawa district, Malawi. Source: Ministry of Agriculture and Food Security, 2001.

Agro-ecological zone	N (%)	P (ppm)	K(mEq/100g)
Chikhwawa Escarpment	>0.12	6 - 18	>0.2

types of eco-san compost had the same level of nutrient content.

Null hypothesis (H$_0$): $\mu_{FA} = \mu_{SL}$
Alternative hypothesis (H$_a$): $\mu_{FA} \neq \mu_{SL}$

Where μ_{FA} represented *Fossa alterna* compost mean values and μ_{SL} represented *Sky loo* compost mean values. The test was carried out at 95% confidence level (p-value = 0.05). If p-value computed was less than 0.05, the null hypothesis was rejected and alternative hypothesis accepted instead. With a p-value < 0.05, it meant there was a significant difference in the means of nutrient levels of *Fossa alterna* compost and *Sky loo* compost.

RESULTS AND DISCUSSION

Results obtained from the analysis of the two types of eco-san compoat are summarised in Table 1. There was a significant difference in levels of nitrogen for *Fossa alterna* and *Sky loo* (1.88% ± 0.08; 0.06% ± 0.00 respectively with $p \leq 0.05$). There was a lower amount of nitrogen in *Sky loo* compost than *Fossa alterna* compost probably because of the urine diversion design in the *Sky loo* latrine. Urine has a high content of readily available nitrogen. Nitrogenous wastes in the body, which result from ingestion of proteins, are excreted as urea after they have been converted from amines and ammonia (Chasseaud, 1970). Moisture content was also significantly different (11.39% ± 0.44, for *Fossa alterna* and 5.03% ± 0.21 for *Sky loo*; with $p \leq 0.05$). High amounts of water in compost from *Fossa alterna* result from urine inclusion because its principal component is water.

Levels of calcium were also significantly different such that *Fossa alterna* compost (0.10% ± 0.03) had lower level as compared to the *Sky loo* compost (0.16% ± 0.01) and p = 0.04. In humid soils, presence of ammonium sulphate may lead to loss of calcium from the soil because clay particles act like ion exchange resins. Calcium ions are exchanged for ammonium ions and are easily leached from the soil (Gardiner and Miller, 2004). This might explain why levels of calcium were lower in *Fossa alterna* compost than *Sky loo* compost.

However, there was no significant difference in levels of phosphorous for *Fossa alterna* compost (0.60% ± 0.00) and *Sky loo* compost (0.05% ± 0.01); $p = 0.17$. No significant difference was observed in levels of potassium for compost from *Fossa alterna* (1.88% ± 0.01) and compost from *Sky loo* (0.04% ± 0.03); $p = 0.34$. There was also no was no significant difference in levels of sodium for *Fossa alterna* (0.007% ± 0.00 and *Sky loo* (0.007% ± 0.00); with $p = 0.94$.

The results obtained in this study are comparable to findings in Zimbabwe (Table 2) which were reported by Morgan (2002) to compare major nutrient levels in samples of naturally occurring topsoil and compost from *Fossa altena* and *Sky loo-toilets*.

The natural soils which are found in the study area (Chikhwawa district, southern Malawi) are classified as eutric cambisols and haplic luviso (Food and Agriculture Organisation, 1999). Table 3 shows major natural occurring plant nutrient levels of these soils.

The levels of plant nutrients in eco-san compost from Chikhwawa district in Malawi are however lower those of eco-san compost from Zimbabwe. This phenomenon underlines the fact that amounts of nutrients in excreta differ from person to person and from region to region (Vinneras, 2002).

It is also important to note that eco-san compost can be a source of microorganisms such as salmonella, especially if the compost is harvested within a period shorter than 4 weeks. Studies done by Lemunier et al. (2005) indicated that compost may support long-term survival of *Escherichia coli and Listeria monocytogenes* but proper long-term (over 12 months) composting may prevent the presence of such microorganisms in compost and therefore pose low health risks to consumers of food plants grown using eco-san compost.

Conclusion

The findings from this study show that *Fossa alterna* produces compost richer in nitrogen than the *Sky loo* therefore more valuable nitrogen based organic fertilizer. The levels of other nutrients such as potassium and phosphorous did not vary between *Fossa alterna* compost and *Sky loo* compost.

The study has increased an understanding on the variations in nutrient levels of human compost depending on the type of eco-san technology used. It also contributes to the existing knowledge on the key factors which influence adoption of eco-san technology to improve soil productivity. Nitrogen is an essential plant nutrient which influences production of most cereals such as maize. Therefore, *Fossa alterna* compost can serve as good alternative for the relatively expensive inorganic fertilizers. However, nitrogen is highly labile and a single analysis is not sufficient to predict its sustainable supply to the crop over the growing season. it is therefore recommended that further extensive studies on crop productivity should be based on long term data of crop response to the two types of eco-san compost.

ACKNOWLEDGEMENTS

Authors would like to thank Water for People (Malawi) and the University of Malawi through the Department of Physics and Biochemical Sciences for financially supporting this study.

REFERENCES

American Public Health Association (APHA) (1998). Standard Methods of Examination of Water and Wastewater, 20th edition. Washington D.C.

AOAC International (1998). Official Methods of Analysis. AOAC Method 978.02. 16th edition. Gaithersburg, Md.

Centre for Community Organisation and Development (2011). Re-use of Eco-san products in Malawi. Experiences from Users in Peri-Urban Areas. http://www.ccodemw.org/phocadownload/malawi.

Chasseaud LF (1970). Processes of absorption, distribution, and excretion. In D. E. Hathway (ed.),Foreign Compound Metabolism in Mammals, Vol. I, The Chemical Society, Burlington House, London. pp. 1–33. http://www.rsc.org/ebooks/archive/free/FC9780851860084/FC9780851860084-00001.pdf.

Esrey S, Anderson I (2001). An Ecosystem Approach to the Management of Human Waste. http:// www. kwaho.eco-sanres.org.

Food and Agriculture Organisation (1999). Integrated Soil Management For Sustainable Agriculture And Food Security In Southern And East Africa. Agritex. Rome.

Gardiner DT, Miller RW (2004). Soils In Our Environment, 10th edition. Pearson Education, Inc. Upper Saddle River, New Jersey.

Lemunier M, Francou C, Rousseaux S, Houot S, Dantigny P, Piveteau P, Guzzo J (2005). Long-term survival of pathogenic and sanitation indicator bacteria in experimental biowaste composts. Appl. Environ. Microbiol. 71(10): 5779-5786.

Ministry of Agriculture and Food Security (2001). Classification, physical and chemical properties of soils in Malawi. Department of Land Survey. www.malawi.gov.org.

Morgan P (2002). Ecological Sanitation in Zimbabwe: A compilation of manuals and experiences Vol. IV. Aquamor Pvt Ltd, Harare.

Morgan P (2010). Ecological sanitation in Malawi - putting recycling into practice putting recycling into practice. http://www.susana.org/docs_ccbk/susana_download/2- 993-ecological-sanitation-in-malawi.PDF.

Nawab B, Nyborg ILP, Esson KB, Jemsen PD (2006). Cultural preferences in designing ecological sanitation system in Northwes, Frontier Province, Pakistan. Transtec Publications Ltd., Switzerland.

Snel M, Smet J (2006). The Value of Environmental Sanitation. IRC International Water and Sanitation Centre, Delft, Netherlands. http://www.lboro.ac.uk/well/resources/factsheets/factsheetstm/Ecological/sanitation.htm

Sugden S (2009). Malawi – One step close to sustainable sanitation. IRC International Water and Sanitation Centre, Delft, Netherlands.

United Nations Development Programme (2012). Human Development Report 2011/2012-Country Fact Sheets –Malawi. http://hdrstats.undp.org/countries/country_fact_sheets_fs_MWI.html

Vinneras B (2002). Possibilities for Sustainable Nutrient Recycling by Faecal Separation Combined with Urine Diversion. Ph.D. Thesis. Agraria 353. Swedish University of Agricultural Sciences, Uppsala.

Water for People (2012). Initiatives. http://www.waterforpeople.org/media-center/publications/fact-sheet/initiatives

Farmers' perception of the effects of soil and water conservation structures on crop production: The case of Bokole watershed, Southern Ethiopia

Kebede Wolka[1] , Awdenegest Moges[2] and Fantaw Yimer[1]

[1]Wondo Genet College of Forestry and Natural Resources, Hawassa University, P.O. Box 128, Shashemene, Ethiopia.
[2]Biosystems and Environmental Engineering Department, Hawassa University, P.O box 5, Hawassa, Ethiopia

Level soil bunds (LSB) and stone bunds (SB) have been widely implemented in Bokole watershed with World Food Programme (WFP) support since 2000-2001. However, the performance of structures against the target has not been studied. This study assessed farmers' opinion on the effect of soil and water conservation (SWC) structures, particularly level soil bunds and stone bunds, in improving agricultural crop production. A household survey was carried out by stratified random sampling. Twenty-seven percent of the farmers who adopted SWC structure (29 households from the upper watershed and 62 households from the lower watershed) were randomly selected and interviewed. Three group discussions were also conducted. Based on their own indicators, a high proportion of those interviewed (79.3% in the upper and 87.1% in the lower watershed) had a positive opinion about LSB and SB on their cropland, in relation to its role in improving soil fertility and crop production. Ninety-three percent of interviewed farmers in both the upper watershed and the lower watershed perceived a change in crop yield within two years after implementation of structures. There is a need for awareness creation and for monitoring the correct management of existing soil and water conservation structures, to ensure that they function as intended, and to improve their efficiency. This can reduce the problem of a 'decrease' or 'fluctuation' in crop yield.

Key words: Crop yield, non-terraced, soil bunds, stone bund, water erosion.

INTRODUCTION

Soil is a critically important resource, the efficient management of which is vital for economic growth and development for the production of food, fibre and other necessities (Troeh et al., 1980). To accommodate the increasing demand for food, either production per unit area must be intensified, or more land must be cultivated. The option of continuously cultivating the same land without appropriate and sufficient management to replenish or maintain nutrients, will likely lead to soil degradation and its consequences. FAO (1985) indicated that the effects of soil degradation, in today's complicated world, can affect nearly every aspect of survival and development.

Ethiopia, one of the developing countries in sub-Saharan Africa, depends on agriculture to satisfy the demand for food, fibre and other goods. Nevertheless, diminishing productivity, resulting from degradation of agricultural land induced by soil erosion, has been and is still a major concern (Admasu, 2005; Aklilu and Graaff, 2006a; Teshome et al., 2012). Soil erosion and consequent land degradation has been recognized as a serious problem in Ethiopia since 1973-74, subsequent to the devastating famine of the time (Bekele and Holden, 1999; Desta et al., 2005; Wagayehu and Drake, 2003; Woldeamlak and Sterk, 2003, Woldeamlak, 2006). Measurements from experimental plots and a micro-watershed estimated that annual soil loss from cropland is about 42 t ha^{-1} $year^{-1}$ (Wagayehu and Drake, 2003).

Tamene et al. (2006) indicated that some 50% of the highlands of Ethiopia were already significantly eroded, and that erosion was causing an annual decline in land productivity of 2.2%. A national-level study in Ethiopia, carried out in the mid-1980s, estimated that soil erosion was removing from use some 20,000-30,000 ha of cropland annually, and projected that *ca.* 10 million highland farmers would have their lands totally destroyed by the year 2010 (Woldeamlak, 2006).

For several decades, an attempt has been made to address the soil erosion problem in Ethiopia by means of different approaches and programmes to ensure the sustainability of agricultural production. The largest soil and water conservation (SWC) activities in the country were those implemented during the 1970s and 1980s, mainly in a food-for-work programme (Woldeamlak, 2006).

In the Ethiopian highlands, the agricultural production system cannot maintain a permanent vegetation cover throughout the year under the given ecological, economic and social circumstances (Herweg and Ludi, 1999; Ludi, 2004). Thus, soil conservation measures are a necessary part of the system for combating erosion during critical times of the year and showed certain effect (Kato et al., 2011; Adimassu et al., 2012).

Among various soil conservation interventions, traditional stone bunds (SB) and diversion ditches are common in the study area, the Bokole watershed in southern Ethiopia. In response to the challenges of climate change and soil degradation, government programmes such as the ‘safety net’, and programmes of non-governmental organizations, such as the World Food Programme (WFP), have promoted agricultural production through environmental rehabilitation since 2000 (OoARD, 2009).

Considerable efforts have been made since then to improve food security by rehabilitating degraded environments and preventing further degra-dation. The construction of level stone bunds (LSB) and SB on cropland has been a major activity of the pro-gramme in the studied watershed (OoARD, 2009).

However, with the exception of simple evaluation, the effects and impacts of watershed management, particularly of SWC, have not been studied. Specifically, no attempt has been made to evaluate how far they have fulfilled their ultimate aim, the attainment of food security through environmental rehabilitation. Therefore, this research was conducted to assess farmers’ opinion on the effect of LSB and SB in improving agricultural crop production.

MATERIALS AND METHODS

The study area

Bokole watershed is situated in Southern Nations’, Nationalities’ and Peoples’ Regional State (SNNPRS) of Ethiopia, in Dawuro zone, Loma *woreda* (Figure 1), between 6°55'-7°01'30"N lat. and 37°15'E-37°19'E long. It is *ca.* 470 km SW of the capital, Addis Ababa. The major landscape type in the study area is undulating and rugged. The watershed drains into the Omo River.

Bokole watershed lies 1160 to 2300 m above sea level (m asl), and has an annual rainfall of 1400 - 1600 mm. The lower watershed receives less rain than the upper. The temperature range in the area is 15.1 - 27.5°C (SNNPRS, BoFED, 2004), with the higher temperature in the lower part of the watershed. The soils of the study area are classified as Orthic Acrisols (SNNPRS, BoFED, 2004). The population of Bokole watershed is estimated at 11,798 (3832 in the upper, and 7936 in the lower, watershed; OoARD, 2008). Mixed agriculture is the main economic activity.

In both the upper and the lower watershed, *Zea mays L.*, *Sorghum bicolor*, *Eragrostis tef*, and *Phaseolus vulgaris* are widely cultivated. *Hordeum vulgare*, *Triticum aestivum*, *Pisum sativum, Ensete ventricosum* and *Vicia faba* are grown in the upper watershed, while *Ipomoea batatas* and *Manihot esculenta* are common in the lower watershed.

Methodology

Data were generated by means of a household survey, carried out by stratified random sampling. For this purpose, the watershed was stratified into the upper (UWS) and the lower watershed (LWS; Figure 2). In the present study, a household was defined as a basic unit of production and consumption, composed of the persons who farm common fields and live under one central decision-maker, the household head (Biruk, 2006). A list of household heads on whose land SWC structures have been implemented as new technology since 2000/01 (during which the programme started), was obtained from the agriculture and rural development office of the three *kebeles* in the *woreda*. The list from Gessa-Chare office covers households which adopted SWC structures in the UWS, whereas that obtained from Ella-Bacho and Subo-Tulama offices covers the LWS area. The list showed that SWC structures were implemented as new technology by 337 households (hh; of which 106 were in the UWS and 231 in the LWS) between 2000-2009. Of farmers who adopted SWC structures, 27% (29 hh in the UWS and 62 hh in the LWS) were selected by lot (the serial number of each hh was written on a piece of paper, wrapped, mixed and withdrawn randomly). For the group discussion, however, participants were selected on two criteria: the geographic location of the hh within the watershed management unit (hh from different corners of the target area were included), and the age of structures on the farmers’ plot of land (range of farmers on whose land the SWC structures have been implemented in different years).

A semi-structured questionnaire was prepared, covering interventions for improving or maintaining cropland productivity, and the effect of SWC structures on cropland productivity, and was pre-tested before the actual interview was conducted. It was therefore possible to modify the questionnaire on the basis of the farmers’ responses. Because the interview was conducted in the field, observation of the structures was important for improving understanding and explanation. Using a checklist, three group discussions were held. Each group was composed of ten heads of households that had adopted the structures. Information obtained from the interview and from the group discussions was checked for consistency and they were also used to supplement one another. The data collected were then analyzed descriptively by means of

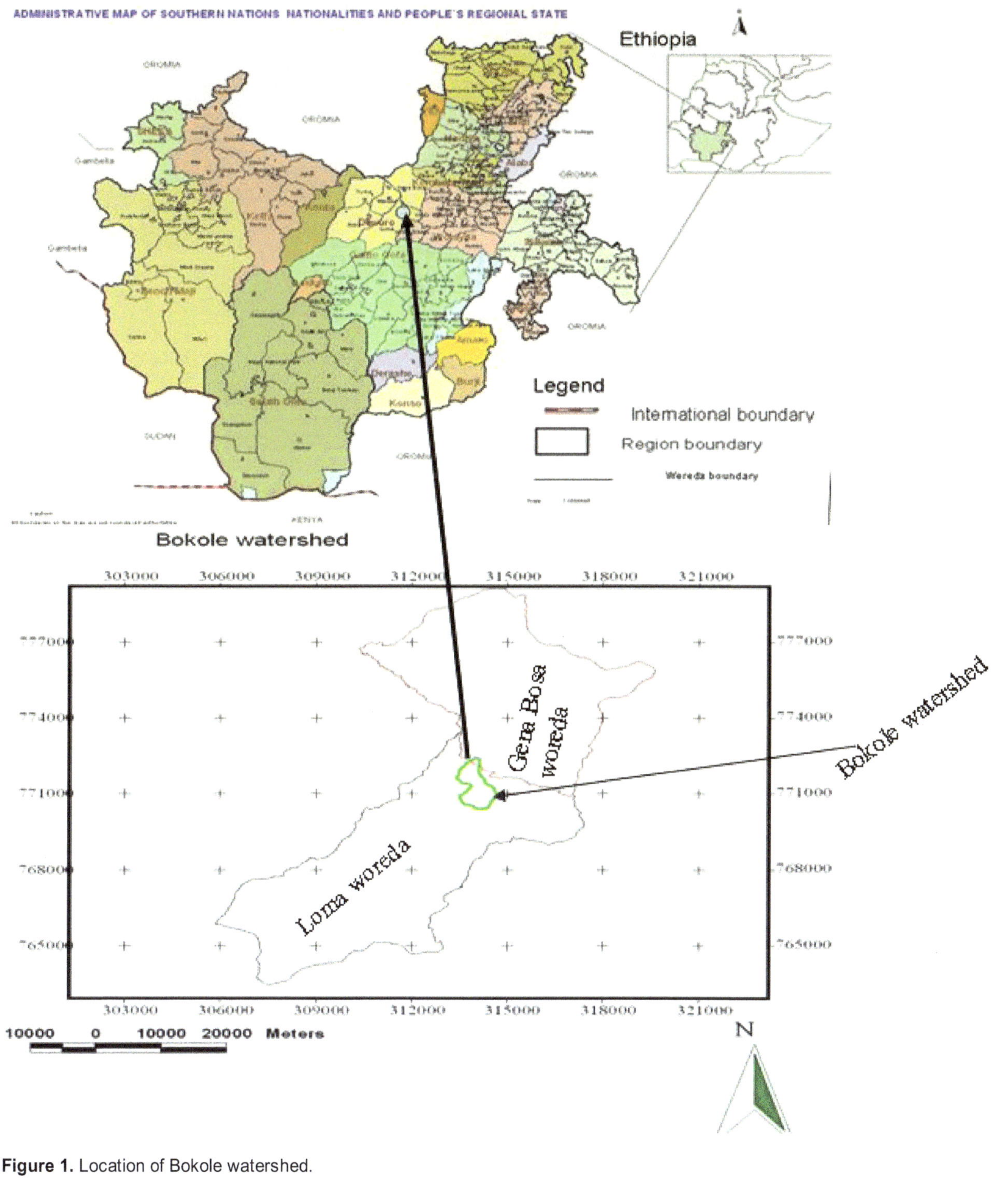

Figure 1. Location of Bokole watershed.

SPSS version 12 (2003).

RESULTS

Socio-economic characteristics in the Bokole watershed

Mean family size, mean age and mean residence period were relatively greater in the UWS as compared to the LWS. The percentage of illiterates and of those who had reached school grade 9 and above, was also higher in the UWS. The socioeconomic situation of the watershed is summarized in Table 1. The primary source of the hh income was the sale of surplus food crops (for 93.1 and 74.2% of respondents in the UWS and LWS, respectively), followed by the sale of livestock products.

Figure 2. Partial view of the topography of the Bokole watershed, Ethiopia; upper watershed (a) and lower watershed (b).

Table 1. Socio-economic characteristics of interviewed households in the Bokole watershed of Ethiopia.

Socio-economic characteristics	Watershed category	
	Upper watershed	Lower watershed
Number of households interviewed	29	62
Mean family size (no.)	9.6	7.9
Mean age of head of family (year)	45.1	41.1
Mean residence period in the area (year)	38.9	31.7
Source of income		
Primarily sale of surplus crop yield	93.1	74.2
Primarily sale of livestock and its product	0.0	14.5
Secondly sale of livestock and its product	82.8	59.7
Education level of family head		
Illiterate (%)	62.1	41.9
First cycle primary (1-4 grades) (%)	13.7	29.1
Second cycle primary (5-8 grades) (%)	13.6	22.6
High school (grade 9 and above) (%)	11.3	6.4
Mean farm area owned per household (ha)	3.0	2.1
Mean number of cattle owned per household	6.6	5.1
Mean number of goats owned per household	0.6	2.0
Mean number of sheep owned per household	2.0	0.3
Mean number of equines per household	0.2	0.3

The sale of livestock products was the primary source of income for 14.5% of LWS residents, but was the second income source for 82.8% in the UWS and 59.7% in the LWS. In general, in Bokole watershed, *ca.* 99% of farmers' primary income was from agriculture (Table 1).

Practices and technologies for crop and soil fertility management

In both the UWS and the LWS, the majority of respondents practice most crop and soil management

Table 2. Exercised practices and technologies for crop and soil fertility management in Bokole watershed of Ethiopia

Crop and vegetation and soil management activities	Respondents practicing the activity (%)	
	Upper watershed	Lower watershed
Crop rotation	100	82.3
Intercropping	86.2	88.7
Fallowing	79.3	38.7
Tree planting	93.1	96.8
Commercial fertilizer	79.3	72.6
Compost	13.7	14.5
Organic matter (animal manure, household refuse)	93.1	90.3
Not using cowdung for fuel	69.0	90.4
Not using crop residue for fuel	3.5	9.7
Adopted level soil bund	93.1	21.0
Adopted stone bund	-	67.7
Adopted both level soil bund and stone bund	6.9	11.3

Table 3. Farmers' indicators to evaluate SWC structures effect in Bokole watershed of Ethiopia

Watershed category	Farmers' indicators/observations	Frequency	Per cent
Upper watershed	Crop performance and yield	23	79.3
	Extent of runoff and erosion	4	13.8
	Both (the above two)	1	3.4
	Sediment accumulation near structures	1	3.4
Lower watershed	Crop performance and yield	30	48.4
	Extent of runoff and erosion	14	22.6
	Both (the above two)	3	4.8
	Sediment accumulation near structures	3	4.8
	Crop performance and yield and sediment accumulation near structures	5	8.1
	Extent of runoff and erosion and sediment accumulation near structures	7	11.3

activities either to improve or to maintain crop yield. In the upper watershed, LSB (Figure 3) was widely practiced, whereas SB (Figure 4) was implemented by a majority of respondents in the LWS. Table 2 presents activities such as crop rotation, intercropping and fallowing, that were practiced in the study area to maintain or enhance agricultural crop yield.

Farmers' indicators, perception of crop yield and effects of SWC structures

In their plot of cropland under their own management, farmers used their own criteria or indicators to evaluate the effect of introduced SWC structures on maintaining or changing crop yield. Observation of crop performance and yield, extent of runoff and erosion, and sediment accumulation near structures, was commonly used for evaluation (Table 3).

In both the UWS and the LWS, crop performance and yield observation were major indicators of whether the introduced SWC structures had decreased, maintained, increased or led to fluctuations in soil fertility and crop yield. The perceptions of farmers concerning crop yield change, and the role of SWC structures that improve soil for better crop yield, are further treated in what follows.

Opinion on crop yield change

In the household survey, interviewed farmers were requested to respond concerning the effects of constructed SWC structures on their cropland, in maintaining or improving soil fertility and thus crop yield. The choices for expressing an opinion were: decrease, maintain, increase and fluctuate. For this opinion, farmers were given the opportunity to compare the situation with other non-terraced cropland, and with conditions on the same land before SWC structures were implemented. A decrease in crop yield was observed only in the LWS,

Figure 3. Typical level soil bunds in upper watershed; (a) newly constructed (b) 6 years old and planted with grass.

Figure 4. Stone bunds in lower watershed; (a) newly built stone bund; (b) series of stabilized stone bunds with grass, six years after construction.

whereas 'maintenance' and 'fluctuation' were perceived by a relatively high proportion of respondents in the UWS as compared to the LWS. In general, in both the UWS (79.3%) and the LWS (87.1%), a majority of respondents had a positive opinion about the effect of SWC structures on crop yield (Table 4).

For the responses 'increase' and 'fluctuate', farmers were requested to give an opinion on how long they perceived that the SWC structures had taken to improve soil fertility and crop yield. The observed effect was immediate, in the opinion of a majority of respondents (74.5%) in the UWS, and took place in not more than a year. But a majority of respondents in the LWS observed a yield increase between one and two years after the construction of structures (Figure 5).

Opinions on the roles of SWC structures

The role played by SWC structures in improving crop yield was in the reduction of runoff and soil loss, as perceived by 27.6 and 54.0% in the UWS and LWS, respectively. The combination of reduced runoff and soil loss and water retention ability, were perceived to improve crop yield by 72.4 and 46.0% of respondents in the UWS and LWS, respectively.

Table 5 presents opinion concerning the effect of structures ('decrease', 'maintain', 'increase' and 'fluctuate')

Table 4. Respondents' opinion on effects of SWC structures on crop yield (in Bokole watershed of Ethiopia).

Watershed category	SWC structures effect on crop yield	Frequency	Percent
Upper watershed	Decrease	-	-
	Maintain	2	6.9
	Increase	23	79.3
	Fluctuate	4	13.8
Lower watershed	Decrease	5	8.1
	Maintain	1	1.6
	Increase	54	87.1
	Fluctuate	2	3.2

Table 5. Opinion on effect of structures on crop yield versus its age in upper watershed (UWS) and lower watershed (LWS); (n = 29 and 62 in upper and lower watershed, respectively).

Age of structure (year)	Decrease (%)		Maintain (%)		Increase (%)		Fluctuate (%)	
	UWS	LWS	UWS	LWS	UWS	LWS	UWS	LWS
1	-	-	-	-	3.4	6.5	-	-
2	-	-	-	-	3.4	9.7	-	-
3	-	-	-	-	10.3	17.7	-	-
4	-	1.6	3.4	-	10.3	12.9	-	-
5	-	-	3.4	1.6	10.3	17.7	-	-
6	-	1.6	-	-	6.9	16.1	-	3.2
7	-	1.6	-	-	20.7	3.2	13.8	-
8	-	3.3	-	-	6.9	3.2	-	-
9	-	-	-	-	6.9	-	-	-
Total	-	8.1	6.9	1.6	79.3	87.1	13.8	3.2

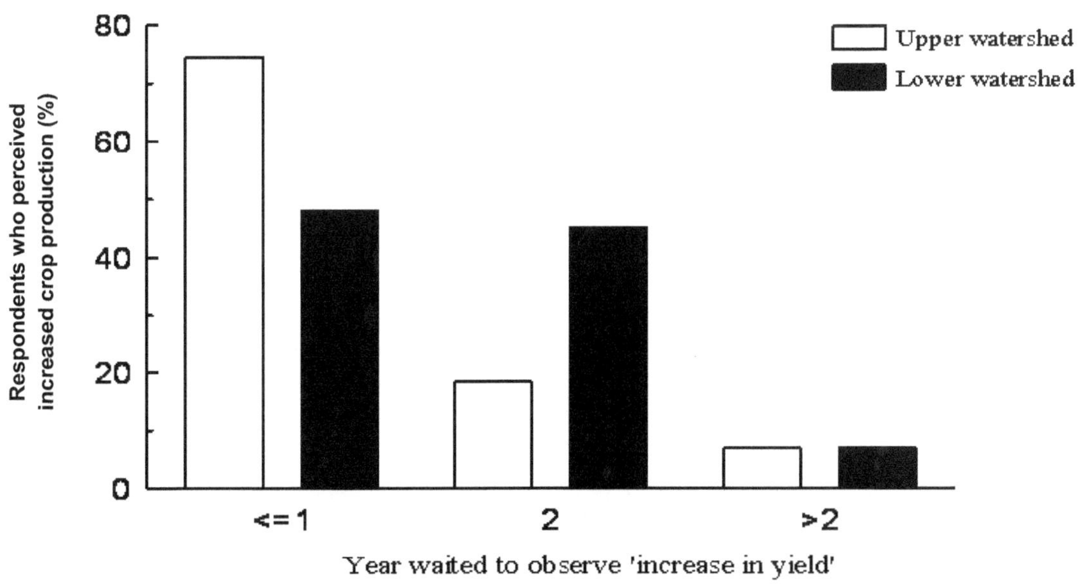

Figure 5. Response on period to observe increased crop yield (n = 27 and n = 56 in upper and lower watershed respectively).

on crop yield after construction, relative to the age of structures. Some of the farmers who owned structures 1-9 years old (in the UWS) and 1-8 years old (in the LWS), perceived an increase as a result of the structures. 'Fluctuation' was observed by farmers who owned structures six and seven years old. In the LWS, a decrease was observed on cropland on which structures were four years old and above.

Does increased yield compensate for the 'lost cropland'?

A majority of the respondents in both parts of the watershed perceived that SWC structures improved cropland and thus crop yield. It is practically observable that the structures, LSB and SB, occupy a proportion of cropland area, and consequently reduce the area available for cultivation and crop production. However, 100% of farmers in the UWS, and 91.0% in the LWS, perceived that increased yield compensated for the 'lost land'.

Increased crop production in the area between structures, as a result of conserved soil and moisture, was a major justification for their use, and was perceived by 66.7 and 60.7% of respondents in the UWS and LWS, respectively, as compensation for the 'lost land'. An alternative assumption was that in the absence of structures, the entire cropland area might be seriously eroded and degraded, and crop yield would be expected to decline, taken as justification by 26.0 and 37.5% of respondents in the UWS and LWS, respectively. In the UWS, 7.3% of farmers considered grass harvesting to be an advantage conferred by LSB as part of the compensation for 'lost land'.

DISCUSSION

Soil and crop management experiences for crop production

Frequent cultivation and soil erosion may reduce the soil fertility of agricultural land in the course of time. To overcome such problems, more than 70% of the interviewed hhs in both the UWS and the LWS practice crop rotation, add organic matter (animal manure, household refuse) to their farms, and use commercial fertilizer (Table 2).

A point raised during the group discussion was that farmers in the LWS appreciate the role of SB for their effect in increasing the efficiency of commercial fertilizer on crop response. This was because SWC structures increase the retention of moisture and soil particles, together with fertilizer, within cropland which might otherwise be washed away by water erosion.

Many farmers (69.0 and 90.4% in the UWS and LWS, respectively; Table 2) did not use cowdung for fuel, but instead used it to improve soil fertility, which increases the possibility of enriching soil with organic matter. In most cases, however, organic matter, including cowdung, is applied around the homestead, hence plots far from the homestead may not benefit. The use of crop residues for fuel by the majority of farmers in both the UWS and the LWS (Table 2), which would otherwise remain on cropland, is expected to affect soil fertility adversely. Aklilu and Graaff (2006b) indicated that the use of crop residues for fuel leads to a deterioration in soil-biological processes which contribute substantially to soil fertility.

Continuous cultivation depletes plant nutrients and makes soil prone to erosion. Farmers appeared to have understood this, and had developed the custom of periodically fallowing cropland as a land-management practice for obtaining better yields. Fallowing was more prevalent in the UWS (79.3%), as compared to the LWS (38.7%; Table 2). The critical factor in the system of fallowing is the length of the fallow period (Morgan, 2005); in the study area, farmers wait (according to the group discussion) three to four years, which was sufficient to restore soil fertility to some extent. However, the fallow land was usually used for intensive free grazing, which would presumably degrade by trampling and removing vegetation cover. The frequency and length of fallowing commonly depend on the area and productivity of the cropland.

In the study, watershed, the average land-holding owned in the UWS was relatively larger (3.0 ha) than that in the LWS (2.1 ha; Table 1), which may have encouraged farmers in the UWS to undertake such practices to a greater extent than those in the LWS. In contrast, as was observed in the field, the trend of constructing SWC structures by investing their own resources, and of maintaining existing structures was 'poor' in the UWS as compared to the LWS. Relatively better climatic conditions in the UWS (especially better moisture and moderate temperature) and the possession of a larger area of farmland, may have contributed to reduced attention to SWC.

Introduced SWC structures and their distribution

LSB were the dominant structures (93.1%) constructed in the UWS, mainly because of a shortage of stone to construct SB. The development agents and some farmers stated during the survey that *Fanya juu* was introduced on a few farmers' cropland, but most of it failed or was broken shortly after construction, owing to the nature of the structure, and its liability to damage on steep land. LSB were relatively 'better functioning', are more accepted than *Fanya juu* in this part of the watershed, and consequently are constructed on the majority of the adopters' land (Table 2). In the LWS, SB was the dominating structure among the introduced technologies, mainly owing to the availability of stones on cropland for building such structures, and to their stability after construction when compared with LSB and *Fanya juu*.

Farmers opinion on LSB and SB effect on soil and crop yield

A core aim was to evaluate the perceptions of the farmers concerning crop yield: whether it decreased, increased or fluctuated. Generally, farmers believed that LSB and SB resulted in an increase in crop yield. The various reasons for this, and the farmers' arguments, are given below.

A decrease in crop yield was perceived in the LWS by only 8.1% of respondents on whose land the SB were four years old and more (Table 5). The reasons given for this were not ineffectiveness of the structures *per se*, but their incorrect implementation and the lack of complementary practices. These include, as conceived by the respondents, a lack of cut-off drains upslope to reduce the volume of incoming surface runoff, and the continuous cultivation of cropland without complementary inputs.

During group discussions in the LWS, a farmer said 'a terrace has a double function for many people in our kebele: the soil becomes fertile due to reduction of erosion, and the grain incentive for the labourer supported the household economy.' A majority of the respondents (79.3% in the UWS and 87.1% in the LWS) perceived that these structures improved crop performance (Table 4). Such an effect was perceived by respondents irrespective of the age of the structures they owned, in both the UWS and the LWS (Table 5). In the group discussion and the interview, crop yield and performance, and reduction of surface runoff and erosion in the inter-structure area, were used as indicators to evaluate SWC structures (Table 3).

In a similar study in northern Ethiopia (Alemayehu, 2007), the majority of the interviewed farmers responded that terraces increased soil fertility, improved moisture status and increased crop yield. Furthermore, in the Gunono area of Wolaita in southern Ethiopia, 80% of the farmers were of the opinion that soil bunds increase yields (Esser et al., 2002). The study conducted by Nyssen et al. (2006) in northern Ethiopia showed that 75.4% of the farmers were in favour of stone-bund building on their land, which is a clear indication that the local community perceives this con-servation measure as beneficial. Another survey in Hagere Selam, Tigray by Esser et al. (2002) also showed that 80% of farmers responded that investments in SWC were profitable, and 68% were of the opinion that conservation practices led to increased yields in normal years. A related study by Woldeamlak (2006) showed that *ca.* 94% of the interviewed farmers in northern Ethiopia believe the physical SWC measures have the potential to improve cropland productivity, and lead to increased crop yield.

Since the gradient of most farmland in the watershed is steep, there is an increasing tendency towards erosion. The channel of the LSB traps and retains surface runoff from the upslope area, which would otherwise erode everything within the cropland. In the LWS, the soil is shallow, and experiences low rainfall and recurrent drought. Thus, the role of structures in reducing runoff, reducing soil loss by water erosion and retaining water, was noted by farmers as improving agricultural production in both the lower and upper watershed areas.

Even though a relatively small proportion of the interviewed farmers perceived the structures negatively, most of their reasons for this were related to management of the structure itself. One of the negative views was 'fluctuation' in crop yield, which was stated by 13.8 and 3.2% in the UWS and LWS, respectively (Table 4). According to the respondents, within three years after construction of the structures, crop yield increased, but subsequently decreased; this was observed by farmers who owned structures six and seven years old. This was predominantly the case in the UWS as compared to the LWS, and was supported with the reason that the channel behind the embankment of the LSB became filled with sediment in the long term (because little attention was given to removing the sediment) with the result that surface runoff overtopped the structure. Thus, in this case, the observed improvement during the earlier life of the structure was reversed or decreased later, owing to the resumption of erosion.

In both the UWS and the LWS, the fluctuating effect of the structures was perceived on structures six and seven years old (Table 5), suggesting that, unless proper management is carried out, their efficiency will decline with time, because they will be broken or sediment will accumulate, leading to overtopping by surface runoff.

Period required to induce increased crop yield and sustainability

The period required to observe the perceived yield increase after the construction of structures varied between the UWS and the LWS. In the UWS, most of the respondents (74.5%) perceived that change was observed after one year, and a few (18.5%) waited for two years. In the LWS improvements observed within a one-year and a two-year period were comparable (48.0 and 45.0% respectively). Araya and Asafu-Adjaye (1999) estimated that a positive net economic benefit as a result of soil conservation measures was obtained after two years from their construction, which is in line with the present study. Natural conditions, such as shallow workable soil, low moisture retention capacity and erratic rainfall, influenced the rate of cropland response after construction of structures in the LWS. Since the channel is empty in a newly constructed LSB, it is more efficient in trapping water and soil. Therefore, the rate of reduction of surface runoff, and thus the retention of soil and water increases, and corresponding subsequent effects accrue. Newly built SB is porous, and their ability to retain water and surface runoff is less than that of LSB (Figure 4a). Eventually, the pore space becomes filled with sediment, more runoff is impounded, and the sediment retained behind bunds enhances vegetation growth (Figure 4b). Thus, the effect of SB was gradualyl compared with LSB.

Since farmers in a subsistence economy accept and use conservation technologies that enhance productivity and provide short-term benefits (Aklilu and Graaff, 2006b), the perceived increase in yield within a few years, for both LSB and SB, may encourage them to continue to adopt it. In both watersheds, only 7.0% of respondents must wait more than two years to observe increased yield. The question is 'how long the increment would be sustained under certain land management conditions?' In the LWS, SB created a fertility gradient in the inter-structure area (as noted in the group discussions), that is, fine soil accumulated in the upslope part of bunds. On the one hand, this was due to tillage erosion related to the surface gradient, and was aggravated by the number of boundaries created (Nyssen et al., 2000; Desta et al., 2005; Li et al., 2006). On the other hand, soil erosion from the upper slopes in the inter-structure area removes fertile soil, whereas excessive accumulation occurs on the lower slope. Such a fertility gradient problem of SB has been reported by other researchers (Nyssen et al., 2006). McConchie and Huan-cheng (2002) also reported that rock terraces trap a greater thickness of soil on the slope, increasing the risk of slope failures, reducing moisture and nutrient availability to plants, and thinning the soil upslope.

To cope with such problems, farmers in the LWS adopted the interesting practice of shifting the position of a SB five to eight years (depending on slope and soil conditions) after its construction. In this practice, they move the position of a SB some distance downslope (about one-third of the distance from the upslope SB on the lower side) and re-build it. The accumulated fine and fertile soil behind the bund and in the spaces between the stones of which the SB is constructed, is spread on the previously denuded area (eroded area). Such cyclic activity interrupts the long-term benefits of a terrace, which is expected to develop progressively to a bench terrace (with a reduced slope gradient), the development of vegetation cover on the bunds themselves and the change in land management. However, the logic, as explained by McConchie and Huan-cheng (2002), which states that a soil several metres deep produces no more than a soil only deep enough to accommodate the roots, reinforces the farmers' practice. Destruction of SB, to redistribute the entrapped soil, and the practice of re-construction, was also reported by Aklilu and Graaff (2006b) from the Baressa area, Ethiopia.

In the case of LSB, the sustainability of increased yield depends on management of the structure, the removal of sediment from the channel and the repair of broken embankments. When the channel accumulates excess sediment, the function of LSB will either cease or become less effective.

Productions and functions of structures to compensate 'lost' land

SWC structures have commonly been condemned, as they occupy a considerable area of productive land (Herweg and Ludi, 1999). In contrast, a majority of farmers in both the UWS and the LWS of the study area did not complain. Instead, their perception (100% in the UWS and 91.0% in the LWS) is that the increased yield that results from LSB and SB compensates for the crop yield from the lost or uncultivated land.

This opinion is in line with that expressed by McConchie and Huan-cheng (2002), that the reduction in potentially productive land may be partly offset by an increase in production from the flatter land created by terracing. The main assumptions of farmers in the study area were that, in the absence of bunds, the entire farmland area would suffer a significant reduction in yield, and that crop production improved as result of their construction. In addition, the cropland occupied by bunds was not considered idle or non-productive on many of the respondents' fields, because planted or naturally growing grass and other plants were harvested.

Conclusions

The household survey and group discussions of the present study revealed that, in the study area, watershed management activities, particularly LSB and SB, have a positive effect on combatting soil erosion, and a potential for sustainable land management towards the improvement of crop productivity, if they are properly managed. The advantages of these introduced structures outweighed their side-effects in the period during which farmers observed their performance. About 79.3% of respondents in the UWS and 87.1% in the LWS recognized that the structures had improved the soil and crop production by reducing soil loss and conserving water.

In both the UWS and the LWS, farmers perceived that, according to their own criteria for evaluating the effect of the structures, in most cases it will not take more than two years to improve land on which the structures are built. The performance of crops or natural grass, the presence or absences of signs of runoff and erosion in the inter-structure area, and the accumulation of sediment near structures, were frequently used evaluation criteria. These criteria in turn indicate a reduction of soil loss from cropland and improved soil moisture retention. Complaints about 'lost land' or land occupied by terraces were few, because of the above-mentioned advantages of the structures.

The criticisms raised concerning introduced SWC structures by a relatively small proportion of farmers- a decrease in crop yield, an increase shortly followed by a decrease (fluctuation) or no change in land and crop yield, were related to incorrect implementation of the structure and watershed management activities. The absence or incorrect construction of supporting structures, such as cut-off drains upslope of cropland, increases the volume of surface runoff on cropland in such a way that structures cannot cope, and thus shortens their life and makes erosion reduction difficult. As observed during field work, neglect of periodic maintenance activities supposed to be carried out by the land user, such as removing sediment from the channel, repairing the embankment and the SB, reduced the effectiveness of the structures.

Effective watershed management activities, including SWC structures, are of substantial benefit for attaining and sustaining food security in smallholder farming, through the successful rehabilitation and management of natural resources. In this context, farmers' perceptions are critical. Therefore, there is a need for awareness creation, and for monitoring the proper management of the existing structures, to ensure that they function as intended, and to improve their efficiency. This can reduce 'decreased' or 'fluctuating' crop yield. In addition, suitable conservation structures, adapted to climatic conditions and slope gradient, need to be implemented. Further research should be carried out to investigate other benefits of the structures, e.g. social benefits. An economic analysis of costs and benefits would provide a better overall insight.

ACKNOWLEDGEMENTS

We are grateful to farmers who willingly participated in the interview and group discussions. The Center for Environment and Society project (at Hawassa University), Food for Work Project (in SNNPRS, Agriculture and Rural Development Bureau) and Wondo Genet College of Forestry and Natural Resources, Hawassa University, are gratefully acknowledged for the financial support.

REFERENCES

Adimassu Z, Mekonnen K, Yirga C, Kessler A (2012). Effect of soil bunds on runoff, soil and nutrient losses, and crop yield in the central highlands of Ethiopia. Land Degradation Dev. DOI: 10.1002/ldr.2182.

Admasu Amare (2005). Study of sediment yield from the Watershed of Angereb reservoir. M.Sc thesis, Department of Agricultural Engineering, Alemaya University, Ethiopia. 98 p.

Aklilu A, Graaff J (2006a). Determinants of adoption and continued use of stone terraces for soil and water conservation in an Ethiopian highland watershed. Ecol. Econ. 61(2-3):294-302.

Aklilu A, Graaff J (2006b). Farmers' views of soil erosion problems and their conservation knowledge at Baressa watershed, central highlands of Ethiopia. Agric. Human Values 23: 99-108.

Alemayehu Assefa (2007). Impact of terrace development and management on soil properties in Anjeni area, West Gojam. MSC thesis, Addis Ababa University, Ethiopia. p. 72.

Araya B, Asafu-Adjaye J (1999). Return to farm level soil conservation on tropical steep slopes: the case of Eritrean Highlands. Department of Economics, University of Queensland, Brisbane Qld 4072, Australia. Discussion paper No 253. (http://www.google.com.et/search?hl=en&source=hp).

Bekele S, Holden S (1999). Soil erosion and smallholders' conservation decisions in the Highlands of Ethiopia. World Dev. 27(4):739-752.

Biruk AH (2006). Woody species composition and socio-economic roles of traditional agroforestry practices across different agro-ecological zones in South Eastern Langano, Oromiya. M.Sc. Thesis, Hawassa University, Wondo Genet, Ethiopia. p. 90.

Desta G, Nyssen J, Poesen J, Deckers J, Mitiku H, Govers G, Moeyersons J (2005). Effectiveness of stone bund in controlling soil erosion on cropland in the Tigray Highlands, northern Ethiopia. Soil Use Manage. 21:287-297.

Esser K, Vågen T, Tilahun Y, Mitiku H (2002). Soil conservation in Tigray, Ethiopia. Noragric Report No. 5, Agricultural University of Norway. Available from: http://www.nlh.no/noragric, soil and water conservation in Tigray. Date accessed: 09/05/2010.

FAO (1985). Protect and produce: soil conservation for development, Food and Agriculture Organization of United Nations, Rome, Italy.

Herweg K, Ludi E (1999). The performance of selected soil and water conservation measures: Case studies from Ethiopia and Eritrea. Catena 36:99-114.

Kato E, Ringler C, Yesuf M, Bryan E (2011). Soil and water conservation technologies: a buffer against production risk in the face of climate change? Insights from the Nile basin in Ethiopia. Agric. Econ. 42:593-604

Li S, Lobb DA, Lindstrom MJ (2006). Tillage translocation and tillage erosion in cereal-based production in Manitoba, Canada. Soil Tillage Res. 94: 164-182.

Ludi E (2004). Economic analysis of soil conservation: Case studies from the Highlands of Amhara region, Ethiopia. African Studies Series A18. Bernee: Geographical Bernensia. pp. 416.

McConchie JA, Huan-cheng MA (2002). A discussion of the risks and benefits of using rock terracing to limit soil erosion in Guizhou Province. J. For. Res. 13(1):41-47.

Morgan RPC (2005). *Soil erosion and conservation*, 3rd edition. Blackwell Science Ltd. pp.304.

Nyssen J, Poesen J, Desta G, Vancampenhout K, Daes M, Yihdego G, Govers G, Leirs H, Moeyersons J, Naudts J, Haregeweyn N, Mitiku H, Deckers J (2006). Interdisciplinary on-site evaluation of stone bunds to control soil erosion on cropland in northern Ethiopia. Soil Tillage Res. 94:151-163.

Nyssen J, Poesen J, Mitiku H, Moeyersons J, Deckers J (2000). Tillage erosion on slopes with soil conservation structures in the Ethiopian highlands. Soil Tillage Res. 57:115 -127.

OoARD (2008). Population of Loma woreda by Kebele, Lomma woreda office of Agriculture and Rural Development, Gessa, Ethiopia.

(OoARD) 2009. Soil and water conservation report. Loma woreda office of Agriculture and Rural Development, Gessa, Ethiopia.

SNNPRS-BoFED (2004). Regional Atlas. Southern Nation, Nationalities and Peoples Regional State, Bureau of Finance and Economic Development, Bureau of Statistics and population. Awassa, Ethiopia.

SPSS Inc. (2003). SPSS for Windows Brief Guide.

Tamene L, Park SJ, Dikau R, Vlek PLG (2006). Reservoir siltation in the semi-arid highlands of northern Ethiopia: sediment yield-catchment area relationship and a semi-quantitative approach for predicting sediment yield. Earth Surface Process. Landforms 31(11):1364-1383.

Teshome A, Rolker D, de Graaff J (2012). Financial viability of soil and water conservation technologies in northwestern Ethiopian highlands. Appl. Geogr. 37:139 -49

Troeh FR, Hobbs AJ, Danahue RL (1980). Soil and water conservation for productivity and environmental protection. Prentice-hall, Inc., Englewood Cliffs. pp.718.

Wagayehu B, Drake L (2003). Soil and water conservation decision behavior of subsistence farmers in the Eastern Highlands of Ethiopia: a case study of the Hunde-Lafto area. Ecol. Econ. 46 (3):437-451.

Woldeamlak B (2006). Soil and water conservation intervention with conventional technologies in northwestern highlands of Ethiopia: acceptance and adoption by farmers'. Land Use Policy 24(2):404-416.

Woldeamlak B, Sterk G (2003). Assessment of soil erosion in cultivated fields using a survey methodology for rills in the Chemoga watershed, Ethiopia. Agric. Ecosyst. Environ. 97:81-93.

Assessing the farmers' perception of climate variability and change: A case study of Miesso-Assebot Plain, Eastern Ethiopia

Solomon Tamiru[1], Kindie Tesfaye[2] and Girma Mamo[3]

[1]Department of Plant Sciences, College of Agriculture and Veterinary Sciences, Ambo University, Ethiopia.
[2]International Livestock Research Institute, Addis Ababa, Ethiopia.
[3]Melkassa Agricultural Research Center, Adama, Ethiopia.

Assessment of farmers' perception and adaptation mechanisms at local level has enormous advantages in Ethiopia, where the driver of the economy is agriculture. This study was conducted to assess the perceptions of farmers to climate change and increased climate variability, and to identify the potential adaptation options. Most farmers noted an increase in temperature and decrease in precipitation in the last fifty years. The farmers' perception of increasing temperature was in accordance with the statistical climate data record; however, the farmers could not differentiate between consistent changes in climate and yearly rainfall fluctuation. Moreover, majority of the farmers were aware of frequent drought occurrence, shift in onset date and early withdrawal of rainfall. As a result, the farmers indicated decreased diversity of cultivated crops, changes in farming practices, new patterns of diseases, crop infestations with new weed species and frequent total crop failures. Furthermore, the majority of the respondents perceived water shortage, great loss of biodiversity and forest resources, and decline in soil fertility as the most serious impact of climate change. Use of different planting dates, on farm soil and water conservation, use of local and early maturing varieties were the major adaptation strategies practiced by farmers to mitigate climate change impacts at Miesso. However, production of drought tolerant cultivars with optimum maturity periods, introduction of new crops, varieties and crop management practices that go in line with the changing climate are recommended to offset climate change impacts on crop production at Miesso-Assebot plain, Eastern Ethiopia.

Key words: Climate variability and change, perception, adaptation options, Miesso-Assebot Plain.

INTRODUCTION

Climate change is rapidly emerging as one of the most serious threats to the totality of human existence. Different studies revealed that global mean surface temperature will be increased by 1.4 to 5.8°C between

1990 and 2100 which is expected to be a much more rapid rate of warming than during the 20th century (Gruza and Rankova, 2004; Majule, 2008). Majule (2008) reported that the average global precipitation was projected to increase but, at regional levels there will be both increase and decrease in intensity of rainfall ranging from 5 to 20°C. Globally, these projected higher temperature and variable precipitation levels will unequivocally reduce crop yields through direct effects as well as indirect impact by triggering insect pests, diseases and weeds (Gadgil et al., 1998). The threats to food security and sustainable growth of developing countries will be much higher as the extent to which the impacts will be felt depends on adaptive capacity of communities (Mendelsohn and Dinar, 2009; Mtambanengwe et al., 2012).

About 66% of the total area of Ethiopia falls within arid and semi-arid climatic zones of the country (MoA, 1998). Nevertheless, agriculture, which is highly sensitive to climate variability and change (Hellmuth et al., 2007; Thornton et al., 2006; Teshome et al., 2008), is the driver of the country's economy as it accounts for half of GDP and 80% of employment (MoARD, 2007). Climate variability, particularly rainfall variability and associated drought, and increased frequency of extreme events could make rainfed agriculture more risky and aggravate food insecurity in Ethiopia (Seleshi and Zanke, 2004; Stern, 2007; Conway and Schipper, 2011). Preliminary projections suggested that climate change can have a sizeable impact that ranges in the order of 7-8% of GDP loss per year in Ethiopia (UNDP, 2011). These will form a serious concern for both researchers and development planners in Ethiopia and elsewhere.

Moreover, in Ethiopia, the risks associated with change in climate patterns that smallholderface is believed to be due to low adaptive capacity of society and limited adaptation options of agricultural sector (Yesuf et al., 2008; Mengistu, 2011). Hence, the livelihood approach of a locality could provide a baseline to probe adaptation options to climate related risks through assessing farmers' perception and local adaptation mechanisms in order to formulate mitigation strategies (Thomas et al., 2007; Stage, 2010; Belaineh et al., 2012). On the other hand, the farmer perception must be integrated with research information and proposed technologies in order to reduce the vulnerability and strengthen the adaptive capacity of communities.

The Miesso-Assebotplain in Eastern Ethiopia is one of the areas where staple food crops are extensively grown under high rainfall variability and unpredictability, strong winds, higher temperature and high evapotranspiration (Mamo, 2005). Therefore, apart from understanding meteorological variability and change on crop production and productivity *per se*, it is important to know the perception, and adaptation mechanisms of communities so as to develop viable climate change and variability adaptation options in a given area. Therefore, this study was conducted to assess the perceptions of farmers to climate change and increased climate variability, and the likely adaptation options used in crop production. The farmers' sources of climate change information and barriers to adaptation were also investigated in this study as these perceptions determine what farmers consider as alternative best adaptation options.

MATERIALS AND METHODS

Study area

The study was conducted in Miesso-Assebot Plain, located in Eastern escarpment of the Central Rift Valley of Ethiopia that forms the heart and corridor of the Ethiopian Rift Valley (Figure 1). The geographical location of the area ranges between 8° 48" 12'- 9° 19" 52'N latitude, 40° 9" 30' and 40° 56'' 44' E longitudes and altitude varying from 1107 to 3106 m above sea level. According to Lemma (2008) and Worku (2006), Miesso is dominated by silty clay loam soil texture with slightly alkaline pH ranging from 7.8 to 8.3. The farming system of Miesso is dominated by crop production. The major crops grown in the area include sorghum (66% of cultivable land), maize (24%), common bean and sesame as staple and cash crops (Kidane et al., 2006).

Miesso-Assebot plain is predominantly categorized under hot and warm sub-moist agro-ecological zone (MoA, 1998), receiving annual average rainfall of 727 mm distributed in a bimodal pattern. The first rainy season extends from March to May while the second (main rainy season) extends from June to September (NMSA, 1996). The annual mean minimum and maximum temperature of the district is 15 and 30.6°C, respectively.

Data collection

Primary data were collected through structured questionnaire using farmers' participatory methodologies which include formal interviews, direct observations and oral discussions. Data were collected on farmer perceptions of magnitude of change in climate patterns along with their underlying causes and consequences as well as the adaptation strategies used in crop production to tackle the risks associated with climate variability and change. Others were perception of farmers on the climate change impacts using indicators such as diversity of livelihood strategies, changes in crop diversity, change in crop management practices, and access to and knowledge of climate related helpful information for their farm level decision that has been in use in the past fifty years. The major constraints to using existing adaptation options were assessed through the questionnaire survey. The questionnaire was pre-tested and improvement made on the results obtained from the pre-test. Long-term temperature and rainfall data of Miesso station were obtained from National Meteorological Service Agency (NMSA, 1996) which was analyzed using Microsoft Office Excel 2007 to present patterns and trends of rainfall and temperature in the form of graphs. The farmers' perceptions were then compared with the meteorological record data analysis.

Sampling techniqueand data analysis

A multi-stage sampling technique was used to select five Peasant Associations (Gorbo, Husse-Mandera, Hunde-Misoma, Odabala and Torbayo) and draw sample farmers for the study. The questionnaire was administered into a total of 75 farmers of the district (15 from each Association). Household heads above the age

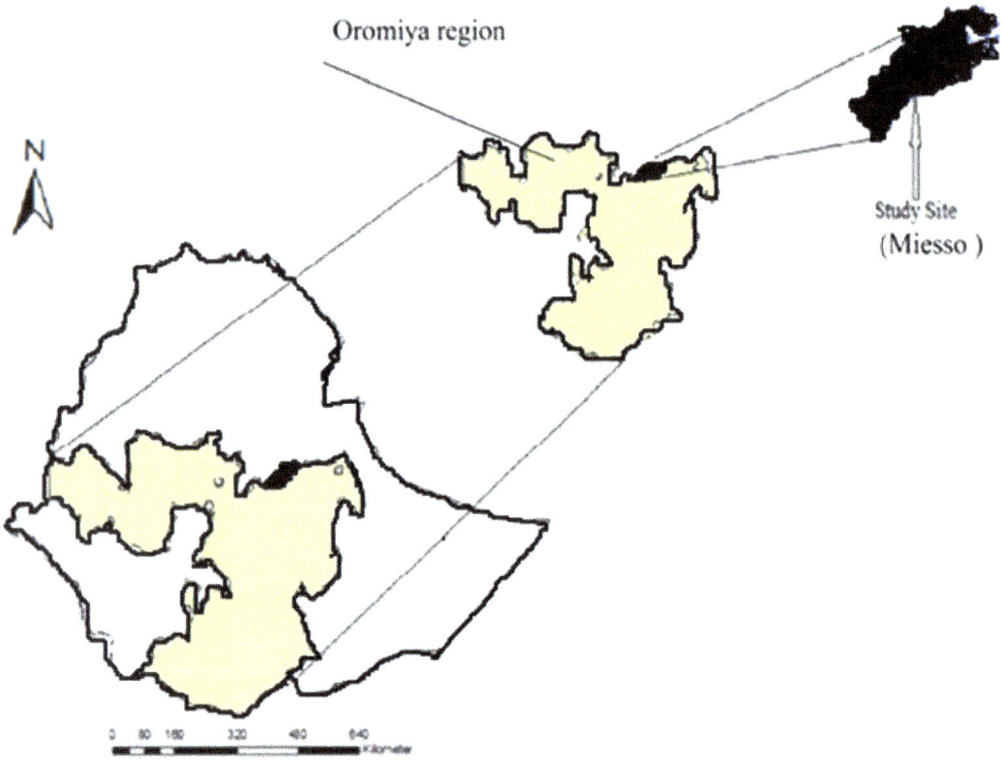

Figure 1. Geographical location of the study area.

of 50 were purposively administered the questionnaires, as more experienced and matured farmers are better at distinguishing climate variability from merely inter-annual variation of weather scenarios (Ishaya and Abeje, 2008). The total sample size was proportionally categorized into primary adopters (50) and non-adopters (25) of new technologies while growing sorghum to make their living. The crucial reason for this category is to compare the perception and adaptation mechanisms between technology adopters and non adopters. Agricultural office workers and development agents of the districts were involved in selecting the farmers.

Qualitative data collected through interview were examined and presented in different forms. Quantitative data were edited, coded and entered in a computer and Statistical Package for Social Science (SPSS) software version 17.0 spread sheet was used for the analysis (Greasley, 2008). Descriptive statistics such as mean, frequency and percentages were used for analysis of the data. Multiple response questions were analyzed so as to give frequencies and percentages. Tables, pie and bar charts were used to present different variables.

RESULTS AND DISCUSSION

Local perception on climate changes

Farmers' perceptions on climate variability and changes over the last fifty years (1960-2009) were assessed (Figure 2) in comparison with measured climate records (Figures 3, 4 and 5). The majority of farmers rated the reduction in rainfall amount as low during the period ranging from 1960-1980, medium during 1981-2000, and high and very high in the 2000s (Figure 1a, b and c). The farmers also opined that the rainfall amount was decreasing progressively every year since 1960s and resulted in drought conditions year after year. The great majority of farmers also indicated that changes which occurred in their ecosystems are related to severe droughts.

In view of that, the farmers ranked the frequency of droughts as low and very low, moderate, high and very high during 1960-1980, 1981-2000 and 2001-2009, respectively. However, the record data on rainfall from 1974-2009 showed that annual and seasonal (March-April-May and June-July-August-September) rainfall trend did not exhibit great reduction in amount except its variability from year to year (Figures 3 and 4). This result is in agreement with previous study on rainfall trend in Ethiopia (Sileshi and Zanke, 2004; Cheung et al., 2008). These indicate that the farmers were not able to differentiate between consistent changes in climate and yearly rainfall fluctuations.

Regarding temperature, most of the farmers interviewed perceived long-term changes in temperature. The farmers are of the view that the rise in temperature is as low (50%) and very low (25%) in the 1960s; moderate (47%) and high (25%) in the 1990s, high (55%) and very high (39%) in the 2000s (Figure 2). By and large, the analysis showed that the farmers' perception of increasing temperature is in accordance with the

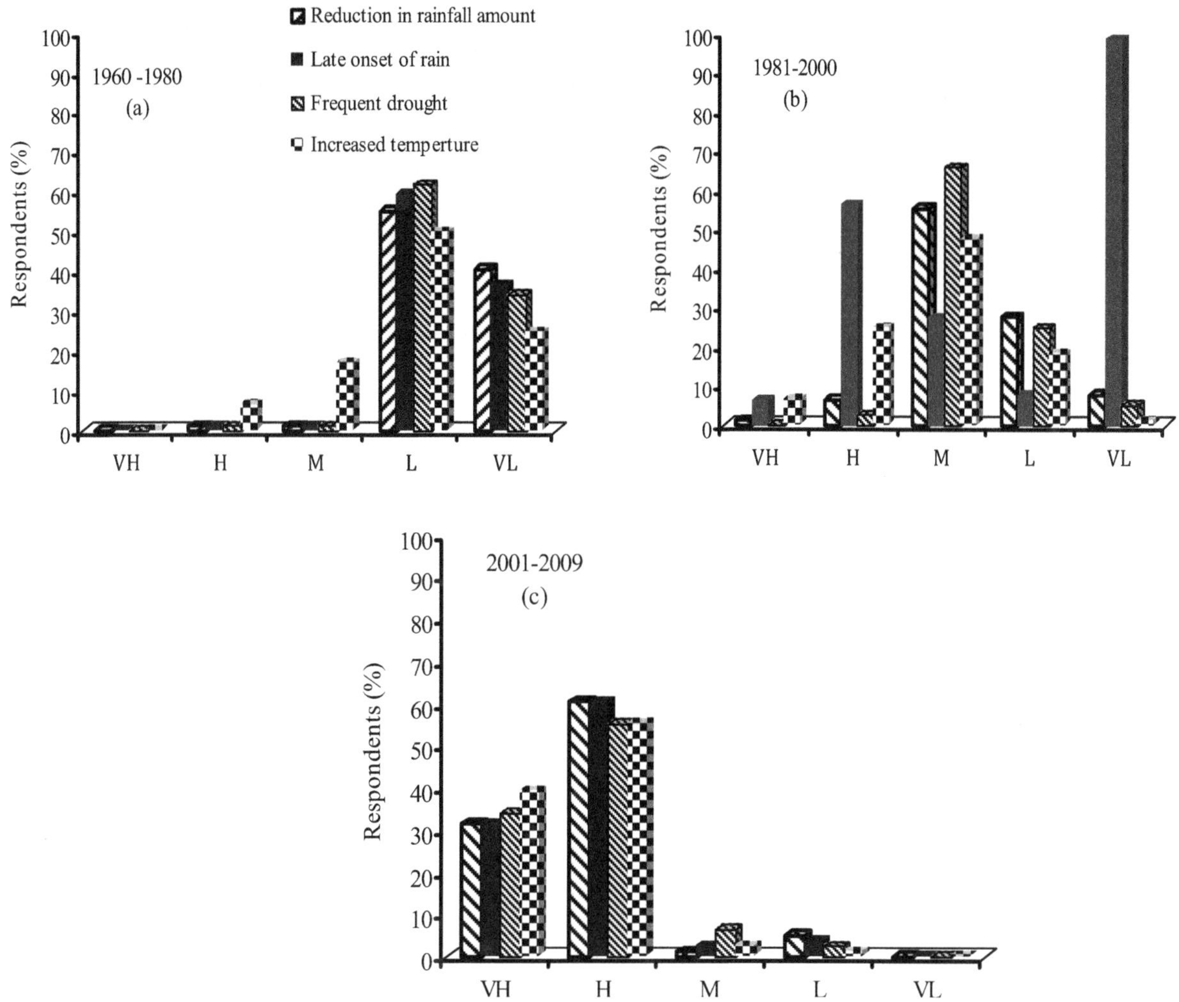

Figure 2. Farmers' perception of extent of change in climate patterns over the last fifty years. VH = very high; H = high; M = moderate; L = low; VL = very low.

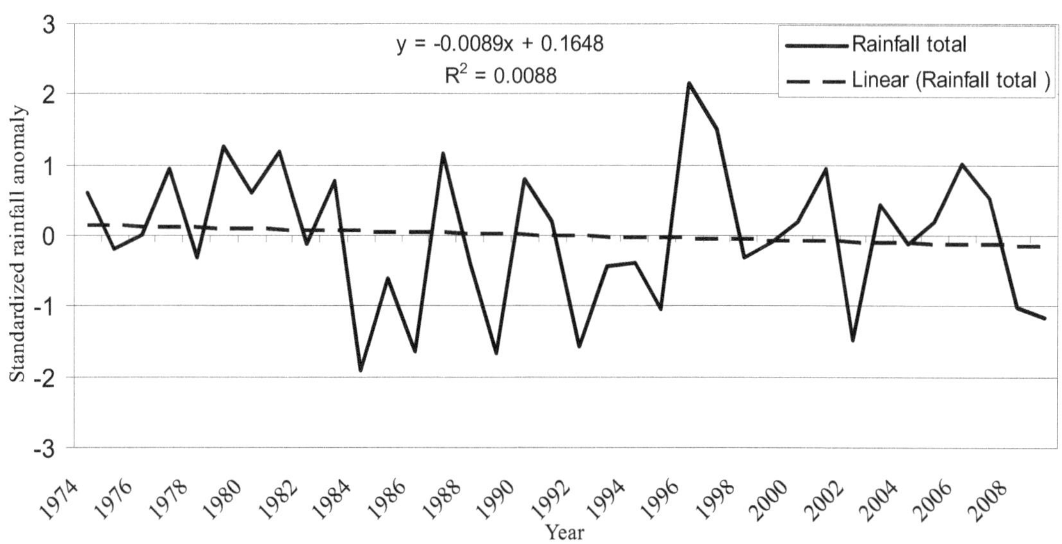

Figure 3. Annual rainfall trend in the study area.

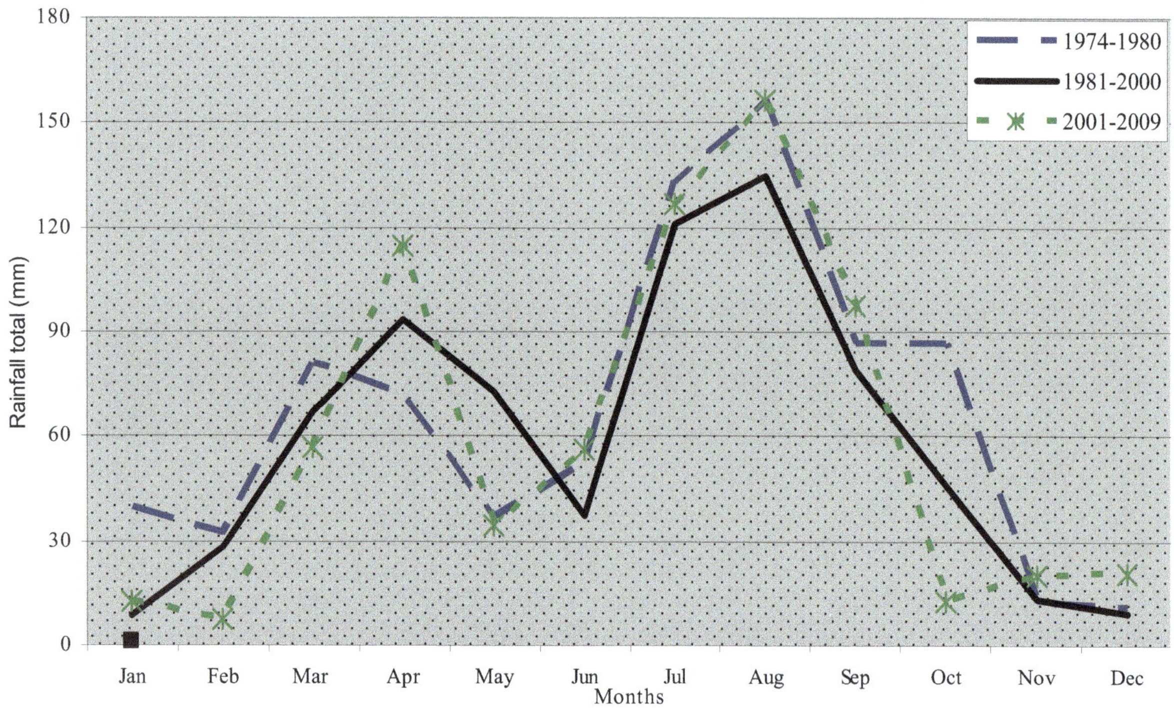

Figure 4. Monthly rainfall trend in the study area.

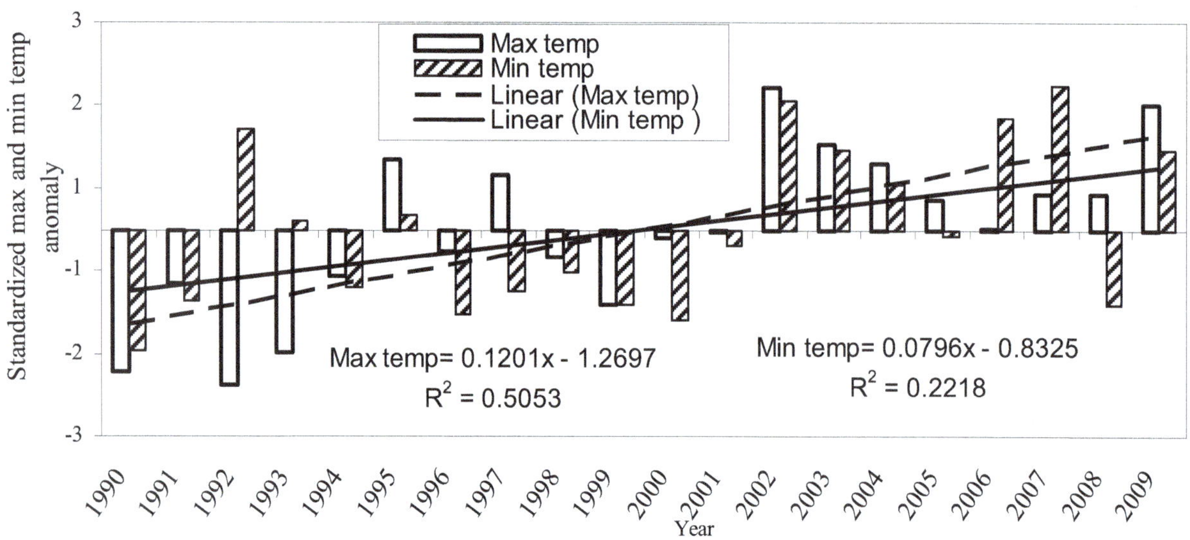

Figure 5. Annual average minimum and maximum temperature trend at Miesso.

statistical record of temperature data (Figure 5). This apart, most of the respondents perceived a shift in onset dates of rainfall, early cessation of rainfall (57%) and erratic rainfall distribution (41%). This corroborated the findings of Mengistu (2011) and Belaineh et al. (2012) which showed awareness of farmers to increased temperature, changes in timing of rains and frequent drought than it was before in central Tigray and Western Hararghe of Ethiopia.

Generally, using their indigenous knowledge and longer years of farming experience, majority of the respondents interviewed had a clear awareness of changing climate. The majority of respondents' awareness on increasing temperature could be attributed to the fact that more experienced sorghum growing farmers were selected as respondents for this study. Similar results were reported

Table 1. Farmers perception of major causes of change in climate patterns.

Variable	Percent of respondents*		Total
	Technology Adopters	Non technology adopters	
Gods anger	61.2	76.9	66.7
Deforestation	69.4	53.8	64.0
Increased population pressure	10.2	23.1	14.7
Burning fossil fuel	4.1	3.8	4.0
Concentration of greenhouse gases in the atmosphere	0	0	0

*Percentages do not add up to 100% because of multiple responses.

Table 2. Farmers' perception of risks (consequences) of climate variability and change.

Variable	Percent of respondents*	
	Serious problem	Very serious
Frequent total crop failures	4.0	96.0
New pattern of diseases and pests of crops	4.0	96.0
Appearance of new weed species	42.7	57.3
Shortage of feed for animals	5.3	94.7
Water scarcity or shortage	25.3	74.7
Loss of biodiversity and forest resources	9.3	90.6
Increased soil degradation (fertility and erosion)	48.0	52.0

*Percentages do not add up to 100% because of multiple responses.

in South Africa and Nigeria in that farmers with more farming experience perceived changes in climate as compared to their less experienced counterpart (Maddison, 2006; Ishaya and Abeje, 2008; Gbetibouo and Ringler, 2009).

Farmers' perception on causes of climate change

About 66.7% of respondents (61% technology adopters and 77% of non adopters) were convinced that the vagaries of climate were a sign of divine anger, that is, punishment from God (Table 1). For example, drought with serious consequences and diseases epidemic to crops and humans, and regular crop infestation by pests were God's punishment. Likewise, most of the respondents (64%) believed that changes in climate patterns were mainly attributed to deforestation (tree cutting and environmental destruction) by the communities, either for short-term economic gains or an attempt to expand arable land. For instance, one of the respondents (Code GR-15) detailed changes in rainfall pattern linked with forest resources as follows: 'Forests attract wind; winds hold (attract) rain; rain comes from mountains and forested areas to other places'. Hence, he concluded that deforestation is the major factor causing climate changes'. Ishaya and Abeje (2008) reported that majority of farmers attributed climate changes to human activities such as deforestation (cutting trees for fuel, roofing and farm land extension).

On the other hand, 14.7% of respondents stated that increased human population that put mounting pressure on natural resources was a major cause of changes in climate patterns. Likewise, 4.1% of technology adopters and 3.8% of non-adopter farmers indicated that fossil fuel burning which comes from other areas or industrialized countries was the cause of climate change at and around the area. This result is in line with the study conducted by Manyatsi et al. (2010) in that a significant number of respondents did not provide any scientifically proven cause of climate change.

Consequences of climate variability and change

Farmers in Miesso areas noted high infestation of crops by new patterns of diseases such as head smut (*Sphacelotheca sorghi*), leaf blights (*Exserohilum turcicum*) and pests such as shoot fly (*Atherigona soccata*) and frequent total crop failures due to reduced rainfall amount, changes in timing of onset of rains and high temperature (Table 2). Increased pest damage may arise from changes in production system and production of crops in warmers climatic regions where plants are

Table 3. Diversity of crops grown by farmers at Miesso over the last 50 years.

Period	Diversity of crops grown (number)		
	Maximum	Minimum	Mean
1960-1980	10	2	5
1981-2000	10+++	1	4
2001-2007	5++	1	3+
2008- recent	4***	1*	2**

*Sorghum (*Sorghum bicolar*); **Sorghum and maize (*Zea mays*); ***Sorghum, maize, sesame (*sesam umindicum*) and soybean (*Glycine max*); + Sorghum, maize and sesame; ++Sorghum, maize, sesame, soybean and tef (*Eragrostis tef*); +++Sorghum, maize, barley (*Hordeum vulgare*), tef, sesame, lentil (*Lens culinaris*), soybean, wheat (*Triticum aestivum*), linseed (*Linum usitatissimum*) and cowpea (*Vigna unguiculata*).

more susceptible to pests (Mary and Majule, 2009). In addition, the local farmers observed various forms of crop infestations with new weed species that reduced the quality and quantity of crops produced. Moreover, many respondents mentioned shortage of feed for animal as a very serious effect of climate variability and change that caused loss and weakness of oxen (Table 2).

Furthermore, the majority of respondents perceived water shortage, great loss of biodiversity and forest resources through excessive de-forestation, and decline in soil fertility as the most important effects of climate variability and climate change. Generally, the outcome shows the sensitivity of social, economic and environment aspects of Meisso farmers (Table 2) to the impacts of climate variability and change. Hence, the interplay between the above climate variability and change consequences and its direct adverse effect on crop yields are the source of vulnerability of production system the farmers at Miesso faced.

Changes in crop diversity (1960-2009)

The majority of farmers revealed that they have switched over to cultivation of only two crops (sorghum and maize) out of diverse types of crops they were earlier cultivating as a result of change in climate patterns (Table 3). Due to losses in climate events that were repeated over time, capacity of crops to maintain productivity has reduced resulting in the withdrawal of crops from production systems. For the different peasant associations assessed, the farmers also recognized that different crops and some species of sorghum (land race or local varieties) are no longer being farmed due to the negative effects of climate variability and change on the productivity of the crops (Table 3).

Changes in farming practices due to climate change

As a result of negative impacts of change in climate patterns, 93% of the respondents changed farming practices. Most of the farmers revealed that decreased tillage frequency, increased frequency of weeding and increased seed rate are practiced in recent periods than before (Table 4). The majority of farmers associate the decrease in tillage frequency with drought and delay in onset of rain as land becomes dry and difficult to plough and feed shortage leads to oxen weakness. Greater use of seed rate is attributed to recover damaged crops due to lack of precipitation which hinders germination of cultivated seeds. Moreover, the highest proportion of households with crop production in their farming portfolio chose not to use farm inputs (fertilizer, herbicides and pesticides). According to the respondents, this situation has resulted in a good opportunity for weeds to stay on cropping land, out compete crops and increased frequency of weeding.

The other farm practice changed include, intercropping, crop rotations and fallowing (Table 4). The increase in fallowing practice is attributed to frequent drought and power shortage. As a result of the impacts of climate change, most of the farmers showed the tendency to allocate more lands to improved sorghum varieties (drought tolerant or early maturing) than local varieties in the 2000s. These findings are in line with a study by Lema and Majule (2009) in semi-arid zone of Tanzania, who reported that farmers adopted tillage methods and other agronomic practices in the face of climate variability and changes risks in order to maximize yield.

Source of climate information

More than 97% of the respondents rated climate information availability from high to very high to their day to day activities. However, the majority of farmers did not have adequate information on climate variability and change for farm level decision making. The limited climate information accessible to farmers was through radio in the form of daily weather forecast (Figure 6). But, the farmers indicated that they did not trust the weather information dispatched through radio. The feeling of the farmers was that the weather information broadcast by radio was not specific to their location or region.The farmers also indicated that at times the season was forecast as good and they invested high in terms of inputs but later it turns out a dry season. The other sources of climate information used include expecting from God (32%), religious leaders and neighbors (4%), and market and radio (9%) (Figure 6).

Adaptations to climate variability and change

Adaptation measures and practices followed by Meisso farmers to combat the adverse effects of climate variability and change over the last fifty years are mostly followed as one or in combination with another measure.

Table 4. Changes to farming practices at Miesso over the last 50 years*

Cropping practice	1960-1980	1981-2000	2001-2007	2008/09
Frequency of tillage	H	H	L	L
Frequency of weeding	L	L	H	H
Fertilizer application	N	N	N	N
Mulches	H	H	H	L
Intercropping	H	H	L	L
Seed rate (kg/hectare)	L	L	H	H
Crop rotation	N	N	L	L
Fallowing	N	N	N	H
Pesticides and herbicides	N	N	N	N
Use of improved seed	N	N	H	H

H = High; L = low; N = Not practiced (used); * = the data are based on the highest percentage of respondents.

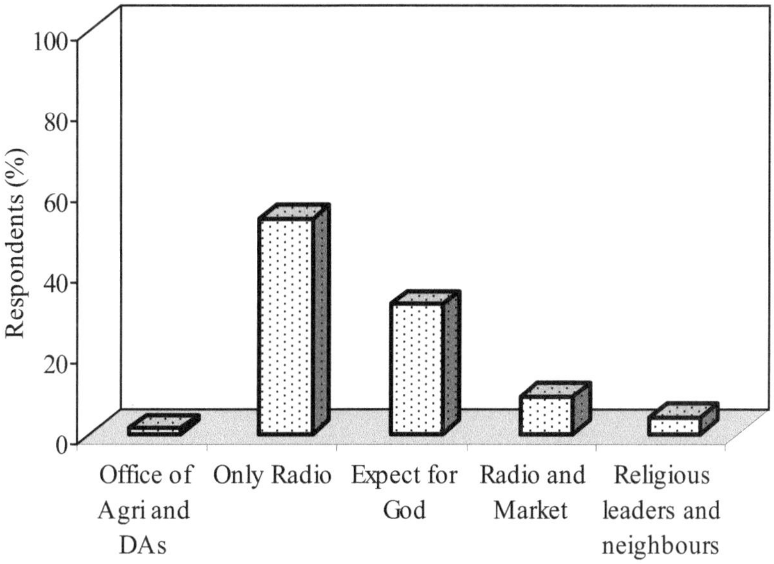

Figure 6. Source of climate information to farmers in Miesso area.

Despite the fact that, livelihood adaptation strategies chosen are not free from constraints a combination of several strategies are used by farmers to cope with the impacts of climate change (Belaineh et al., 2012). They are grouped as agronomic or crop management, technology adoption and promotion activities, and livelihood adaptation strategies (Table 5).

The dominant adaptation systems are using different planting dates (86.7%) and on farm soil and water conservation practices (80%). The greater shifting of planting dates was for the purpose of reducing the risks of crop damage during germination due to dry spells soon after sowing. Likewise, Komba and Muchapondwa (2012) reported that large size owned farmers preferred crop and variety diversification than planting dates to easily spread climate change risks. The use of on farm soil and water conservation practices as an adaptation method is associated with the efficient use of scarce rainfall received over the area. Other adaptation strategies practiced by farmers include application of fertilizers and chemicals (4%) and water harvesting techniques for supplemental irrigation (34.7%). The low use of fertilizers, chemicals and water harvesting techniques as on farm adaptation mechanism could be due to the fact that these practices require high capital investment. The results of farmers agronomic adaptation strategies practiced by Miesso farmers are similar to autonomous adaptation strategies reported by FAO (2007) and Hassan and Nhemachena (2008), which were carried out in response to or in anticipation to changing climate patterns.

Since crop cultivation is the dominant livelihood strategy of farmers, growing of local/land race cultivars

Table 5. Types of climate variability and change adaptation strategies used by farmers in Meisso.

	Adopter (%)*	Non-adopter (%)	Total (%)
A. Agronomic or crop management			
Use of different planting dates	89.8	80.8	86.7
Use of fertilizers and chemical	6.1	-	4.0
Use of on farm soil and water conservation practices	81.6	76.9	80
Use of water harvesting techniques for supplemental irrigation	38.8	26.9	34.7
B. Technology adoption			
Use of local/land race varieties	81.6	92.3	85.3
Use of drought resistant varieties	59.2	19.2	45.3
Use of early maturing varieties	73.5	26.9	57.3
C. Livelihood adaptation options			
Changing from farming to non-farming	10.2	38.5	20.0
Move to different site	10.2	7.7	9.3
Use of credit	18.4	7.7	14.7

*Percentages do not add up to 100% due to multiple responses.

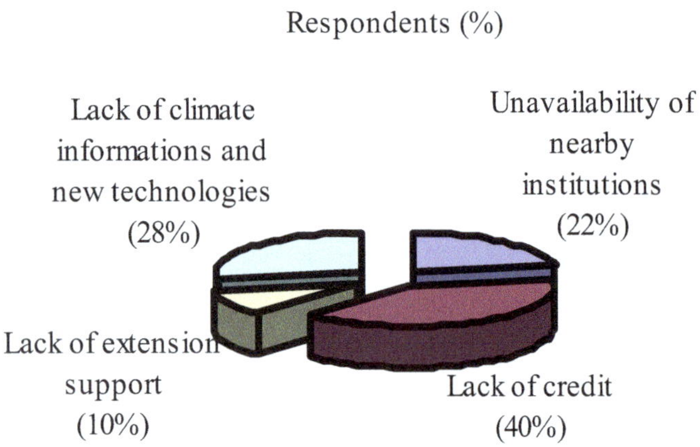

Figure 7. Constraints that hinder climate variability and change adaptation of farmers in Miesso area.

(85.3%) is one of the adaptation option for climate variability and change. Greater use of local varieties is associated with their use as animal feed resources during severe droughts because of their good fodder quality and productivity. The use of drought resistant varieties (45.3%) and early maturing crop varieties (57.3%) are other adaptation strategies practiced by farmers. Planting short season and drought resistant crop varieties increases the chances of successful harvests despite adverse climatic conditions (Mano and Nhemachena, 2006). These adaptation mechanisms are in line with the adaptation practices reported by Cox et al. (2008). Livelihood diversification (shift from farming to non-farming such as charcoal making, firewood sale) was one of the most commonly used adaptation methods by the farmers (Table 5), whereas temporary migration (9.3%) to other places was the rarely practiced adaptation method during severe conditions. Likewise, 14.7% of respondents used credit (from government or local lenders) as an adaptation strategy against the shocks of changing climate patterns to recover and develop resilience for the next event (s). Similar adaptation strategies were reported from various studies conducted in Ethiopa, Zimbabwe, South Africa and Nigeria (Mano and Nhemachena, 2006; Hassan and Nhemachena, 2008; Deressa et al., 2009; Mengistu, 2011; Belaineh et al., 2012). The proportion of technology adopter farmers that use the adaptation mechanisms is almost equal to the proportion of non-adopter farmers except the high proportion of technology adopters in using early and drought resistant crop varieties.

Major constraints to adaptation

The survey results indicate that, there are four major constraints to climate variability and change adaptation by sorghum farmers in Miesso area. These are unavailability of nearby supporting organizations (that provide technologies, farm inputs, information related to farm activities and climate), lack of credit, lack of support from extension workers and lack of climate information and new technologies (Figure 7).

Lack of credit for input purchase (improved seed, fertilizer, chemicals) is the prominent constraint to adapting climate variability and change effects. This result is in line with the survey carried out across Africa by Maddison (2006), where 33% of respondents in Ethiopia reported lack of credit as the main constraint to adapting to climate change. Lack of information on climate and new technologies is the second main barrier to adaptation in Miesso. This is also related to week support from extension and research in the provision of up-to-date information and technologies.

Summary and recommendation

The analysis of perception of farmers on climate variability and change indicated that the majority of farmers were aware of a decline in rainfall amount, increasing temperature, shift in onset dates of rainfall, early withdrawal of rains and frequent drought occurrence. As a result, majority of farmers indicated decreasing diversity of cultivated crops, changes in farming practices and reduction in crop yields due to changes in climate patterns in their area. Production of drought or heat tolerant crop cultivars with optimum maturity periods are recommended to offset the adverse effects of increasing temperature. Moreover, introduction of new crops, varieties and crop management practices that goes in line with the changing climate patterns should be the prior agenda for research and development planners in order to arrest declining diversity of crop grown in the area due to climate change. Additional advantage can be achieved by the distribution of pamphlets containing weather and climate information prepared by Ethiopian weather service. The farmers should be encouraged to stabilize their family size.

Conflict of Interests

The author(s) have not declared any conflict of interests.

REFERENCES

Belaineh L, Yared A, Woldeamlak B (2012). Smallholder farmers' perceptions and adaptation to climate variability and change in Doba district, West Hararghe, Ethiopia. Asian J. Empir. Res. 3(3):251-265.

Cheung WH, SenayGB, Singh A(2008). Trends and spatial distribution of annual and seasonal rainfall in Ethiopia. Int. J. Climatol. 28(13):1723-1734.

Conway D, Schipper LF (2011). Adaptation to climate change in Africa: challenges and opportunities identified from Ethiopia. Glob. Environ. Change 21:227-237.

Cox HW, Hammer GL, Robertson MJ (2001). Opportunities for crop modeling in barley. pp. 21-28. Proceedings of the 10th Australian Barley Technical Symposium, 22-24 December 2001, Australian.

Deressa T, Hassan RM, Ringler C, Alemu T, YesufM (2009). Determinants of Farmers' Choice of Adaptation Methods to Climate Change in the Nile Basin of Ethiopia. Glob. Environ. Change 19:248-255

FAO (2007). Adaptation to climate change in agriculture, forestry and fisheries: Perspective framework and priorities. Food and Agricultural Organization of the United Nations. Rome, Italy.

Gadgil S, Seshagiri PR, Sridhar S (1998). Modeling impacts of climate variability on rain fed groundnut. Indian Institute of Science. Bangalore, India. 11p.

Gbetibouo GA, Ringler C (2009). Mapping South African farming sector vulnerability to climate change and variability. International Food Policy Research Institute (IFPRI) discussion paper.

Greasley P (2008). Quantitative data analysis using SPSS: an introduction for health and social sciences. Open University press, Berkshire, England.

Gruza G, Rankova E (2004). Detection of changes in climate states, climate variability and Climate extremity. Meteorology and Hydrology number 4, 2004. Institute for Global Climate and Ecology (IGCE), Russia.

Hassan R, Nhemachena C (2008).Determinants of African farmers' strategies for adapting to climate change: Multinomial choice analysis. Afr. J. Agric. Res. 2(1):83-104.

Hellmuth ME, Moorhead A, Thomson MC, Williams J (2007). Climate Risk Management in Africa: Learning from practice. New York: International Research Institute for Climate and Society, Columbia University.

Ishaya S, Abaje IB (2008). Indigenous people's perception on climate change and adaptation strategies in Jema'a local government area of Kaduna State, Nigeria. J. Geogr. Reg. Plan. 1(8):138-143.

Kidane G, Abebe T, Degefie T (2006). Estimating crop water use and simulating yield reduction for maize and sorghum in Adama and Miesso districts using the Cropwat model. Ethiopian Agricultural Research Institute. 1-14p.

Komba C, Muchapondwa E (2012). Adaptation to Climate Change by Smallholder Farmers in Tanzania. ERSA working paper 299, School of Economics, University of Cape Town, South Africa.

Lema MA, Majule AE(2009). Impact of climate change, variability and adaptation strategies on agriculture in semi-arid areas of Tanzania: The case of Manyoni district in Singida region, Tanzania. Afr. J. Environ. Sci. Technol. 3(8):206-218.

Lemma W (2008). Physicochemical studies of soils of Miesso areas, Eastern Hararhge, Oromiya region. MSc Thesis presented to the School of Graduate Studies of Haramaya University. Haramaya, Ethiopia.

Maddison D (2006). The perception of and adaptation to climate change in Africa. CEEPA discussion Paper No.10. Centre for Environmental Economics and Policy in Africa, University of Pretoria. Pretoria, South Africa.

Majule AE (2008). Climate change and variability impacts on agriculture and water resource and implications for livelihoods in selected basins.Institute of Resource Assessment (IRA), University of Dares Salaam. Dares Salaam, Tanzania.

Mamo G (2005). Using seasonal climate outlook to advice on sorghum production in the Central Rift Valley of Ethiopia. PhD thesis, Blomefontein, Republic of South Africa.

Mano R, Nhemachena C (2006). Assessment of the economic impacts of climate change on agriculture in Zimbabwe: A ricardian approach. CEEPA Discussion Paper No. 11, University of Pretoria, South Africa.

Manyatsi AM, Mhazo N, Masarirambi MT(2010). Climate variability and change as perceived by rural communities in Swaziland. Res. J. Environ. Earth Sci. 2(3): 164-169.

Mary AL, Majule AE (2009). Impacts of climate change, variability and adaptationstrategies on agriculture in semi-arid areas of Tanzania:The case of Manyoni District in Singida region,Tanzania. Afr. J. Environ. Sci. Technol. 3 (8):206-218.

Mendelsohn R, Dinar A (2009). Climate Change and Agriculture. An Economic Analysis of Global Impact, Adaptation and Distributional Effects. Cheltenham, Elgar

Mengistu DK (2011). Farmers' perception and knowledge of climate change and their coping strategies to the related hazards: Case study from Adiha, central Tigray, Ethiopia.

MoA (1998). Agro ecological zones of Ethiopia. Ministry of Agriculture. Addis Ababa, Ethiopia.

MoARD (2007). Ministry of Agriculture and Rural Development crop development crop variety register. Issue No.7. Addis Ababa, Ethiopia. 104p.

Mtambanengwe F, Mapfumo P, Chikowo R, Chamboko T (2012). Climate hange and variability: Small holder farming communities in Zimbabwe portray a varied understanding. Afr. Crop Sci. J. 20 (Issue Supplement s2):227-241.

NMSA (1996). Climatic and agroclimatic resources of Ethiopia. Vol.1: No. 1. National Meteorology Service Agency of Ethiopia, Addis Ababa.

Seleshi Y, Zanke U (2004). Recent changes in rainfall and rainy days in Ethiopia. Int. J. Climatol. 24:973-983.
Stage J (2010). Economic Valuation of Climate Change Adaptation in Developing Countries. Ann. N Y Acad. Sci. 1185: 150-163.
Stern NH (2007). The Economics of Climate Change: The Stern Review. Cambridge, UK, Cambridge University Press.
Teshome W, Peterson N, Gebrekirstos A, Muniappan K (2008). Micro insurance demand assessment in AdihaTabia. Progress Report, Tigray Regional State, Ethiopia.
Thomas DG, Twyman C, Osbahr H, Hewitson B (2007). Adaptation to climate change and variability: farmer responses to intra-seasonal precipitation trends in South Africa. Clim. Change 83(3):301-322.
Thornton PK, Jones PG, Owiyo T, Kruska RL, Herrero H, Kristjanson P, Notenbaert A, Bekele N, Omolo A (2006). Mapping climate vulnerability and poverty in Africa. Reportto department for international development. Nairobi, Kenya: International Livestock Research Institute (ILRI).
UNDP (2011). Framework for UNDP Ethiopia's climate change, environment and disaster risk management portfolio. Addis Ababa, Ethiopia.
Worku AW (2006). Quantifying rainfall-runoff relationship on selected benchmark ecotopes in Ethiopia: A primary step in water harvesting research. PhD Dissertation University of Free State, Bloemfontein, South Africa.
Yesuf M, Salvatore DF, Temesgen DR, Gunnar K (2008). The impact of climate change and adaptation on food production in low-income countries: Evidence from the Nile Basin, Ethiopia. IFPRI discussion paper 2008. pp. 1-16.

Sediment and nutrient lost by runoff from two watersheds, Digga district in Blue Nile basin, Ethiopia

Alemayehu Wudneh[1] , Teklu Erkossa[2] and Prhaba Devi[3]

[1]Natural Resource Management Course Team, Madawalabu University, P.O. Box 247, Robe-Bale, Ethiopia,
[2]International Water Management Institute, Regional office for East Africa and Nile Basin, P. O. Box 5689, Addis Ababa, Ethiopia.
[3]Ambo University, Biology Course Team, Ambo, Ethiopia.

The study was conducted in two sub watersheds in the Upper Blue Nile Basin in Ethiopia to determine the quantity, quality of sediment lost and its onsite costs in terms of crop yield. Two monitoring stations at the outlets were selected. Discharges were estimated and depth integrated daily runoff samples were collected during the rainy season in 2011. The sediment concentration and nitrogen (N) and phosphorus (P) content was analyzed and related to crop yield using a nutrient response equation for maize (*Zea mays*). The result shows statistically significant differences in sediment concentration of the two watersheds. Suspended sediment concentration was strongly correlated with the discharge from Chekorsa River (R^2 = 0.7) but it was very weak for Dapo. There was higher concentration of nutrients in the sediment than the surface soil in both watersheds. Particularly, sediment enrichment ratio of available P was 2.7 and 9 for Dapo and Chekorsa Rivers, respectively. The estimated yield reduction of maize due to N and P loss was about 950 and 1420, and 1015 and 665 kgha^{-1} from Dapo and Chekorsa catchments, respectively. Such results are equivalent to an onsite cost to farmers of about 190 and 285 USD ha^{-1} for Dapo, and 203 and 133 USDha^{-1} for Chekorsa catchments in that order. This monetary value can alert farmers and decision makers to take soil and water conservation measures.

Key words: Soil loss, enrichment ratio, nutrient depletion, Blue Nile Basin, sediment.

INTRODUCTION

Soil erosion is a worldwide problem and has become a major global concern since its severe impacts on agriculture and environment (Le Bas and Kozak, 2007; Tenaw and Bekele, 2009). Pimentel et al. (1995) indicated that about 80% of the world's agricultural land suffers moderate to severe erosion. Soil erosion is among the major causes of nutrient depletion and, shortage of basic plant nutrients (Wortmann, 2006; Pimentel et al., 1995) with a devastating impact on overall soil organic matter levels (Van-Camp et al., 2004). According to FAO (1999), Ethiopia is among the countries with high rate of erosion and nutrient depletion rate. Studies showed soil erosion in Ethiopia is more severe in the highlands. Henao and Baanante (1999) reported NPK loss of more than 60 kgha^{-1} based on their study between 1993 and 1995 in Ethiopia and losing fertile topsoil that exacerbated impacts of dry spells and drought in the Upper Blue Nile Basin (Getnet et al., 2009).

Water is the main cause of soil erosion in the highlands of Ethiopia (Dubale, 2001) removing about 1493 Mtyr^{-1} (Hurni, 1993) whereas in the Abay Basin (the Ethiopian part of the Blue Nile Basin) loss of soil was about 303 Mtyr^{-1}(Hagos et al., 2009). Nearly 50 and 34% of the total soil loss was estimated to come from cultivated fields in the central highlands of Ethiopia and Abay Basin, respectively (Hagos et al., 2009; Hurni, 1993).

According to Gruhn et al. (2000), and Desta et al., (2000) soil fertility loss by erosion are the main ways for nutrient outflow from a watershed. Eroded soil was often significantly enriched by plant nutrients than the soil left behind with enrichment ratios ranging from 1.2 to 2.3 (Fenli et al., 2005). It is the characteristics of watershed that influence river nutrient enrichment ratio. Eroded soil by runoff constitutes a large proportion (>60%) of the P transported in surface runoff from surface soil, with enrichment ration greater than 1 (Scalenghe et al., 2002). The enrichment ratio greater than 1 is related to the high P content in the top 5 cm of the surface soil and the fact that runoff preferentially erodes finer particles including clays with which P is associated (Sharpley et al., 2000).

The losses of the nutrients can limit crop growth and productivity. Soils in all major maize growing regions of Ethiopia are depleted of NPK nutrients, thus demanding high soil amendments (Kebede et al., 1993). Decline in soil fertility due to depletion of NPK nutrients in the country is therefore eradicating production including maize production. Nitrogen is the most limiting nutrient for maize productivity (Kogbe and Adediran, 2003; Alley et al., 2009). As a result of water erosion, organic matter content is highly reduced since its largest part is found in the fertile topsoil and cause decline of production because of low physical structure (Helmecke, 2009).

All physical and economic evidence shows that agricultural land productivity decline is a severe problem in Ethiopia (Berry et al., 2003) and in Africa in general, for example, about 86% of the countries in Africa lose more than 30 $kg^{-1}ha^{-1}yr$ of NPK (Henao and Baanante, 1999). This severe erosion is reducing agricultural land productivity (Berry et al., 2003) particularly, in the Blue Nile basin (Tenaw and Bekele, 2009; Dubale et al., 2009) in which soil erosion is among the major threat (Desta et al., 2000; Bekele and Holden, 1998; Ashagrie, 2009).

Most soil erosion research in Ethiopia focused on its physical, chemical and biological effects (Gorlach et al., 2004) and limited to farm plots (Berry et al., 2003). In Ethiopia, farmers and other decision makers at grass-roots level cannot easily understand estimated soil and nutrient losses by runoff in terms of scientific units. However, converting the lost soil nutrients by runoff from watersheds into monetary value can easily make them aware of the severity of soil erosion. In other words, the monetary value of lost nutrients can induce and initiate farmers and policy makers to implement land and water management practices. The cost of soil erosion to farmers can be related to loss of productivity due to the lost of nutrients or the cost of fertilizer to compensate for the lost nutrients (Gruhn et al., 2000).

Yield reduction cannot be because of soil nutrient depletion alone but also affected by other physical and biological factors and conditions such as land management, nutrient status, soil microbs, soil depth, etc. To estimate financial cost of lost soil nutrients, it is necessary to consider those factors and conditions prevailing in the specific area (Pimentel et al., 1995). As a result, productivity lost method was preferred to replacement cost for this study.

Depending on the characteristics of watershed, suspended sediment concentration related to discharge can produce different shapes of hysteresis loops (Baca, 2002; Sander et al., 2011). These relationships are normally not homogenous in time, neither within nor between events (Baca, 2002). The most common shapes of the hysteresis effects are clockwise and counter clockwise loop (Sander et al., 2011; Ahanger et al., 2013). The characteristics of watershed that produce hysteresis effects are source of sediment which could be from distant hill slope, or channel or nearby sources, higher rainfall intensities at the beginning of storms, sediment availability, increasing of base flow supply to the river, the duration and interval of storm events (Eder et al., 2010; Baca, 2002; Slattery et al., 2002; Lefrançois et al., 2007)

In this context, the objectives of the research were to estimate the sediment concentration at the outlets of Dapo and Chekorsa watersheds and patter of sediment hysteresis loop, to analyze soil nutrients lost by runoff and consequently river nutrient enrichment ratio, and to estimate the financial cost due to soil erosion at the study

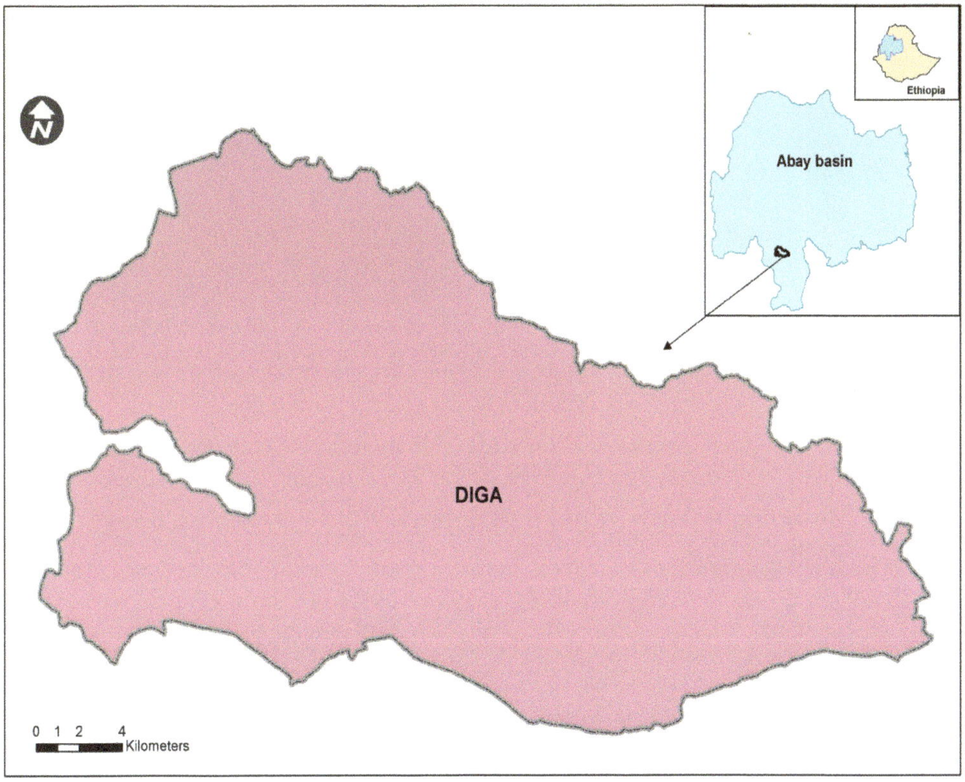

Figure 1. Study location map.

region.

MATERIALS AND METHODS

Description of the study area

The study was conducted in two sub watershed located in Diga district in East Wollega Zone of Oromia Regional State, Ethiopia (Figure 1) from March 2011 to October 2012. The area is characterized by an average annual mono-modal rainfall of 1707 mm which begins late April, and ends in early September, with annual minimum and maximum temperature of 14 and 30°C, respectively (Joshua et al., 2010). Information from Diga District Water Resource Office in 2010 shows that the major livelihood is mixed crop-livestock rain fed agriculture. In middle altitude of the study area, the dominant soil color is red whereas black in the low land (Birhanu Z, 2012) and according to soil map developed by Seleshi et al. (2009) which are classified as Acrisols and Alisols.

Two sub watersheds (Dapo and Chekorsa) were selected by the Nile Basin Development Challenge Program (NBDC) for rainwater management research in the Blue Nile Basin. Zemadim et al. (2010) indicated that the two catchments have an area of 18 and 5.60 km^2, respectively with their altitude ranging between 1,347 - 2,011 m (Dapo) and 1,266 - 1,430 m (Chekorsa).

Both Dapo River (DR) and Chekorsa River (CR) are tributaries of Didesa River, the largest tributary to the Blue Nile River, in terms of water volume (MWRE, 2010). Both Rivers are supplied by numerous first and second order streams. Similar to Chekorsa River Catchment (CRC), the physiographic, land use and land cover condition in the downstream of Dapo River Catchment (DRC), around the water level gauging site, consists of mango trees and sparse natural vegetation, maize and grazing areas(Figure 2) (Zemadim et al., 2012). Coupled with the intensive rainfall, Dapo River has cultivation in steep slopes that aggravates soil erosion and plant nutrient depletion (Zemadim et al., 2010). Depleted soil fertility caused by soil erosion and crop uptake is a major limiting factor for crop production and productivity in both catchments (Negassa et al., 2005). In response to the productivity decline, farmers open a new agricultural land which increases deforestation. Maize, sesame and finger millet are the dominant crops in DRC while about one third of the watershed area is covered by forest, mainly located at the most upper land (Figure 3), but no significant forest is left in Chekorsa catchment.

Methods

Selection of runoff sampling site

Hydrologists of the International Water Management Institute (IWMI) selected the bridge on the main highway that goes from Diga to Ghimbi for DRC and the bridge from Arjo Gudatu to Lalisa Dimtu for CRC, where flow monitoring stations where established. The bridges are 4.5 and 3.5 m wide through which Dapo and Chekorsa river flows respectively, and all flows were contained inside the culvert of the bridges.

Discharge measurement

Discharge measurement and sampling was started from the beginning of major runoff in July, 2011 and ended in early September, 2011 when the rainfall ended. Based on Gierke (2002), the flow rate of the river at the outlet was determined using current

Figure 2. Dapo watershed outlet (a) and land use and land cover condition around the outlet (b), at the Diga district, Ethiopia. Photo credit (Brihanu Zemedin, 2011).

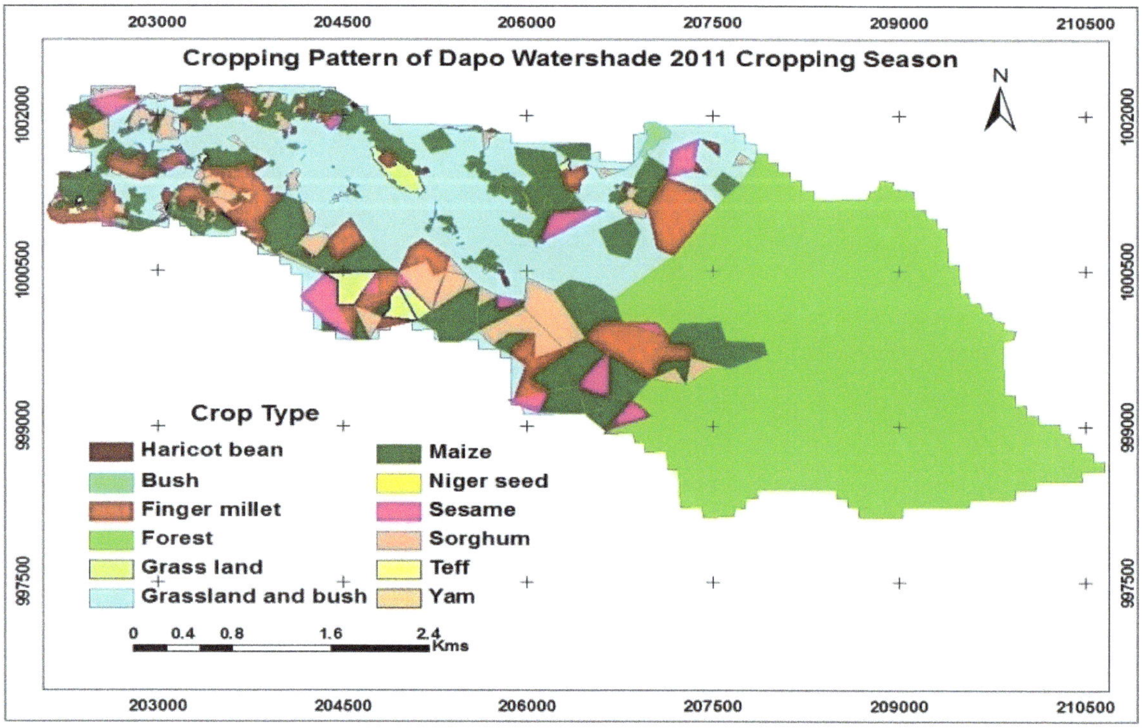

Figure 3. Land use and land cover of Dapo watershed. Source: IWMI (2012).

meter (Model 0012B Surface Display Unit and Model 002 Flow Meter (Figure 4A and B)); the depth of the rivers at the cross-section was measured using 1.5 m wading rod (Figure 4E), and the flow rates and depths (h) were measured across the rivers simultaneously at 9 points spaced at 0.5 m intervals for DR and at 5 points at 0.75 m interval for CR. The cross sectional area of the flow was calculated using the depth records and the average flow rates at each point (Equation 1). Discharge (Q) passing the outlet of the watersheds was calculated using Equation 2.

$$qi = vi*ai \ (a) \quad Q = \sum_{i=0}^{n} qi \ (b) \qquad (1)$$

Where:

qi = discharge at each cross sectional area (m^3sec^{-1}); vi = flow velocity at each cross sectional area ($msec^{-1}$); ai = cross sectional area at each point (m^2); Q = total discharge (m^3sec^{-1}).

$$Q = c\,(h + a)^b \qquad (2)$$

Where: Q = discharge (m^3sec^{-1}); h = measured water level (m); a = water level (m) corresponding to Q = 0; c = coefficients derived for the relationship corresponding to the station characteristics; b = coefficient derived for the power in relation to the station characteristics.

Discharge rating curves were developed by fitting the measured gauge to discharge into power curve (Equation 4) for the two rivers. And having water levels measured throughout the study period by the staff gauge (Figure 4E), the discharge for each was calculated from the equations of the curves.

Runoff sampling and storage

Depth integrated runoff was collected manually from catchments using one liter plastic bottle three times a day. The daily samples were mixed and two litters were sub sampled and bulked in a 20 L plastic jar for 10 consecutive days. The bulked samples were kept

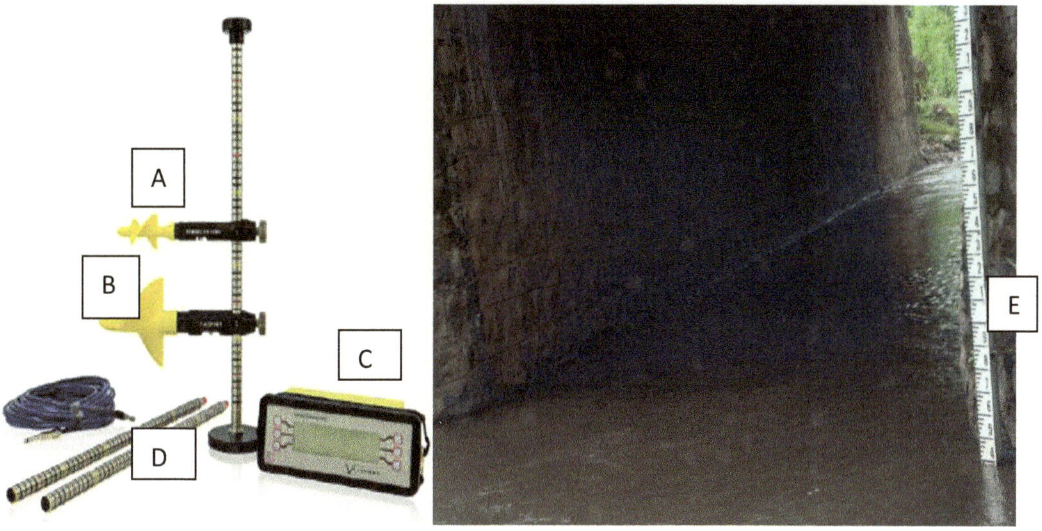

Figure 4. Current Meter model used for measuring river flow and staff gauge (E) at DR monitoring station. Photo Credit: Brihanu Zemedin (2011).

Table 1. Methods and procedure used for the chemical analysis sediment and water.

Sample	Parameter	Method budget	Reference
Soil	OM	Wet oxidation/Walkley-Black	Jackson (1967)
	TKN	Modified Kjeldahl digestion	Dalal et al (1984)
	NO_3-N NH_4-N	Magnesium oxide-Devrda's alloy	Maiti (2004)
		Magnesium oxide-Devrda's alloy	Maiti (2004)
	P_2O_5	Alkaline Extraction of Olsen Method	Olsen and Sommers 1982
	Texture	Hydrometer	Bouyoucos (1962)
Water	Dissolved ammonia	Phenate method using Spectrophotometer; Modele Eleco SL-160 double beam UV	Patnaik (2010)
	Dissolved nitrate and phosphorus	Spectrophotometer; Modele Eleco SL-160 Double beam UV	Patnaik (2010)

in the refrigerator at 4°C in a soil laboratory to minimize further chemical and physical changes.

Estimation of sediment load and chemical analysis

Sediment in water sample collected for each consecutive 10 days was allowed to settle down before the clear water at the top was decanted into laboratory beakers. The turbid water remaining at the bottom containing most of the sediment was filtered using filter papers and air dried. Soil loss was calculated based on the sediment concentration and discharge per 10 consecutive days at the gauged sites. The water collected after filtration and that which was stored after decantation were mixed proportionally for chemical analysis (Table 1).

Estimates of nitrogen and phosphorus loss

The quantity of N and P delivered to the outlet of the watersheds with water and suspended sediment was estimated using Equations 3 and 4, respectively. The total N and P lost was estimated by summation of the amount lost with water in dissolved form and by the quantity lost with suspended sediments (Equation 5). The maize yield loss due to the two essential plant nutrients was estimated according to FAO (1999) method.

$$\mathrm{Nwi} = \mathrm{Ncwi} * \mathrm{qi(a)} \text{ and } \mathrm{TNw} = \textstyle\sum_{i=1}^{9} \mathrm{NLi} \quad \text{(b)} \tag{3}$$

$$\mathrm{Nsi} = \mathrm{Ncsi} * \mathrm{SSL} \text{ and } \mathrm{TNs} = \textstyle\sum_{i=1}^{9} \mathrm{Nsi} \quad \text{(b)} \tag{4}$$

$$(\mathrm{TNw} + \mathrm{TNs})/\mathrm{AGTN} = (\mathrm{TNw} + \mathrm{TNs})/\mathrm{A} \tag{5}$$

Where, Nw: nutrient loss with water per ten consecutive days (gm/d) (nitrogen or phosphorus); Ns: nutrient loss with suspended sediment per ten consecutive days (gm/d) (nitrogen or phosphorus); Ncw: nutrient concentration in water (gm/L) ((nitrogen or phosphorus); Ncs: nutrient concentration in suspended sediment (gm/kg) (nitrogen or phosphorus); SSL: suspended sediment loss (Kg); q: discharge of the rivers per ten consecutive days (L/d); A: area of the catchments (ha); i: ten consecutive days; TN: total nutrient loss (Kg) (nitrogen or phosphorus); grand total nutrient loss (Kg/ha) (nitrogen or phosphorus).

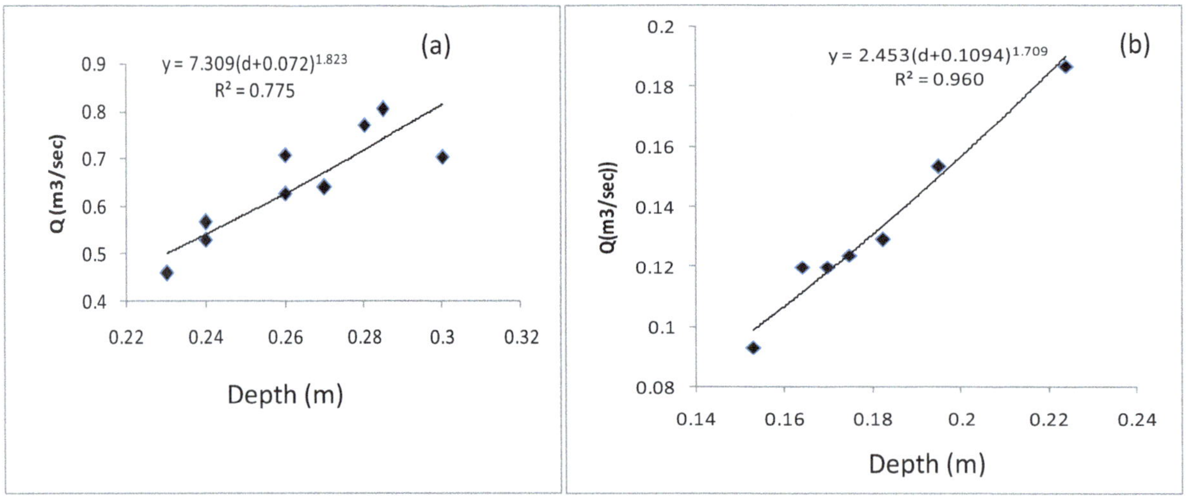

Figure 5. Discharge rating curve for DR (a) and CR (b).

Table 2 Discharge, total suspended sediment and sediment concentration from Dapo River and Chekorsa River.

Parameter	Dapo River				Chekorsa River				t value	p-value
	Mean	SD	CV	R^2	Mean	SD	CV	R^2		
Discharge (Mm^3/d)	0.64	0.17	26.7	1	0.24	0.043	18.09	1	6.83	0
TSS (tons/d)*	747.86	379	50.7	0.7	434	217	49.82	0.85	2.16	0.05
SSC (gm/L)	1.12	0.36	32	0.3	1.74	0.597	34.29	0.73	2.66	0.02
Soil loss (t/ha/d)	0.42	0.21	50.7	0.7	0.75	0.335	44.82	0.79	2.52	0.023

*TSS: Total suspended sediment loss in tons per 10 consecutive days.

Crop yield reduction due to the loss of nitrogen and phosphorus

Since maize is a major crop in both watersheds, reported research data of maize grain yield response to N and P under similar agro-ecological conditions were obtained from Bako Agricultural Research Center. The yield response curve for N and P was developed by fitting the data into quadratic relation (Rosas, 2011; Naher et al., 2011). The yield reduction due to loss of available N and P were estimated using their respective response curves (Equation 6).

$$\mathrm{Y} = ax^2 + \mathrm{bx} + \mathrm{C} \quad (6)$$

Y: Maize yield per hectare; X: nutrient (N and P) per hectare; a, b and c: constant of quadratic equation.

The local market price of maize which was 3.5 BirrHa^{-1} (1birr = 17.50USD)) was used to convert the reduction in grain yield to financial loss incurred by the farmers.

Data analysis

The data was analyzed using SPSS and presented using Sigma-plot software (version 10). Linear regression analysis was carried out for suspended sediment load (SSL) and suspended sediment concentration (SSC) with hydrological parameters. T-test at 95% confidence limit was used to analyze the differences in sediment load between the two monitoring stations. Pearson correlation analysis between soil texture percentage and nutrients were also determined.

RESULTS AND DISCUSSIONS

Discharge

Water levels or depth across the width of the rivers and corresponding flows of water resulted in power curves (Figure 5). According to Braca (2008), continuous measurement of flow past a river section is usually time consuming, impractical during flood event and prohibitively expensive. The average discharge estimated using the stage-discharge relationship curve was 0.64 DR and 0.24 Mm^3/d for DR and CR, respectively (Table 2).

Suspended sediment and its interaction with discharge

Eroded soil by runoff from watersheds is mostly

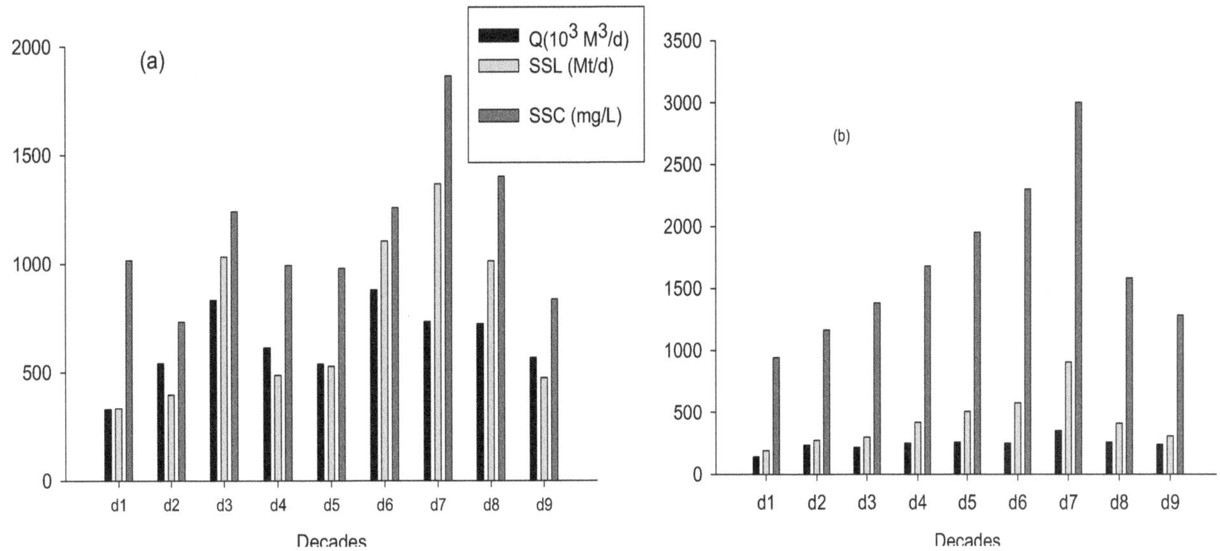

Figure 6. Average discharge and SSC, SSL of Dapo (a) and Chekorsa Rivers (b).

transported by river flow in suspension (Sileshi et al., 2008). The sediment concentration increased with increasing discharge throughout the measurement period for CR (Figure 6b). For DR, the maximum concentration occurred during day 7 after the maximum discharge occurred (Figure 6a).The regression analysis showed SC is strongly related to discharge at CR (R2=0.73) but it was weak at DR weak (R2=0.29) (Table 2). SSC of DR did not remain high with all the discharges throughout the study period. This may be because of the quantity of sediment on a river bed, limited and less sediment availability from the watershed (Ongley, 1996).

Consequently, although the total discharge from DR was about three times greater than that from CR, the amount of sediment lost from the two catchments was similar.

Some possible causes of these variations are soil erodibility, proximity to the waterways, and the difference in riparian zones vegetation (Figure 8), catchments dimensions, land use and cover (GLCUS, 2008), sediment availability (Baca, 2002) and population density, turbulent fluctuations of stream velocity. As shown in Figure 6a, sediment becomes readily available for transportation following the peak discharge that occurred during day 6.

Temporal variability of suspended sediment with discharge

There is significant difference in discharge and sediment transport between ten consecutive days. The discharge to SC (Q-SC) hysteretic loops identified for the two catchments were different (Figure 7a and b) such as it was characterized by anti-clockwise for DR but clockwise for CR. However, since short-term storm events are important determinants of sediment concentration (Eder et al., 2010), and single events hysteresis analysis must be conducted to confirm this interpretation.

The counter clockwise hysteresis in SC for Dapo watershed may be related to the reduced rainfall and the attendant decrease in discharge up to the fifth ten consecutive days, the delay in sowing time of the major crops in the upstream part as compared to those in the downstream part of the watershed. The eroded sediment from cultivated field had been trapped to its maximum at the riparian zone and then detached and transported after the peak discharge event when the supply of sediment was limited (Figure 8) corroborating the findings of Baca (2012) who conducted a similar study in Slovakia.

The clockwise hysteresis is attributed to the rapid displacement of sediment from source close to the stream. The size of CRC is less than a third of Dapo watershed. Similar result where found by Vanmaercke et al. (2006) in Geba River Catchment of Northern Ethiopia. Clockwise loops most commonly occur when the sediment peak occurs before the water discharge peak and when there is a source of easily erodible sediment which can be rapidly depleted (Hudson, 2002; Picouet et al., 2001; Baca, 2010).

Plant nutrient loss and enrichment ratio

In Chekorsa River, N loss increased towards the end of the rainy season up to three times as compared to at the beginning of the study, whereas P loss increased up to 2 times (Figure 9) although increase in discharge was not significant (Figure 9). On the other hand, it is vice versa

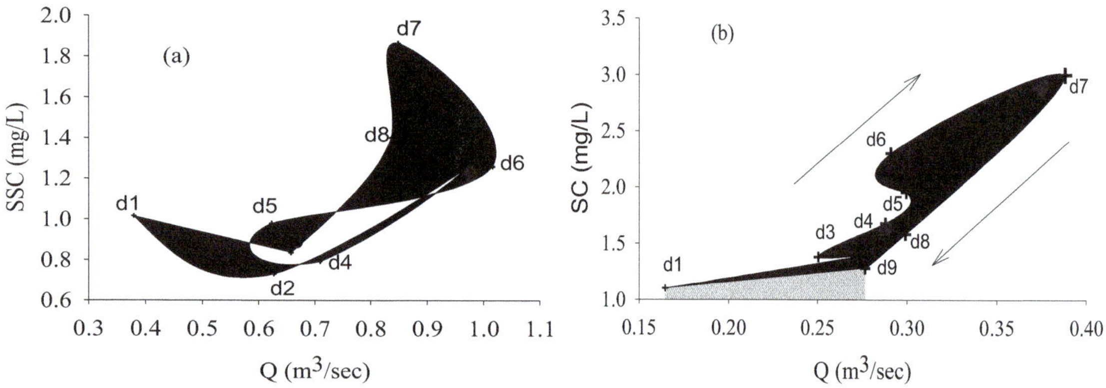

Figure 7. Changes of SSC with decade average discharge of Dapo (a) and Chekorsa Rivers (b).

Figure 8. Accumulated sediment at the edge teff field (a) and vegetation along DR (b).

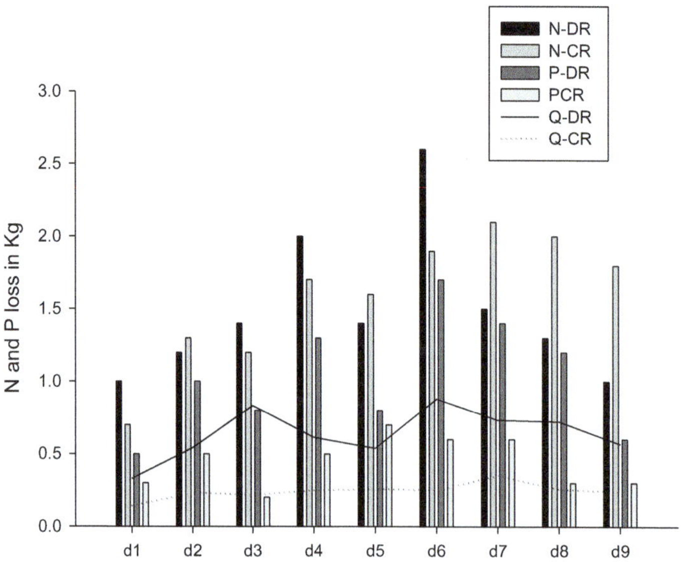

Figure 9. Amount of N and P loss (kg) and discharge ((Q) in Mm^3/d) in each decade from DR and CR during the study period.

Table 3. Mean nutrient content of the surface soil of the catchments, and sediment (kg/ton) and enrichment ratio.

Sub areas	Surface soil			Enrichment ratio (: s)		
	Total N	OC	P	Total N	OC	P
Dapo downstream	4.19	57.06	0.013	0.89	1.18	2.69
Dapo upper stream	2.74	26.49	0.022	1.36	2.54	1.62
Chekorsa catchment	2.58	20.47	0.002	0.95	1.5	8.99
DR (Suspended sediment)	3.73	67.2	0.04			
CR (Suspended sediment)	2.44	30.78	0.02			

Table 4. R^2 regration analysis of nutrient with particle size distribution percentage in the sediment for CR and DR.

TKN	OC		P		Sand (%)		Silt (%)		Clay (%)	
	CR	DR	CR	DR	CR	DR	CR	DR	CR	DR
TKN	0.485	0.43	0.621	0.278	0.074	0.39	0.997	0.14	0.672	0.936
OC			0.981	0.989	0.251	0.1	0.428	0.34	0.964	0.539
P					0.142	0.11	0.565	0.35	0.997	0.523
Sand (%)							0.107	0.91	0.107	0.17
Silt (%)									0.617	0.016
Clay (%)										

TKN, Total Kjeldal nitrogen; OC, organic carbon; CR, Chekorsa River; DR, Dapo River; P, phosphate.

for Dapo River, which is quite similar result found by Liu et al. (2012). The loss of P and particularly N from CR is as higher as that of DR (Figure 9) though the average discharge of CR is as about three times less than DR (Figure 9). It was probably because of an effect remarked by Arheimer and Lide (2000), that is, contribution from rural point sources such as the traditional manuring, the availability of nutrients in the lower parts of a catchment (Chekorsa River catchment is more arable land and dense population), and water has easy access to the rivers and dilution effects by water from upper part of the catchment.

Similarly, the ground water which came as base flow from the upper region of DRC (covered by natural dense forest increase infiltration of rain water into ground water) probably diluted the river. Clay particles have been washed away from both catchments in greatest percentage. For example, if washing of clay particles continues in such a way (Table 4), it would exacerbate pressure on production, since soils with higher clay content have more favorable chemical properties (Hutton et al., 2008) than coarser textured soils.

The concentration of OM, TKN and available P in the eroded sediment were greater than the surface soil. The concentration of available P reaches up to greater than 2.7 and 9 time its initial concentration in some areas of the watershed in the DRC and CRC (Table 3), respectively. This indicates that surface runoff is washing P in large amount to the water body. This may be because of the largest concentration of P in the soil surface layers, and concentrated hydrologic energy on the soil surface (Zaimes and Schultz, 2002). This high P ER coincides with high clay transport by the rivers, (for example: r^2 of 0.997) (Table 4). Scalenghe et al. (2002) indicated the highest accumulation of clay has strong relation with P ER because of the preferential loss of P with finer clay sized particle.

Soil erosion selectively washes fine particles such as clay and organic matter leaving coarse sand behind. Sand and silt are comparatively inert and act only as diluents to the more active clay (Page, 1950). Therefore, the higher the clay fraction the greater is the surface area of the soil available for sorption (Hutton et al., 2008).

Surface soil of DRC has greater soil fertility and hence loses much more nutrient than CRC (Table 3). This result also supports why soils with higher fertility status lose much more nutrients relative to those with a lower fertility status. Evans (2006) showed that the amount of nutrient lost was found to be strongly dependent on the nutrient status of the soil.

The severity of nutrient loss

The classifications were based on FAO (1999) calcification for available nutrient loss. The result of classification signifies how much soil erosion alone is contributing for the very high nutrient loss (FAO, 1999; Henao and Baanante, 1999). However, the classification in the report had been based on the nutrient removal including other major means of nutrient removal such as crop residue removal, leaching evapo-transpiration, grazing, etc.

Regarding nitrogen, Table 5 indicates only 35 and 30%

Table 5. Severity classes of the loss available nutrients.

Available	Dapo catchment		Chekorsa catchment		Loss left behind to be classified as high nutrient loss class (%)	
	Total loss	Severity class	Total loss	Severity class	DRC	CRC
N	13.58	Moderate	14.3	Moderate	35.4	30.6
P	9.31	High	4.20	Moderate	---*	47.5

Table 6. Estimated monitory value of available nutrient loss by the two Rivers.

Step	Estimated	Dapo catchment		Chekorsa Catchment	
		N	P	N	P
1	Total Lost/ha	13.6	9.3	14.3	4.1
2	Potential grain yield response (Kg/Ha)	3338.1	3905.1	3402.4	3147.6
3	Mean grain yield with no P and N fertilizer*	2389.3	2483.7	2389.3	2483.7
4	Net yield (Kg/ Ha)	948.8	1421.4	1013.1	663.9
5	Total price (Birr/ha)**	3320.8	4974.9	3545.9	2323.7

*Using Figure 12 and 13 equations accordingly. **Since market price of maize at Nekemte is 3.5 birr Ha^{-1} (1 birr = 17.50 USD).

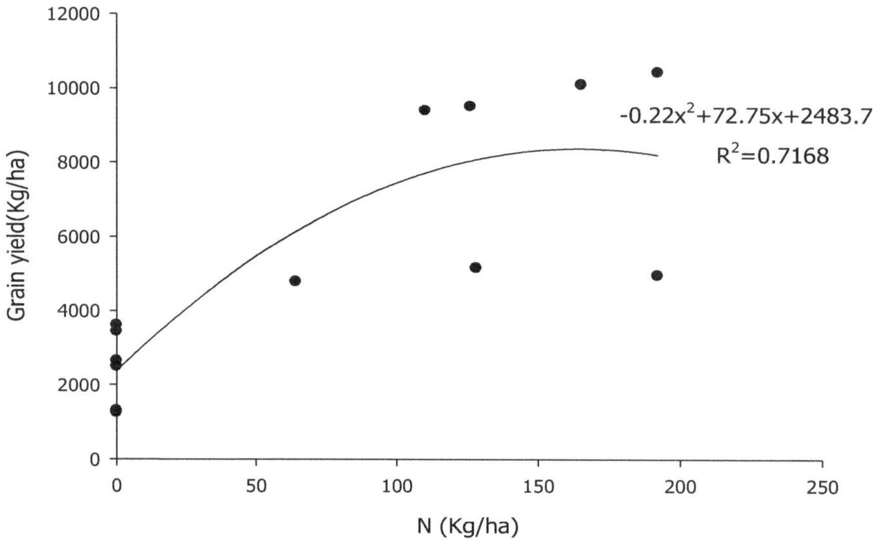

Figure 10. Response of maize grain yield to nitrogen application rate.

of high nutrient loss rate class stated by FAO (1999) was left behind in DRC and CRC, respectively, to fall in the high nutrient loss rate class. The soil nutrient loss due to soil erosion alone contributed about 65 and 69% high nutrient loss rate class by FAO (1999). According to FAO (1999), if P_2O_5 losses from cultivated land are between 4 and 7 $kg^{-1}ha^{-1}yr$, such value are considered as high losses. Only by soil erosion, P_2O_5 has already attained the high soil nutrient depletion rate class for Dapo (Table 5). On the other hand, in the CRC, 52.53% has already attained the minimum amount of P_2O_5 loss to be classified as high losses.

Costs of nutrient loss

The estimated macro nutrients (Table 6) and the yield response curve (Figures 10 and 11) were used as a tool to the estimated monetary value of onsite economic cost of the lost nutrients in Table 6. However, estimating total monetary cost of erosion presents many difficulties (GLCUS, 2008) since it involves many different factors (Derek, 1996).

R^2 of Figure 10 shows a wide variation of yield response to the almost equivalent amount of fertilizer rate; that is, 2.3 to 6.5 tha^{-1}. Similarly, Negassa et al.

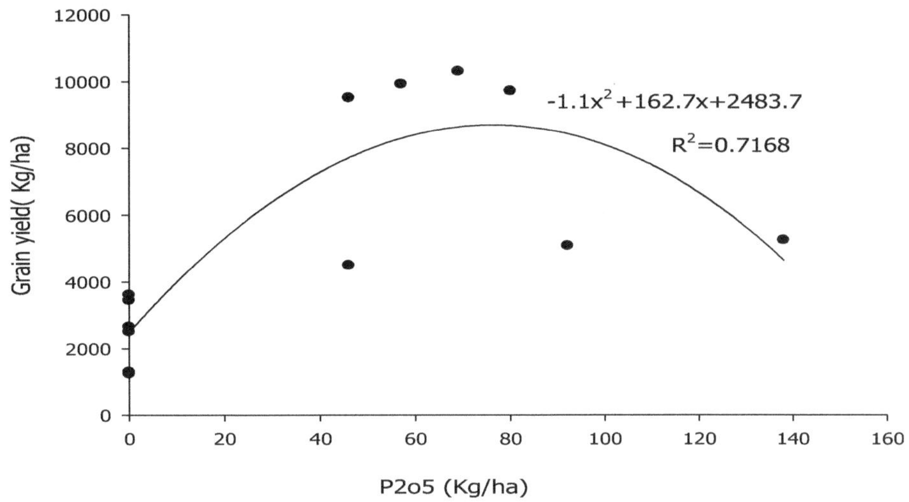

Figure 11. Grain yield response curve of maize to phosphorus rate.

(2005) indicated high variation of maize yields on farmers' fields in the study areas ranging from <1.0 to almost 6.0 t ha^{-1}.

Equations of Figures 10 and 11 represents the yield response curves showing the trend of yield increment for different rates of additional N and P application. Mean grain yield with no N and P fertilizers were 2,389 and 2,484 $Kgha^{-1}$, respectively (Table 6). Therefore, lost net maize grain yield due to the loss of available N and P_2O_5 by runoff were about 949 and 1,421 $kgha^{-1}$ from DRC whereas 1,013 and 664 $kgha^{-1}$ from CRC (Table 6). Farmers in the study area lost about 190 and 284 USD/ha from Dapo catchment and 203 and 133 USD/ha from Chekorsa catchment due to the loss of N and P, respectively.

Yield decline due to erosion follows a curvilinear, negative exponential (FAO, 1999). Therefore, eventually the current decline of yield can gradually reach a worst period in the future where there is no observation in yield decline. Negassa et al. (2005) also found that the relatively common practice of sole application of low rate of N and P fertilizers has not sustained maize production and productivity in the region because of severe nutrient loss. Therefore, combined application of these nutrients with manure/compost gives a better yield.

Conclusion and recommendations

The results revealed that suspended sediment concentration is strongly related to discharge or flood events for both catchments. The quality of the sediment lost from both sites is similar, though the catchment size and total discharge from Dapo River is about three times greater than Chekorsa, once sediment concentration of the Chekorsa River (varies from 942 to 3,002 mg/L) is higher than one at the Dapo River (ranging from 731 to 1,865 mg/L).

Suspended sediment concentration increases directly coincide with the surge in discharge for Chekorsa River but not for Dapo River. Discharge-suspended sediment concentration produced two types of hysteretic loops, an overlapped anticlockwise hysteresis for Dapo River Catchment and clockwise hysteresis relationships for Chekorsa River. The study reveals the loss of productivity as a result of both dissolved and sediment-sorbed N and P nutrient transported in overland flow delivered to the monitoring station.

In addition, from the result of the study, it is possible to conclude that:

1. Nitrogen and phosphorus losses due to soil erosion are elevated;
2. Clay has eroded more than silt and sand;
3. There is a strong correlation between percentage of clay with phosphate and TKN;
4. The rate of phosphorus loss was nine times in the sediment than that of the surface soil;
5. The monetary value of the lost nutrient may persuade farmers to implement sustainable nutrient practices and soil conservation activities;
6. As recommendation, similar studies must continue in the catchment including event based sediment hysteresis assessment in order to compare it with the weekly based sediment hysteresis.

Conflict of Interests

The author(s) have not declared any conflict of interests.

ACKNOWLEDGEMENTS

The authors gratefully acknowledge the International Water Management Institute (IWMI) for funding this study.

Their great appreciation goes to the International Water Management Institute staff members especially to Dr Charlotte McAlister previous Nile 4 Project leader and Dr Brihanu Zemedin (Ph.D.) who supported the work technically, and all the laboratory technicians of Nekemte Soil Research Institute.

REFERENCES

Ahanger M, Asawa G, Lone M (2013). Hysteresis effect on sediment rating curves. J. Acad. Ind. Res. 1(8)

Alley M, Marvin E, Martz J, Paul H, Davis J, Hammons L (2009). Nitrogen and phosphorus fertilization of corn. Communications and Marketing, College of Agriculture and Life Sciences, Virginia Polytechnic Institute and State University. Virginia State, Petersburg Management Institute. 87p. (IWMI Working Paper 131)

Arheimer B, Liden R (2000). Nitrogen and phosphorus concentrations from agricultural catchments —influence of spatial and temporal variables. Swedish Meteorological and Hydrological Institute. J. Hydrol. 227 (2000) 140-159SE-601 76

Ashagrie T (2009). Modeling rainfall, runoff and soil loss relationships in the northeastern highlands of Ethiopia, AnditTid watershed. A Thesis Presented to the Faculty of the Graduate School of Cornell University.

Baca P (2002). Temporal variability of suspended sediment availability during rainfall-runoff events in a small agricultural basin.ERB and Northern European FRIEND Project 5 Conference, Demänovskádolina, Slovakia.

Baca P (2012). Hydrological Sciences Journal: Hysteresis effect in suspended sedimentconcentration in the Rybárik basin, Slovakia. Taylor & Francis. London W1T 3JH, UK.

Bekele S, Holden ST (1998). Resource degradation and adoption of land conservation technologies in the Ethiopian Highlands: A case study in Andit Tid, North Shewa. Agricultural Economics 18:233-47

Berry L, Olson J, Campbell D (2003). Assessing the extent, cost and impact of land degradation at the national level: findings and lessons learned from seven pilots. World Bank

Bouyoucos G (1962). Hydrometer method improved for making particle size analysis of soils. Agron. J. 54:464-465.

Braca G (2008). Stage-discharge relationships in open channels: Practices and problems. FORALPS Technical Report, 11.UniversitàdegliStudi di Trento, Dipartimento di IngegneriaCivile e Ambientale, Trento, Italy. 24 pp.

Dalal R, Sahrat K, Myers R (1984). Inclusion of nitrate in the Kjeldahl nitrogen determination of soil and plant materials using sodium thiosulphate. Commun. Soil Sci. Plant Anal. 15:14531461.

Derek E (1996). The Economics of Soil Erosion: a Model of Farm Decision-Making. Environmental Economics Programme, 2502 LS Thee Hague The Netherlands

Desta L, Menale K, Benin S, Pende J (2000). Amhara National Regional State Bureau of Agriculture. IRLI. Addis Ababa, Ethiopia.

Dubale G, Yasir A, Mohamed, Ann van Griensven I. Popescu, Arthur M (2009). Modeling of soil erosion and sediment transport in the Blue Nile Basin using the Open Model Interface Approach in; Improved Water and Land Management in the Ethiopian Highlands: Its Impact on Downstream Stakeholders Dependent on the Blue Nile IWMI Subregional Office for East Africa and Nile Basin, Addis Ababa, Ethiopia.

Dubale P (2001). Soil and Water Resources and Degradation Factors Affecting Productivity in Ethiopian Highland Agro-Ecosystems.Michigan State University Press. Northeast African Studies, Volume 8, Number 1, 2001 (New Series). pp. 27-51.

Eder A, Strauss P, Krueger T, Quinton J (2010). Comparative calculation of suspended sediment loads with respect to hysteresis effects (in the Petzenkirchen catchment, Austria). J. Hydrol. 389:168-176.

Evans R (2006). Sustainable practices to limit soil erosion: a review and discussion. CABI Publishing. Department of Geography, Anglia Ruskin University, East Road, Cambridge CB1 1PT, UK.

FAO (1999). Integrated soil management for sustainable agriculture and food security in southern and east Africa.Proceedings of the expert consultation.8-12 December, 1997 Harare, Zimbabwe. FAO, Rome

Fenli Z, Xiubin H, Xuetian G, Chang Z, Keli T (2005). Effects of erosion patterns on nutrient loss following deforestation on the Loess Plateau of China. Agric. Ecosyst. Environ. 108:85-97.

Getnet D, Yasir A, Ann van Griensven I, Arthur M (2009). Modeling of soil erosion and sediment transport in the Blue Nile Basin using the Open Model Interface Approach in; Awulachew, S. B.; Erkossa, T.; Smakhtin, V.; Fernando, A. (Comp.). 2009. Improved water and land management in the Ethiopian highlands: Its impact on downstream stakeholders dependent on the Blue Nile. Intermediate Results Dissemination Workshop held at the International Livestock Research Institute (ILRI), Addis Ababa, Ethiopia, 5-6 February 2009. Summary report, abstracts of papers with proceedings on CD-ROM. Colombo, Sri Lanka: International Water Management Institute. doi:10.3910/2009.201

Gierke J (2002). Measuring River Discharge: ENG5300 Engineering Applications in the Earth Sciences. Department of Geological and Mining Engineering and Sciences Michigan Technological University1400 Townsend Drive Houghton, MI 49931-1295.

GLCUS (Great Lakes Commission for the U.S. Army Crops of Engineers Great Lakes and Ohio River Division) (2008).The Economics of Soil Erosion and sedimentation in the Great Lake Basin. USA.

Gorlach B, Trinkunaite R, Interwies E, Bouzit M, Darmendrail D, Rinaudo D (2004). Assessing the Economic Impacts of Soil Degradation. Volume IV: Executive Summary. Study commissioned by the European Commission, DG Environment, Berlin.

Gruhn P, Francesco G, Yudelman M (2000). Integrated Nutrient Management, Soil Fertility, and Sustainable Agriculture: Current Issues and Future Challenges. International Food Policy Research Institute 2033 K Street, N.W. Washington, D.C. 20006 U.S.A.

Hagos F, Amare H, Seleshi B (2009). Assessment of Local Land and Water Institutions in the Blue Nile and their Impact on Environmental Management in; Improved Water and Land Management in the Ethiopian Highlands: Its Impact on Downstream Stakeholders Dependent on the Blue Nile IWMI Sub regional Office for East Africa and Nile Basin, Addis Ababa, Ethiopia.

Helmecke M (2009). Assessment of soil erosion in selected micro-watersheds in koraput District, Orissa. Foundation for Ecological Security Anand - 388 001 (India).

Henao J, Baanante C (1999). Estimating Rate nutrient depletion in Soils of Agricultural Lands in Africa, IFDC, Muscle Shoals, Alabama.

Hurni H (1993). Land degradation, famine, and land resource scenarios in Ethiopia. In: Pimentel, D. (Ed.), World soil erosion and conservation. Cambridge University Press, Cambridge. pp. 27-62.

Hutton S, Harrison S, Halloran J (2008). An evaluation of the role of forests and forest practices in the eutrophication and sedimentation of receiving waters. Department of Zoology, Ecology and Plant Science Environmental Research Institute University College Cork.

Jackson M (1967). Soil chemical analysis. Presntice Hall of India, Pvt Ltd. New Delhi, India.

Joshua O, Bharat R, Matthew P, Biranu Z (2010).Biophysical Aspects of landscape soil quality improvement for sustainable Agricultural productivity in Ethiopian highlands (draft).

Kebede M, Gezahegne B, Benti T, Mosisa W, Yigzaw D, Assefa A (1993). Maize producti on trends and research in Ethiopia. In Benti Tolessa and J.K. Ransom (ed.), Proceedings of the First Nati onal Maize Workshop of Ethiopia. May 5-7, 1992. IAR/CIMMYT, Addis Ababa, Ethiopia. pp. 4-12.

Kogbe J, Adediran J (2003). Influence of nitrogen, phosphorus and potassium application on the yield of maize in the savanna zone of Nigeria. Institute of Agricultural Research and Training, Obafemi Awolowo University, PMB 5029, Ibadan, Nigeria.

Le Bas, Kozak J (2007). Soil Erosion. European commition Joint Research.http://eusoils.jrc.ec.europa.eu/esbn/Plenary_esbn_2007/Workshop/ESBN_SoilErosion

Lefrançois J, Grimaldi C, Gascuel C, Gilliet N (2007). Suspended sediment and discharge relationships to identify bank degradation as a main source on small agricultural catchments. Hydrol. Process. 21:2923-2933.

Maiti S (2004). Handbook of Methods in Environmental Studies Vol. 1: Water and Wastewater Analysis. ABD Publishers. Jaipur, India.
MWRE (2010). Environmental and Social Impact Assessment of the Ribb Irrigation and Drainage Project Volume 2/2 Ministry of Water and Energy, Ethiopia.
Naher U, Mohammed A, Qurban A, Othman R, Shamsuddin J (2011). Techniques of efficient fertilizer management for wetland rice- a review. Aust. J. Crop Sci. 5(12):1661-1669.
Negassa W, Heluf G, Friesen DK (2005). Integrated Use of Farmyard Manure and NP fertilizers for Maize on Farmers' Fields. J. Agric. Rural Dev. Trop. Subtropics 106(2):131-141.
Olsen S, Sommers L (1982). Phosphorus; In Page, Miller, Keeny, Leeds.Method of soil analysis II. Chemical and micro biological properties, agronomic, monograph, vol.9, Second ed. SSSA, Madisen, Wisconsin.
Ongley E (1996). Water Quality Monitoring - A Practical Guide to the Design and Implementation of Freshwater Quality Studies and Monitoring Programmes. United Nations Environment Programme and the World Health Organization.
Page J (1950). Role of physical properties of clays in soil science.Agricultural and Mechanical College of Texas, College Station, Texas.
Patnaik P (2010). Hand book of Environmental analysis chemical pollutants in air, water, soil and solid wastes, CRC press Boca Raton London, New York.
Picouet C, Hingray B, Olivry JC (2001). Empirical and conceptual modelling of the suspended sediment dynamics in a large tropical African river: the Upper Niger River basin. J. Hydrol. 250:19-39.
Pimentel D, Haway C, Resosudarmo K, Kurz D, McNair M, Crist S, Shpritz L, Fitton L, Saffourl R, Blair R (1995). Environmental and Economic Costs of soil Erosion and Conservation Benefits. Science 267(5201):1117-1123.
Rosas R (2011). World Fertilizer Model; The World NPK Model. Working Paper 11-WP 520.Center for Agricultural and Rural Development Iowa State University, Ames, Iowa 50011-1070.
Sander G, Zhenga T, Heng P, Zhong Y, Barry D (2011). Sustainable soil and water resources: Modelling soil erosion and its impact on the environment. 19th International Congress on Modelling and Simulation, Perth, Australia.
Scalenghe R, Barberis E, Edwards A (2002). The Phosphorus Enrichment rationOverfertilized Soils. 17th WCSS August 2002, Thailand.
Seleshi BA, Fitsum H, Godswill M,Regassa EN (2009).Importance of Irrigated Agriculture to the Ethiopian Economy: Capturing the Direct
Net Benefits of Irrigation.IWMI Research Reports.
Sharpley N, Foy H, Withers P (2000). Practical and Innovative measures for the control of agricultural phosphorus losses to water: An overview. J. Environ. Qual. 29:1-9.
Sileshi A, McCartney S, Abdalla A (2008). A review of hydrology, sediment and water resource use in the Blue Nile Basin. Colombo, Sri Lanka.
Slattery MC, Gares PA, Philips JD (2002). Slope-channel linkage and sediment delivery on North Carolina Coastal Plain cropland. Earth Surf. Process. Landforms 27:1377-1387.
Tenaw M, Seleshi BA (2009). Soil and Water Assessment Tool (SWAT)-Based Runoff and Sediment Yield Modeling: A Case of the Gumera Watershed in Lake TanaSubbasinSeleshi B. Awulachew, Teklu Erkossa, Vladimir Smakhtin and Ashra Fernando Improved Water and Land Management in the Ethiopian Highlands: Its Impact on Downstream Stakeholders Dependent on the Blue Nile IWMI Subregional Office for East Africa and Nile Basin, Addis Ababa, Ethiopia.
Van-Camp L, Bujarrabal B, Gentile A, Jones R, Olazabal C, Selvaradjou S (2004). Reports of the Technical Working Groups Established under the Thematic Strategy for Soil Protection.EUR 21319 EN/3, 872 pp. Office for Official Publications of the European Communities, Luxembourg.
Vanmaercke M, Zenebe A, Poesen J, Nyssen J, Verstraeten G, Deckers G (2006). Magnitude and Dynamics of Runoff and Sediment Transport in the Geba River Catchment, Northern Ethiopia. Physical and Regional Geography Research Group, Catholic University of Leuven, Celestijnenlaan 200E Tower 3 Floor 3, 3001 Heverlee.
Wortmann C (2006). Agricultural Nitrogen Management for Water Quality Protection in the Midwest EPA Region VII.The Board of Regents of the University of Nebraska, USA.
Zaimes G, Schultz R (2002). Phosphorus in Agricultural Watersheds Department of Forestry, Iowa State University, Ames, Iowa.
Zemadim B, Matthew M, Bharat S (2012). Establishing Hydrological and Meteorological Monitoring Networks in Jeldu, Diga and Fogera Districts of the Blue Nile Basin, Ethiopia. IWMI, Addis Ababa, Ethiopia
Zemadim B, Matthew M, Bharat S, Joshua O, Gerba L, Seleshi B (2010). Integrated rain water management strategies in the Blue Nile Basin of the Ethiopian highlands. IWMI, Addis Ababa, Ethiopia.

Land degradation in Northern Nigeria: The impacts and implications of human-related and climatic factors

B. M. Macaulay

Natural Resources Institute, University of Greenwich, Medway Campus, United Kingdom.

Northern Nigeria despite its promising agricultural potential, is currently not enjoying economic bloom like its Southern counterpart as a result of political, religious, ethnic and socio-economic factors. The degradation of land and its resulting impact on other agricultural resources may have further exacerbated the economic and social conditions of the region. The perceived encroachment of Nigeria's Savannah into its rainforest zone indicated possible land degradation. Some authors attributed this degradation to anthropogenic sources whilst others reported climatic variability (which is nature-driven) as the singular culprit. Therefore, this study reviews the impacts and implications of both human-related and climatic factors on land degradation in Northern Nigeria. Human-related activities such as agricultural/pastoral expansion, agricultural intensification and fuel-wood extraction as well as climatic/physical factors such as rainfall variability and land-atmosphere feedback mechanisms were suggested by various authors as agents of land degradation in Northern Nigeria. The remediation strategies to reduce the impacts of anthropogenic factors include the practice of agroforestry, rainwater harvesting, local irrigation techniques, utilization of wetter sites, contour ridging and terracing to conserve nutrient and water run-off, cautionary expansion of cultivated sites, and the maintenance of a viable seed stock well-suited to variable climatic conditions. The recommendations made on the resulting conflict between crop farmers and pastoralists include a federal government-assisted water project to build more waterpoints at strategic areas in the region and the practice of agropastoralism. However, the controversy over the cause of land degradation in Africa has led to a number of research questions itemised in this paper on the main cause of land degradation, the elements of climate inducing the dryness and wetness of the region, and the extent of land recovery when wet seasons return.

Key words: Land degradation, impacts and implications, human-related factors, climatic factors, northern Nigeria.

INTRODUCTION

Nigeria, a West African country located in the tropical zone of the world has a land area of about 923,769 km^2 with the Northern region covering about 79% of the entire land mass (FOS, 1989; Salako, 2003; Aregheore, 2005). The country comprises about five broad ecological zones: Swamp forest, Tropical rainforest, Guinea Savannah,

Sudan Savannah and Sahel (Okpara et al., 2013). Of the five zones, Northern region is made up of three: Guinea Savannah, Sudan Savannah and the Sahel. Northern Nigeria is composed of 19 of the country's 36 states. It is inhabited by over 50% of the country's 167 million people (Pate and Dauda, 2013) sparsely distributed across 79% of the country's total land mass. It is home to over two-third of the country's 250 ethnic groups (Pate and Dauda, 2013).

According to Cleaver and Shreiber (1994), the surface area of Nigeria is about 91.07 million hectares and 57% of this land surface is suggested to be used for crops and pasture production (commonly practiced in Northern Nigeria) whilst the remaining 43% is divided amongst forests, water bodies and other uses such as construction and human settlements (Aregheore, 2005). Iortim (2012) also reported that agriculture contributes about 42% of the country's GDP and the Northern part of the country is responsible for most of these; yet the region is suffering from both economic and infrastructural deficiency as indicted by the country's Human Development Index (UNDP, 2010).

There are six geo-political zones in Nigeria: the South-South, South-West, South-East, North Central, North-West and North-East. Of the six geopolitical zones in Nigeria, the most developed in the North, is the North Central, possessing economically-viable cities like Kano (the centre for commerce in the North) and Abuja (the Federal Capital Territory) whilst other northern geo-political zones appear to be economically less-attractive. The north-west and north-east, according to the 2012 National Bureau of Statistics (NBS) data, have the highest poverty rates of about 77.7 and 76.3% respectively (UNESCO, 2010); and alarming illiteracy levels of about 86% each in the country (UNESCO, 2010). The North also has the least annual per capita income, below the national average of twenty thousand naira (approximately $127). The industrial deficiency in the North has increased the unemployment levels and youth restiveness in the region (Pate and Dauda, 2013). The region thus, experiences more violence than other parts of the country as a result of the poor management of its diverse resources (Pate and Dauda, 2013). The violence could manifest in the form of religious, ethnic, economic, political or value-based crisis, which makes the region "increasingly stigmatised as a theatre of violent clashes and as a parasite that cannot survive without monthly federal allocations" (Pate and Dauda, 2013). As a result of the socio-economic structure of the region, most families practice subsistent farming and/or nomadic, pastoralism for economic sustenance.

Northern Nigeria (as one advances north-ward) is characterised by low rainfall and drought-like conditions (Xue and Shukla, 1993). Despite the perceived harsh weather conditions of this region, many plant species such as *Acacia* species, baobab tree, mango, orange and *Moringa* species, with soils suitable for the production of crops such as cereals (millet, rice, corn, sorghum and maize) and legumes (soybeans and cowpea) inhabit the area. The vast 'fertile' land of this region has the potential for agricultural revolution, however, most farming practices are not mechanised (Pate and Dauda, 2013) probably because the states in the north are not economically-fit to commercialise agriculture.

Human activities such as agricultural expansion/-intensification, pastoralism and fuel-wood extraction, and climatic/physical factors such as rainfall variability and land-atmospehere feedbacks have been reported by many authors to have a significant impact on land degradation. Therefore, this study intends to review the impact of the aforementioned human-related and climatic factors on land quality in northern Nigeria, including the underlying implications and possible misconceptions and debates. The knowledge would improve the understanding of how land and other agricultural resources can be properly managed so that the potential of the region to lead the 'green revolution' crusade in the country can be actualised.

HUMAN-RELATED IMPACTS AND IMPLICATIONS ON LAND DEGRADATION IN NORTHERN NIGERIA

According to the WRI (1992) and Barbier (1999), land degradation and conversion in Africa is mostly caused by agricultural activities. Land uses such as tree felling for fuel-wood and timber production, crop and pastoral land expansion and agricultural intensification have been suggested to be part of the causes of land degradation in Africa (including Nigeria). The possible loss of natural vegetation cover due to changes in land use, influenced by man, could expose the soil surface and render it vulnerable to the elements of weather such as rain and wind. Lanly (1982) and Skoupy (1987) reported that for the past 30 years, changes in human activities have resulted in the continuing loss of vegetation cover and soil fertility in West Africa. Land use change has been suggested to impact the northern ecosystems of the country over the years leading to a perceived creeping of the savannah into the tropical rain forest zone (Badejo, 1998). Changes in crop and livestock production methods and increasing demand for forest products (such as fuel-wood and timber) will accelerate the rate of land degradation which if not checked, would have a negative implication on food and livestock quality and yield.

The total annual rainfall in the ecological zones of Northern Nigeria is lowest in the Sahel and increases southward. This variation in rainfall distribution greatly affects the soil type and biodiversity in each ecological zone in the country. The Northern region is also characterised by high intensity solar radiation in the area (Hekstra, 1985). Evapo-transpiration is therefore, higher in the Northern part of the country than in the South. Despite the increased temperature and decreased precipitation in Northern Nigeria, many vegetation species (generalists with a broad ecological niche) have

Table 1. The soil types in the ecological zones of Nigeria.

Ecological zone	Major soil types (FAO classification)
Humid forest	Ferralsols, nitosols and gleysols.
Derived savannah	Ferralsols, luvisols, lithosols, arenosols, nitosols and acrisols.
Guinea savannah	Luvisols, acrisols, ferrasols and lithosols.
Sudan savannah	Luvisols, ferrasols.
Sahel	Ferralsols and nitosols

Source: Salako (2003).

been found to survive and thrive in the area as a result of their resilience and ability to grow on a variety of soil types. In spite of the low precipitation in the Northern region, the soil types mostly iron-rich (Table 1) are well-suited for crop production under the right application of fertiliser and soil management practices.

About 80% of the inhabitants of the Northern region are involved in crop production, pastoral farming and nomadic pastoralism (Schaefer, 1998). These agricultural activities require the use of land, which if not managed properly may lead to the over-exploitation of the natural resource and consequently, degradation. Barbier (2000) reported that many poor pastoralists and farmers in Africa (including Nigeria) often respond to declining land fertility by moving on to a new piece of land for cultivation and grazing. There are cases when farmers and pastoralists will seek new lands for the purpose of expansion and not as a result of declining soil fertility. This practice is known as agricultural/pastoral land expansion.

Agricultural expansion takes place when uncultivated or unused lands are utilized for crop production. The aggressive acquisition of unused or virgin lands could mean the indiscriminate destruction of forest lands. Therefore, land fallowing is suggested by some authors as a means of reducing the incessant need by farmers and pastoralists to occupy virgin lands. Taylor et al. (2002) reported that if used land is abandoned for a period of 20 years, the land will replenish itself. However, farmers in the North have started reducing the fallow periods of used lands due to the rising demand on food crops in order to meet natural population growth and this may affect crop quality and production rate over time. For example, the reduction of fallow period from 6 to 2 years led to a decline in cassava yield from 11 to 2 tonnes/ha whilst maize yield reduced from 3 to 0.7 tonnes/ha in sub-Saharan Africa (Sanginga et al., 1995). Honlonkou (1999) suggested the use of inorganic fertilisers to restore degraded farmlands, although noted that it might be too expensive for farmers in Northern Nigeria to afford. However, Mortimore and Adam (2001) discouraged the incessant application of chemical fertilisers for environmental and health reasons whilst Tiffen et al. (1994) suggested that the local farmers be encouraged to maximize agricultural output using the resources at their disposal.

Both agricultural and pastoral land expansion may lead to severe conflicts between farmers and pastoralists in Northern Nigeria. This is because two or more interdependent resource users, in this case, between crop farmers and pastoralists, are very likely to have a clash over common environmental resources such as land, pasture, crop-residue, livestock routes and water-points (Ime, 2013). Farm encroachment, that is, crop damage caused by animal belonging to herdsmen as a result of cattle routes or waterpoints on farmlands (Gefu and Gills, 1990) and even the straying of livestock into the grazing land of other pastoralists are possible sources of conflict (Traore, 1996). For example, Amzat (2013) reported the violent clashes between the Fulani herdsmen and the Berom natives of Jos in Plateau State, Northern Nigeria. Pastoralists engage in seasonal rotation; migrating to southern Nigeria (which is usually moist) during the dry season in search of pasture and waterpoints, and returning to their major base in the north during the wet season (Shettima and Tar, 2008). This means that a possible face-off with pastoralists is not confined to the North as cases of bloody clashes have been recorded in southern states such as Enugu, Osun and Delta (Amzat, 2013).

The Nigerian Senate discussed a bill in 2012 seeking for the establishment of a National Grazing Reserves Commission, which will possess the authority to cease or/and acquire land that will serve as grazing reserves and routes for herdsmen in any part of the country. However, the majority of the senators expressed their dissatisfaction with the bill, pointing out that it negates the Land Use Act and the concept of federalism. It was then decided that the matter be handled by the State House of Assembly in each of the 36 states of the country (Odebode et al., 2012). Currently, the matter is yet to be addressed by the State House of Assemblies.

Taylor et al. (2002) measured the percentage land use of areas in Northern Nigeria using the Sahelian Land use Model (SALU) and selecting three reperesentative years (1961, 1996 and 2015) which span through a 54 year period (Table 2). The result indicated a continuous decrease in the percentage of forest areas but showed a continuous increase in the percentages of land area used for pasture and cropland. This inverse relationship suggests the direct impact of agricultural/cropland

Table 2. The percentage land use of areas in Northern Nigeria using the SALU for 1961, 1996 and 2015.

Land	1961	1996	2015
Forest	72	40	35
Pasture	14	16	17
Cropland	5	14	22
Fallow	9	16	15
Unused	0	13	11

Source: Taylor et al. (2002).

expansion (and probably fuel-wood demand) on forest resources which may likely have grave environmental/climatic consequences over time. Forest components such as trees act as nutrient pumps in ecosystems (Kang, 1993); provide shade; and preserve the little soil water available in the case of a low rainfall.

Trees also make the soil less-susceptible to erosion. Therefore, the roles of forest resources in the presser-vation of land quality cannot be overemphasized. However, Mortimore and Adams (2001) presented a contrasting idea that indigenous intensification has the potential to promote the stability of high densities of multi-purpose trees on farmlands. They considered agricultural expansion into virgin forested land as an adaptive strategy of the local farmers and pastoralists to climate variation.

In the Sahel and Savannah, nutrient deficiency due to the absence of vegetation cover poses a more serious threat to crop/pastoral productivity than insufficient water supply (Badejo, 1998). An study by Bremen and De wit (1983) revealed that the problem of insufficient soil nutrients in the Sahel and Savannah is a serious limitation to crop productivity as compared to low rainfall. The application of fertiliser in the right proportion was observed to promote vegetation growth in the area. They also added that water is the limiting resource in the desert but in the semi-arid ecosystems (such as the Sahel and Savannahs found in northern Nigeria), nutrient is a more limiting resource. The need to increase fertiliser inputs has led to agricultural intensification.

Agricultural intensification refers to the application of very high amounts of fertilisers, pesticides and other agricultural products, even including labour inputs, with the sole aim of maximizing crop production. Krogh (1997) reported that agricultural intensification may improve soil performance in sites intensely cultivated, however, there may be a net loss of soil fertility and quality in the surrounding extensive land areas. A better approach of restoring soil fertility is to combine nutrient amendments with fertiliser application. Pieri (1989) reported that the use of fertilisers alone has often, in the long run, had negative impacts on soil productivity. The choice of nutrient amendment depends on soil organic matter, phosphorus availability and the pH of the soil (Breman et al., 2001).

Agricultural intensification may increase crop produc-tion to a break-even point where incessant nutrient addi-tions will have little or no impact on soil quality and fertility. This is because the soils in Northern Nigeria gene-rally comprise coarse-textured aggregates which bind loosely and therefore do not retain sufficient water and nutrients. This implies that even in the midst of sufficient nutrients, if the soil physical condition is not managed, soil productivity might still be low. Therefore, the addition of organic matter like animal dungs or crop residues may be required to safe-guard nutrient or water run-off/-leaching, thereby, restoring the soil's fertility and quality. According to Batterbury and Warren (2001) agricultural intensi-fication is an improvised means by which local farmers cope with the changing climatic conditions of the region as successfully practised in Northern Nigeria (precisely Kano), but stated that the underlying challenges can be managed effectively by adopting an environmentally-healthy indigenous land restoration methods such as rainwater harvesting, local irrigation techniques, utilization of wetter sites, contour ridging and terracing to conserve nutrient and water run-off, cau-tionary expansion of cultivated sites, and the main-tenance of a viable seed stock well-suited to variable climatic conditions.

CLIMATIC FACTORS INFLUENCING LAND DEGRADATION IN NORTHERN NIGERIA

A study in 1998 by Badejo (1998) showed that the savannah is creeping into the rainforest in Nigeria. Similarly, a study by Nobre et al. (1991) predicted that climate change due to deforestation may cause tropical savannahs to creep into tropical rainforests stretching 500 to 1000 km into the zone whilst the savannah will in turn be encroached by the sahara. Many authors termed this 'southward creep' as desertification.

Desertification has been identified to be one of the processes that degrades land, however, there has been an unending debate about the use of the concept to describe the environmental changes occurring in the African terrain. The process of desertification was described by the UN as human-induced with climatic factors increasing susceptibility (Nicholson, 2001). One of the definitions of desertification is 'the expansion of desert-like conditions and landscapes to areas where they should not occur climatically' (Graetz, 1991). The UN later modified their definition to 'land degradation in arid, semi-arid and dry sub-humid areas resulting from various factors, including climate variations and human activities' (Puigdefabregas, 1995; Warren, 1996). The concept of the 'advancement of the desert' was first proposed by Lamprey (1975) using maps of 1950s and aerial surveys of 1975. He concluded that between 1958 and 1975, the Sahara desert had advanced southward into Western Sudan by 90 to 100 km (a desert expansion of about 5.5 km per year). Similarly, in 1983, Skoupy

(1987) reported that about 4.125 million km^2 of land area in Africa's Savannah was affected by desertification. However, in 1984, a field study by Hellden (1984) at the same location found no proof to support Lamprey's claims.

A 10 year study by Tucker et al. (1991) from 1980 to 1990 using satellite-derived vegetation index to determine the annual variations in rainfall and the rate of expansion or retraction of the Saharan-Sahelian boundary, found that there was a progressive decrease in rainfall between 1980 and 1984. However, a steady improvement in rainfall was observed between 1985 and 1990, which led to the restoration of the native vegetation of the Saharan-Sahelian boundary. This variation in rainfall was found to determine the direction of movement of the boundary. Between 1980 and 1984 (during the drought), there was an overall southward movement of the Saharan-Sahelian boundary (the expansion of the desert) whilst between 1985 and 1990 (during the wet season), there was an overall northward movement of the boundary (the retreat of the desert). This study revealed that the average estimation of the overall direction of movement for the 10 year study period was southward (130 km) but most importantly, it should be noted that the movement is not progressive but fluctuating based on rainfall patterns in the region. This discovery rendered the desertification estimates problematic because Lamprey and Skoupy did not put into consideration the rainfall decline of about 50% that occurred during the period of assessment (Nicholson, 1990). Thus, while 'desertification' itself was defined as anthropogenic, evidence shows that such land degradations could have been a product of climatic variability.

In a similar vein, a group of scientists from the University of Lund, Sweden, carried out studies in Sudan in 2001 and revealed through a combination of satellite images and field work that there was neither a systematic advancement of the desert or other vegetation zones, nor a reduction or disappearance of vegetation cover, although the replacement of forage with woody species was observed which indicated declining soil quality (Nicholson, 2001). They reported no evidence of wide-spread removal of vegetation cover in the villages. However, the study revealed that vegetation changes was as a result of drought and noted that there was a full recovery of the land as soon as the drought ended. Therefore, the researchers added that the perceived 'advances of the desert' was as a result of fluctuations in rainfall pattern in the Sahelian region with no large-scale reduction or disappearance of biological entities in the region (Prince et al., 1998) which is in support of Tucker et al (1991) propositions. In other words, it might be technically wrong to refer to the environmental changes in the Sahelian as a product of desertification since these changes are not irreversible neither do they lead to the sterility of land. A better term could be 'land degradation' which explains the replacement of diverse and nutrient-rich plant species by vegetation of poorer quality (Nicholson, 2001) due to reduced soil quality caused by drought.

Many studies (Bonan et al., 1992; Xue and Shukla, 1993; Zhang et al., 1996) have shown that the climate is altered when deforestation (mostly caused by human activities) occurs, due to land-atmosphere feedbacks. Charney (1975) and Garratt (1993) reported that rainfall over a semi-arid region may be affected if there is a change in the surface albedo (the reflectivity of solar radiation). Charney (1975) argued that bare soils have a higher albedo than vegetation surfaces, therefore, a decline in rainfall following increased deforestation is a possibility. This is because an increase in albedo can reduce the heat flux within the troposphere (atmosphere closest to the Earth's surface) thereby, weakening convection mechanisms and limiting cloud formation and precipitation (Hoffmann and Jackson, 2000). As a result, an increase in albedo may promote a positive feedback mechanism leading to a drier environment, unable to sustain natural vegetation recovery and may cones-quently cause prolonged drought (Wang and Eltahir, 2000). Similarly, Woodward (1987), Zeng et al. (1999) and Taylor et al. (2002) reported that land-atmosphere interactions may cause inter-decadal climate variability in the semi-arid regions of the world.

Zheng and Eltahir (1997) reported that the West African monsoon which fed rain to the Sahel regions of Africa is sensitive to vegetation removal. They added that changes in land use such as the overgrazing, agricultural expansion and increased fuel-wood extraction may collectively alter the behaviour of the West African monsoon over time. Clark et al. (2001) pointed out that the climate in semi-arid regions is very sensitive to land degradation whilst Nicholson et al. (1998) referred to the region as the most "ecologically unstable" in the world. The ecological instability of the Sahel and Savannah regions was explained by Zeng and Neelin (2000) who revealed that these "grassland ecosystems" try to smoothen out the large climatic variability between the desert and the rainforest, thus creating a state of unsteadiness in-between. Nevertheless, human population in the region is doubling every 20 years which has caused a proportional increase in demand for agroforestry resources (Taylor et al., 2002), therefore an accelerated loss of vegetation cover from anthropogenic sources may increase the albedo of the area, thus exacerbate climatic conditions. However, Taylor et al. (2002) argued that rainfall conditions have been generally poor since the 1960s and this would have caused the loss of vegetation cover by biophysical feedbacks even without changes in land use which are not large enough to cause the persistently low rainfall observed in the region over the decades. Taylor et al. (2002) argument may not hold water because it was reported that there was persistent drought between the 1950s and 1970s (Nicholson, 2001) but rainfall improved in the late 1980s

(Tucker et al., 1991) restoring the native vegetation of the region. Therefore, there should not be a progressive loss of vegetation as a result of a temporary drought that occurred many decades ago.

CONCLUSION AND RECOMMENDATIONS

This study reviewed the human-related and climatic factors inducing land degradation in northern Nigeria. Some authors believe that the human-related activities such as agricultural/pastoral expansion and agricultural intensification are localised adaptation measures by the natives of northern Nigeria in response to the changing climatic condition of the area, an idea referred to as the 'optimists view'. Others believe that the so-called adaptation measures are further destroying the integrity of the ecosystem, an idea referred to as the 'pessimists view' (Batterbury and Warren, 2001). Various studies in literature show that neither the optimists view nor the pessimists view can be totally discarded. Only when the uncertainties in the arguments presented by both parties are well-clarified with supporting evidences, will a common ground be found.

Conflicts between pastoralists and crop farmers may be reduced by the siting of more boreholes or waterpoints in the region by granting Northern states experiencing such crisis, a special federal-supported water project fund as a result of the political structure of the country and the financial limitations of most Northern states. Ime (2013) recommended 'agro-pastoralism' (a form of farming that combines the growing of crops with the rearing of livestock) as a way of managing pasture lands more effectively, thereby, reducing pastoral expansion and face-off with crop farmers.

Owing to the various debates on the cause of land degradation in Africa, it would be imperative for further studies to be carried out. The following controversial questions are to be further researched for the purpose of clarity of knowledge and a better understanding of land degradation in Nigeria and West Africa:

1. Is the perceived creeping of the Savannah in Nigeria, a result of either climatic variations which could be reversed during better weather conditions or human-related activities which some authors claimed are insufficient to cause large-scale land degradation?
2. It has been suggested that the African Monsoon and the InterTropical Convergence Zone (ITCZ) may be responsibile for the rainfall variablility in West Africa (Nicholson, 2001), therefore, what are the elements in these meteorological entities forcing the dryness and wetness of the region?
3. What is the extent of land recovery (whether total or partial) of the West African region each time rain is restored?

It is important for the inter-decadal measurement of the current Savannah-Rainforest boundary as well as the current Saharan-Sahelian boundary to be carried out as this would help project a possible expansion or retreat of Nigeria's Savannah. The proper management of land and its resources including the full understanding of the climatic forces at work, may go a long way to help Northern Nigeria reach its full potential as a pace-setter for agricultural revolution in the country. The State Governments in Northern states are advised to utilize the potentials in agriculture as a reliable route to economic emancipation.

REFERENCES

Amzat A (2013). Menace of herdsmen: Like Sudan like Nigeria. The Guardian Nigeria. Published Saturday 27th April, 2013.

Aregheore EM (2005). Country pasture/forage resource profiles: Nigeria. www.fao.org/ag/GP/AGPC/doc/Counprof/nigeria/nigeria.htm. Updated May/June, 2005. Accessed on the 10th May, 2013.

Badejo MA (1998). Agroecological restoration of savannah ecosystems. Ecol. Eng. 10:209-219.

Barbier EB (1999). The economics of land degradation and rural poverty linkages in Africa *In* Proceedings of the UNU/INRA Annual lecture on natural resource conservation and management in Africa, United Nations University press, Tokyo, Japan.

Barbier EB (2000). The economic linkages between rural poverty and land degradation: Some evidence from Africa. Agric. Ecosyst. Environ. 82:355-370.

Batterbury S, Warren A (2001). The African Sahel 25 years after the great drought: Assessing progress and moving towards new agencies and approaches. Glob. Environ. Change 11:1-8.

Bonan GB, Pollard D, Thompson SL (1992). Effects of boreal forest vegetation on global climate. Nature 359:716-718.

Breman H, Rob-Groot JJ, van Keulen H (2001). Resource limitations in Sahelian Agriculture. Global Environ. Change 11:59-68.

Bremen H, De wit CT (1983). Rangeland productivity and exploitation in the Sahel. Science 221 (4618): 1341-1347.

Charney JG (1975). Dynamics of deserts and drought in the Sahel. Quart. J. Roy. Meteor. Soc. 101:193-202.

Clark DB, Xue YK, Harding RJ, Valdes PJ (2001). Modeling the impact of land surface degradation on the climate of tropical northern Africa. J. Clim. 14:1809-1822.

Cleaver KM, Shreiber AG (1994). Reversing the spiral: The Population, Agriculture and Environment nexus in sub-Sahara Africa. The world bank, Washington DC, USA.

FOS (1989). Annual report of the Federal Office of Statistics (1989), Nigeria.

Garratt JR (1993). Sensitivity of climate simulations to land-surface and atmospheric boundary-layer treatments - A review. J. Clim. 6:419-449.

Gefu JO, Gills JE (1990). Pastoralist ranchers and the states in Nigeria and North Africa: A comparative analysis. Nomad. People 25 - 27: 34 - 49

Graetz RD (1991). Desertification: A tale of two feedbacks In: Mooney HA *et al.* (eds) Ecosystem experiments. John Wiley and Sons, Chichester, pp. 59-87.

Hekstra GP (1985). Productivity and land use in semi-arid regions - lessons from the Sahel. Ecoscript 27. Foundation for ecological development alternatives (Stiching mondiaal Alternative). The Netherlands.

Hellden U (1984). Drought impact monitoring: a remote sensing study of desertification in Kordofan, Sudan. Lunds Universitets, Naturgeografiska Institution, Avhandlingar, Sweden.

Hoffmann WA, Jackson RB (2000). Vegetation-climate feedbacks in the conversion of tropical savannah to grassland. J. Clim. 13:1593-1602.

Honlonkou AN, Manyong VM, Tchetche N (1999). Farmers perceptions and the dynamics of adoption of a resource management technology: The case of *Mucuna* fallow in Southern Benin, West Africa. Int. For. Rev. 1:228-235.

Ime E (2013). Resolving the conflicts involving cattle herdsmen. 2nd

Asian-Nias Joint Annual meeting, September 8-12, 2013. Abuja, Nigeria.
Iortim HT (2012). How North can surmount the poverty scourge. Retrieved from http://blueprintng.com/2012/03/how-north-can-surmount-the-poverty-scourge/. Uploaded March 22, 2012. Accessed January 23, 2014.
Kang BT (1993). Sustainable agroforestry systems for the tropics: Concepts and examples. IITA research guide 26:24.
Krogh L (1997). Field and nutrient balances in millet cultivation in northern Burkina Faso: A village level case study. J. Arid Environ. 35(6):147-159.
Lamprey H (1975). Report on the desert encroachment reconnaissance in Northern Sudan. National Council for Research/Ministry of Agriculture, Food and Natural Resources, Khartoun, Sudan.
Lanly JP (1982). Tropical forest resources. FAO Forestry paper, 30 FAO, Rome, Italy. pp. 106.
Mortimore MJ, Adams WM (2001). Farmer adaptation, change and 'crisis' in the Sahel. Glob. Environ. Change 11:49-57.
Nicholson SE (1990). The need for a reappraisal of the question of large-scale desertification: some arguments based on consideration of rainfall fluctuations. Report of the SAREC-Lund International Meeting on Desertification in Lund University, Sweden.
Nicholson SE (2001). Climatic and environmental change in Africa during the last two centuries. Clim. Res. 17:123-144.
Nicholson SE, Tucker CJ, Ba MB (1998). Desertification, drought and surface vegetation: An example from the west African Sahel. Bull. Amer. Meteor. Soc. 79:815-829.
Nobre CA, Sellers PJ, Shukla J (1991). Amazonian deforestation and regional climate change. J. Clim. 4:957-988.
Odebode N, Alechenu J, Soriwei F, Owuamanam J, Aborisade S, Attah D (2012). Senators fight over grazing land for Fulani herdsmen. Punch Newspaper, Nigeria. Published on the 21st July, 2012.
Okpara JN, Tarhule AA, Perumal M (2013). Study of climate change in Niger River Basin, West Africa: Reality not a Myth in climate change-realities, impacts over ice cap, sea level and risks (Chapter 1), BR Singh (Ed.). January 16, 2013. ISBN: 978-953-51-0934-1.
Pate UA, Dauda S (2013). Media and Socio-economic development in Northern Nigeria. Malay. J. Commun. 29 (1):1-19.
Pieri C (1989). Fertilite des terres de savanes. Bilan de trente ans de recherche et de deHveloppement agricol au Sud du Sahara. Ministe`re de la CoopeHration et CIRAD-IRAT, Montpellier, France.
Prince SD, Brown de Colstoun E, Kravitz LL (1998) Desertification of the Sahel? Evidence from remotely sensed rainuse efficiencies. Glob. Change Biol. 4:359-374
Puigdefabregas J (1995) Desertification: stress beyond resilience, exploring a unifying process structure. Ambio 24:311-313
Salako FK (2003). Soil physical conditions in Nigerian savannas and Biomass production. Department of Soil Science and Agricultural Mechanisation, University of Agriculture, Abeokuta, Nigeria. Lecture given at the college of soil physics, Trieste, Italy, 3-21 March, 2003.
Sanginga N, Vanlauwe B, Danso SKA (1995). Management of biological N_2 fixation in alley cropping systems: Estimation and contribution to nitrogen balance. Plant Soil 174:119-141.
Schaefer C (1998). Pastorale Wiederkauerhal-tung in der sudan-savanne: Eine untersuchung im zamfara forstschutzgebiet im Nordwesten Nigerias. Cuvillier, Goettingen, The Netherlands. pp. 217.
Shettima AG, Tar UA (2008). Farmer-Pastoralist Conflict in West Africa: Exploring the Causes and Consequences. Info. Soc. Justice 1(2):163-184
Skoupy J (1987). Desertification in Africa: Agricultural and meteorological program. Proc. Regional training seminar on drought and desertification in Africa, Addis Ababa, WMO 33-45.
Taylor CM, Lambin EF, Stephenne N, Harding RJ, Essery RLH (2002). The influence of land use change on climate in the Sahel. J. Clim. 15:3615-3629.
Tiffen M, Mortimore MJ, Gichuki F (1994). More people, Less Erosion: Environmental Recovery in Kenya. Wiley, Chichester, United Kingdom.
Traore M (1996) Problems In Pastoral Land Management Related To Tenure: Policies Versus Basis Practice. In: Managing Land Tenure And Resource Access In West Africa GRET and IIED (eds). Proceedings of a Workshop held in Goree, Senegal.
Tucker CJ, Dregne HE, Newcomb WW (1991). Expansion and contraction of the Sahara Desert from 1980 to 1990. Science 253:299-301.
UNDP (2010). Nigeria human development index, 2007-2009. Abuja: United Nations Development Programme (UNDP). Abuja, Nigeria.
UNESCO (2010). Education for all (EFA) global monitoring report 2010: Reaching the marginalised. Paris: United Nations Educational, Scientific and Cultural Organisation (UNESCO), Paris, France.
Wang GL, Eltahir EAB (2000). Ecosystem dynamics and the Sahel drought. Geophys. Res. Lett. 27:795-798.
Warren A (1996). Desertification; In: Adams WM, Goudie AS, Orme AR (Eds.), The Physical Geography of Africa. Oxford University Press, Oxford, United Kingdom. pp. 342-355.
Woodward FI (1987). Cambridge studies in ecology: Climate and plant distribution. Cambridge University press, United Kingdom. pp. 174. ISBN: 0-521-28214-4.
WRI (1992). World resources (1992-1993). World Resources Institute (WRI), Oxford University, press, New york, USA.
Xue Y, Shukla J (1993). The influence of land surface properties on Sahel climate. Part I: Desertification. J. Clim. 6:2232-2245.
Zeng N, Neelin JD (2000). The role of vegetation-climate interaction and interannual variability in shaping the African savannah. J. Clim. 13:2665-2670.
Zeng N, Neelin JD, Lau KM, Tucker CJ (1999). Enhancement of inter-decadal climate variability in the Sahel by vegetation interaction. Science 286:1537-1540.
Zhang H, Henderson-Sellers A, McGuffie K (1996). Impacts of tropical deforestation. Part I: Process analysis of local climate change. J. Clim. 9:1497-1517.
Zheng XY, Eltahir EAB (1997). The response to deforestation and desertification in a model of west African monsoons. Geophys. Res. Lett. 24:155-158.

Natural radioactivity and environmental risk assessment of Sokoto phosphate rock, Northwest Nigeria

Kolo, M. T.

Department of Physics, Federal University of Technology, Minna, Niger State, Nigeria.

Phosphate rock samples collected from the phosphate bearing formations of Sokoto State, Northwest Nigeria were assessed for their radiological contents by gamma spectrometric technique (sodium iodide (NaI (Tl) detector). The mean activity concentrations of ^{226}Ra, ^{232}Th and ^{40}K in all the studied samples were 720.1±4.2 $Bqkg^{-1}$, 33.5±1.4 $Bqkg^{-1}$, and 315.3±6.7 $Bqkg^{-1}$ respectively. Contributions from radium contents to the overall radionuclide activity concentrations were quite significant. The calculated mean value of the absorbed dose rate was found to be 3.6 times the maximum value for the world background level, which shows the severity of radiation risk the public is exposed to in the study areas. The computed values for radiation hazard indices were above the world's proposed safety limits by UNSCEAR, which place the study areas among high background radiation zones. Some restrictive measures and precautionary actions are therefore obligatory for the local farmers and the general public from the view point of radiation protection.

Key words: Natural radioactivity, Sokoto, phosphate rock, radiation hazard, NaI (Tl) detector, radiation dose.

INTRODUCTION

Phosphate rocks, which can be of sedimentary, volcanic or biological origin (Sam and Holm, 1995), generally refer to rocks with high concentration of phosphate minerals (Cevik et al., 2010). Phosphate rocks are the starting raw material for all phosphate products and fertilizer production (Khater et al., 2001; Ogunleye et al., 2002). They are also used in the preparation of food and animal feeds, and as a source of phosphorus for agricultural purposes as well as for the manufacturing of certain industrial phosphorus-based chemicals and detergents (Abbady et al., 2005; Khan et al., 1998). Phosphate rocks contain relatively high amounts of naturally occurring radioactive materials (NORMs) from ^{238}U and ^{232}Th and their respective descendants, and ^{40}K in concentrations that depend on geographical and geological origin. Though the concentrations of NORMs observed in phosphate rocks of all types are similar to those observed normally in soils, uranium and its decay products tends to occur at elevated concentrations in sedimentary phosphates (Harb et al., 2008; Khater et al., 2001). This uranium enrichment makes phosphates one of the major sources of technologically enhanced natural radiation (TENORM), (Abbady et al., 2005), which contributes to human exposure. It is widely believed that the radioactivity associated with phosphate rocks of sedimentary origin is formed by the adsorption and co-precipitation of uranium with calcium (Khater et al., 2001). Phosphate ores are found to occur as phosphorites, $[Ca_3(Po_4)_2]$, which are very old marine deposits associated with fossils; and apatite, $Ca_5[(Po_4)_3(F)]$, which is of igneous origin (Schmidt, 1993; OSPARCOM, 1997; Abbady et al., 2005). van Straaten (2002), reported that sedimentary marine phosphate rock deposits accounts for about 75% of the world's phosphate source, which according to Okosun (1989), makes it a great economic base that provides most of the raw materials for the manufacture of phosphate fertilizers and phosphorus based chemicals

Mining and processing of phosphate rock and its use in the manufacturing industry, redistributes the radioactive trace elements and ^{238}U and its decay products throughout the environment and also in the final products, thereby enhancing the concentrations of natural radionuclides in the end products and wastes of the phosphate industry. This constitutes ways in which the workers, the public and the environment are exposed to enhanced radioactivity (IAEA, 2004; UNSCEAR, 1993). Significant radiological investigations have been conducted on phosphate rocks and their products in different countries of the world, which has led to establishment of standards and regulations for the protection of man and his environment from the radiation effects that will arise due to the usage of the rock. Sam et al. (1999) assessed the radiological and chemical constituents of Uro and Kurun rock phosphates from Sudan. Khater et al. (2001), investigated the radiological impacts of natural radioactivity in Abu-Tartor phosphate deposits, Egypt. Saueia et al. (2005) undertook similar investigation for phosphate rock, phosphogypsum and phosphate fertilizers in Brazil. Abbady et al. (2005), studied the natural radioactivity and dose assessment for phosphate rocks from Wadi El-Mashash and El-Mahamid Mines in Egypt. Cevik et al. (2010), reported the radiological, structural and chemical characteristics of Mardin-Mazidagi phosphate rock in Turkey.

As a result of increasing interest by the Nigerian local farmers in direct application of phosphate rocks to agricultural soils as a substitute for and ready alternative to chemical fertilizer, which is becoming increasingly expensive and out of reach, it becomes important to assess the potential radioactivity exposures of phosphate rocks.

In the present work, gamma spectrometric technique is used to quantify the activity concentrations of ^{226}Ra, ^{232}Th and ^{40}K of Sokoto phosphate rocks. The human radiation exposure from the radioactive content of phosphate rock samples was estimated, along with possible environmental implications of direct application of Sokoto phosphate rocks for agricultural purposes.

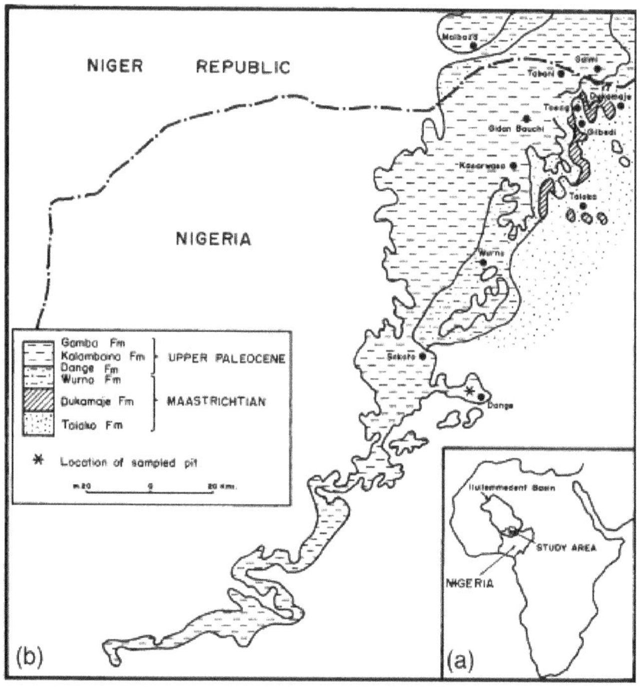

Figure 1. (a) Map of Africa, showing the Sokoto sub-basin, Northwest Nigeria I relation to the Iullemmenden basin. Inset: Location of (b). **(b)** Simplified geological map of the southeastern sector marine and marginal facies of Illemmenden basin, Sokoto sub-basin (Okosun, 1997).

Sokoto phosphate deposit

The Sokoto basin, Northwest Nigeria is Nigeria's sector of extensive Iullemeden Basin (Figure 1). A phosphate deposit of great significance is located in Sokoto state, northwestern Nigeria, in a paleocene sedimentary sequence. Marine sediments were deposited in Sokoto basin during the maastrichtian-paleocene trans-Saharan transgressions.

The phosphates, whose basic components are from vertebrate remains, are present in the Dange formation which from Paleocene age occur as nodular beds and nodular disseminations in the shales and siltstone of the formation. The Dange formation is phosphate bearing in the entire basin though the concentration differs from place to place. Phosphate rocks are found in commercial deposits in Dukamaje, Dange, Gilbadi, Dankilo-Dange, Wurno, Kindiru, Dillingu and Gada in Sokoto State. Ogunleye et al. (2002) also reported occurrence of phosphate nodules and pellets in Gidan Bauchi, Illela and Kalambiana. Etu-Efeotor (1998), Okosun (1997) and Ogunleye et al. (2002), reported that phosphate deposit thickness in Dukamaje formation range between 1 to 5 m,

Table 1. Mean activity concentrations of ^{226}Ra, ^{232}Th and ^{40}K ($Bqkg^{-1}$) in Sokoto phosphate rock samples.

Sample ID	Number of samples	Activity concentrations ($Bqkg^{-1}$)		
		^{226}Ra	^{232}Th	^{40}K
DNG	3	650.7±3.8 (602.5 - 732.0)	35.7±1.2 (17.0 - 58.6)	264.3±6.4 (111.2 - 392.3)
KDR	3	1073.2±5.9 (1055.7 - 1087.3)	55.9±1.7 (39.2 - 85.3)	380.3±8.3 (310.3 - 459.0)
DLG	3	804.5±3.7 (854.7 - 883.6)	28.9±1.0 (17.7 - 40.8)	394.6±5.6 (375.8 - 428.4)
WRN	3	510.1±4.1 (443.7 - 617.8)	19.5±1.8 (13.2 - 27.7)	272.2±7.7 (202.7 - 362.8)
DNK	3	562.0±3.3 (491.0 - 606.3)	27.4±1.3 (24.7 - 32.2)	265.2±5.3 (110.7 - 404.7)
Av. range	15	720.1±4.2 (443.7 - 1087.3)	33.5±1.4 (13.2 - 85.3)	315.3±6.7 (111.2 - 459.0)

while the phosphate nodules/pellets occur in sizes of 0.1-1cm. Amapu (1998), investigated the agronomic potential of direct application of Sokoto phosphate rock as an alternative phosphate fertilizer for Nigeria's sub-humid Savannah region. Phosphate nodules from Sokoto basin are presently mined superficially by the natives especially women and children during the non-farming season and transported, as a major raw material to feed the Crystal Talc Nigeria, Kagara, Niger State and Federal Superphosphate Fertilizer company plant at Kaduna, Kaduna State, for the production of phosphate fertilizer. They are also blended by the locals and applied directly in the agricultural farms as substitute for fertilizer.

MATERIALS AND METHODS

The Sokoto phosphate deposits occur generally as nodules on the surface, with nodular beds measuring to between 0.2-0.3 m. The phosphate nodules are superficially mined on daily basis mostly by women and children. To measure the natural radioactivity, a total of 15 phosphate rock samples, three each, were collected from five villages which represents the Sokoto phosphate bearing formations. The villages are Dange (DNG), Dankilo (DNK), Wurno (WRN), Kindiru (KDR) and Dillingu (DLG). Phosphate nodules were picked at random from four different points and mixed together thoroughly to give a true representation of each sampling point throughout all the sample locations. The samples, each about 1 kg in weight, were packed in well labelled sampling polyethylene bags, properly tied and sealed to avoid cross contamination of samples, and transported to the laboratory for analysis at the Center for Energy Research and Training (CERT), Ahmadu Bello University, Zaria, Kaduna State.

In the laboratory, the samples were opened and all weeds, stones and other organic and non-phosphate materials, were picked out. The samples were sundried under ambient temperature to almost constant weight for 72 h. The dried samples were crushed into fine powder and sieved through a 2-mm mesh. The dried samples were homogenized, weighed and packed into identical 7.2 cm diameter and 6.0 cm height radon impermeable marinelli beakers which holds between 300-350g of sample. The beakers were then subjected to a triple stage sealing process to ensure the containment of radon-222 (Jamok, 2014), and stored at room temperature for a period of 5 weeks to allow for radioactive equilibrium between the parents and their daughter nuclides (Amin et al., 2013; Ghose et al., 2012; Ibeanu, 2002; Kumar et al., 2003).

The radiometric analysis of the samples was carried out using a gamma spectrometric technique comprising a 7.62 x 7.62 cm NaI (Tl) detector coupled to a computer based multichannel analyzer (MCA) for data acquisition. To effectively reduce the gamma-ray background radiation from the acquired spectrum, the detector is housed in a 6 cm thick lead shield lined with cadmium and copper sheets. The system was set at operating energy range of 0-3000 keV and energy resolution of 661.6 keV peak of ^{137}Cs. For accurate quantitative determination of ^{226}Ra, ^{232}Th and ^{40}K in the samples, the detector calibration is necessary to ensure a good relationship between the respective peak positions of the spectrum and their corresponding gamma-ray energies. RGU-1, RGTh-1 and RGK-1 gamma spectrometric reference materials (gamma sources of known energies) supplied by the International Atomic Energy Agency (IAEA), whose activity is distributed homogenously within the same volume and container geometry as that of the samples, was used for the energy calibration and absolute photopeak efficiency evaluation of the detector. Each sample was counted for 29,000 s. Knowing that through the period of sample storage, uranium attained circular equilibrium with its daughters, the gamma energy line of 1764 keV (15.6%) for ^{214}Bi used to represent the activity concentration of ^{226}Ra, while the activity concentration of ^{232}Th was evaluated from the average activity of photopeaks of 911.2 keV (25.8%) for ^{228}Ac and 238.6 keV (43.6%) for ^{212}Pb. The ^{40}K activity was determined from its 1460.8 keV (10.66%) single characteristic gamma line. An empty sealed beaker in the same geometry as the samples was first counted in the same manner for 29,000 s to determine the radiation background around the detector environment. The obtained background was used to correct the net peak area of gamma rays of measured radionuclides (Uosif and El-Taher, 2008). The activity concentrations were computed using the equation (Cevik et al., 2010; Khandakeret al., 2012):

$$A(Bq/kg) = \frac{N}{\varepsilon_\gamma \times I_\gamma \times M \times t} \quad (1)$$

Where A is the activity concentration of the nuclide, N is the total net count of a specific gamma emission, ε_γ is the detection efficiency for the specific gamma emission, I_γ is the emission probability, M is the mass of the sample (kg) and t is the counting time.

RESULTS AND DISCUSSION

Phosphate samples collected from the villages visited were analysed for their natural radioactivity concentrations using gamma ray spectrometry. The measured specific activity values for ^{226}Ra, ^{232}Th and ^{40}K along with their statistical uncertainties are summarized in Table 1. These values are actually taken as representing

Table 3. Activity concentration of ^{226}Ra, ^{232}Th and ^{40}K ($Bqkg^{-1}$) and Ra_{eq} ($Bqkg^{-1}$) in phosphate samples from different countries.

Country	Activity concentration ($Bqkg^{-1}$)			Ra_{eq} ($Bqkg^{-1}$)	Reference
	^{226}Ra	^{232}Th	^{40}K		
Pakistan (Hazara)	440	50	207	527.4	Khan et al. (1998)
Egypt (El-Mahamid)	567	217.3	217.3	921	Abbady et al. (2005)
Egypt (El-Mashash)	666	329.4	329.4	1182	Abbady et al. (2005)
Tanzania	5022	717	286	6069	Makweba and Holm (1993)
Algeria	619	64	22	712	Olszewska-Wasiolek (1995)
Sudan (Kurun)	393	6.9	141.3	414	Sam et al. (1999)
Tunisia	250	28	167	303	Cevik et al. (2010)
Nigeria (Sokoto)	720.1	33.5	315.3	792.3	Present work

the NORMs level in Sokoto phosphate rocks. The activity concentration levels for NORMs in phosphate rock samples from the five villages range from 44.37 to 1087.3 $Bqkg^{-1}$, with a mean value of 720.1±4.2 $Bqkg^{-1}$ for ^{226}Ra, from 13.2 to 85.3 $Bqkg^{-1}$ with an average value of 33.5±1.4 $Bqkg^{-1}$ for ^{232}Th and from 111.2 to 459.0 $Bqkg^{-1}$ with a mean value of 315.3±6.7 $Bqkg^{-1}$ for ^{40}K. The results show that the activity levels are higher for phosphate rocks from Kindiru (KDR) village. Generally however, the ^{226}Ra activity concentration values are higher than those of ^{232}Th and ^{40}K indicating that Sokoto phosphates are of sedimentary origin. It can be clearly seen that ^{238}U and its daughters plays a prominent role in contributing to the overall radioactivity content of phosphate rocks (Abbady et al., 2005; Khater et al., 2001). Although the values of radium activity concentration varies within the phosphate rock samples from the five villages, which according to Guimond (1990) and Ivanovich and Harmon (1992), may be due to uranium solubility under oxidation conditions and the percentage of P_2O_5 present, they are generally high. This can pose external radiation risk especially to women and children who mined this commodity on daily basis.

Table 3 give a comparison of activity concentration of natural radioactivity of phosphate rocks from different parts of the world. It can be seen from this table that the activity concentration of ^{226}Ra in this present study is relatively higher than those reported for many countries, except for Tanzania, while the average activity concentration value for ^{40}K falls within the same range as that reported in literature, except for Algeria that is relatively lower. The activity concentration value of ^{232}Th reported for Egypt, Tanzania and Algeria are however higher that that reported in this present study.

Numerous propositional criteria abound in literature for estimating environmental and human hazard due to radiation exposure (Bashir et al., 2013; Sabiha-Javied et al., 2010; Tufail, 2012). ^{226}Ra, ^{232}Th, and ^{40}K are not distributed uniformly within the soils and other environmental samples. A common radiological index, the radium equivalent activity (Ra_{eq}), which therefore, compare the activity concentrations of the above three radionuclides and their associated radiation hazards, has been defined. This definition is consequent upon an assumption that 370 $Bqkg^{-1}$ of ^{226}Ra, 259 $Bqkg^{-1}$ of ^{232}Th and 4810 $Bqkg^{-1}$ of ^{40}K produce the same gamma dose rate (Abbady, 2005; Cevik et al., 2010; Saueia et al., 2005). Ra_{eq} is calculated using the equation (Beretka and Mathew, 1985; Yu et al., 1992):

$$Ra_{eq}\ (Bqkg^{-1}) = A_{Ra} + 1.43A_{Th} + 0.077A_K \quad (2)$$

Where, A_{Ra}, A_{Th} and A_K are the respective activities of ^{226}Ra, ^{232}Th and ^{40}K in $Bqkg^{-1}$.

The total air absorbed dose rate, D ($nGyh^{-1}$), 1.0 m above the ground, from outdoor external exposure to natural radioactivity for the population living in the phosphate bearing areas is calculated using the formula (UNSCEAR, 2000):

$$D = 0.462A_{Ra} + 0.604A_{Th} + 0.0417A_K \quad (4)$$

Where, A_{Ra}, A_{Th} and A_K are the respective average activity concentrations ($Bqkg^{-1}$) of Ra, Th, and K in the phosphate rock samples.

Jibiri et al. (2007), reported that absorbed dose rates does not give a direct representation of exposure risk of an individual. To estimate the annual effective dose equivalent to an individual from outdoor terrestrial gamma radiation, UNSCEAR (2000), proposed two conversion factors: 0.7 $Sv.Gy^{-1}$ as conversion coefficient from absorbed dose in air to effective dose received by adults, and 0.2 as outdoor occupancy factor, signifying that 80% of time is spent indoors across the world. The annual effective dose equivalent (AE) is estimated using the equation (UNSCEAR, 2000):

$$AE\ (mSvyr^{-1}) = D \times T \times \xi \quad (5)$$

Where, D ($nGyh^{-1}$) is the absorbed dose rate, T(= 0.2 x 24hr x 365) is the outdoor occupancy time, and ξ (= 0.7 x 106 $SvGy^{-1}$) is the conversion factor.

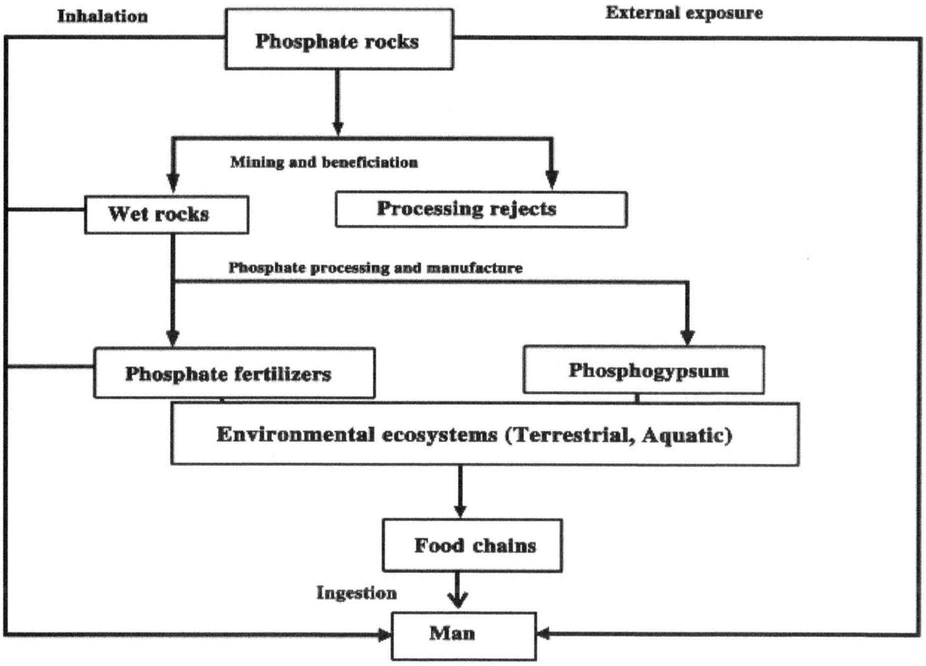

Figure 2. Environmental pathways of natural radionuclides from phosphate rocks.

Table 2. Mean values of total absorbed dose rate (D) and annual effective dose equivalent (AE) of Sokoto phosphate rock sample.

Sample ID	Ra(eq) (Bq/kg)	D (nGy/h)	AE (mSv/yr)
DNG	722.2 (656.3-778.9)	334.5 (303.2-361.5)	0.4 (0.4-0.4)
KDR	1182.4 (1135.6-1234.0)	547.5 (526.1-570.6)	0.7 (0.6-0.7)
DLG	876.1 (748.3-942.3)	406.9 (347.9-436-9)	0.5 (0.4-0.5)
WRN	558.9 (502.6-664.6)	259.6 (233.1-309.4)	0.3 (0.3-0.4)
DNK	621.7 (545.6-672.8)	288.3 (252.0-312.9)	0.4 (0.3-0.4)
Av. Range	792.3 (502.6-1234.0)	367.3 (233.1-570.6)	0.4 (0.3-0.7)

The two basic environmental pathways by which radiation dose from phosphate rocks are imparted to the general public and phosphate workers in particular are the external and internal exposures represented in Figure 2 (Khater et al., 2001).

The calculated results for radium equivalent activity (Ra_{eq}), total absorbed dose rate in air (D) and the annual effective dose equivalent (AE) are presented in Table 2. Phosphate rock samples from Kindiru (KDR) recorded the maximum value of Ra_{eq} activity (1182.4 $Bqkg^{-1}$), while phosphate samples from Wurno (WRN) had the minimum Ra_{eq} activity value of 558.9 $Bqkg^{-1}$. The mean value of Ra_{eq} across the entire study areas is 792.3 $Bqkg^{-1}$ which is higher than the safety limit of 370 $Bqkg^{-1}$ recommended by the Organization for Economic Cooperation and Development (NEA-OECD, 1979).

The total absorbed doses calculated from the measured activities of phosphate samples varies from 259.6 $nGyh^{-1}$ for WRN samples to 547.5 $nGyh^{-1}$ for KDR samples, with a mean value of 367.3 $nGyh^{-1}$. These high dose rate values witnessed across the entire study areas may not be unconnected with the high radium contents of the phosphate rocks. The mean value of the total absorbed dose rate as seen in Table 2 is almost 3.6 times the maximum value of the world's soil background level of 102 $nGyh^{-1}$ proposed by UNSCEAR (2000), which shows the severity of radiation risk the inhabitants of the study areas are exposed to. The computed values for the annual effective dose equivalent in the studied phosphate rock samples as presented in Table 2 varies from 0.3 $mSvyr^{-1}$ for WRN sample to 0.7 $mSvyr^{-1}$ for KDR samples, with a general mean value of 0.4 $mSvyr^{-1}$. These values are found to be lower than the safety limit of 1 $mSvyr^{-1}$ recommended by the International Commission on Radiological Protection (ICRP-60, 1990). A plot of radium equivalent activity and average dose rates with their res-

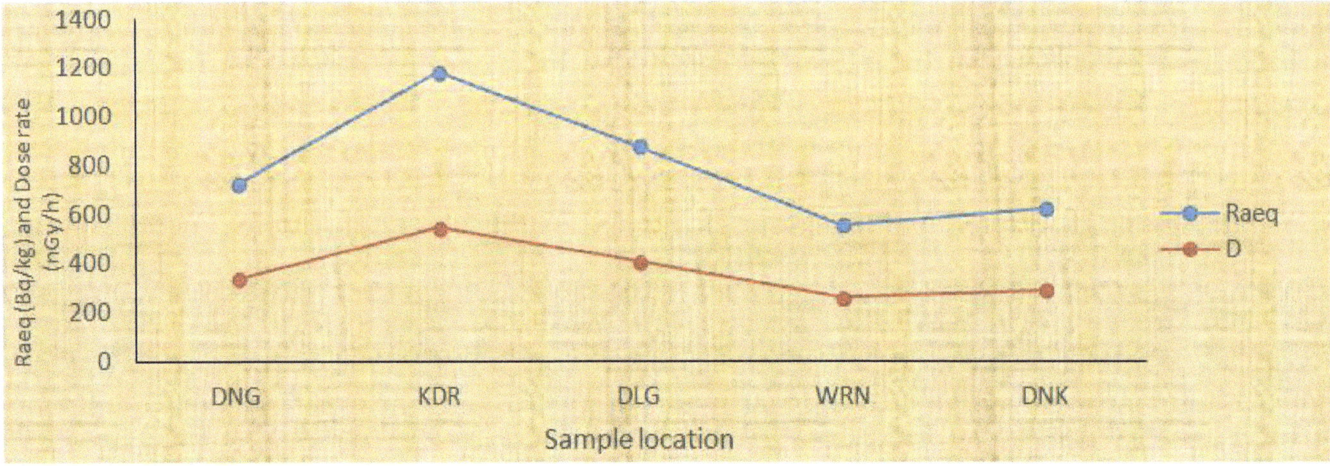

Figure 3. Radium equivalent and average dose rates for phosphate rock samples from different locations.

pective sample locations are shown in Figure 3. These values are higher than the respective recommended safety limits, which presents a serious radiation risk to the inhabitants of the study areas, especially when the untreated phosphate rocks are used generally by the local farmers as substitute for fertilizer. Women and children within the study areas are at greater radiation risk from internal dose due to inhalation of radon and dust during mining and blending of phosphate rocks for direct application to plants. Detail study of this exposure pathway is therefore recommended so as to estimate the occupational radiation dose in these phosphate bearing areas of Sokoto state and to enforce restrictions and precautionary measures where necessary, from the view point of radiation protection.

Conclusion

The results of this preliminary radiological investigation of Sokoto phosphate rocks clearly demonstrates the radiological burden an average member of phosphate bearing communities are exposed to. The obtained results show that ^{238}U and its decay daughters are the primary contributors to the radioactivity level of phosphate bearing formations of Sokoto State. The computed values of external radiation exposure from natural radioactivity of phosphate rocks show that the radioactivity of Sokoto phosphate rock samples is above the proposed safety limit of 1 (370 $Bqkg^{-1}$), which demonstrates the severity of radiation risk the local farmers are exposed to especially when using the untreated ground phosphates as fertilizer. It is also obvious that phosphate industry, among other geological materials' processing industries, play a significant role in the radiation dose received by the public. The data obtained from this investigation underscores the need for in-depth radiological assessment of all the phosphate bearing communities in Sokoto State. Underground water supplies and local building materials from these communities should be investigated for their contribution to the radiation dose incurred by the general public.

Conflict of Interests

The author(s) have not declared any conflict of interests.

ACKNOWLEDGEMENTS

I like to acknowledge the support of the Director and the entire staff of Centre for Energy Research and Training (CERT), Ahmadu Bello University, ABU, Zaria, Nigeria.

REFERENCES

Abbady A (2005). Assessment of the natural radioactivity and its radiological hazards in some Egyptian rock phosphates. Indian J. Pure Appl. Phys. 43(7):489-493.

Abbady AGE, Uosif M, El-Taher, A (2005). Natural radioactivity and dose assessment for phosphate rocks from Wadi El-Mashash and El-Mahamid Mines, Egypt. J. Environ. Radioact. 84(1):65-78.

Amapu IY (1998). Potentials of Sokoto phosphate rocks as alternative phosphate fertilizer for the sub-humid Savannah of Nigeria. Unpublished PhD thesis, Ahmadu Bello University, Zaria.

Amin Y, Uddin Khandaker M, Shyen A, Mahat R, Nor R, Bradley D (2013). Radionuclide emissions from a coal-fired power plant. Appl. Radiat. Isot. 80:109-116.

Bashir M, Ibeanu I, Zakari Y, Sadiq U (2013). Assessment of Radiological Risk in Flooded Soil Samples of Kudenda, Kaduna State Nigeria. Int. J. Eng. Sci. Invent. 69-74.

Beretka J, Mathew P (1985). Natural radioactivity of Australian building materials, industrial wastes and by-products. Health Phys. 48(1):87-95.

Cevik U, Baltas H, Tabak A, Damla N (2010). Radiological and chemical assessment of phosphate rocks in some countries. J. Hazard. Mater.

182(1):531-535.

Etu-Efeotor J (1998). A review of the mineral resources of Sokoto basin, Northwest Nigeria. J. Min. Geol. 34(2):171-180.

Ghose S, Asaduzzaman K, Zaman N (2012). Radiological significance of marble used for construction of dwellings in Bangladesh. Radioprotection 47(01):105-118.

Guimond R (1990). Radium in fertilizers. Technical Report No. 310 International Atomic Energy Agency (IAEA), The environmental behaviour of radium pp. 113-128.

Harb S, El-Kamel A, El-Mageed AA, Abbady A, Negm H (2008). Natural Radioactivity Measurements in Soil and Phosphate Samples from El-Sabaea, Aswan, Egypt. Paper presented at the IX Radiation Physics and Protection Conference. Nasr City-Cairo.

IAEA (2004). International Atomic Energy Agency, Safety Standard Series No. RS-G-1. Application of the concepts of exclusion, exemption and clearanace safety guide.

Ibeanu I (2002). Tin mining and processing in Nigeria: cause for concern? Journal of environmental radioactivity, 64(1):59-66.

ICRP-60(1990).Recommendations of the International Commision on Radiological Protection, Pergamon Press, Oxford, 1990.

Ivanovich M, Harmon RS (1992).Uranium-series disequilibrium: applications to earth, marine, and environmental sciences. Oxford: Carendon Press.

Jamok EJ, Jonathan Y, Abdullahi M, Matthew K (2014). Radiological and X-Ray Fluorescence Spectrometric Analyses of Selected Inorganic Fertilizers in Zaria Kaduna State, Nigeria: Source of Possible Environmental Pollution. Int. J. Sci. Res. 3(4):86-93.

Jibiri N, Farai I, Alausa S (2007).Estimation of annual effective dose due to natural radioactive elements in ingestion of foodstuffs in tin mining area of Jos-Plateau, Nigeria. J. Environ. Radioact. 94(1):31-40.

Khan K, Khan H, Tufail M, Khatibeh A, Ahmad N (1998). Radiometric analysis of Hazara phosphate rock and fertilizers in Pakistan. J. Environ. Radioact. 38(1):77-84.

Khandaker M, Jojo P, Kassim H, Amin Y (2012). Radiometric analysis of construction materials using HPGe gamma-ray spectrometry. Radiati. Prot. Dosim. 152(1-3):33-37.

Khater AE, Higgy R, Pimpl M (2001). Radiological impacts of natural radioactivity in Abu-Tartor phosphate deposits, Egypt. J. Environ. Radioact. 55(3):255-267.

Kumar A, Kumar M, Singh B, Singh S (2003). Natural activities of 238U, 232Th and 40K in some Indian building materials. Radiat. Meas. 36(1):465-469.

NEA-OECD (1979). Exposure to Radiation from Natural Radioactivity in Building Materials. Report by NEA Group of Experts. OECD, Paris

Ogunleye P, Mayaki M, Amapu I (2002).Radioactivity and heavy metal composition of Nigerian phosphate rocks: possible environmental implications. J. Environ. Radioact. 62(1):39-48.

Okosun EA (1989). A review of stratigraphy of dange formation (Paleocene) Northwestern Nigeria, Nigeria. Newsl. Stratigr. 21(1):39-47.

Okosun E (1997). The potential application of Sokoto phosphate for the manufacture of fertilizer. J. Agric. Technol. 5(2):59-64.

Sabiha-Javied M, Tufail M, Asghar M (2010).Hazard of NORM from phosphorite of Pakistan. J. Hazard. Mater. 176(1/3):426-433.

Sam AK, Ahamed MM, El Khangi F, El Nigumi Y, Holm E (1999). Radiological and chemical assessment of Uro and Kurun rock phosphates. J. Environ. Radioact. 42(1):65-75.

Sam AK, Holm E (1995). The natural radioactivity in phosphate deposits from Sudan. Sci. Total Environ. 162(2):173-178.

Saueia CH, Mazzilli BP, F·varo, DIT (2005). Natural radioactivity in phosphate rock, phosphogypsum and phosphate fertilizers in Brazil. J. Radioanalyt. Nucl. Chem. 264(2):445-448.

Tufail M (2012). Radium equivalent activity in the light of UNSCEAR report. Environ. Monit. Assess. 184(9):5663-5667.

UNSCEAR (1993). Sources and Effects of Ionizing Radiation. Report to General Assembly with Scientific Annexes. United Nations, New York.

UNSCEAR (2000). Sources and Effects of Ionizing Radiation. Report to General Assembly, with Scientific Annexes. United Nations, New York.

Uosif M, El-Taher A (2008). Radiological assessment of ABU-TARTUR phosphate, western desert Egypt. Radiat. Prot. Dosim. 130(2):228-235.

Yu K, Guan Z, Stokes M, Young, E (1992). The assessment of the natural radiation dose committed to the Hong Kong people. J. Environ. Radioact. *17*(1):31-48.

Unusual rainfall shift during monsoon period of 2010 in Pakistan: Flash flooding in Northern Pakistan and riverine flooding in Southern Pakistan

Muhammad Arslan[1,3] , Muhammad Tauseef[2], Mehtab Gull[1], Mujtaba Baqir[1], Iftikhar Ahmad[1], Usman Ashraf[1] and Bassam Shafiq Al-Tawabini[3]

[1]College of Earth and Environmental Sciences University of the Punjab, Pakistan.
[2]Department of Physics, COMSATS Institute of Information Technology, Pakistan.
[3]Department of Earth Sciences, King Fahd University of Petroleum and Minerals, Dhahran 31261, Saudi Arabia.

Floods due to "blocking event" in the jet stream during 2010 caused intense rainfall and flash floods in northern Pakistan which resulted to riverine flooding in southern Pakistan. In the beginning of July 2010, changes in summer monsoon rainfall patterns caused the most severe flooding in Pakistan history. Process control charts suggest that monsoon pattern was not normal which made one-fifth of the country to be inundated. In this study, our main concern was to check the upward shifts (floods) in the rainfall pattern of all provinces of Pakistan. Results indicate that there was significant and sudden shift in the rainfall pattern of monsoon in 2010 which might be due to prolong "blocking event" in the jet stream. In late July, rainwater from the highlands entered major rivers which affected nearby areas of the Indus River. More than 250 mm of rain fell over a 36-h period in late July. Abeyant policies by the Pakistan Irrigation Department (PID) caused destruction in Jacobabad which was not a normal Indus waterway. The first week of August marked the worst week of extreme flooding in southern Pakistan. Flood simulation overylay technique showed the affected areas of the country in comparison with normal waterways by using vector and raster data images.

Key words: Indus River, monsoon, flooding in 2010, rainfall pattern, Climate Change, Floods.

INTRODUCTION

Natural disasters are common in today's world. They happen as a result of sudden changes in the state of natural elements. Flooding is one of the catastrophic natural hazards that take place every year around the globe. Pakistan is a country built around the Indus- the green swathe that cuts north-south across the deserts of central and southern Pakistan and essentially defined by Indus River and its tributaries (Figure 1). There are two major hydro-climatic factors which are responsible for flooding in Pakistan (Indus River basin). The first factor is the melting of snow cover in northern Pakistan while second factor of flooding is attributed to monsoon period (Shamshad, 2011). Flooding in Pakistan during 2010 was due to unusual shifts in the rainfall pattern during monsoon season. Monsoon is the periodic rainfall season which is responsible for the heat stabilization and land-mass warming by series of precipitation and/or rain-storms. In Indian subcontinent, climate is being controlled by tropical monsoon period which is the seasonal reversal of winds between late June and late September. During early July to late August of 2010, northern Pakistan and northwestern India faced one of the biggest natural

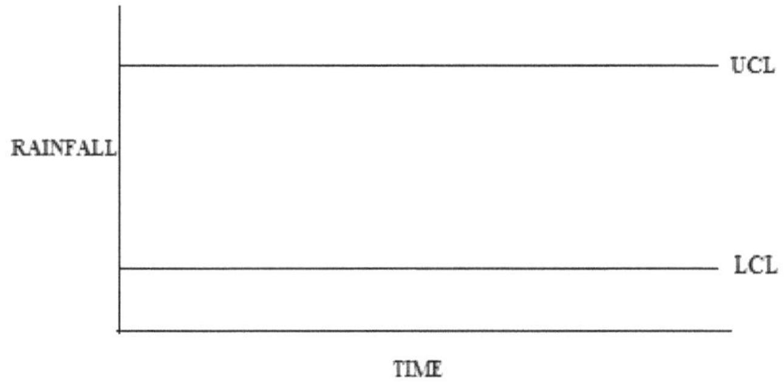

Figure 1. Basic theme of control chart with fixed upper and lower limits for floods and droughts.

disasters in the form of flash floods (Chi-Cherng et al., 2011), while riverine flooding were observed in southern Pakistan (Webster et al., 2011). Indus River, the longest river of Pakistan, was over-flooded during this period and was the main cause of riverine flooding in southern Pakistan as rainfall water receded downward towards Indian Ocean. Normal waterways of Pakistan along with the length of Indus River are shown in Figure 2. Tropical rainfall measuring mission satellites rainstorm structures also explained that rainfall behavior was not normal during 2010 in Indian Subcontinent (Houze Jr. et al., 2011).

In addition to this, the phenomenon of high rainfall was also correlated with another disastrous natural event called Russian heat wave-wildfire due to "blocking event" in subtropical Jet stream (William and Kyu-Myong, 2012). Jet streams are great rivers of air that ring earth at levels in the atmosphere ranging from 7 to 8 miles (11 to 13 km) above the surface. The subtropical jet stream is a permanent feature flowing west to east. Blocking event froze the meander of jet stream for approximately, 36 days because jet stream stops when an area of high pressure air forms, which is stronger enough to stop the normal flow of the jet stream from west-to-east as shown in Figures 4 and 5. This halting of jet stream allowed more fuel recharge (moisture content) from Arabian Sea and finally, resulted into unusual shift in the monsoon pattern and high rainfall in northern Pakistan because the upward movement of trade winds was to balance the effects of climate change in terms of pressure gradient (Figure 5). The whole picture reflects the intense precipitation and flash flooding in northern areas which, further, affected the downstream areas in terms of riverine flooding as rain water rolled across the length of country.

The study was conducted to show the variations and unusual shifts (upward and downward) during 2010 in the normal rainfall pattern of monsoon. Upward shifts explain flooding periods while downward shifts elucidate drought seasons in rainfall pattern. Permanence in downward shifts explains the desertous environment in some regions of southern Pakistan, that is, Thar Desert. Moreover, affected inundated areas were illustrated in polygons.

MATERIALS AND METHODS

In this study, rainfall data of 900 days for all the four provinces of Pakistan including Azad Jammu Kashmir (at each rainfall station) was collected from Pakistan Meteorological Department (PMD). Average rainfall rate, in millimetre, for each month was computed by taking average of daily measured rainfall of each station and the process was repeated for all the provinces viz: Azad Jammu Kashmir (AJK), NWFP, Punjab, Baluchistan and Sindh. Finally, rainfall data was summarized for all the stations to generate a single value for the representation of rainfall behavior. Data source was Pakistan Meteorological Department (PMD) which is located at Flood Forecasting Division, 46-Jail Road Lahore, Pakistan. Finally, data was plotted in excel sheet with upper and lower limits defined according to definition of floods and droughts. On average basis, rainfall rate for floods was fixed to 200 mm/month (> 100 mm/h) (Robert, 2004) while rainfall rate for droughts was fixed to 20 mm/month (< 25 cm/year). If rainfall rate remains less than 20 mm for longer periods of time, more than a year, succession of droughts will turn the green land into desert (Robert Maliva and Thomas Missimer, 2012). Basic ideology behind the experimentation was process control with fixed upper and lower control limits whose illustration is shown below:

The second objective of this research work was to integrate flood simulation model, remotely sensed data with topographic and spatial data in a GIS environment for flood risk mapping in the flood plain of Indus River in Pakistan. GIS based overlay analysis was done by using software MapPro Maker and Arc View 3.1. Data source for waterways of Pakistan was http://pakgis.blogspot.com/ which was further compiled in Arc View and adobe photoshop to mark the places (Figure 2). MapMaker Pro, compared to other packages, was more convenient and easy to use for analysis regarding GIS overlay technique based on raster and vector data images. Vector data images were used to illustrate normal water ways (red lines) which were merged with topographic raster data images showing abeyant behavior of water flow in Indus River and inundation in Pakistan (blue polygons) as shown in Figure 3. Lastly, layers were merged together based on themes union and MapMaker Pro was used for the impact analysis mainly for inundation.

RESULTS AND DISCUSSION

Daily rainfall data of 900 days from all the provinces elucidate more than 250 mm of rain fell over northern Pakistan in July 2010. Gradient coloring of line chart for each province is divided into red, yellow, green and brown zones. Red zone is the representation of intense rainfall which is beyond the upper control limit (UCL) whereas brown zone is the indication of droughts which are lower than lower control limit (LCL). Results illustrate that there was intense rainfall in three provinces, that is, Azad Jammu Kashmir, NWFP and Upper Punjab including federal (Figure 6). Maximum rainfall is observed in Azad Jammu Kashmir which is, approximately, at 4900 feet elevation and ultimately resulted into flash as shown in Figure 7 (Houze Jr. et al., 2011). Punjab also get intense rainfall (>250 mm) and faced riverine flooding in lower Punjab due to settlement of people on floodplain areas (Figure 8). Similar results are observed for NWFP

Figure 2. Normal waterways of Pakistan along with the length of Indus River.

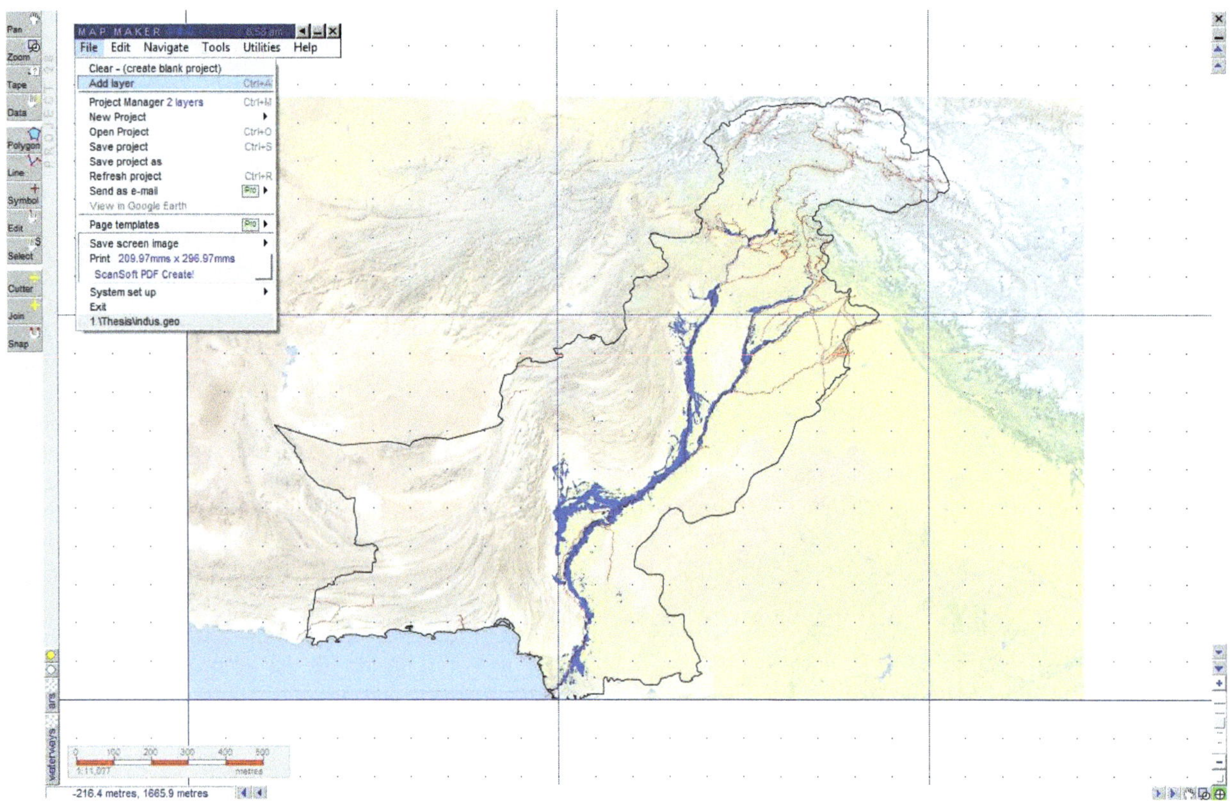

Figure 3. Water Inundated areas (blue polygons) with normal waterways (red lines).

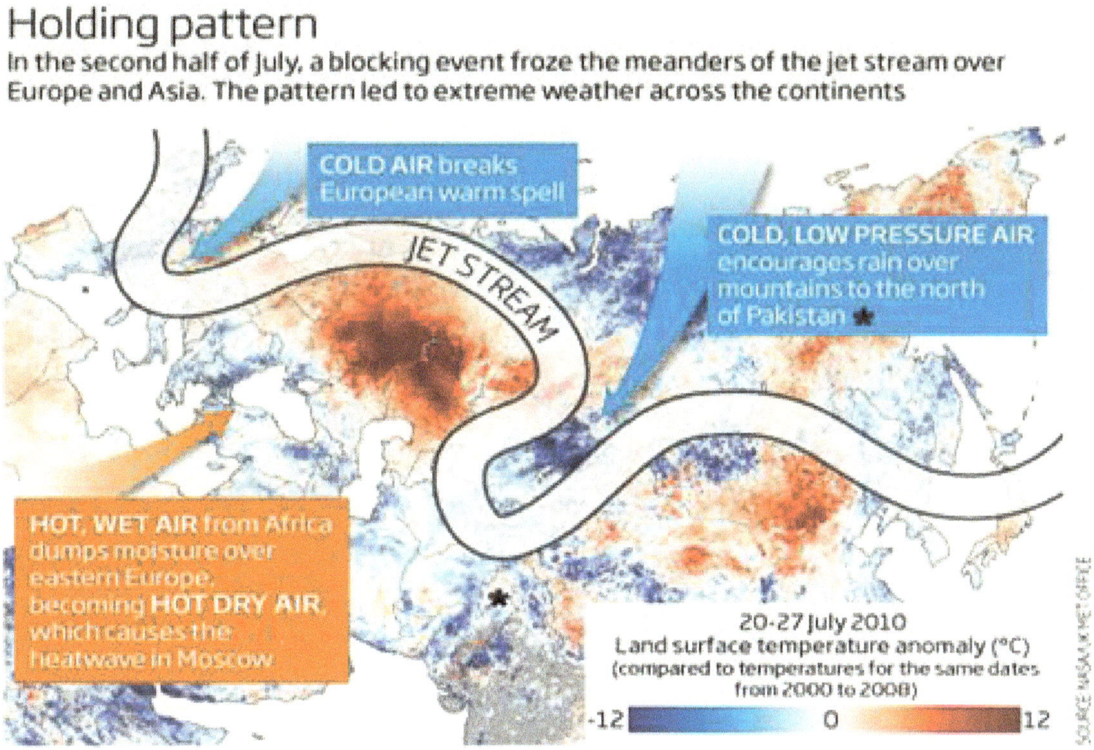

Figure 4. Frozen meander of jet stream showing blocking event
(Source: http://www.newscientist.com/article/mg20727730.101-frozen-jet-stream-leads-to-flood-fire-and-famine.html)

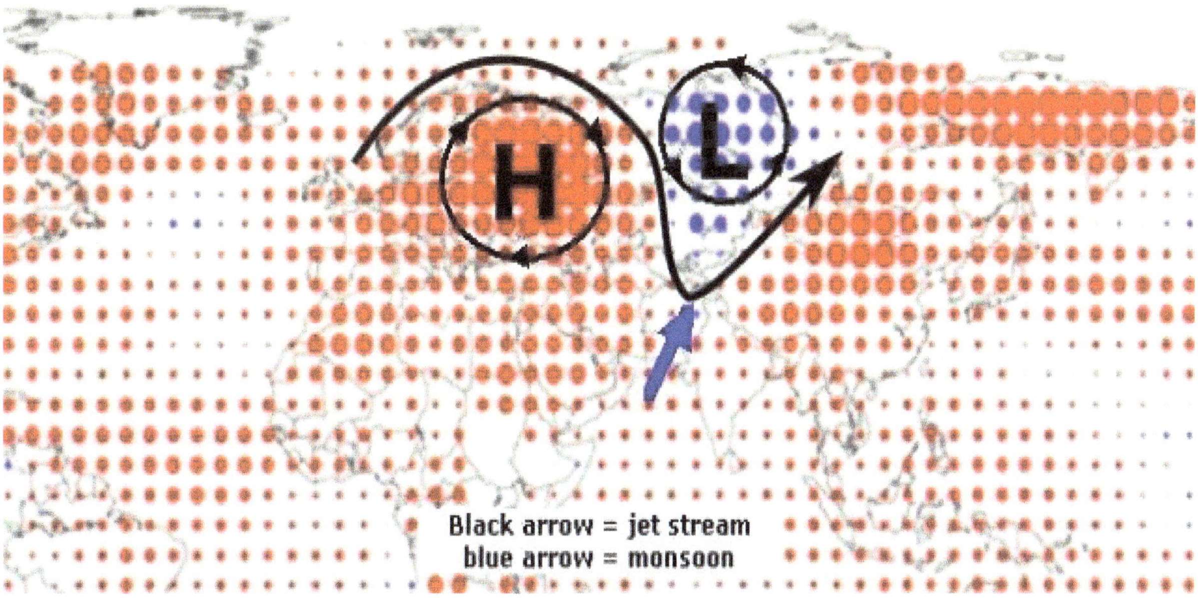

Figure 5. Pressure gradient created in subtropical jet stream during blocking event
(Source: http://www.newscientist.com/article/mg20727730.101-frozen-jet-stream-leads-to-flood-fire-and-famine.html)

as shown in Figure 9. Upward shift during 2010 was very abrupt and sudden as compared to monsoon periods of 2008 and 2009. The sudden, significant and unusual shift in July 2010 is attributed to changes in rainfall pattern which lead to extreme rainfall events at Gilgit-Baltistan and AJK as explained by Lau et al. (2012). Lau et al. (2012) have also correlated this phenomenon of extreme rainfall with excitation of Rossby wave train of Russian

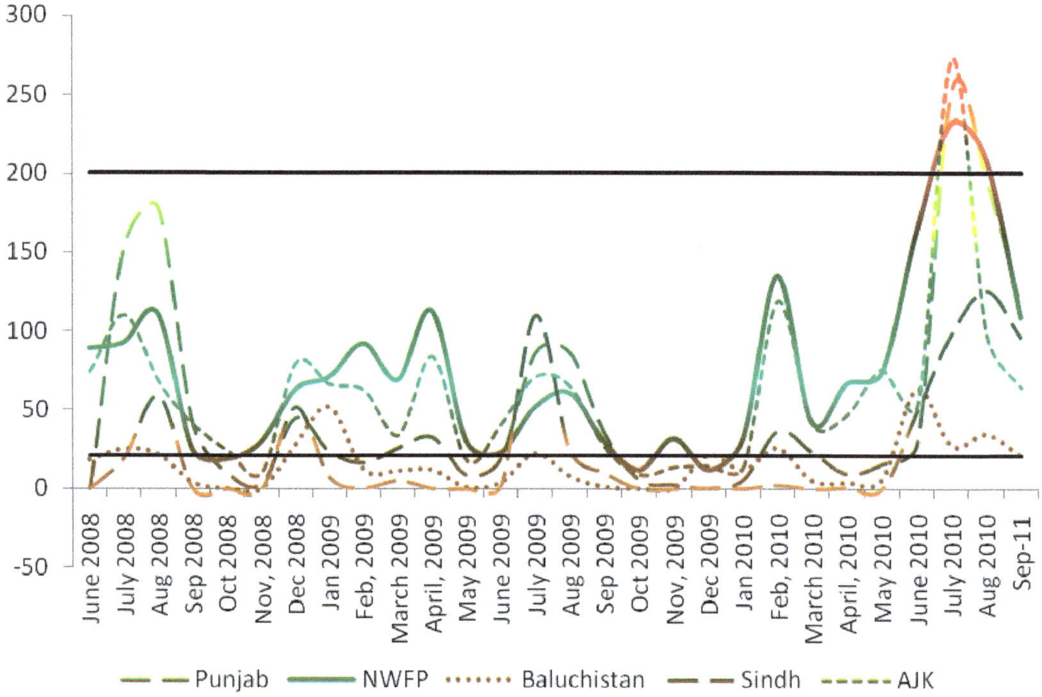

Figure 6. Rainfall data of all of the provinces for 900 days showing high rainfall in July 2010

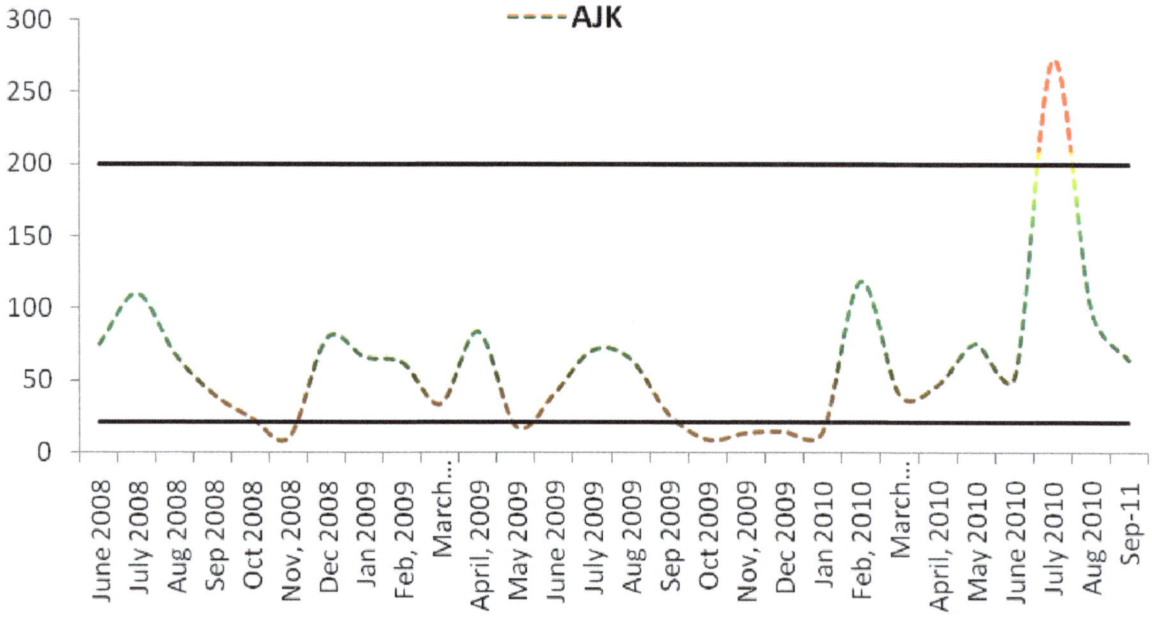

Figure 7. Rainfall data of Azad Jammu Kashmir for 900 days showing high rainfall in late July to early August 2010.

wildfires. Rossby wave train is meandering over Gilgit-Baltistan and northwestern China-Tibetan Pleatue region which reflects the association of two natural disasters occurring at the same time. Due to these reasons, intense rainfall and extreme flooding was observed in northern Pakistan which led to a big calamity in the history of Indian Subcontinent. According to previous records, extreme rainfall events of flooding have happened between 26th and 30th August 1929 and 31st July to 3rd August 1976 (Shamshad, 2011) but their impacts were not as much as in 2010 (OCHA, 2010). Monsoon rainfall season of 2010 is reported to be greatest natural disaster in previous 110 years history of the continent (UN Report).

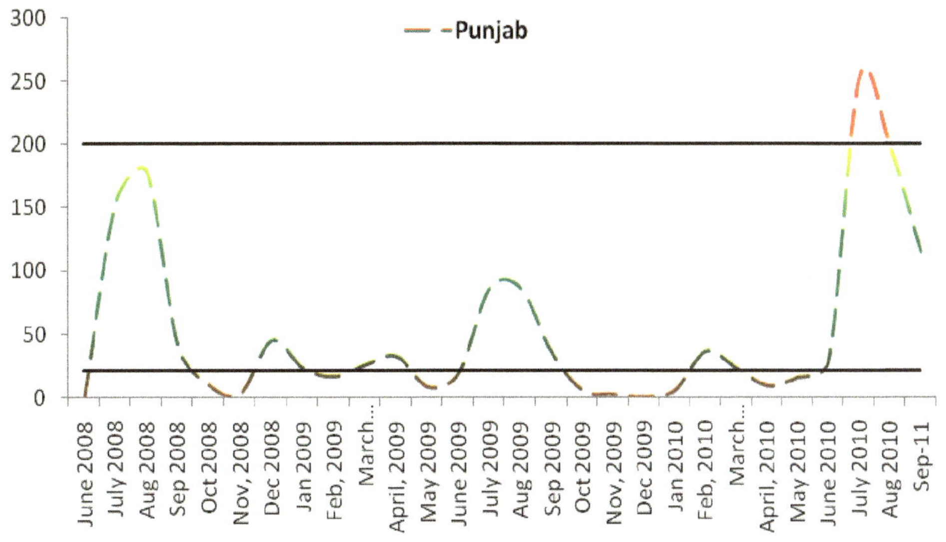

Figure 8. Rainfall data of Punjab for 900 days showing high rainfall in late July to early August 2010

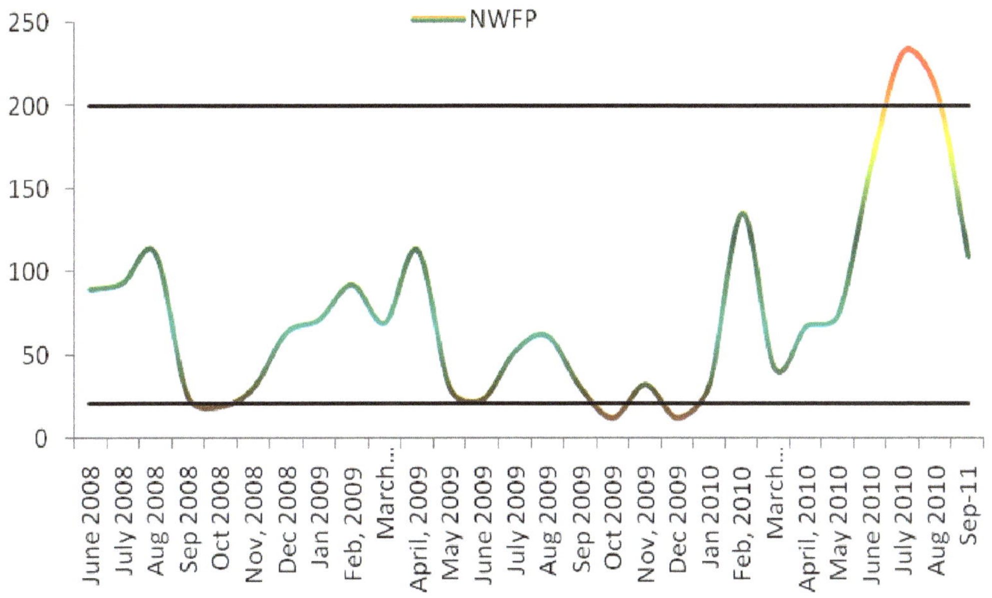

Figure 9. Rainfall data of NWFP for 900 days showing high rainfall in late July to early August 2010

In southern Pakistan, Baluchistan and Sindh, rainfall was not as much intense as it was in the northern areas but losses were reported in term of riverine flooding. Hereby, rainfall data show that there are consecutive periods of droughts in Baluchistan and Sindh which have converted the land of these provinces into desert (Figures 10 and 11). In Baluchistan, very low rainfall was observed even in monsoon periods due to its north-western geography which has very low influence on monsoon. Whereas, province Sindh faces seasonal rainfall as compared to Baluchistan but at lower extent though it is also desertous arid environment due to droughts in rest of the periods. However, in 2010, high rainfall in highland mountainous areas caused flood water to recede into the low lying areas and finally rolled across the length of the country. This created high water level in Indus River with the passage of time. Prior to destruction in floodplain areas, GIS based raster data images and GIS overlay technique illustrates the destruction of those areas which are not along normal waterways. This was due to poor water management policies of Pakistan government which lead to the destruction as in the case of Jaccobabad as shown in Figures 3 and 12. By overlaying the waterways of vector data image on inundated areas of raster data image, flood affected areas have been shown in blue polygons as compared to normal waterways in red lines. Intense monsoon rainfall season affected one fifth of the total area of Pakistan affected due to primary and

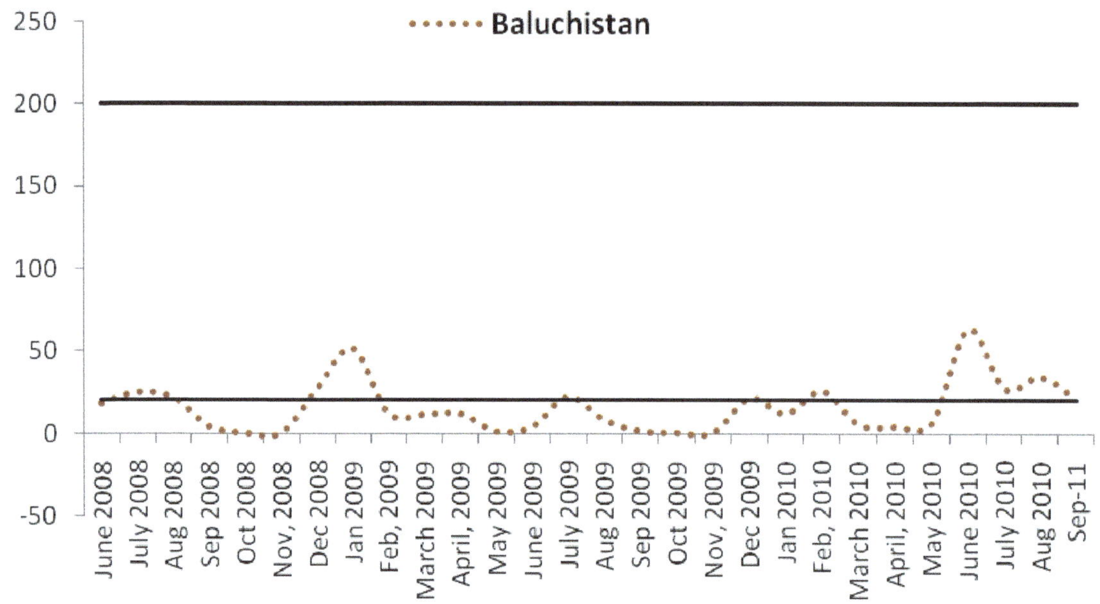

Figure 10. Rainfall data of Baluchistan for 900 days showing continuous droughts in rainfall pattern elucidating the desertous nature of the region

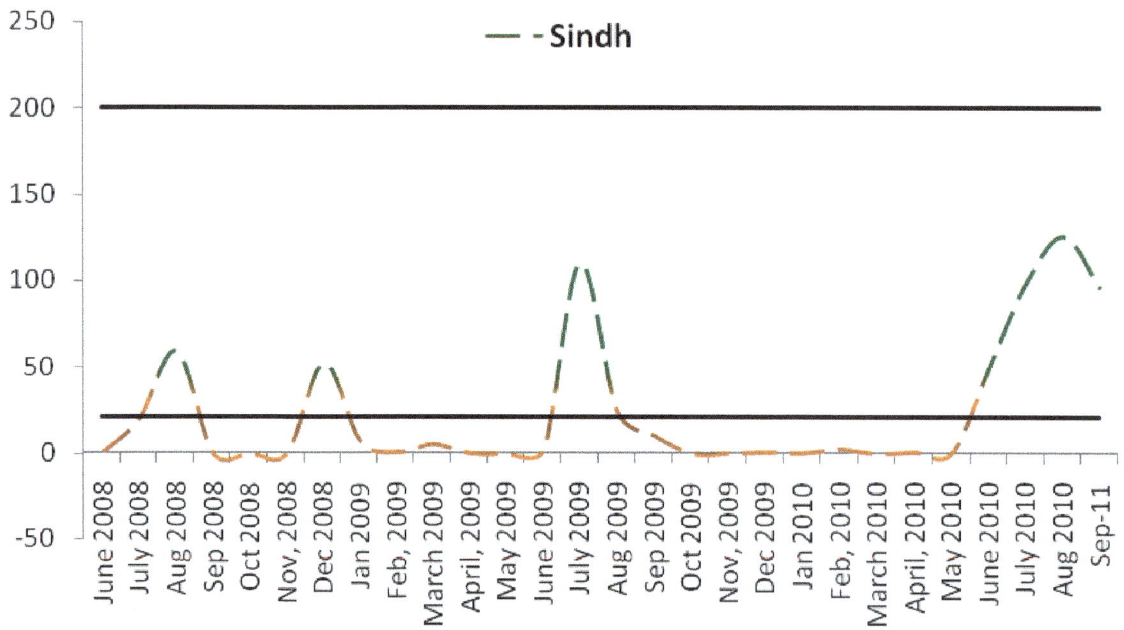

Figure 11. Rainfall data of Sindh for 900 days showing droughts in general rainfall pattern with three consecutive monsoon periods.

secondary effects (OCHA, 2010).

In early August 2010, two weeks of devastating monsoon rains transformed the landscape of Pakistan by pushing rivers over their banks, inundating villages, washing away bridges and roads, destroying crops and killing livestock. According to reports by the United Nations Office for the Coordination of Humanitarian Affairs (OCHA, 2010), more than 14 million people were affected in the middle of August. By comparing the water level in Indus River during monsoon season of 2009 with monsoon 2011, the Indus River appears as a thin river towards north of Sukkur with a skinny meandering waterway as compared to water level of Indus River in 2010 (Figures 13 and 14). In 2010, the river completely filled the river valley, merged the braided streams north of Sukkur, and pushed water over riverbanks in places.

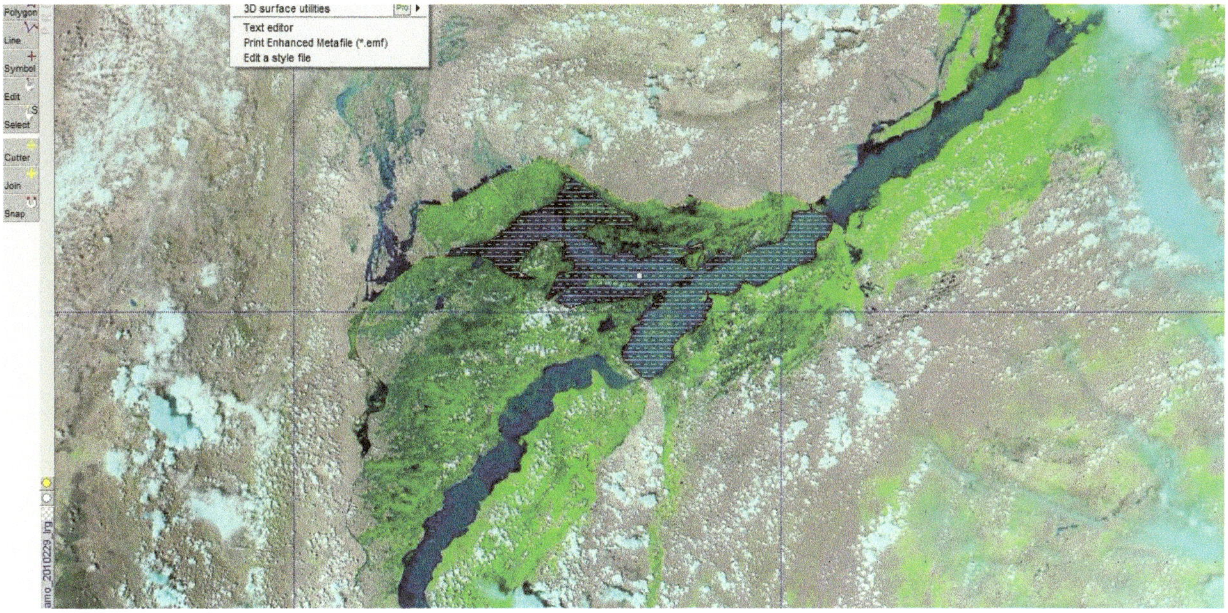

Figure 12. Water Inundation and destruction in Jaccoabad during late August 2010.

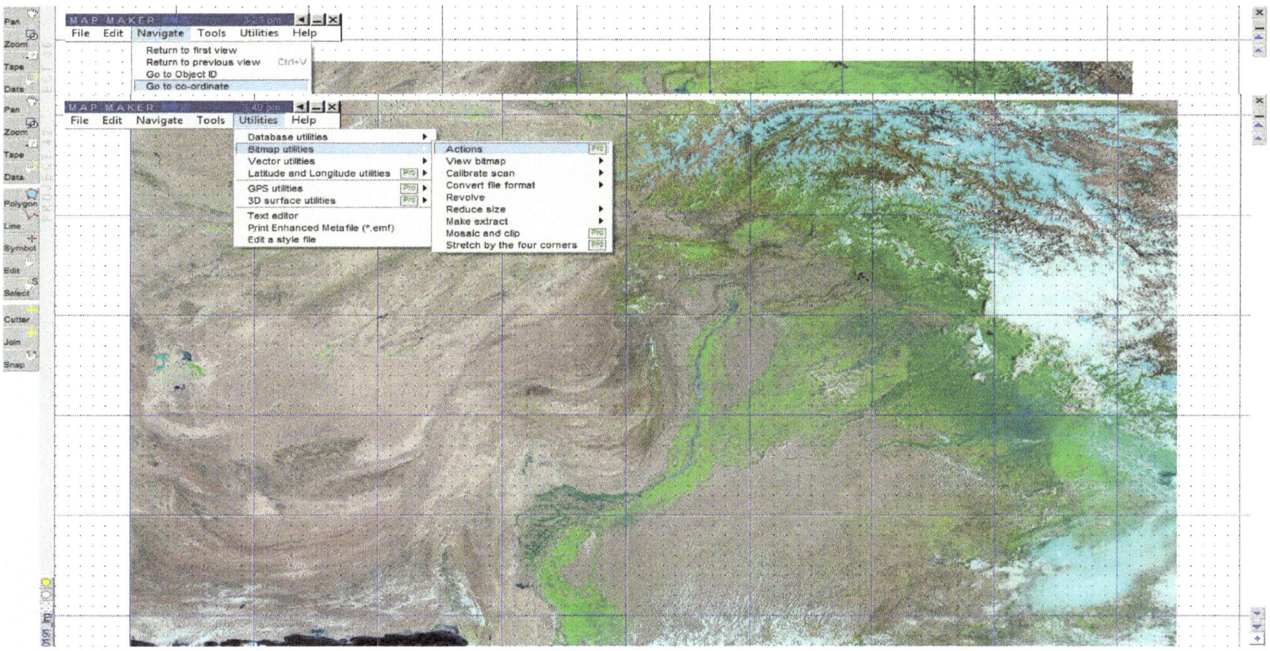

Figure 13. Monsoon Period of 2009 (18 August) showing water in Indus River. Source: NASA Earth observatory.

Northwest of the river valley, around the city of Sibi, standing water is evident on what appears to be normally arid ground as shown in Figure 14. Rainfall and flooding was also observed in 1929 and 1976, but the effects of flooding were not as much as they were in 2010. This might be due to the fact that population settlement on flood plains was not as much as it was in 2010. With the increase of population, people started living on flood plains to acquire the available resources but heavy rainfall heavy and flooding in 2010 destroyed their infrastructure, property, homes and lives.

Conclusion

In environmental studies, climate change is of prime importance because it can directly and indirectly affect the environment. Climate studies always played a major

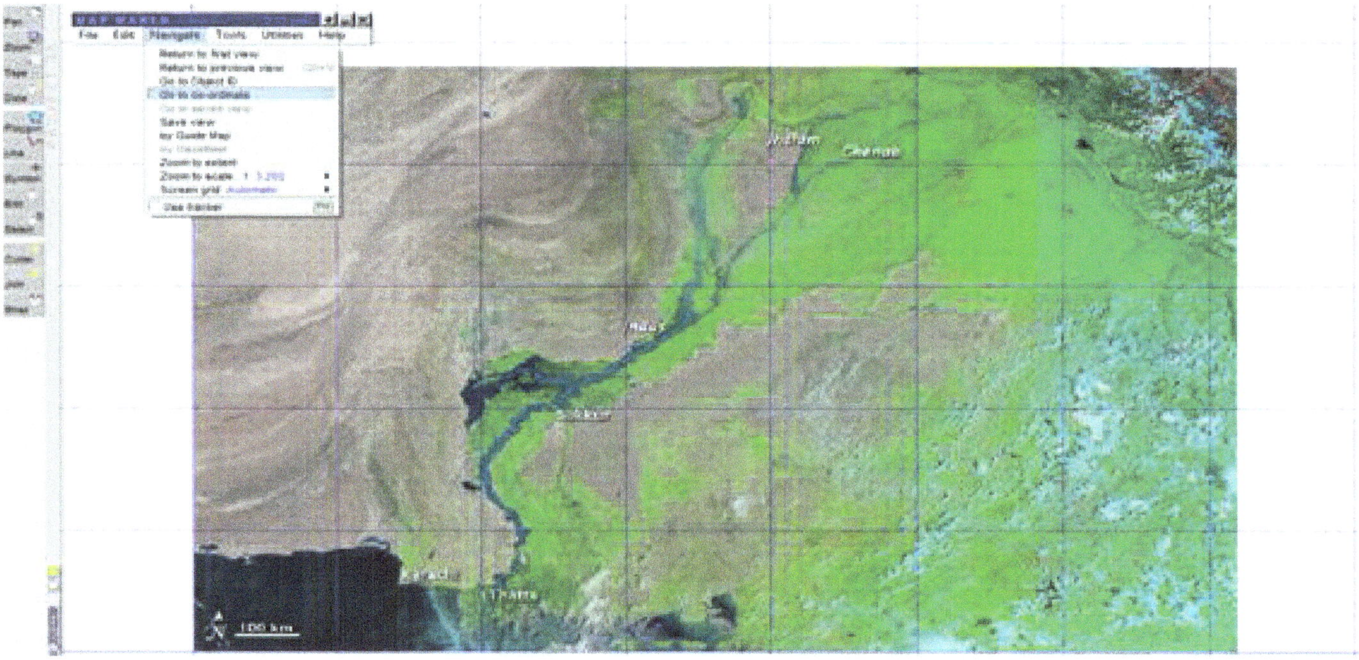

Figure 14. Monsoon period of 2010 (30 July) showing water in Indus River. Source: NASA Earth observatory.

role in the interpretation of hydrological cycle especially when changes are in rainfall behavior. Changes in monsoon of 2010 affected major portion of Pakistan due to flash and riverine flooding. In this study, we have tried to investigate the rainfall behavior for 900 days with three consecutive monsoon periods in which third monsoon period was very significant. By fixing the upper and lower control limits according to the definitions of floods and droughts, major shift in rainfall pattern was detected in 2010 which was due to blocking event in the jet stream. Jet stream halting created low pressure which allowed more fuel recharge from Arabian Sea and, hence, lead to severe rainfall in northern Pakistan. On the other hand, lower control limit for droughts supported the desertous environment of southern Pakistan which was affected due to riverine flooding. From the GIS based approaches, we can say that riverine flooding in Sukkur and Jacobabad was mainly due to heavy monsoon rainfall in highland areas of northern Pakistan. In conclusion, flash flooding in northern Pakistan was due to freezing of subtropical jet stream whereas riverine flooding in southern Pakistan was attributed to high water level in Indus River as rain-water moved across the length of the country. Poor water management policies by Pakistan government also contributed to destruction in major cities of the country.

ACKNOWLEDEGMENTS

We would like thank College of Earth and Environmental Sciences and Pakistan Meteorology Department for providing Resources, for different solutions and motivation towards this research.

REFERENCES

Shamshad A (2011). The South Asiatic Monsoon and Flood Hazards in the Indus River Basin, Pakistan. J. Basic Appl. Sci. 7:101-115.

Robert DS (2004). The impact of tropical rainfall rates on flash flood detection, Preprints 22nd Conference on Severe Local Storms.

Chi-Cherng H, Hsu, Lin HH, Chiu H (2011). Roles of European blocking and tropical-extratropical interaction in the 2010 Pakistan flooding. Geophys. Res. Lett. 38.

Houze Jr, Rasmussen R, Medina K, Brodzik S, Romatschke SU (2011). Anomalous atmospheric events leading to the summer 2010 floods in Pakistan. Bull. Am. Meteorological Society. 92, 291-298.

Lau WK, Kim K-M (2012). The 2010 Pakistan flood and Russian heat wave: Teleconnection of hydrometeorological extremes. J. Hydrometeorol. 13:392-403.

Maliva, R., Missimer, T., 2012. Aridity and Drought, Arid Lands Water Evaluation and Management. Springer, pp. 21-39.

Webster P, Toma V, Kim HM (2011). Were the 2010 Pakistan floods predictable? Geophys. Res. Lett. 38.

Pakistan Meteorology Department (PMD), Flood Update . http://www.pakmet.com.pk/FFD/cp/floodpage.asp [Accessed 02/01/2011]

Office for the Coordination of Humanitarian Affairs Pakistan (OCHA) (2010). www.unocha.org/where-we-work/pakistan [Accessed 02/03/2011]

National Aeronautics and Space Administration NASA [Electronic version]. Retrieved Dec 02, 2010, from http://earthobservatory.nasa.gov/NaturalHazards/event.php?id=44986

33

Studies on the effect of simulated acid rain of different pH on the growth response of TMS 419 cultivar of *Manihot esculenta* (Crantz)

Bridget O. Odiyi[1] and Joseph F. Bamidele[2]

[1]Department of Biology, Federal University of Technology, PMB 704, Akure, Ondo State, Nigeria.
[2]Department of Plant Science and Biotechnology, University of Benin, Edo State, Nigeria.

The growth response of TMS 419 cultivar of cassava *Manihot esculenta* Crantz exposed to simulated acid rain of different pH was studied. The plant was exposed to simulated acid rain of pH 2.0, 3.0, 4.0, 5.0, 6.0 and 7.0 (control), respectively. Simulated acid rain induced morphological changes including chlorosis, senescence, necrosis, leaf abscission, leaf folding and death. Plant height, leaf area, fresh weight, dry weight, relative growth rate, the chlorophyll content of the leaf and the harvest index was highest at pH 7.0 (control) and significantly ($p<0.05$) decreased with increasing acidity. The result shows that the TMS 419 cultivar of *M. esculenta* exhibited growth suppression at low pH level.

Key words: Simulated acid rain, *Manihot esculenta*, senescence, harvest index, leaf abscission.

INTRODUCTION

Acid rain is a major polluting agent harmful to terrestrial and aquatic ecosystems (Brimblecombe et al., 2007). It is the wet deposition that occurs when pollutants such as oxides of sulphur and nitrogen contained in power plant emission, factory smoke and car exhaust, react with the moisture in the atmosphere (Kita et al., 2004). In natural conditions, atmospheric precipitation is slightly acidic due to the dissolution of atmospheric carbon dioxide (Liu et al., 2010).

Rain that presents a concentration of H^+ ions greater than 2.5 μeq^{-1} and pH values lower than 5.6 is considered acid (Reshma and Manju, 2011). When the air becomes more polluted with nitrogen oxide and sulphur oxide, the acidity can increase to a pH value of 3. Occasional pH readings in rain and fog water of well below 2.4 have been reported in industrialized areas. Acid deposition may cause decline in health and growth of trees as well as other plants (Wyrwicha and Sklodowska, 2006). Several experiments have been carried out in the field and in greenhouses to investigate the effect of acid rain in plants (Silva et al., 2005). Exposure of plants to acid rain results in characteristic foliar injury symptoms, modified leaf anatomy (Stoyanova and Velikova, 2004), structural changes in the photosynthetic pigment apparatus and a decrease in the chlorophyll concentrations (Sant' Anna-Santos et al., 2006).

Cassava (*Manihot esculenta* Crantz) is a woody shrub that belongs to the family Euphorbiaceae (Nweke et al.,

2002). Since the introduction into Nigeria it has become very popular throughout the country. It is grown in 70% of the Nigerian states, available from the swamp forest area to the Guinea savannah area of Nigeria (Remison, 2005). The cassava cultivar TMS 419 is about 3-4 m tall and usually has no branching. The rain forest belt in Southern Nigeria is potentially susceptible to problems related to acid rain because of the increase in the consumption of petroleum oil products such as diesel, gasoline and coal used to produce energy for different economic sectors of the economy.

There is a need to examine the potential effects of acidic precipitation on cassava since it is a staple food in southern Nigeria for millions of Nigerians. There are over 2,000 cultivars of cassava in Nigeria (Ekanayake, 1994). TMS 419 cultivar was used for this research because it is a common cultivar found in the southern part of Nigeria. In view of the importance of this plant in human diet and the adverse effect of acid rain, the present study was carried out to assess the effect of simulated acid rain on this plant cultivar.

MATERIALS AND METHODS

Source of plant material

Disease free stem cuttings from matured plant of TMS 419 cultivar of *M. esculenta* were collected from the International Institute of Tropical Agriculture (IITA) Ibadan in Oyo State.

Planting procedure

A field experiment was carried out in the school farm of the Federal University of Technology, Akure, Ondo state, Nigeria. Stem cuttings of 30 cm long each of TMS 419 cultivar of *M. esculenta* were planted horizontally with a spacing of 100 cm and four stem cuttings were planted on each row. Each pH treatment of 2.0, 3.0, 4.0, 5.0, 6.0 and 7.0 had four replicates and was arranged in a completely randomized design (CRD). The plants were watered every other day and allowed to grow for a week before the application of the simulated acid rain treatment (Reshma and Manju, 2011). Simulated acid rain was sprayed to the planted cassava cultivars every three days according to their pH values. The solutions were applied using a medium size pressurized sprayer on the plants. The plants grew for 25 weeks before the experiment was terminated.

Preparation of simulated acid rain

The acids used were a mixture of concentrated sulphuric acid (H_2SO_4) and concentrated nitric acid (HNO_3) in a ratio 2:1. This is because the most important gas which leads to acidification is sulphur dioxide. The acidic solution was then calibrated using distilled water with a Deluxe pH meter to get the desired pH and cross checked with pH pen. Distilled water was used as the control of pH 7.0.

Several parameters were used in assessing the growth and productivity of the plant. The height of shoots was measured (cm) from the soil level to the terminal bud. The measurements were taken two weeks from the day the acid rain treatment commenced to the day of harvest at twenty four weeks. Leaf area was determined by the proportional method of weighing a cut-out of traced area of the leaves on graph paper with standard paper of known weight to area ratio. The fresh and dry weights were determined after twenty four weeks of treatment. The tuber dry weight was determined by cutting the tuber into smaller pieces and drying before weighting. Relative growth rate (RGR) was calculated following the methods of Hunt (1990) and the fresh weight of the whole plant was used to determine the relative growth rate.

$$RGR = (W_2 - W_1)/(T_2 - T_1)$$

Where, W_2 = final weight, W_1 = initial weight, T_2 = final time and T_1 = initial time.

The chlorophyll content of the leaves was determined and the harvest index was determined by the method of Ekanayake (1994):

Harvest Index (HI) = Tuber dry weight/total plant dry weight

Statistical analysis

Data obtained were subjected to analysis of variance (ANOVA) using the Statistical Package for Social Sciences, Version 15.0 (SPSS, 2003). Treatment means were separated using the Duncan multiple range test (Zar, 1984).

RESULTS

Morphological changes were observed in TMS 419 cultivar of *M. esculenta* throughout the period of the experiment. Table 1 shows the effects of simulated acid rain on the morphology of TMS 419 cultivar of *M. esculenta* at 24 weeks after simulated acid rain treatment. Leaves turned brownish, withered with 60% leaf abscission at pH 4.0. The falling and eventual collapse of leaves stretched over a period of 16-20 weeks. At 4.0 pH treatment, leaf abscission started with the leaves at the base of the shoot falling with long petiole. Leaves were chlorotic and necrotic. At 2.0 pH treatment, the plants die from the base of the shoot and leaves had 40% leaf abscission.

The results of the plant height, leaf area, fresh and dry weights are presented in Table 2. There was a significant ($p<0.05$) decrease in the plant height, leaf area, fresh and dry weights of the cultivar with increasing acidity. The plant height, leaf area, fresh weight and dry weight were significantly higher ($p< 0.05$) at the control (pH 7.0) as compared to the other acidity treatments.

The effect of simulated acid rain on the relative growth rate (RGR), the chlorophyll content and the harvest index of TMS 419 of *M. esculenta* is presented in Table 3. The cultivar had relative growth rate, chlorophyll content and harvest index at pH 7.0 as compared to the other acidity treatments. There was a significant reduction in the relative growth rate, the chlorophyll content and the harvest index with decreasing pH level.

DISCUSSION

Symptoms of plants polluted with simulated acid rain

Table 1. Morphological changes observed in TMS 419 cultivar of *M. esculenta* polluted with simulated acid rain 24 weeks after treatments.

pH treatment	Observed effects
7.0 (Control)	Luxuriant growth
6.0	Plant had good growth and the leaves had necrotic dots on its surface.
5.0	Production of new leaves on the stem and at the leaf apex. Growth was retarded. There was 90% survival. Older leaves showed signs of curling and chlorosis.
4.0	The leaves were curled and chlorotic. Plant growth was stunted. Leaves were showing signs of burnt surfaces from the tip. Had 60% leaf abscission. There was 70% survival of the plant
3.0	Plants had stunted growth. The new leaves became folded, chlorotic followed by necrosis and eventually some of the plants died. 40% survival. 60% leaf abscission.
2.0	Had stunted growth. 30% survival. All the surfaces of the leaves were necrotic and curled. Had 40% leaf abscission from the base of the shoot.

Table 2. Effect of simulated acid rain (SAR) on the plant height (cm), leaf area (cm), fresh weight (g), dry weight (g) of TMS 419 cultivar of *M. esculenta*, 24 weeks after treatment.

pH of SAR	Plant height (cm)	Leaf area (cm)	Fresh weight (g)	Dry weight (g)
7.0 (Control)	274.20 ± 4.16^{a}	288.24 ± 3.25^{a}	782.11 ± 0.16^{a}	326.24 ± 0.21^{a}
6.0	96.12 ± 4.02^{b}	270.11 ± 3.14^{b}	638.30 ± 0.14^{b}	276.14 ± 0.20^{b}
5.0	90.13 ± 3.20^{b}	224.34 ± 3.13^{b}	420.16 ± 0.09^{c}	216.22 ± 0.16^{b}
4.0	84.30 ± 3.12^{b}	218.24 ± 3.10^{b}	354.20 ± 0.09^{d}	168.40 ± 0.14^{c}
3.0	62.34 ± 2.21^{c}	184.10 ± 1.24^{c}	$228 .10 \pm 0.05^{e}$	102.36 ± 0.10^{d}
2.0	50.02 ± 1.10^{d}	166.32 ± 1.16^{c}	114.27 ± 0.02^{f}	60.14 ± 0.04^{e}

Means followed by the same letter vertically are not significantly different at 5% level of significant by New Duncan's multiple range test for the parameters tested.

Table 3. Effect of simulated acid rain (SAR) on the relative growth rate ($gg^{-1}d^{-1}$), chlorophyll content (mg/g) and harvest index of TMS 419 cultivar of *M. esculenta*, 24 weeks after treatment.

pH of SAR	Relative growth rate	Chlorophyll content	Harvest index
7.0 (Control)	4.2 ± 0.22^{a}	4.4 ± 0.20^{a}	0.6
6.0	3.4 ± 0.20^{a}	4.1 ± 0.18^{a}	0.5
5.0	2.6 ± 0.17^{b}	3.2 ± 0.14^{b}	0.4
4.0	1.8 ± 0.14^{b}	2.4 ± 0.10^{b}	0.3
3.0	1.4 ± 0.11^{b}	1.6 ± 0.07^{c}	0.2
2.0	0.6 ± 0.05^{c}	0.8 ± 0.03^{c}	0.1

Means followed by the same letter vertically are not significantly different at 5% level of significant by New Duncan's multiple range test for the parameters tested.

include chlorosis, necrosis, stunted growth, lesion, suppression of leaf production, leaf curling, withering of leaves, leaf abscission and even death of plants. Silva et al. (2006) found that plants exposed to low pH rain (pH 3.0) are generally slow-growing with leaf chlorosis and necrotic spot coupled with dehydration of the plants. In this study, simulated acid rain exposure caused chlorosis, necrotic lesions and leaf tip injuries at different pH levels (Table 1). Necrosis progressed from nodal region to the adjacent inter-nodal region leading to large scale leaf abscission. Marked chlorotic and marginal necrotic symptoms were observed at pH 4.0 and 5.0. However, this was less pronounced in comparison with pH 3.0 and 2.0. Similar symptoms were also observed by Johnston and Shriner (1985) on wheat at pH 4.3 and 2.3.

The TMS 419 cultivar of *M. esculenta* showed a decrease in growth parameters. Simulated acid rain at pH 2.0 caused burned irregular lesions on the plant leaves. It is well reported by many workers that plants sensitive to acid rain can present changes in their morphology, anatomy, physiology and biochemistry (Neufeld et al., 1985). All the plant growth parameters studied, including plant height, leaf area, fresh weight and dry weight were reduced significantly at all acidity levels as compared to the control. The highest reduction was observed at pH 2.0 level (Table 2). The adverse effects of simulated acid rain on plant growth parameters on several crops were also observed by Evans et al. (1997), Banwart et al. (1990), Chevone et al. (1984) at pH 2.0

Photosynthetic pigments were also inhibited with respect to acidity levels. Chlorophyll content was significantly reduced by simulated acid rain treatment relative to the control at pH 2.0 and 3.0 (Table 3). The greater foliar injury noticed in plants exposed to pH 2.0 is associated with the decreased chlorophyll content and the damage to the photosynthetic apparatus. This is similar to the earlier results of Sheridan and Rosenstreter (1973) and Evans (1984). Chlorophyll reduction has been attributed to the removal of magnesium ion from the tetrapyrol ring of the chlorophyll molecules by hydrogen ion (Foster, 1990) or increased transpiration due to acid rain (Evans et al., 1997). Recently, similar reductions of chlorophyll content by simulated acid rain were observed on many crops like mustard, radish, potato (Agrawal et al., 2005; Kausar et al., 2005; Khan and Deopura, 2005; Varshney et al., 2005).

The highest relative growth rate and harvest index that was recorded at pH 2.0 and 3.0 (Table 3) is similar to other results for cassava reported by a number of authors (Seinfield et al., 1998; Ekanayake, 1994; Cock et al., 1977; Kawano, 1978). Harvest index is the fraction of total dry matter in the economically useful parts. In cassava, storage roots are the economic yield component. According to Iglesias et al. (1994), harvest index of 0.5-0.6 is the optimum level because at higher values of harvest index, root production decreases due to reduced leaf area, light interception and photosynthesis.

Conclusion

The present results show that simulated acid rain with pH 2.0 and 3.0 has negative effect on the growth and yield of TMS 419 cultivar of *M. esculenta* due to reduction of photosynthesis as a result of chlorosis, necrosis and leaf abscission.

Conflict of Interests

The author(s) have not declared any conflict of interests.

REFERENCES

Agrawal S, Raghav D, Khan AA (2005). An evaluation of the impact of simulated acid rain on the growth of mustard in pots. Test of Agrochemicals and Cultivars 21:25-26.

Banwart WL, Finke RL, Porter PM, Hassett JJ (1990). Sensitivity of twenty soybean cultivars to simulated acid rain. J. Environ. Qual. 19:339-346.

Brimblecombe P, Hara H, Houle D, Novak M (2007). Acid rain: Deposition to recovery. Lavoisier Publ., Paris. France. 350pp.

Chevone BI, Herzfeld K, Chappelka AH (1984). Effects of acid aerosol and ozone on yellow poplar seedlings. Can. J. Res. 14:820-891.

Cock JH, Wholey DW, Casas OG (1977). Effects of spacing on cassava (*Manihot esculenta* Crantz). Exp. Agric. 13:289-299.

Ekanayake IJ (1994).Terminology for growth analysis of cassava. Tropical Root and Tuber Crops Bulletin 8(1):2.

Evans LS (1984). Botanical aspects of acidic precipitation. Botanical Review 50: 449-490.

Evans LS, Gmor NF, Dacosta F (1997). Leaf surface and histological perturbations of leaves of *Phaseolus vulgaris* and *Helianthus annuus* after exposure to simulated acid rain. Am. J. Bot. 4: 304-313.

Hunt R (1990). Basic Growth Analysis: Plant Growth Analysis for Beginners. Academic Press. London, UK. 41pp.

Iglesias CA, Calle F, Hershey C, Jaramillo G, Mesa E (1994). Sensitivity of cassava (*Manihot esculenta* Crantz) clones to environmental changes. Field Crops Res. 36: 213-220.

Johnston JW Jr., Shriner DS (1985). Response of three wheat cultivars to simulated acid rain. Environ. Exp. Bot. 25: 349-353.

Kausar S, Khan AA, Rahman F (2005). Impact of simulated acid rain on potato.In: Herbal Drug and Environmental Pollution. A satellite Session of Third International Conference on Plants and Environmental Pollution, JamiaHamdard, New Delhi, 56pp.

Kawano K (1978). Genetic improvement of cassava (*Manihot esculenta* Crantz) for productivity. Trop. Agric. Res. 11: 9-21.

Khan TI, Deopura S (2005). Physiological and biochemical effects of simulated acid rain on *Phaseolus vulgaris* Var HUR 15. Environmentalist 24: 224-226.

Kita I, Sato T, Kase Y, Mitropoulos P (2004). Neutral rains at Athens, Greece: a natural safeguard against acidification of rains. Sci. Total Environ. 327: 285-294.

Liu KH, Mansell RS, Rhue RD (2010). Cation removal during application of acid solution into air dry soil columns. Soil Sci. Soc. J. 4: 1747 – 1753.

Neufeld HS, Jernstedt JA, Haines BL (1985). Direct foliar effect of simulated acid rain. I. Damage, growth and gas exchange. New Phytol. 99: 389-405.

Remison SU (2005). Arable and vegetable crops of the Tropics. Gift Prints Association, Benin City. 248pp.

Reshma B, Manju M (2011). Impact of simulated acid rain of different ph on the seeds and seedlings of two commonly cultivated species of legmes in Kerala, India. Plant Arch. 2:607-611.

Sant' Anna-Santos BF, Silva LC, Azevedo A, Deque-Brasil R (2006). Effects of simulated acid rain on photosynthesis, Chlorophyll fluorescence and antioxidative enzymes in *Cucumis sativus* L. Photosynthetica 40:331-335.

Seinfield JH, Pandis SN, Spyries N (1998). Atmospheric chemistry and physics from air pollution to climate damage. AVI publishing Co, West Print Connecticut. 224pp.

Sheridan RP, Rosenstreter R (1973). The effect of hydrogen ion concentration in simulated acid rain on the moss *Tortula ruralis*(Hedgw). Bryologist 76:168-173.

Silva LC, Azevedo AA, Silva EAM, Oliva MA (2005). Effects of simulated acid rain on the growth and anatomy of five Brazilian tree species. Aust. J. Bot. 53:789-796.

Silva LC, Oliva MA, Azevedo AA, Araujo JM (2006). Responses of resting plant species to pollution from an iron pelletization factory. Water Air Soil Pollut. 175(1-4): 241-256.

SPSS (2003). Statistical package for social sciences (SPSS 15.0 Inc., Chicago, U.S.A).

Stoyanova D, Velikova V (2004). Effects of simulated acid rain on chloroplast ultrastructure of primary levels of *Phaseolus vulgaris.* Biol. Plant 40:589-598.

Varshney P, Kausar S, Khan AA (2005). An evaluation of the susceptibility of radish in pots to simulated acid rain. Test of Agrochemical Cultivar 27: 29-30.

Wyrwicha A, Sklodowska M (2006). Influence of repeated acid rain treatment on antiocidative enzyme activities and on lipid peroxidation in cucumber leaves. Environ. Exp. Bot. 56:198-204

Zar JH (1984). Biostatatical Analysis, 2nd edition, Prentice- Hall International, Englewood Cliffs, N.J.

Permissions

All chapters in this book were first published in AJEST, by Academic Journals; hereby published with permission under the Creative Commons Attribution License or equivalent. Every chapter published in this book has been scrutinized by our experts. Their significance has been extensively debated. The topics covered herein carry significant findings which will fuel the growth of the discipline. They may even be implemented as practical applications or may be referred to as a beginning point for another development.

The contributors of this book come from diverse backgrounds, making this book a truly international effort. This book will bring forth new frontiers with its revolutionizing research information and detailed analysis of the nascent developments around the world.

We would like to thank all the contributing authors for lending their expertise to make the book truly unique. They have played a crucial role in the development of this book. Without their invaluable contributions this book wouldn't have been possible. They have made vital efforts to compile up to date information on the varied aspects of this subject to make this book a valuable addition to the collection of many professionals and students.

This book was conceptualized with the vision of imparting up-to-date information and advanced data in this field. To ensure the same, a matchless editorial board was set up. Every individual on the board went through rigorous rounds of assessment to prove their worth. After which they invested a large part of their time researching and compiling the most relevant data for our readers.

The editorial board has been involved in producing this book since its inception. They have spent rigorous hours researching and exploring the diverse topics which have resulted in the successful publishing of this book. They have passed on their knowledge of decades through this book. To expedite this challenging task, the publisher supported the team at every step. A small team of assistant editors was also appointed to further simplify the editing procedure and attain best results for the readers.

Apart from the editorial board, the designing team has also invested a significant amount of their time in understanding the subject and creating the most relevant covers. They scrutinized every image to scout for the most suitable representation of the subject and create an appropriate cover for the book.

The publishing team has been an ardent support to the editorial, designing and production team. Their endless efforts to recruit the best for this project, has resulted in the accomplishment of this book. They are a veteran in the field of academics and their pool of knowledge is as vast as their experience in printing. Their expertise and guidance has proved useful at every step. Their uncompromising quality standards have made this book an exceptional effort. Their encouragement from time to time has been an inspiration for everyone.

The publisher and the editorial board hope that this book will prove to be a valuable piece of knowledge for researchers, students, practitioners and scholars across the globe.

List of Contributors

E. O. Oladele
Department of Cell Biology and Genetics, University of Lagos, Nigeria

P. G. C. Odeigah
Department of Cell Biology and Genetics, University of Lagos, Nigeria

I. A. Taiwo
Department of Cell Biology and Genetics, University of Lagos, Nigeria

Maud Kamatenesi-Mugisha
Division of Medical Ethnobotany and Product Development, Department of Biological Sciences, College of Natural Sciences, Makerere University, P. O. Box 7062, Kampala, Uganda

John Paul Buyungo
Division of Medical Ethnobotany and Product Development, Department of Biological Sciences, College of Natural Sciences, Makerere University, P. O. Box 7062, Kampala, Uganda

Patrick Engeu Ogwang
Natural Chemotherapeutics Research Institute, Ministry of Health, Uganda

Alex Kasibante
Natural Chemotherapeutics Research Institute, Ministry of Health, Uganda

Arop L. Deng
Egerton University, P. O. Box 536, Egerton, Kenya

Joshua O. Ogendo
Egerton University, P. O. Box 536, Egerton, Kenya

Matobola J. Mihale
Open University of Tanzania, P. O. Box 31608, Dar es salaam, Tanzania

Arhab Rabah
Laboratoire de Génie Microbiologique et Applications, Université Mentouri, BP 325 route de Aïn-El-Bey 25 017 Constantine, Algérie

Khenaka Karima
Département des Sciences de la Nature et de la Vie, Faculté des Sciences Exactes et des Sciences de la Nature et de la Vie, Université Larbi Ben Mhidi, Oum El Bouaghi, Algérie

Leulmi Nassima
Laboratoire de Génie Microbiologique et Applications, Université Mentouri, BP 325 route de Aïn-El-Bey 25 017 Constantine, Algérie

Belaidi Hakim
Laboratoire de Génie Microbiologique et Applications, Université Mentouri, BP 325 route de Aïn-El-Bey 25 017 Constantine, Algérie

Harzallah Besma
Département des Sciences de la Nature et de la Vie, Faculté des Sciences Exactes et des Sciences de la Nature et de la Vie, Université Larbi Ben Mhidi, Oum El Bouaghi, Algérie

Bousseboua Hacene
Laboratoire de Génie Microbiologique et Applications, Université Mentouri, BP 325 route de Aïn-El-Bey 25 017 Constantine, Algérie

Yapo Toussaint Wolfgang
Laboratory of Environmental Sciences, Research Group of Water Chemistry, University of Abobo-Adjamé (UAA), 02 BP 801 Abidjan 02, Côte d'Ivoire
National Institute of Public Hygiene (INHP), BPV 14 Abidjan, Côte d'Ivoire

Mambo Veronique
Laboratory of Environmental Sciences, Research Group of Water Chemistry, University of Abobo-Adjamé (UAA), 02 BP 801 Abidjan 02, Côte d'Ivoire

Yapo Ossey Bernard
Laboratory of Environmental Sciences, Research Group of Water Chemistry, University of Abobo-Adjamé (UAA), 02 BP 801 Abidjan 02, Côte d'Ivoire

Seka M'bassiguié Arsène
Laboratory of Environmental Sciences, Research Group of Water Chemistry, University of Abobo-Adjamé (UAA), 02 BP 801 Abidjan 02, Côte d'Ivoire

Houenou Pascal Valentin
Laboratory of Environmental Sciences, Research Group of Water Chemistry, University of Abobo-Adjamé (UAA), 02 BP 801 Abidjan 02, Côte d'Ivoire

P. I. D Kinyua
Department of Land Resource Management and Agricultural Technology, University of Nairobi, P.O. Box 29053 - 00625 Nairobi, Kenya

J. T. Njoka
Department of Land Resource Management and Agricultural Technology, University of Nairobi, P.O. Box 29053 - 00625 Nairobi, Kenya

Chris O. Magero
Department of Land Resource Management and Agricultural Technology, University of Nairobi, P.O. Box 29053 - 00625 Nairobi, Kenya

R. A. Wuana
Department of Chemistry, University of Agriculture, Makurdi, 970001, Benue State, Nigeria

S. G. Yiase
Department of Chemistry, Benue State University, Makurdi, 970001, Benue State, Nigeria

P. D. Iorungwa
Department of Chemistry, University of Agriculture, Makurdi, 970001, Benue State, Nigeria

M. S. Iorungwa
Department of Chemistry, University of Agriculture, Makurdi, 970001, Benue State, Nigeria

Mohamed Seid
Department of Soil and Water Resources Management, Wollo University, Dessie, Ethiopia

Tessema Genanew
Department of Soil and Water Resources Management, Wollo University, Dessie, Ethiopia

P.O Youdeowei
Department of Environmental Management, Institute of Geosciences and Space Technology, Rivers State University of
Science and Technology, P.M.B 5080, Nkpolu-Oroworukwo, Port Harcourt, Nigeria

H.O Nwankwoala
Department of Geology, University of Port Harcourt, P.M.B 5323, Choba, Port Harcourt, Nigeria

Javaid Ahmad Shah
Centre of Research for Development (CORD), University of Kashmir, Srinagar- 190006, Kashmir, India

Ashok K. Pandit
Centre of Research for Development (CORD), University of Kashmir, Srinagar- 190006, Kashmir, India

I. B. M. Kosamu
University of Malawi Polytechnic, Private Bag 303, Blantyre 3, Malawi

A. A. Mkandawire
University of Malawi Polytechnic, Private Bag 303, Blantyre 3, Malawi

W. Utembe
University of Malawi Polytechnic, Private Bag 303, Blantyre 3, Malawi

H. W. T. Mapoma
University of Malawi Polytechnic, Private Bag 303, Blantyre 3, Malawi

Arijit Das
Centre for the Study of Regional Development, School of Social Science, Jawaharlal Nehru University, New Delhi-110067, India

E. G. Olumayede
Department of Chemical Sciences, Faculty of Science, Ondo State University of Science and Technology, Okitipupa, Ondo state, Nigeria

J. M. Okuo
Chemistry Department, University of Benin, Benin City, Edo State, Nigeria

P. C. Okonkwo
Chemical Engineering Department, Ahmadu Bello University, Zaria, Kaduna State, Nigeria

E. E. Ikurekong
Department of Urban and Regional Planning, University of Uyo, PMB 1017 Uyo, Akwa Ibom State, Nigeria

Atser Jacob
Department of Urban and Regional Planning, University of Uyo, PMB 1017 Uyo, Akwa Ibom State, Nigeria

A. E. Majule
Institute of Resource Assessment (IRA), University of Dar es Salaam, Box 35097, Dar es Salaam, Tanzania

A. S Kauzeni
Institute of Resource Assessment (IRA), University of Dar es Salaam, Box 35097, Dar es Salaam, Tanzania

M. Mujwahuzi
Institute of Resource Assessment (IRA), University of Dar es Salaam, Box 35097, Dar es Salaam, Tanzania

Abdullahi Hussaini
Nigerian Meteorological Agency, Mallam Aminu Kano International Airport, Nigeria

Padmaja R. Jonnalagadda
Food and Drug Toxicology Research Centre (FDTRC), National Institute of Nutrition, Indian Council of Medical Research, Department of Health Research, Hyderabad-500007, (A.P), India

A. Y. E. Prasad
Food and Drug Toxicology Research Centre (FDTRC), National Institute of Nutrition, Indian Council of Medical Research, Department of Health Research, Hyderabad-500007, (A.P), India

U. V. R. Vijaya Sarathi
National Centre for Mass Spectrometry, Indian Institute of Chemical Technology [IICT], Tarnaka, Hyderabad-607, India

G. Manohar
Field Division, National Institute of Nutrition, Indian Council of Medical Research, Department of Health Research, Hyderabad-500007, (A.P), India

K. Venkaiah
Department of Biostatistics, National Institute of Nutrition, Indian Council of Medical Research, Department of Health Research, Hyderabad-500007, (A.P), India

A. I. Obasi
Department of Geological Sciences, Nnamdi Azikiwe University, Awka, Anambra State, State

A. U. Okoro
Department of Geology, Ebonyi State University, Abakaliki, Ebonyi State, Nigeria

O. M. Nweke
Department of Geological Sciences, Nnamdi Azikiwe University, Awka, Anambra State, State

A. Chukwu
Department of Geological Sciences, Nnamdi Azikiwe University, Awka, Anambra State, State

Syed Sana Mehraj
Department of Environmental Science, University of Kashmir, J&K 190006, India

G. A. Bhat
Department of Environmental Science, University of Kashmir, J&K 190006, India

Henah Mehraj Balkhi
Department of Biotechnology, University of Kashmir, J&K 190006, India

Taseen Gul
Department of Biotechnology, University of Kashmir, J&K 190006, India

O. Adisa, Banji
Department of Agriculture Extension and Rural Development, Obafemi Awolowo University, Ile-Ife, Osun State, Nigeria

Ikusemoran, Mayomi
Department of Geography, University of Maiduguri, Borno State, Nigeria

J. O. Olokor
Hydrology Unit, National Institute for Freshwater Fisheries Research, New Bussa, Niger State, Nigeria

A. N. Mutune
Jomo Kenyatta University of Agriculture and Technology, P.O. Box 62000 - 02000, Nairobi, Kenya

M. A. Makobe
Jomo Kenyatta University of Agriculture and Technology, P.O. Box 62000 - 02000, Nairobi, Kenya

M. O. O. Abukutsa-Onyango
Jomo Kenyatta University of Agriculture and Technology, P.O. Box 62000 - 02000, Nairobi, Kenya

C. Wariboko
Department of Crop Production Technology, Niger Delta University, Bayelsa State, Nigeria

I. A. Ogidi
Department of Crop Production Technology, Niger Delta University, Bayelsa State, Nigeria

Aasifa Gulzar
Department of Botany, Aligarh Muslim University, Aligarh, India

M. Badruzzaman Siddiqui
Department of Botany, Aligarh Muslim University, Aligarh, India

Bing-Hua Liao
The Key Laboratory of Ecological Restoration in Hilly Areas, Forestry Department of Henan Province, Ping-ding-shan University, Ping-ding-shan, Henan Province 467000, China
Department of Environment and Geography, Ping-ding-shan, Henan Province 467000, China
Institute of Ecological Science and Technology, College of Life Sciences, Henan University, Kaifeng 475001, China

I. B. M. Kosamu
Department of Physics and Biochemical Sciences, The Polytechnic, University of Malawi, Private Bag 303, Blantyre 3, Malawi

D. Mlelemba
Department of Physics and Biochemical Sciences, The Polytechnic, University of Malawi, Private Bag 303, Blantyre 3, Malawi

H. W. T. Mapoma
Department of Physics and Biochemical Sciences, The Polytechnic, University of Malawi, Private Bag 303, Blantyre 3, Malawi

P. Masache
Department of Physics and Biochemical Sciences, The Polytechnic, University of Malawi, Private Bag 303, Blantyre 3, Malawi

C. C. Kaonga
Department of Physics and Biochemical Sciences, The Polytechnic, University of Malawi, Private Bag 303, Blantyre 3, Malawi

K. Harawa
Water for People, P.O. Box 1207, Blantyre, Malawi

Kebede Wolka
Wondo Genet College of Forestry and Natural Resources, Hawassa University, P.O. Box 128, Shashemene, Ethiopia

Awdenegest Moges
Biosystems and Environmental Engineering Department, Hawassa University, P.O box 5, Hawassa, Ethiopia

Fantaw Yimer
Wondo Genet College of Forestry and Natural Resources, Hawassa University, P.O. Box 128, Shashemene, Ethiopia

Solomon Tamiru
Department of Plant Sciences, College of Agriculture and Veterinary Sciences, Ambo University, Ethiopia

Kindie Tesfaye
International Livestock Research Institute, Addis Ababa, Ethiopia

Girma Mamo
Melkassa Agricultural Research Center, Adama, Ethiopia

Alemayehu Wudneh
Natural Resource Management Course Team, Madawalabu University, P.O. Box 247, Robe-Bale, Ethiopia

Teklu Erkossa
International Water Management Institute, Regional office for East Africa and Nile Basin, P. O. Box 5689, Addis Ababa, Ethiopia

Prhaba Devi
Ambo University, Biology Course Team, Ambo, Ethiopia

B. M. Macaulay
Natural Resources Institute, University of Greenwich, Medway Campus, United Kingdom

M. T. Kolo
Department of Physics, Federal University of Technology, Minna, Niger State, Nigeria

Muhammad Arslan
College of Earth and Environmental Sciences University of the Punjab, Pakistan
Department of Earth Sciences, King Fahd University of Petroleum and Minerals, Dhahran 31261, Saudi Arabia

Muhammad Tauseef
Department of Physics, COMSATS Institute of Information Technology, Pakistan

Mehtab Gull
College of Earth and Environmental Sciences University of the Punjab, Pakistan

Mujtaba Baqir
College of Earth and Environmental Sciences University of the Punjab, Pakistan

Iftikhar Ahmad
College of Earth and Environmental Sciences University of the Punjab, Pakistan

Usman Ashraf
College of Earth and Environmental Sciences University of the Punjab, Pakistan

Bassam Shafiq Al-Tawabini
Department of Earth Sciences, King Fahd University of Petroleum and Minerals, Dhahran 31261, Saudi Arabia

Bridget O. Odiyi
Department of Biology, Federal University of Technology, PMB 704, Akure, Ondo State, Nigeria

Joseph F. Bamidele
Department of Plant Science and Biotechnology, University of Benin, Edo State, Nigeria